GEOLOGY OF
SELECTED AREAS IN NEW JERSEY AND
EASTERN PENNSYLVANIA
AND GUIDEBOOK OF EXCURSIONS

And some rin up hill and down dale, knapping the chucky stanes to pieces wi' hammers, like sae many road makers run daft. They say it is to see how the warld was made ~

1824 SIR WALTER SCOTT
ST. RONAN'S WELL

GEOLOGY OF SELECTED AREAS IN NEW JERSEY AND EASTERN PENNSYLVANIA AND GUIDEBOOK OF EXCURSIONS

SEYMOUR SUBITZKY, *Editor*

Prepared for the 1969 Annual Meeting of
THE GEOLOGICAL SOCIETY OF AMERICA
and ASSOCIATED SOCIETIES
Atlantic City, New Jersey, 1969

RUTGERS UNIVERSITY PRESS
New Brunswick, New Jersey

CONTENTS

The organizer of the field trips described in this volume is
Harold E. Gill
Chairman, Field Trip-Committee

FOREWORD

This volume, *The Geology of Selected Areas in New Jersey and Eastern Pennsylvania and Guidebook of Excursions*, contains reports on the latest concepts of geo-scientific knowledge pertaining, in part, to the Middle Appalachian region and areas immediately adjacent to the Delaware Valley in New Jersey and eastern Pennsylvania. The book is far more than a guide to field excursions. It is designed to provide background material and guides to six field trips of the 1969 Annual Meeting of the Geological Society of America and associated societies in Atlantic City, New Jersey, and also included are selected topical reports which will be valuable references for all geologists interested in this region.

Within the reports are descriptions of the geology of the Precambrian and folded Paleozoic rocks of the middle Appalachian region and the Triassic rocks of the Newark basin, the paleoenvironment of the Cretaceous-Tertiary deposits of the Coastal Plain of New Jersey, and the geology of Quaternary rocks of New Jersey including the glacial deposits of the northern part of the State and, locally, fluviatile and marine deposits of the unglaciated southern part of the State. The significance of the engineering geology and hydrogeologic problems of the Tocks Island dam site and Yards Creek pumped storage reservoir are presented. The Pleistocene geology of the Tocks Island dam site is discussed in connection with foundation problems and its influence on the design of engineering structures. The geology of the ilmenite deposits in the New Jersey Coastal Plain is described and the exploratory work of the New Jersey Geological Survey leading to the delineation of four ore bodies is summarized.

The efforts of the authors and field trip leaders who fitted into their work schedules the time necessary to prepare these reports is gratefully acknowledged.

Seymour Subitzky

Trenton, New Jersey
August, 1969

GEOLOGY OF
SELECTED AREAS IN NEW JERSEY AND
EASTERN PENNSYLVANIA
AND GUIDEBOOK OF EXCURSIONS

THE PENSAUKEN FORMATION—
A PLEISTOCENE FLUVIAL DEPOSIT
IN NEW JERSEY

JAMES F. BOWMAN, II
Department of Geology
Lehman College of the City University of New York, N. Y. 10031

WILLIAM LODDING
Department of Geology
Rutgers University, New Brunswick, N. J. 08903

ABSTRACT

A detailed study of the texture, mineralogy, and structure of the Pensauken Formation strongly indicates a fluvial environment of deposition with paleoslope toward the southwest.

A bauxitic laterite-type of weathering is revealed by X-ray diffraction analysis. This weathering, in conjunction with other evidence, suggests that the Pensauken was deposited during a relatively warm, interglacial period.

INTRODUCTION

Along the southeastern margin of the Triassic Lowlands of New Jersey and in scattered patches throughout the Coastal Plain, including southern Staten Island, the southern two-thirds of New Jersey, and the southeastern edge of Pennsylvania, occur a series of surficial Pleistocene formations that are distinct from the deposits of the Wisconsin and earlier glaciers.

From oldest to youngest, the formations of the Columbia Group are the Bridgeton, Pensauken, and the Cape May. The absolute age and precise stratigraphic position of the Pensauken are unknown but the Wisconsin glacial deposits unconformably overlie it. This relationship is well shown at Woodbridge, N. J., and Staten Island, N. Y.

The Pensauken Formation consists of ferruginous, quartzose sand and gravel, and contains a major feldspathic facies. The Bridgeton is similar in general aspect but may be somewhat more weathered. The Cape May is less weathered than both the others and has coloration of lighter shades than the typical yellows and browns of the Bridgeton and Pensauken Formations. Average thickness is about 20 feet.

The major portion of the Pensauken is confined to a 20-mile-wide belt whose axis extends from Staten Island southwestward to Salem, N. J. The Geologic Map of New Jersey (Johnson, 1950) provides the best presentation of the areal distributions of the Bridgeton, Pensauken, and Cape May Formations. Figure 1 generalizes the areal distribution of the Pensauken Formation.

A brief account of the early work done on the Columbia Group in New Jersey is given by Wilmarth (1957). A more up-to-date, annotated bibliography appears in Bowman (1966b). The stratigraphic nomenclature of Salisbury and Knapp (1917) remains in use.

TEXTURE AND COMPOSITION

Pensauken sedimentation units are generally coarse textured. Bimodality of particle size distribution is a common attribute of many gravel units. The minor mode is often in the sand range. Sand units are also frequently bimodal. Bowman (1966b) made size distribution analyses of samples from more than 100 sedimentation units. No systematic

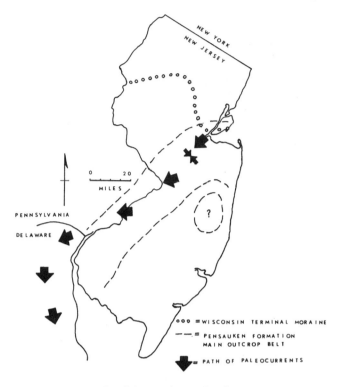

Figure 1. Generalized Pensauken Paleoslope Map (Tentative Correlation with Jordan (1964), Delaware).

TABLE 1. SAND SIZE MINERALS IN THE PENSAUKEN FORMATION
(After Bowman, 1966b)

Light Minerals

Quartz
Chert
Feldspar, potash
Feldspar, plagioclase
Glauconite, single grains
Glauconite, aggregates
Gibbsite, vermiform grains
Muscovite (some)
Shale fragments

Heavy Minerals

Ilmenite	Muscovite (some)
Magnetite	Biotite
Leucoxene	Chlorite
Limonitic aggregates	Chloritoid
Zircon	Tremolite
Tourmaline	Actinolite
Rutile	Hornblende, blue
Anatase	green
	brown
	basaltic
Andalusite	Enstatite
Kyanite	Hypersthene
Sillimanite	Augite
Staurolite	Diopside
Epidote	Topaz
Zoisite	Monazite
Garnet	Sphene
Spinel	Corundum

geographic variation of median grain diameter was found.

The base of the Pensauken Formation is, in most places, defined by a conglomerate. Pebbly lenses occur scattered throughout the section at most localities. Quartzose rocks, chert, granite and other crystallines, and Triassic sandstone and shale are the most abundant constituents of Pensauken gravel. The greatest variety of rock types, within a given section, occurs at Staten Island, N. Y., and Turkey Hill, Pa. The general trend is a reduction in variety of rock types from northwest to southeast.

A great variety of mineral species occurs in the light and heavy fractions of Pensauken sands. The minerals found in the sand fraction of the Pensauken are shown in table 1.

Of the sedimentation units studied in detail, 80 percent were classified as feldspathic sands and 20 percent as glauconitic sands. Some of the feldspathic sands are true arkoses. However, the distinction between feldspathic and glauconitic facies is not well defined. Of samples containing feldspar, 40 percent also contained some glauconite. Only 3 percent contained neither feldspar nor glauconite.

Geographic trends in the heavy mineral fractions of the sands, with certain exceptions, are not well defined. Samples along a northeast to southwest line passing from the southwestern corner of Staten Island through Jamesburg, N. J., Turkey Hill, Pa., Deacon's Station, N. J., Pensauken Township, N. J., and Swedesboro, N. J., have high feldspar and amphibole content. The heavy fractions of these samples also tend to contain maximum variety of mineral species. There appears to be a tendency for abundant amphibole content to occur in samples high in feldspar.

With respect to the northeast-southwest line just mentioned, roundness of Pensauken sand grains increases towards the southwest. Samples from Blackwood and Swedesboro, N. J., near the southern extreme of the main outcrop belt, contain well-rounded heavy minerals in their fine sand grades. Many zircon and amphibole grains are strikingly well rounded.

Maturity (determined according to the ZTR index of Hubert, 1962) of Pensauken sands is also greatest near the southwest extremity of the main outcrop belt.

STRUCTURES

Campbell and Bascom (1933) show that the Pensauken Formation, along the main outcrop belt, occupies a trough-shaped depression. This trough is mapped by means of structure contours drawn on the base of the formation (p. 302, Plate 1). They conclude that the Pensauken of the main outcrop belt is an alluvial deposit formed by the combined "original Delaware" and "original Hudson Rivers" and that the Pensauken was later folded into a syncline with its axis coincident with the central axis of the outcrop belt.

If this structure has been caused by post-depositional flexure, it is probably related to isostatic rebound following retreat of one of the Pleistocene glaciers. If not, it must be regarded as having been coexistent with, and having exercised a structural control upon Pensauken sedimentation.

The association of the development of the presently occupied valley of the Hudson River with Wisconsin and Holocene events (Newman and others, 1969) must be examined relative to the hypothesis of Campbell and Bascom. In this connection we emphasize the occurrence of "typical" Pensauken, unconformably overlain by Wisconsin till, on southwestern Staten Island.

Bedding in the Pensauken Formation is seldom of uniform thickness. Virtually everywhere examined, the bedding consists of a series of coalescing lenses. Individual lenses range from a few tens of feet to a few hundred feet wide at the maximum. The lenticular nature of the bedding appears, in large part, to be the result of channel cut-and-fill.

Cross-bedding is abundant. Bowman (1966b) measured the orientation of more than 400 foreset beds distributed throughout the Pensauken of the main outcrop belt. All foresets measured were of the *planar* type according to the definition of Potter and Pettijohn (1963, p. 69). Since the average dip of the measured foresets is 25°, no stereonet correction for tilt was made (Potter and Pettijohn, 1963, p. 79). Tilt of foresets imposed by isostatic flexures of the crust would not alter dip azimuths by more than a few degrees. The complication of interpretation that would occur with the presence of anti-dunes (Harms and Fahnestock, 1965, p. 84–115) did not arise since no structures of this kind were encountered. Anti-dunes, with laminae dipping upstream at angles of 0 to 10°, are characteristic of high gradient ephemeral streams. The average 25° dip of Pensauken foresets is in the range cited for tabular-planar cross-stratification of intermediate angle (generally 18 to 28° but locally less) by McKee (1964, p. 286). This type of cross-bedding is characteristic of alluvial plains (valley-bottom flood plains of aggradation) and delta cones.

A generalized Pensauken paleoslope map is shown in figure 1. A tentative correlation is made with data generalized from the work of Jordan (1964).

Secondary structures include concretions of the voidal type and cementation of thin layers which frequently do not parallel bedding planes and may be of highly irregular shape. The cementing material probably is mostly limonite.

In exposures of the central part of the outcrop belt, the upper few feet of the Pensauken are observed to have little stratification or the layers are much disturbed and contorted. At least part of this disturbance of bedding is the result of periglacial deformation. Near Collingwood Park, N. J. (adjacent to the Naval Depot), the section is deformed to a depth of 12 feet from the surface.

WEATHERING

Much of the silt and most of the clay in the Pensauken are believed to be the result of weathering. All sand size vermiform grains of gibbsite (Lodding, 1961) are formed by weathering processes.

X-ray diffraction analyses of weathered materials were carried out and reported upon by Bowman (1966a). Samples obtained from sections at Columbus and Jamesburg, N. J., show a feldspar to kaolinite to gibbsite trend with increasing depth from the top of the weathered zone. Although ferruginous laterites have been found to form in cold climates, bauxitic laterites are believed to form only in warm, humid climates. Not only is this zone of weathering of the bauxitic laterite type, it appears to be intraformational in the Pensauken.

Floral remains found by Berry and Hawkins (1933) resemble the suite native to the warm Carolinas of today.

CONCLUSIONS

The criteria of Salisbury and Knapp (1917), for distinguishing between the Bridgeton, Pensauken, and Cape May, are based chiefly upon the topographic elevations and degree of weathering of these Formations. These criteria were assumed to be valid for the purposes of this paper. No better map has yet been made of their outcrop patterns than that of Johnson (1950). The Bridgeton and Pensauken are virtually indistinguishable from one another on the outermost reaches of the Coastal Plain—perhaps because they are part of a continuous sequence in this region. Possibly this part of the Coastal Plain, sloping oceanward from the drainage divide separating it from the southeastern limb of the trough of Campbell and Bascom (1933), was relatively unaffected by isostatic flexures. If this were the case, the elevation criteria of Salisbury and Knapp would here be meaningless. Figure 1 reflects these stratigraphic uncertainties. Although the southern part of the Cape May Formation contains a marine facies, the longer period of erosion of the Bridgeton and Pensauken Formations would account for their apparent lack of marine facies. In any case, erosion has removed all but small patches of Bridgeton and Pensauken from the region south of Raritan Bay to Forked River, N. J. The Pensauken Formation has not been identified south of Forked River (outside the main outcrop belt).

The bimodality of Pensauken sediments, abundant channel cut-and-fill structure, lack of marine fossils, and cross-bedding data all demonstrate a fluvial environment of deposition. The main outcrop belt must have been the locus of a stream or stream system flowing southwest. Tributaries, flowing down the slopes on either side and into the trough, are suggested by the trends of cross-beds in the vicinity of Jamesburg. The correlation of the major trend of cross-bedding with that of Jordan appears excellent.

The Pensauken cannot be directly dated by current age-determination methods. However, the floral suite of Berry and Hawkins and lateritic-type weathering suggest that the Pensauken was deposited during a relatively warm, interglacial period.

A tentative synthesis of climatic cycle including Pensauken time is set forth:

Climate (Impending glaciation)	*Evidence*
Aridity	Loess in vicinity of Jamesburg
Cold	Periglacial deformation, tops of sections
-----------	-----------
	(Pensauken time)
Warm	Lateritic-type weathering
Cold	Ice rafted boulders near base of formation
(Glacial recession)	Fluvial environment of deposition

REFERENCES CITED

Berry, E. W., and Hawkins, A. C., 1933, Flora of the Pensauken Formation in New Jersey: Geol. Soc. America Bull., v. 46, p. 245–252.

Bowman, James F., II, 1966a, Lateritic Weathering in the Pensauken Formation: Abstracts, Geol. Soc. America, First Annual Meeting of the Northeastern Section, Philadelphia, Pennsylvania, February, 1966.

Bowman, James F., II, 1966b, Petrology of the Pensauken Formation: unpublished PhD. Thesis, Rutgers University, 155 p.

Campbell, M. R., and Bascom, F., 1933, Origin and structure of the Pensauken gravel: Amer. Jour. Sci., v. 26, p. 300–318.

Harms, J. C., and Fahnestock, R. K., 1965, Stratification, bedforms, and flow phenomena (with an example from the Rio Grande): *in* Middleton, G. V. (ed.), 1965, Primary Sedimentary Structures and their Hydrodynamic Interpretation: S.E.P.M., Sp. Pub. no. 12, p. 84–115.

Hubert, J. F., 1962, A zircon-tourmaline-rutile maturity index and the interdependence of the composition and texture of sandstones: Jour. Sed. Petrology, v. 32, p. 440–450.

Johnson, M. E., 1950, Geologic Map of New Jersey: Atlas Sheet No. 40, Department of Conservation and Development, State of New Jersey.

Jordan, R. R., 1964, Columbia (Pleistocene) Sediments of Delaware: Delaware Geological Survey, Bull. No. 12, Newark Delaware.

Lodding, W., 1961, Gibbsite vermiforms in the Pensauken Formation of New Jersey: Amer. Mineralogist, v. 46, p. 394–401.

McKee, E. D., 1964, Inorganic sedimentary structures: *in* Imbrie, J., and Newell, N. (ed.), 1964, Approaches to Paleoecology: John Wiley and Sons, Inc., New York, p. 275–295.

Newman, W. S., Thurber, D. H., Zeiss, H. S., and Rokach, A., 1969, Late Quaternary geology of the Hudson River estuary: A preliminary report, to be presented at the meeting of the Geol. Soc. Amer., Northeastern Section, Albany, N. Y., March 13–15, 1969.

Potter, P. E., and Pettijohn, F. J., 1963, Paleocurrents and Basin Analysis: Academic Press, Inc., New York.

Salisbury, R. D., and Knapp, G. N., 1917, The Quaternary Formations of Southern New Jersey: The Final Report Series of the State Geologist, v. viii Department of Conservation and Economic Development, Trenton, New Jersey.

Wilmarth, M. G., 1957, Lexicon of Geologic Names of the United States: U. S. Geol. Survey Bull. 896.

FACIES CHANGES
IN SEDIMENTS OF MIOCENE AGE
IN NEW JERSEY

W. C. ISPHORDING
University of South Alabama, Mobile, Alabama 36608

WILLIAM LODDING
Rutgers University, New Brunswick, New Jersey 08903

INTRODUCTION

Miocene sediments crop out in a northeast-southwest trending belt in the New Jersey Coastal Plain and can be traced, almost continuously, from Delaware Bay to Raritan Bay (fig. 1). Two formations have been recognized for this time interval, (1) the middle Miocene Kirkwood Formation and (2) the upper Miocene Cohansey Sand. Both of these formations dip seaward at less than 50 feet per mile and are of interest not only because lithologic changes within them reflect clearly the final advance and withdrawal of the late Tertiary seas but also because of the controversial nature of their contact.

KIRKWOOD FORMATION

The Kirkwood is the second largest formation in the New Jersey Coastal Plain and was deposited, disconformably, on a highly eroded surface of lower Tertiary greensands and limesands. This surface, known as the Schooley peneplane, resulted from prolonged weathering during the late Eocene through the lower Miocene when the region was largely emergent. The maximum outcrop thickness is found in the central portion of the State and is on the order of 100 feet; downdip, in the vicinity of Atlantic City, the thickness increases to nearly 1,000 feet. Recent investigation suggests that the Kirkwood Formation may be divided into three members: (1) a basal marine clay unit, the Alloway Clay Member, which is

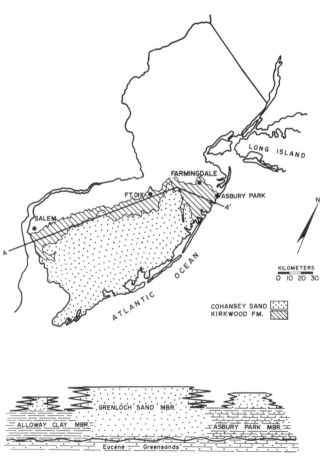

Figure 1. Map showing areal distribution of Kirkwood and Cohansey Formations. Cross-section shows lateral lithologic variation within Kirkwood Formation along strike.

found in the southern portion of New Jersey in Salem and Cumberland Counties, (2) a finely laminated, clayey-silt, known as the Asbury Park Member, which is found resting on Eocene greensands in the northeastern portion of the State, and (3) a very fine-grained, silty, orange, yellow and white sand, which has been designated the Grenloch Sand Member. This unit is, in part, contemporaneous with the Asbury Park and Alloway Clay Members but also overlies them. The generalized lithologic variation along strike is shown in figure 1. All three members of the Kirkwood Formation, though easily distinguished in outcrop, rapidly lose their identity downdip and cannot be differentiated, lithologically, in deep wells.

ALLOWAY CLAY MEMBER

The Alloway Clay Member consists of light tan to dark brown or gray, silty-clay which, frequently, is encrusted with iron oxides and surfates formed by oxidation of disseminated grains of pyrite. Its marine origin is indicated by the reported presence of abundant mollusk fossils found in marl pits in the upper portion of this member near the turn of the century. Richards and Harbison (1942) correlated fauna collected from the fossiliferous zone (known locally as the "Shiloh Marl") with the lower middle Miocene Calvert Formation of Virginia and Maryland and indicated that the fauna was of a middle-neritic nature. Fossils are generally rare or absent in present outcrops, however, and the fossiliferous localities described in earlier literature have, long since, been covered by slumping sands and gravels of the overlying Bridgeton Formation of Pleistocene age.

Mineralogically, the clays of this member are remarkably homogeneous both laterally and vertically in the section (table 1). The higher percentage of montmorillonite and the finer texture of the sediments, both of which reflect marine depositional conditions, easily allow this member to be differentiated from the other two, in outcrop. Locally the Alloway Clay Member has undergone extensive diagenetic alteration, however, with the most striking change involving a transformation of illite and montmorillonite clays near the updip base of the member to kaolinite with concurrent epitaxial growth. This has resulted in the formation of an enriched kaolinitic horizon, possessing abundant "macro-kaolinite" grains, which appears to be unique in the geological literature. So unusual is this unit that, for years, its kaolinitic nature went unrecognized and it was mistakenly called a "micaceous talc-like clay." This horizon ranges up to 10 feet or so in thickness and may, discontinuously, cover an area of about 20 square miles. Textural analyses indicate that the largest kaolinite grains are on the order of 65 mesh (0.2mm) but that most are between 100 mesh (0.149 mm) and 200 mesh (0.074 mm) in size. Figure 2 shows locations where the macro-kaolinite was found by augering. At location 200, the enriched macro-kaolinite zone was found in a well drilled by the U. S. Geological Survey at a depth of 184 feet and persisted down to a depth of 196 feet. Here, as in all other locations, the horizon is located in close proximity to the base of the formation and apparently has formed as a result of upward leaching by ground waters in the underlying Eocene Vincentown Formation. Few surface exposures

TABLE 1. REPRESENTATIVE X-RAY ANALYSES FROM ALLOWAY CLAY MEMBER OF THE KIRK-WOOD FORMATION (IN PERCENT)

Sample Number	Montmorillonite	Mixed Layer	Illite	Kaolinite	Quartz	Feldspar	Gibbsite
862A	15	—	35	45	5	—	—
826B	10	5	30	50	5	—	—
826C	15	5	45	35	5	—	—
851A	25	—	40	30	5	—	—
851B	20	—	30	40	5	—	—
851C	20	—	40	30	5	—	—
858A	10	10	30	35	10	5	—
855B	10	10	30	40	5	5	—
852A	15	10	25	35	10	2	1
852B	15	10	30	40	5	—	—
852C	10	10	30	35	10	trace	—
852D	25	—	30	35	10	2	—
1817A [*]	15	10	20	50	5	1	trace
1817B [*]	5	5	10	70	5	—	2
1897B [*]	3	5	10	70	10	—	2

[*] Samples 1817 and 1897B are from the Macro-kaolinite horizon

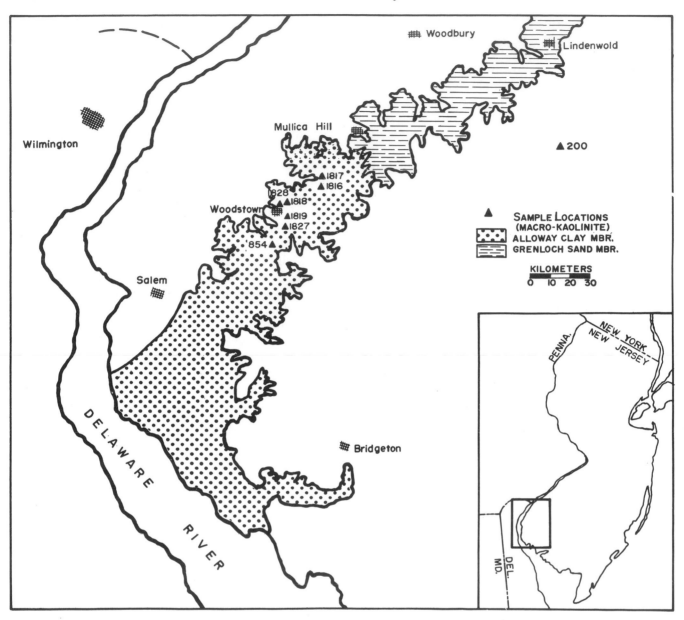

Figure 2. Outcrop map of the Alloway Clay Member of the Kirkwood Formation showing locations where macro-kaolinite was found.

are known of this horizon, as it weathers rapidly on exposure, but several have been found in the vicinity of Woodstown, New Jersey, where the Alloway Clay thinly overlies the Manasquan Formation. A more detailed account of the origin and mineralogy can be found in the discussion by Isphording and Lodding (1968).

The Alloway Clay Member is interpreted as having been deposited during the initial transgression of the middle Miocene seas and can be traced laterally, in outcrop, to the vicinity of Ewan, New Jersey, where it merges with the lower portion of the Grenloch Sand Member and loses its identity. As with the other two members, this unit retains its identity, downdip, for only a few miles and, thereafter, contemporaneous strata can only be identified by using fossils.

ASBURY PARK MEMBER

In the vicinity of Asbury Park and extending west toward Farmingdale, New Jersey, sediments which are equivalent in age to the Alloway Clay Member are again encountered. These sediments form the Asbury Park Member of the Kirkwood Formation and appear to represent material deposited in transitional marine environments, such as marshes, lagoons, or estuaries. This unit consists of dark brown, organic-rich, finely laminated, silty, clayey sands whose clay mineralogy is strikingly different from that of the Alloway Clay Member. This can be seen in table 2 which shows the average mineral content for the minus 2-micron fraction for samples from each of the two members. The difference in mineralogy coupled with the marked textural difference indicates that even though this unit is contemporaneous with the Alloway Clay Member, as indicated by faunal correlation, that they are environmentally dis-similar and do not form a continuous unit. South of Point Pleasant, New Jersey, well logs indicate that the Asbury Park Member is completely lacking, having given way to thick beds of marine silt and fine sand. West of Famingdale, the Asbury Park Member becomes interbedded with, and overlain by, the clayey sands that make up the lower portion of the Grenloch Sand Member and can no longer be traced.

TABLE 2. COMPARISON OF AVERAGE MINERAL CONTENT OF MINUS 2-MICRON FRACTION FOR ALLOWAY CLAY AND ASBURY PARK MEMBERS (IN PERCENT)

	Alloway Clay Member	Asbury Park Member
Montmorillonite	17	5
Mixed Layer	5	4
Illite	31	35
Kaolinite	39	44
Quartz	7	11
Feldspar	trace	—
Gibbsite	trace	—
Lepidocrocite	—	trace
Goethite	—	trace

GRENLOCH SAND MEMBER

The name Grenloch Sand Member was proposed by Isphording (1968) for the structureless, fine grained, silty, sands that make up the bulk of the Kirkwood Formation in the central portion of the New Jersey Coastal Plain. This unit is the largest of the three members in areal extent and it is found resting on Paleocene and Eocene greensands from Sewell, New Jersey, northward to the vicinity of Farmingdale. Stratigraphically, it grades laterally into both the Alloway Clay and Asbury Park Members and also overlies them. It consists, typically, of lignitic, clayey, fine sands near the base, with occasional lenses of fine gravel, which grade upward into pastel yellow, orange, and white silty, fine sands displaying prominent Liesegang banding (iron diffusion banding). Near the top the sediments generally coarsen and reflect a final regres-

sion of the sea in late middle Miocene time. The coarsening of the sediments in the upper part of the Grenloch Sand Member has made the placing of the Kirkwood-Cohansey contact difficult in wells and has been responsible for the common mis-identification, in outcrop, of this unit for that of the overlying upper Miocene Cohansey Sand. It has also resulted in a controversial argument regarding the conformable or unconformable nature of the Kirkwood-Cohansey contact. This controversy has raged for almost 50 years with the contact being described as everything from a "slight disconformity," involving only a minor regression, to a major unconformity which involved a complete withdrawal of the Miocene sea to beyond the present shoreline. In order to shed more light on this problem the senior author, in 1966, subjected numerous samples from over 30 different locations to detailed mineralogical and textural analysis to determine if a means existed which would permit differentiation of the Kirkwood and Cohansey Formations both in outcrop and downdip and which might show, more clearly, the exact nature of the contact.

DIFFERENTIATION OF THE KIRKWOOD AND COHANSEY FORMATIONS

The Cohansey Sand, in contrast to the Grenloch Sand Member of the Kirkwood Formation, was found to consist chiefly of medium to coarse grained white to orange-brown sand which, more often than not, displayed well developed bedding or was cross-bedded. Clay lenses were also common in the Cohansey and ranged from a few inches to almost 10 feet in thickness. Richards and Harbison (1942) correlated the Cohansey with the Yorktown Formation of Virginia (upper Miocene) and considered it the non-marine equivalent of the Kirkwood Formation. Though part of the Cohansey undoubtably represents fluvial deposition, recent investigation now indicates that the Cohansey was also deposited in transitional marine environments, namely swamps, deltas, mudflats, and lagoons.

Detailed mineralogical analysis of the Kirkwood and Cohansey Formations led to the conclusion that it would be difficult, if not impossible, to differentiate the formations, either in outcrop or downdip, solely on this basis. The provenance areas for both are identical, hence, the mineralogy is similar. Table 3 shows a comparison of their respective heavy mineral suites and indicates that only minor differences occur; these were not found

TABLE 3.   COMPARISON OF AVERAGE NUMBER PERCENT OF HEAVY MINERALS FOR THE KIRKWOOD AND COHANSEY FORMATIONS

	Kirkwood Formation	Cohansey Sand
Andalusite	VR	VR
Chloritoid	R	VR
Corundum	VR	VR
Epidote	VR	VR
Garnet	VR	VR
Kyanite	VR	VR
Monazite	VR	VR
Rutile	R	R
Sillimanite	R	R
Staurolite	R	R
Tourmaline	R	R
Zircon	C	C
Black Opaques	VA	VA
White Opaques	C	C

>75%   Flood (F)
50–75%   Very Abundant (VA)
25–50%   Abundant (A)
15–25%   Very Common (VC)
5–15%   Common (C)
1–5%   Rare (R)
<1%   Very Rare (VR)

significant enough to be used as a criterion. Similar results were obtained with the light mineral suite. Feldspar, which is essentially absent in the Cohansey Sand, is common in the lower and middle part of the Grenloch Sand Member but is rare to absent near the top. The clay minerals, which occur as lenses, matrix, and coating on sand grains, at first appeared to offer more promise as a means of differentiating the two formations but the differences were restricted to outcrop samples only and did not persist in wells only a few miles downdip. In outcrop, the Cohansey was found to consistently have lesser amounts of illite and greater amounts of kaolinite than the Kirkwood and was generally lacking in lepidocrite, a common mineral in the sediments of the Grenloch Sand Member.

The conclusion was reached that the only reliable means of differentiating the upper Kirkwood from the Cohansey, in outcrop, was on the basis of texture. The Cohansey Sand is coarser, will usually display well developed bedding or cross-bedding, and often contains lenses of white, almost pure, kaolinitic clay. The Grenloch Sand, in contrast, is generally finer grained, except near the contact, lacks bedding, and clay, if present, occurs as thin lenses which are often made up of alternating laminae of fine silt and clay-sized material, highly stained by iron oxide. Downdip, the placing of the contact is difficult, at best, and

for reasons to be discussed later, a range of several feet or more often must be used.

CONTACT RELATIONSHIP

Investigation of the contact relationship indicated, initially, that a disconformity did, in fact, exist. This could be seen by the development of as much as several feet of relief on the upper surface of the Kirkwood Formation, in outcrop. The unconformable nature was also indicated by increases or decreases of certain heavy mineral percentages and by a decrease in the amount of clay and silt-sized material as the contact was crossed (fig. 3). Detailed examination of well samples across the contact zone, however, did not yield any textural or mineralogical information that would indicate a disconformable relationship. Changes in texture were of a gradual nature or any minor abrupt changes noted could not be correlated with other wells in the area. This incongruity between well and outcrop samples was most baffling until an outcrop was found in the central part of the State, near Ft. Dix, and another further north, in the vicinity of Farmingdale, where the Cohansey Sand seemed to rest with apparent conformity on the upper sands of the Kirkwood. Textural analyses from both of these locations revealed that the change from a low to a high energy environment (marine to transitional marine) was of a gradual nature and began well below the position of the contact as mapped by the U. S. Geological Survey. In addition, the sediments of the Grenloch Sand Member at both locations were distinctly cross-bedded and clearly reflected a change in depositional environment from moderate depth to shallow water (fig. 4).

The Cohansey Sand, at least toward its base, represents a continuation of transitional marine deposition which, following a brief transgression, was later coupled with a gradual eastward migration of the shoreline. The local disconformity, seen between the Kirkwood and Cohansey Formations in some outcrops, probably resulted from a minor regression of, at best, a few miles and was of short duration. The conclusion was thus reached that there is no justification for postulating a widespread withdrawal of the sea from the present Coastal Plain, following deposition of the Kirkwood sediments, and that the Cohansey Sand lies, essentially, conformably on the Kirkwood Formation.

Well samples were made available by Dr. Kemble Widmer, State Geologist, New Jersey.

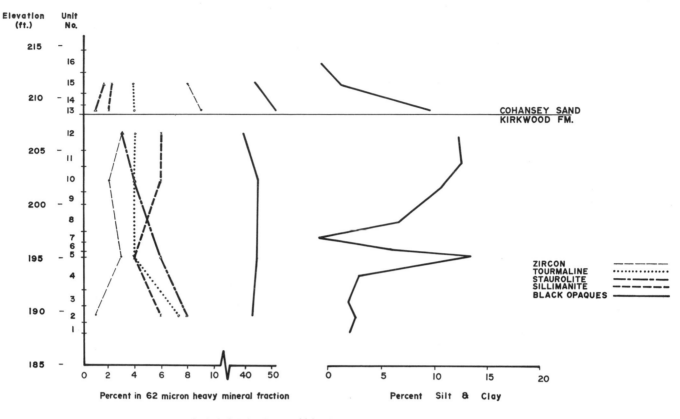

LOCATION KC-9

Figure 3. Vertical variation of relative mineral percent (62 micron fraction) and silt-clay percentages.

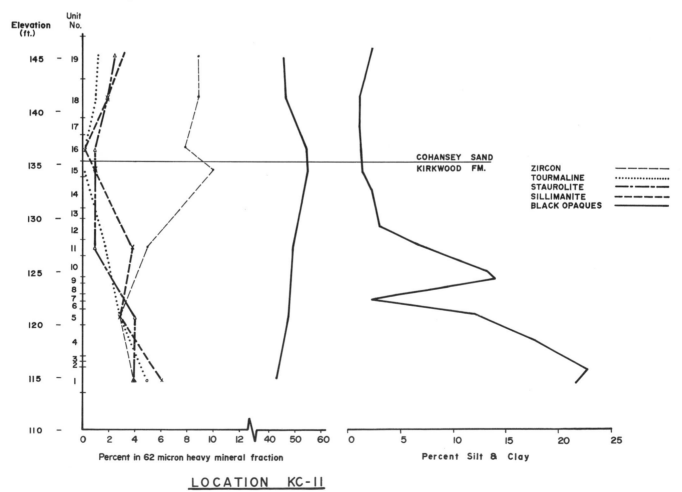

LOCATION KC-11

Figure 4. Vertical variation of relative mineral percent (62 micron fraction) and silt-clay percentages.

REFERENCES CITED

Isphording, W. C., Upper Tertiary paleoclimatology of eastern United States: submitted to Amer. Assoc. Petroleum Geol. Bull., October, 1968.

Isphording, W. C., and Lodding, W., 1968, Origin of the Woodstown, New Jersey macro-kaolinite: Clays and Clay Minerals, v. 16, No. 3, p. 257–264.

Richards, H. G., and Harbison, A., 1942, Miocene invertebrate fauna of New Jersey: Phila. Acad. Nat. Sci. Proc., v. 94, p. 167–250.

THE STRATIGRAPHY AND CLAY MINERALOGY OF THE EOCENE SEDIMENTS OF THE NORTHERN NEW JERSEY COASTAL PLAIN

RICHARD ENRIGHT

Rutgers University, New Brunswick, New Jersey 08903

INTRODUCTION

Early works (table 1) on the New Jersey Coastal Plain indicated a three-fold subdivision of the Upper Marl Bed of Cook (1868). Clark (1894, 1898) included the lower two members in the Manasquan Formation and the upper member was named the Shark River Formation. Since then, other investigators have attempted to lump the two formations in a single lithologic unit (e.g., Richards, Olmsted and Ruhle, 1962). Recent detailed investigation of the type localities and numerous subsurface samples indicate that this single lithology approach is unwarranted. The Manasquan Formation is seen to consist of two members (figs. 1 and 2) as originally described. The upper or Deal Member is the dominant Manasquan lithology in the subsurface. In addition, the Shark River Formation is redefined to include a new subsurface unit (Toms River Member).

Planktonic foraminiferal zonation (table 2) of the Eocene sediments (Olsson, 1967; Enright, 1969) indicates that deposition of the Manasquan and Shark River Formations spanned the interval from the latest Paleocene to approximately middle middle Eocene. Although deposition was apparently continuous, the condensed section and spotty distribution of well samples makes it impossible to ascertain whether or not there are any minor breaks in the Eocene section.

STRATIGRAPHY

MANASQUAN FORMATION

FARMINGDALE MEMBER

The lower member of the Manasquan is herein designated the Farmingdale Member for its good exposures in and near Farmingdale, Monmouth County, New Jersey. The type section is located

TABLE 1. STRATIGRAPHIC NOMENCLATURE OF THE EOCENE SEDIMENTS OF THE NEW JERSEY ATLANTIC COASTAL PLAIN

Cook (1868)		Clark (1894, 1898)		This Report	
Not recognized		Not recognized		Shark River Formation	Toms River Member [*]
Upper Marl Bed	Blue Marl	Shark River Formation			Squankum Member
	Ash Marl	Manasquan Formation	Upper Member	Manasquan Formation	Deal Member
	Green Marl		Lower Member		Farmingdale Member

[*] Subsurface only

14

on the south bank of the Manasquan River 2,500 feet upstream from Southard Avenue. The Farmingdale Member is in general a mottled, slightly clayey, medium to coarse, quartzose glauconite sand. It is coarse grained, slightly feldspathic, and is more quartzose in the lower portion and more clayey in the upper portion, where it grades into the overlying Deal Member. Molds, casts, and occasional shell materials occur near outcrop in a layer approximately 10 to 15 feet above the base of this unit. This layer seems to persist along strike and is a good stratigraphic marker.

Contact relations with the underlying Vincentown Formation wherever observed are sharp. In the northern part of the state, the basal Farmingdale contact is in several places marked by the presence of quartz granules and small pebbles. The abrupt change to a glauconite sand from the underlying slightly glauconitic quartz sand of the Vincentown Formation is quite distinctive. South of New Egypt, the Farmingdale lies directly upon the cemented layers of the upper Vincentown Formation. Minard, Owens, and Nichols (1964) and Minard (written communication, March, 1966) found pieces of cemented Vincentown in the basal few feet of the Farmingdale Member just north of Medford, which suggests an unconformable relationship. Hence, the presence of small pebbles and granules and, in other places, pieces of the Vincentown in the basal Manasquan support some reworking of the Vincentown during deposition of the basal beds of the Farmingdale. Olsson (1967), noted the poor representation of the upper Paleocene *Globorotalia velascoensis* zone in the northeastern part of the coastal plain and believed that possibly a disconformity existed between the Vincentown and Manasquan. He also noted, if a disconformity existed, it was probably of very short duration.

Approximately eight miles downdip, the Vincentown Formation undergoes a rapid facies change to a micaceous, glauconitic, silty, fine grained, quartz sand. This facies appears to grade upward into the basal Farmingdale Member. The disconformable relationship observed between the Vincentown and Manasquan near outcrop would therefore seem to be the result of local submarine erosion or winnowing.

Outcrop thickness of the Farmingdale Member averages about 40 feet in the northern part of the State. To the southwest, along strike as far as Medford there is no appreciable thinning. Southwest of Clementon exposures and well-log data are very sparse, hence, the extent of the Farmingdale cannot be ascertained at this time. Jordan

(1962) does not report the unit from Delaware. Downdip, the Farmingdale becomes more clayey and thinner, and shows facies relations with the underlying Vincentown Formation and the overlying Deal Member. At distances of 20 miles downdip, the Farmingdale appears to pinch out.

In the vicinity of Asbury Park, it is apparent (Olsson, 1967) that the basal 10 or 11 feet of the Farmingdale contain a latest Paleocene planktonic foraminiferal fauna. The rest of the unit contains a *Globorotalia subbotinae* (=*rex*)-*Globorotalia aequa* zone fauna indicating an early Eocene age. In bore-holes drilled by the author southwest of New Egypt, the lowest fossiliferous

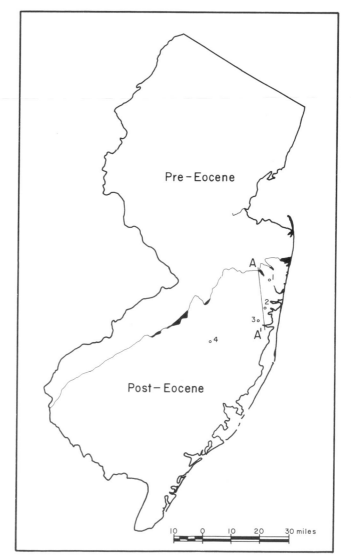

Figure 1. Map of New Jersey showing Eocene outcrops in black. Line A-A' is the section line for figure 2. Small circles are important wells referred to in text: 1, Allaire State Park Well; 2, Bricktownship Intermediate School Well; 3, Toms River Chemical Company Well no. 84; 4, United States Geological Survey Tests Well at Butler Place, Lebanon State Forest.

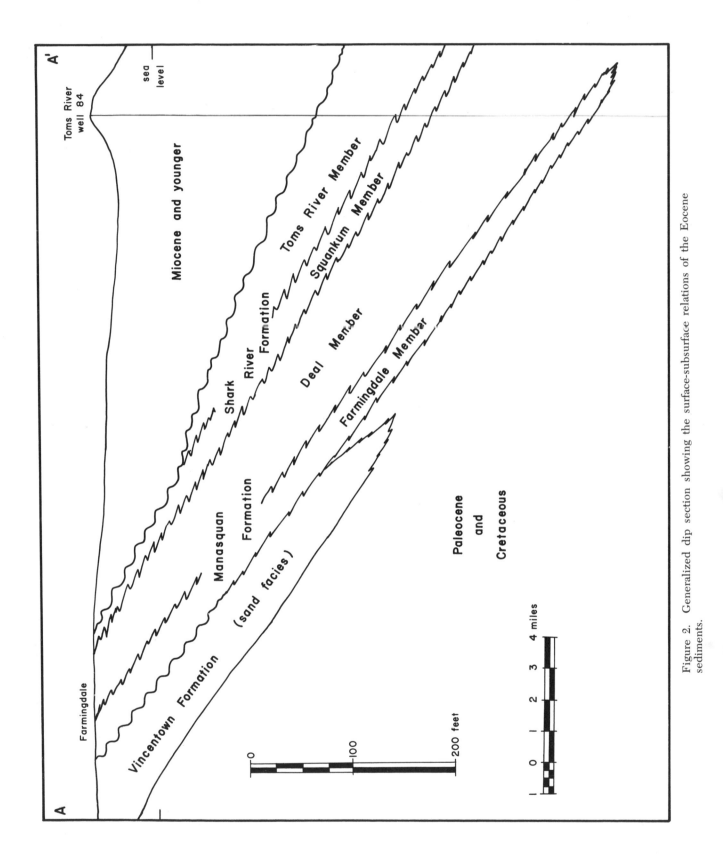

Figure 2. Generalized dip section showing the surface-subsurface relations of the Eocene sediments.

TABLE 2. PLANKTONIC FORAMINIFERAL BIOSTRATIGRAPHY OF THE EOCENE SEDIMENTS OF THE
NORTHEASTERN PART OF THE NEW JERSEY ATLANTIC COASTAL PLAIN (AFTER OLSSON (1967)
AND ENRIGHT (1969))

Age	Planktonic Foraminiferal Zones	Lithologic Units updip downdip	
Middle Eocene	*Turborotalia centralis* Zone	Shark River Formation	Toms River Member (subsurface)
	Subbotina frontosa Zone		Squankum Member
Lower Eocene	*Pseudohastigerina sharkriverensis* Zone	Manasquan Formation	Deal Member
	Subbotina inaequispira Zone		
	Globorotalia formosa formosa Zone		
	Globorotalia subbotinae-G. aequa Zone		Farmingdale Member
Pal- eocene	*Globorotalia velascoensis* Zone		Vincentown Formation

core of the Farmingdale was obtained about 10
to 15 feet above the base of the unit. The fauna of
this sample consisted of typical *Globorotalia subbotinae* (=*rex*)-*Globorotalia aequa* zone forms thus
substantiating the zonal assignment.

DEAL MEMBER

The upper member of the Manasquan Formation (Ash Marl of Cook, 1868) is herein designated
the Deal Member for the good exposures in the
vicinity of the Boro of Deal, Monmouth County,
New Jersey. The type section is located on the
grounds of the Hollywood Golf Club on the north
bank of Hog Swamp Brook. This 14 foot section
is exposed approximately 100 feet downstream from
the Roseld Avenue bridge. The Deal Member
ranges from very slightly glauconitic, clayey,
fine grained, quartz sand to a clayey sandy silt.
Downdip the unit ranges from a slightly sandy
clayey silt to a silty clay. The unit is finely laminated and also contains scattered cristobalitic
nodules which in many cases are porcellanite.
When wet the Deal is dark to medium gray green
and yellowish-gray (5Y 7/2) to pale greenish-yellow
(10Y 8/2), or greenish-gray (5GY 6/1) (ashen) when

dry. Another good exposure is to be found on the
Manasquan River just south of Preventorium
Road. This outcrop appears to be the southwestern most exposure of the Deal Member.
Submarine exposures have been cored directly
offshore from Deal and six miles to the southeast
by the Lamont Geological Observatory research
vessel, R/V Goldberger (Fray and Ewing, 1960).
After storms, many of the porcellanite nodules
can be found along the beach in the general area
of Deal. Thickness of this unit ranges from approximately 40 feet near outcrop to about 140 feet in
the Toms River area.

This member grades into the underlying Farmingdale Member and the overlying Squankum
Member of the Shark River Formation. As the
Farmingdale and Squankum become more clayey
and less glauconitic downdip they are replaced
by the Deal Member. The upper and lower boundaries of the Deal are placed where coarse glauconite
sand grains appear in more than trace amounts.
Downdip the Deal Member is a distinctive lithology and can be traced a long distance. Owens
(1967), Parker and others (1964), and other investigators have apparently misidentified the Deal

Member in the subsurface and either called it an unnamed formation or placed it in the Vincentown. This misidentification appears to be due to the lack of knowledge of the Deal Member exposures and the absence of faunal data.

An extensive, well preserved microfauna is contained in the Deal Member well samples. Foraminifera, radiolarians, diatoms, coccoliths, discoasters, hystrichospheres, and sponge spicules are common. In some samples the entire sand fraction is composed of foraminifera. Outcrops are generally barren except for siliceous fossils. The enclosed planktonic foraminiferal fauna indicates that deposition of this member spanned the uppermost part of the *Globorotalia subbotinae* (=*rex*) zone to the basal *Subbotina frontosa* zone (lowermost lower Eocene to lowermost middle Eocene).

SHARK RIVER FORMATION

SQUANKUM MEMBER

The Shark River Formation, as redefined, is expanded to include the Squankum Member and a new unit—the Toms River Member, which is recognized only in the subsurface. The Squankum Member is herein designated for the Blue Marl of Cook (1868) which is excellently exposed on the Manasquan River 1.1 miles northwest of Squankum, Monmouth County and 100 yards east of the New Jersey Central Railroad bridge. The Squankum Member is an argillaceous, glauconite sand. Very coarse sand size glauconite grains are common, and they seem to be suspended in a cream to gray brown matrix. An indurated layer one to three feet thick commonly caps the formation in outcrop and serves to preserve the numerous molluscan molds and casts. Occasionally one can find traces of bedding but in general extensive burrowing has eliminated them. The clay content of this unit increases with depth, and hence it grades downward into the Deal Member. In the subsurface, it grades rapidly upward into the Toms River Member by an increase in quartz sand. Thickness of the Squankum ranges from about 40 feet near outcrop to 20 feet or less in the subsurface.

The geographic extent of the Squankum is difficult to determine as pre-Kirkwood erosion has removed the member in many places. Facies relations also tend to obscure the true extent of the Squankum Member in the subsurface. Typical Squankum exposures are limited to a few areas in Monmouth County. The Squankum outcrop, Clark's (1894) type locality of the Shark River,

and the exposures along an unnamed tributary to Deal Lake, are the only exposures known to the writer. A 14-foot exposure approximately 300 yards downstream from the type locality of the Deal Member on Hog Swamp Brook is probably Squankum. The lithology of this exposure is somewhat intermediate between the typical Squankum and typical Deal. It is a creamy, clayey, fine sand which contains many large glauconite sand grains. Also present are shark teeth and casts and molds of *Vernericardia antiquata* (Whitfield). South of this exposure a water well at Whitesville penetrated a similar lithology near the surface. The presence of many large glauconite sand grains suggests this lithology can be correlated with the Squankum. Typical Squankum lithology is present a short distance updip (at the Asbury Park Traffic circle). The absence of typical Squankum lithology at Whitesville and on Hog Swamp Brook is thought to be indicative of a rapid facies change. In the subsurface, the Squankum is present in the U.S.G.S. Butler Place test well, Toms River Chemical Company well no. 84, Bricktownship Intermediate school well and the Allaire State Park well and is found immediately below the Toms River Member.

The Squankum Member in the U.S.G.S. test well at Butler Place contains the *Subbotina frontosa* planktonic foraminiferal zone and in the Bricktownship well it contains the *Pseudohastigerina sharkriverensis* zone fauna as well as the basal part of the *Subbotina frontosa* zone faunas, thus indicating migration of the Squankum facies in a southwest direction with time.

TOMS RIVER MEMBER

The upper member of the Shark River Formation is restricted to the subsurface and is designated the Toms River Member from studies of core samples obtained from well no. 84 of the Toms River Chemical Company (altitude 60 feet) Toms River, Ocean County. In the type well, this unit occurs from approximately 160 feet to 240 feet below sea level. It is a micaceous, slightly clayey, slightly glauconitic, fine to medium quartz sand. In general, samples contain many shell fragments. Most of the sand grains are angular to subangular and the percentage of sand and grain size increase as the upper limit of the member is reached. Cuttings from upper samples of other wells are quite coarse grained. This member grades into the underlying Squankum Member and is unconformably overlain by the basal clays of the Kirkwood Formation of Miocene age. The Toms River Member has been found in sample cuttings obtained from wells at the Allenwood School and Allaire State Park, Monmouth County; Butler Place in Lebanon State Forest and Atsion, Burlington County; and the type well and Bricktownship Intermediate School well, Ocean County. Thickness of this member ranges from a featheredge to over 100 feet in a downdip direction. In the area studied, this member contains the upper part of the *Subbotina frontosa* zone and the *Turborotalia centralis* zone and is therefore of middle Eocene age.

Core samples of the type well are on file at the Geology Department, Rutgers — The University of New Jersey. Duplicate samples are also on file at the New Jersey Geological Survey and the Water Resources Division District Office of the U. S. Geological Survey in Trenton, New Jersey. The well was drilled by the Layne–New York Drilling Company and completed in March, 1968.

To the southwest in Maryland and Delaware, a somewhat similar lithology overlies the Nanjemoy Formation of Eocene age underlies the Miocene in the subsurface. It was described as glaucontic sands and interspersed shell beds of Jackson age and named the Piney Point Formation by Otton (1955). Rasmussen, Groot, and Depman (1958) traced this unit in the subsurface to northeastern Delaware, and Richards, Olmsted, and Ruhle (1962), and Parker, and others (1964) have traced it as far north as Atlantic City. So far as known (Richards, and others and Parker and others) it has not been found north of Atlantic City. The original description and subsequent usage of the term Piney Point has placed too much emphasis on its supposed "Jackson age" and not enough on its lithologic and stratigraphic relationships with other subsurface units in Maryland, Delaware and New Jersey. The general lithologic description of the Piney Point is similar to the original description of the Shark River lithology (Cook, 1868) and seems identical to the Toms River lithology. The Toms River may therefore subsequently prove equivalent to the unit identified in New Jersey as the Piney Point by Richards, Olmsted and Ruhle (1962).

CLAY MINERAL DISTRIBUTION

X-ray diffraction analysis of the minus two micron fraction of the Eocene sediments indicates the existence of three distinct clay mineral assemblages (lower, middle and upper). Oriented slides of the magnesium saturated minus two micron fraction were prepared for diffraction analysis. The slides were also run after glyceration and heating to 300° and 550°C.

Quartz was present in all clay fractions. Calcite content varied greatly and appears to be related to the presence of calcareous nannofossils. It is conceivable that some of the calcite could be finely comminuted shell material of larger organisms. Small amounts of feldspar were identified in the lower and upper clay mineral assemblages. In addition, trace amounts of the zeolite, clinoptilolite, were identified in the middle assemblage. This identification was substantiated by petrographic examination of the fine silt fraction.

For the purpose of this report, only gross clay mineral assemblages need be discussed. The lower clay assemblage consists of montmorillonite, kaolinite and illite as major constituents. Small amounts of glauconite and feldspar also are present. This assemblage is found in the Farmingdale Member and is very similar to the clay mineral assemblage of the underlying Vincentown Formation.

The middle assemblage is quite different in that it contains no kaolinite and is almost entirely composed of montmorillonite. Illite and glauconite are also present. This assemblage also contains small amounts of clinoptilolite. Gill (oral communication 1968) has also found what he considers to be clinoptilolite in equivalent sediments from the subsurface. The middle assemblage is confined to the Deal Member and the lower part of the Squankum Member.

Kaolinite, montmorillonite and illite are again present in substantial amounts in the upper part of the Squankum Member and the Toms River Member. Feldspars are present in trace amounts in the lower part of this assemblage. In addition, glauconite is present in more than trace amounts in the upper part of this assemblage.

Owens, Minard, and Blackmon (1961) examined the clay minerals from four samples of the Manasquan Formation southwest of New Egypt. Three of the four samples, NE.67, PEm. 35, and PEm. 11, are apparently from the Farmingdale Member. The clay minerals, in general, correspond to the clay minerals of the lower assemblage of this report. The fourth sample, PEm. 3, appears to be from strata transitional between the Farmingdale and the Deal Member (oral communication, Minard, 1966). This assignment is substantiated by the montmorillonitic character of the clay fraction which indicates that this sample belongs to the middle clay assemblage.

Analysis of several Manasquan sections has indicated that the depth of weathering and acid leaching is about 40 feet. As the above four samples are from within 40 feet of the surface, the more complex clay minerals reported by Owens, Minard and Blackmon are probably alteration products.

Both the Farmingdale and the Toms River clay mineral assemblages are similar to assemblages from the Paleocene sediments of the New Jersey Coastal Plain and detrital marine clays in general. On the other hand, the Deal assemblage is unique. Reynolds (1966), and Deffeyes (1959) have concluded that the combined occurrence of clinoptilolite, montmorillonite, and cristobalite in clay deposits is the result of the diagenesis of rhyolitic ash. The occurrence of a montmorillonite—clinoptilolite clay fraction and porcellanite nodules would therefore seem to indicate that acidic volcanic sources contributed fine sediments to the Deal Member. Although no direct evidence of volcanic activity was found in these Eocene sediments, Jordan and Adams (1962), have found altered shards in a thin montmorillonite bed from the lowermost Tertiary of Delaware. The possible sources of these volcanics is at present indeterminate.

CONCLUSION

Examination of well samples and outcrops of the Eocene sediments of the northeastern New Jersey Coastal Plain indicate that sedimentation was continuous from the latest Paleocene to at least the middle middle Eocene. These Eocene sediments are divided into two formations—the Manasquan Formation and Shark River Formation.

The Manasquan Formation consists of the newly named Farmingdale and Deal Members. Along strike to the southwest, the Farmingdale persists as far as Medford. In the subsurface it appears to wedge out at distances of 20 miles downdip. On the other hand the Deal Member thickens downdip and is an important unit in the subsurface. Its unique clay mineral assemblage substantiates its lithologic continuity in the subsurface and also indicates that volcanic sources contributed significant amounts of volcanic ash to this member.

The Shark River Formation as redefined includes the Squankum Member and the overlying Toms River Member, which is restricted to the subsurface. The quartzose Toms River Member rapidly thickens downdip where it too becomes a major unit in the subsurface.

REFERENCES CITED

Clark, W. B., 1894, Origin and classification of the greensands of New Jersey; Jour. Geol., v. 2, p. 161–177.

—— 1898, Report upon the Upper Cretaceous formations, 1897; New Jersey Geol. Surveys Rept. State Geol. 1897, p. 161–210.

Cook, G. H., 1868, The geology of New Jersey: New Jersey Geological Survey.

Deffeyes, K. S., 1959, Zeolites in sedimentary rocks: J. Sed. Pet., v. 29, no. 4, p. 602–609.

Enright, R., 1969, Eocene planktonic foraminiferal zonation of the New Jersey Atlantic Coastal Plain: Am. Assoc. Petroleum Geologist Bull., Abstracts, Dallas Meeting, 1969.

Fray, C. T., and Ewing, J., 1961, Project 555-Monmouth County "Offshore borings," Report no. 1: unpublished, Columbia University.

Jordan, R. R., 1962, Stratigraphy of the sedimentary rocks of Delaware: Delaware Geol. Survey, Bull., No. 9.

—— and Adams, J. K., 1962, Early Tertiary bentonite from the subsurface of central Delaware: Geol. Soc. America Bull., v. 73, p. 395–398.

Minard, J. P., Owens, J. P. and Nichols, T. C., 1964, Pre-Quaternary geology of the Mount Holly Quadrangle, N. J.: U. S. Geol. Survey. Geologic Quadrangle Map, GQ-272.

Olsson, R. K., 1967, Early Tertiary planktonic foraminiferal zonation of New Jersey: in Proceedings at the International Planktonic Conference, Geneva, in Press.

Otton, E. G., 1955, Ground-water resources of the southern Maryland Coastal Plain: Maryland Dept. Geol., Mines and Water Res. Bull., 15, 347 p.

Owens, J. P., 1967, Map I-514-B, Sheet 2, in Engineering geology of the northeast corridor Washington, D. C., to Boston, Massachusetts: Coastal plain and surficial deposits: U. S. Geol. Survey, Misc. Geol. Inv. Map I-514-B.

—— Minard, J. P. and Blackmon, P. D., 1961, Distribution of clay-sized sediments in the coastal plain sediments near Trenton, New Jersey: U. S. Geol. Survey, Prof. Paper 424-C, p. c317–c319.

Parker, G. G., Hely, A. G., Keighton, W. B., Olmsted, F. H., and others, 1964, Water resources of the Delaware River Basin; U. S. Geol. Survey Prof. Paper, 381, 200 p.

Rasmussen, W. C., Groot, J. J., and Depman, A. J., 1958, High capacity test well developed at the Air Force Base, Dover, Delaware: Delaware Geol. Survey Rept. of Investigations No. 2, 36 p.

Reynolds, W. R., 1966, Stratigraphy and genesis of clay mineral and zeolite strata in the lower Tertiary of Alabama, p. 26–37, in Copeland, C. W., Editor, Facies changes in the Alabama Tertiary: Ala. Geol. Survey Guidebook no. 4, 103 p.

Richards, H. G., Olmsted, F. H. and Ruhle, J. L., 1962, Generalized structure contour maps of the New Jersey Coastal Plain: New Jersey Geol. Survey, Geologic Report Series No. 4, 38 p.

PETROLOGY AND ORIGIN OF THE HARDYSTON QUARTZITE (LOWER CAMBRIAN) IN EASTERN PENNSYLVANIA AND WESTERN NEW JERSEY [1]

JOHN M. AARON

U. S. Geological Survey, San Juan, Puerto Rico 00936

ABSTRACT

The Hardyston Quartzite contains a varied lithic assemblage that consists principally of feldspathic sandstone, arkose, and orthoquartzite, with lesser quartz pebble conglomerate, silty shale, and, locally, jasper. The unit is about 100 feet thick and forms the base of the Paleozoic section in eastern Pennsylvania and western New Jersey, where it overlies Precambrian crystalline rocks of the Reading Prong and is in transitional contact with the superjacent Leithsville Formation of Early to Middle Cambrian age.

The textural fabric of the Hardyston is broadly graded. Discontinuous, lenticular beds of quartz pebble conglomerate and coarse, poorly sorted arkose are common in the lower part of the Hardyston. The upper part is composed of finer, better sorted arkosic sandstone, orthoquartzite, and silty shale.

Field and petrographic studies show that the Hardyston detritus was derived from a high-grade igneous and metamorphic terrane essentially similar to the Reading Prong. Mineralogy and textures indicate that the source was quite local, probably the Reading Prong itself.

The Hardyston probably is both alluvial and marine in origin. Poorly-sorted angular arkosic detrital material (represented by the lower part) formed an alluvial apron on the low-lying flank of a shield of Precambrian rocks (the Reading Prong). These deposits were partly reworked and augmented by better sorted orthoquartzitic marine deposits (upper Hardyston) of an advancing Early Cambrian sea.

INTRODUCTION

The Hardyston Quartzite is about 100 feet thick and forms the base of the Paleozoic section in eastern Pennsylvania and western New Jersey, where it unconformably overlies Precambrian crystalline rocks of the Reading Prong. It has, of course, been recognized and mapped since the beginning of geologic work in the area, but all the early work and much of that done more recently, was concerned with problems of stratigraphic definition and mapping and the resultant structural and tectonic implications. Little attention has been given to the Hardyston as a unit of sedimentation that records important aspects of its history not normally available simply by mapping.

This paper summarizes briefly the current state of knowledge concerning the Hardyston, with particular emphasis on petrography and petrology. It results from some of the writer's recent work in the area. Most of the data were obtained from exposures in the Nazareth quadrangle, Pa. (fig. 1), and the quadrangles immediately east and south (Easton, Bloomsbury, Hellertown, and Riegelsville), but important localities in other quadrangles also were studied.

I am grateful to A. A. Drake, Jr., and R. B. Mixon, U. S. Geological Survey, for their very constructive technical reviews of this paper, and particularly to Drake, my coworker in this area, for several years of penetrating discussion of the structural, stratigraphic, and petrologic problems.

[1] Publication authorized by the Director, U. S. Geological Survey.

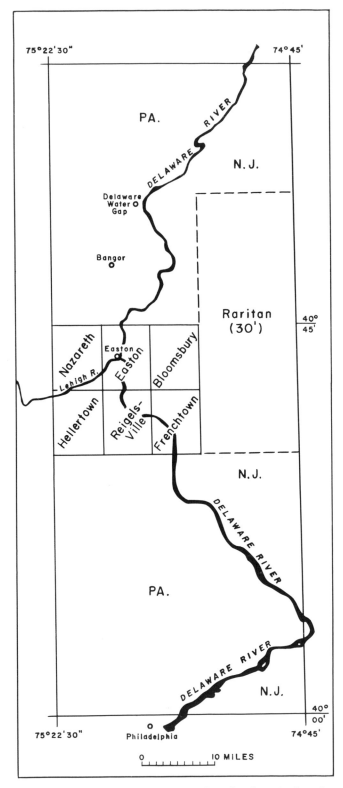

Figure 1. Location of area and quadrangles described in this report.

NAME

The Hardyston Quartzite was named for Hardistonville (Wolff and Brooks, 1898), a village near Franklin Furnace, Sussex County, N. J., where it is particularly well exposed. The name was shortened to Hardiston (Kümmel and Weller, 1901), the township name, which later was changed to Hardyston.

Formerly, the unit has been designated as Formation I of the lower Secondary rocks (Rogers, 1838), the Primal White Sandstone (Rogers, 1858), and the Potsdam sandstone (Prime, 1883). The discovery that the Potsdam sandstone in New York State (Van Ingen, 1902; Weller, 1903) was much younger than rocks called Potsdam in New Jersey and Pennsylvania led to the abandonment of the term in the latter areas.

DISTRIBUTION

The Hardyston Quartzite commonly is closely associated with Precambrian rocks, not only for obvious stratigraphic reasons but for structural reasons as well. Mapping in the area has shown that even where Precambrian rocks are allochthonous and far-traveled, the Hardyston usually is carried along. The same is rarely true for the younger carbonate rocks that overlie the Hardyston. Thus, Hardyston is exposed or present as float on one or both flanks of most of the Precambrian ridges in the area. Figure 2 summarizes the distribution of the Hardyston in the quadrangles on which the bulk of this report is based.

LITHOLOGY

The Hardyston Quartzite contains a rather varied assemblage of lithic types, principally feldspathic sandstone, arkose, and orthoquartzite, with lesser quartz-pebble conglomerate, silty shale, and, locally, jasper. Carbonate rocks occur in the Hardyston in New Jersey. There is a broad textural gradation within the Hardyston; conglomerate and coarse sandstone are more common at or near the base and shale is more common at or near the top. These are not generally mappable as stratigraphic subdivisions, however.

Arkose and feldspathic sandstone are light gray to very light gray and orthoquartzite is grayish orange-pink; all weather to moderate brown or moderate yellowish-brown (all rock-color designations determined from Rock-color chart (Goddard, 1948). Conglomerates are composed largely of iron-stained light- to dark-gray granules and pebbles of quartz and feldspar in a dark, poorly sorted, argillaceous matrix of finer quartz and feldspar.

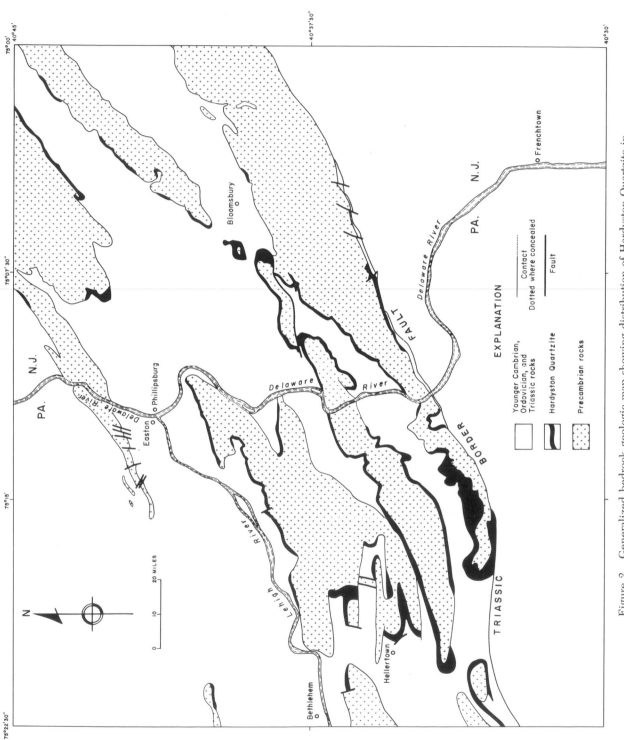

Figure 2. Generalized bedrock geologic map showing distribution of Hardyston Quartzite in part of eastern Pennsylvania and western New Jersey (in the quadrangles outlined in fig. 1). Data compiled from: Drake, 1967a, b; Drake and others, 1961, 1967; Miller and others, 1939, 1941; Willard and others, 1959 (Hellertown quadrangle); and J. M. Aaron, unpub. data (Nazareth quadrangle).

The granules and pebbles range in maximum diameter from 2 mm to 100 mm but most are less than 50 mm. They are subangular to subround, and the larger ones are platy to elongate. Shales are pale red or pale yellowish-brown, weather moderate yellowish-brown, and are silty and sandy. In all rock types, quartz grains are vitreous light to dark gray, but vitreous blue quartz is abundant locally. Pyrite is commonly a conspicuous constituent; it occurs as a fine "dust" (0.1 mm diam.) widely disseminated throughout individual hand specimens, or as larger (0.2 mm–0.5 mm) blebs and cubes.

Compositional stratification in sandstone and conglomerate typically ranges in thickness from 5 cm to 1.5 m, sometimes more, and in small exposures may be very difficult to locate. Stratification in all Hardyston rock types is irregular in thickness and rather discontinuous laterally, especially in the lower part. Cross-strata, small scale and planar, are present locally but the relative paucity of outcrops and the complexity of folding preclude meaningful systematic statistical study of them.

PETROGRAPHY

COMPOSITION

Modes, based on detailed examination of thin sections of representative Hardyston lithic types, are presented in table 1. Each mode represents approximately 400 counts. The total percentage of accessories is based on the point count, but the variety and relative abundance of these species is based on estimates by visual scan of the entire section and by examination of heavy-mineral separates from crushed rock.

Arkose and orthoquartzite are the end-member rock-types and all gradations exist between these rock-types, which have much the same character over the entire area.

MINERALOGY

Quartz. Quartz is the most abundant mineral in all Hardyston rock types, constituting from 33 to 80 percent of the rock (table 1). Most quartz is sand size and angular to subangular. Well-rounded quartz grains are rare.

Microscopic examination of thin sections reveals that almost without exception quartz in the Hardyston occurs as single grains with strongly undulose extinction (fig. 3A). The rare exceptions are composite grains with the same extinction. The predominance of this type of quartz strongly suggests that there was but a single source of the detrital quartz, most probably a high-grade metamorphic terrane. Moreover, in terms of microscopic morphology, the quartz in the Hardyston in no way differs from that of the several quartz-bearing gneisses and metamorphosed granitic

TABLE 1. MODES (VOLUME PERCENT) OF TYPICAL HARDYSTON ROCK TYPES

Field Number [1]	Arkose		Arkosic sandstone		Orthoquartzite	
	166-21	B-O-7-2	152-9	166-66	R-K³-32-C	166-48
Quartz	56.4	33.0	75.3	60.4	73.3	78.5
Potassium feldspar	20.4	15.0	10.3	12.0	3.3	1.0
Perthitic feldspars (all varieties)	4.8	16.7	4.7	4.8	—	1.3
Plagioclase	—	0.5	—	—	—	—
Silica cement	11.7	20.6	6.2	6.2	19.0	13.4
Clay and micromica	4.3	13.7	1.6	8.6	3.4	2.5
Accessories [2]	2.4	0.5	1.9	8.0	1.0	3.3
Zircon	C	tr	tr	tr	tr	—
Monazite	C	—	C	C	tr	tr
Sphene	tr	—	tr	tr	—	—
Tourmaline	tr	—	C	C	—	—
Garnet	—	tr	—	—	—	—
Rutile	—	—	tr	—	—	—
Topaz	—	—	tr	—	—	—
Hornblende	—	tr	—	—	—	—
Pyrite	tr	—	tr	C	tr	—
Magnetite	A	A	A	A	A	C
Limonite	C	C	C	C	—	C

[1] Nazareth quadrangle: 166-21, 152-9, 166-66, 166-48
Bloomsbury quadrangle: B-O-7-2 (collected by A. A. Drake, Jr.)
Reigelsville quadrangle: R-K³-32-C (collected by A. A. Drake, Jr.)
[2] Visual estimates of relative abundance of species in the detrital accessory fraction are limited as follows: A, 50 percent or greater; C, 10 to 50 percent; tr, 10 percent

rocks that underlie a major part of the Reading Prong in this area.

Feldspar. Feldspar is common to abundant in most of the Hardyston. The most abundant variety is potassium feldspar, mostly microcline with prominent grid twinning (fig. 3A). Feldspars having microperthitic, mesoperthitic, or micro-antiperthitic structure are also common, and in some areas these collectively constitute the principal feldspar in the rock (for example B-O-7-2, table 1). Plagioclase, chiefly oligoclase, is rare.

All of the above minerals are commin in the Precambrian rocks presently exposed in the Reading Prong. Surprisingly, however, plagioclase is much less abundant in the Hardyston than one might expect if the Reading Prong as it stands today were the ultimate source of the detritus, as plagioclase, chiefly albite-oligoclase-andesine, is quite abundant there. The dominant feldspars in the Hardyston are typical only of a small fraction of rocks in the Reading Prong. The predominance of the potassium feldspars must be due either to the relative unavailability of the plagioclase-bearing gneisses to erosion and transport, an unlikely possibility, or to the superior resistance and stability of the potassium feldspars. Differentiation of the latter sort is considered quite likely as there are clear examples of it in the coastal plain of New Jersey (J. P. Owens, oral commun., 1967). The possibility that the Reading Prong is not the source of the Hardyston detritus is considered highly unlikely in the light of other data to be presented.

The feldspar, like the quartz, is chiefly angular to subangular and most is of sand size not significantly different from the quartz. Although feldspars in transport should undergo a greater size reduction than quartz, the fact that quartz and feldspar are in the same size range in any given Hardyston sample is not significant because the feldspars probably were much larger initially (fig. 4D, for example).

The feldspars are fresh to moderately altered, but on the whole remarkably fresh. Alteration of microcline consists of patchy areas of dusty brown cloudiness or turbidity, which conventionally is ascribed to kaolinitization, but which also may be due to vacuolization or bubble formation (Folk, 1961). In no case, however, are microcline grains reduced to a mass of micromica or clay. The alteration is always of minor volumetric importance. Perthitic feldspars are altered differentially according to the composition of the exsolution lamellae (fig. 3B). Sodic lamellae are weakly to moderately altered to sericite; the potassic lamellae are quite fresh or are altered weakly in the fashion described for microcline. Plagioclase is weakly to moderately altered to sericite.

Clay and micromica. Clays and micromica are minor constituents of the Hardyston, constituting less than 5 percent of the sandstones and as much as 50 percent of the shales. These minerals are predominantly sericite-illite, principally derived by weathering of feldspars in the source area, and secondarily by diagenetic alteration of feldspars after deposition. There is a significant correlation between feldspar and clay content (table 1).

Magnetite and limonite. Magnetite is the most abundant heavy accessory mineral; it composes from 35 to 95 percent (by volume) of the heavy fraction, but averages about 50 percent of that fraction. Limonite, used here for poorly crystalline hydrated iron oxides of uncertain identity, is commonly associated with magnetite. Both occur as irregularly shaped grains and rare euhedral and subhedral crystals in the finer fraction of the sandstones and conglomerates. Limonite also occurs, though rarely, as very round opaque grains from 0.5 to 1 mm in diameter, and both the enclosing sand grains and the round limonite grains are very well sorted. These could possibly be pseudomorphs after glauconite as glauconite readily oxidizes under subaerial conditions. The magnetite is detrital, but much of the limonite is authigenic.

Pyrite. Pyrite is present in conspicuous amounts in some Hardyston sandstones but volumetrically it is a minor constituent. It occurs as cubes and irregularly shaped grains widely scattered throughout the sediment. Undoubtedly it is authigenic in origin.

Monazite. Monazite, an unusually common accessory in the Hardyston, is second in abundance only to magnetite and limonite. It occurs as colorless to pale yellow, angular to subround, subequant to slightly elongate sand- and silt-size grains. Subhedral and euhedral detrital grains are not at all uncommon (fig. 3D).

Although monazite is quite resistant to chemical destruction (Pettijohn, 1957), it is not notably resistant to abrasion (Friese, 1931; Cozzens, 1931). The angular, subhedral, and euhedral detrital grains suggest an extremely local source. Monazite is a common accessory in granites and pegma-

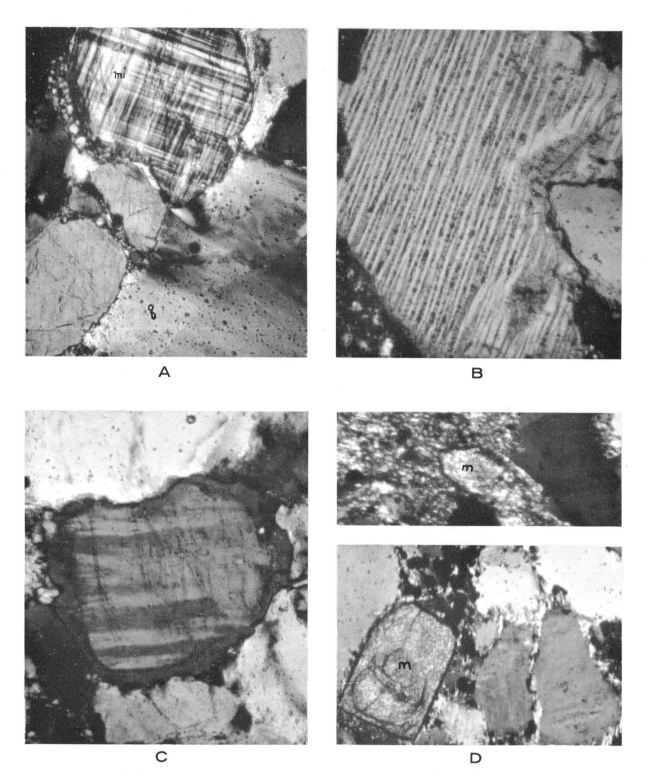

Figure 3.   Photomicrographs showing some petrographic features of the Hardyston Quartzite. A, Arkose with quartz showing undulose extinction (q) and microcline (mi).   Crossed polarizers, ×100, B, Differentially weathered detrital mesoperthite grain.   Lighter lamellae are potassium feldspar, darker are altered sodic feldspar.   Crossed polarizers, ×100.   C, Authigenic feldspar overgrowth on detrital feldspar grain.   Crossed polarizers, ×100.   D, Euhedral (above) and sub-hedral detrital monazite grains (m) with angular quartz grains.   Crossed polarizers, ×100.

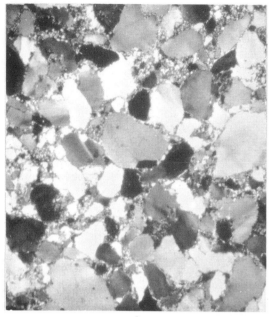

A

B

C

D

Figure 4. Photomicrographs and photographs of Hardyston Quartzite. A, moderately sorted, subangular to subround quartz and feldspar grains in clay and silica cement. Note nature and variety of grain-to-grain contacts and compare with figure 4B. Crossed polarizers, ×30. B, Orthoquartzite showing sutured grain boundaries. Crossed polarizers, ×30. C, *Scolithus* tubes in orthoquartzite. Pencil points to well-developed tube emerging from plane of outcrop. D, Typical conglomerate from lower part of Hardyston. Large clasts are potassium feldspar.

tites, both of which occur in the Reading Prong. Moreover, the prong is known to have at least one monazite-rich belt several miles long and several hundred feet wide within the gneissic terrane in New Jersey (A. A. Drake, Jr., oral commun., 1967); therefore, a local source for the monazite in the Hardyston is all the more probable.

Zircon. Trace amounts of zircon occur throughout the Hardyston. The grains are small (0.125–0.05 mm), colorless, extremely well rounded (in marked contrast to the monazite), and contain very few inclusions. Very probably the zircon is multicyclic and was derived from older metasedimentary rocks similar to those of the Reading Prong, the Canadian Shield, or even the Piedmont rocks to the east.

Tourmaline. Common in some specimens, virtually absent from others, tourmaline occurs in subequant to elongate, angular to rounded grains of medium to fine sand. The grains are strongly pleochroic but the colors are variable; typically, they range from very pale brownish-yellow to pale green (E) and from dark brownish-green to dark green (O). Neither inclusions nor zoning is present; the grains are clear. These color varieties would appear to be part of the dravite (Mg-rich)-schorlite (Fe-rich) series that occurs in granitic and metamorphic or metasomatic rocks (Deere and others, 1966). In the Reading Prong tourmaline is present in some gneisses and granites, and in Precambrian pegmatites.

Other accessories. Garnet, hornblende, rutile, sphene, and topaz are found in trace amounts in some heavy separates of the crushed rock. They are very rarely seen in thin section and are not common enough to generalize or discuss their optical properties or occurrence.

TEXTURE

The microscopic textural fabric of most Hardyston specimens is generally homogeneous. The commoner exceptions are streaks and stringers of magnetite, clay, or mica, and larger grains of quartz and feldspar.

The degree of compaction or tectonic compression is indicated by the nature of grain-to-grain contacts. Most contacts range from straight to concavo-convex, suggestive of minor to moderate pressure solution (Adams, 1964). However, the entire range of possibilities, from sand grains "floating" in silica or clay cement to highly sutured grains, is present at one place or another in the Hardyston (figs. 4A, B).

Grain enlargement by secondary overgrowth on quartz and feldspar is common (fig. 3C). The material forming the overgrowths presumably is the result of diagenetic precipitation but it is no less conceivable that it was generated by pressure solution. Some of the most striking overgrowths, particularly on feldspar, occur in samples with abundant highly sutured grain boundaries. Abraded overgrowths have not been observed in the Hardyston.

Results of grain-size analyses using the microscope and sieve-size conversion methods developed by Friedman (1958; 1962) are summarized for four representative samples (table 2). Grain-size frequency distributions were determined by measuring the apparent long axes of grains in a thin-section and converting this apparent size to the equivalent sieve size according to Friedman's

TABLE 2.   SUMMARY OF TEXTURAL PARAMETERS OF TYPICAL HARDYSTON ROCK TYPES

Field No.	Arkose 166-21	Arkosic sandstone 152-9	Orthoquartzite R-K³-32-C	166-48
Mean grain size (mm)	0.34 Medium sand	0.7 Coarse sand	0.37 Medium sand	0.5 Medium-coarse sand
Sorting	Poor	Very poor	Moderate	Moderate-well
Standard deviation (φ)	1.21	2.2	0.79	0.53
Skewness	Coarse (−0.23)	Strongly coarse (−0.32)	Nearly symmetrical (0.07)	Fine (0.23)
Kurtosis	Very platykurtic (0.61)	Platykurtic (0.72)	Leptokurtic (1.20)	Leptokurtic (1.25)
Textural maturity	Submature	Submature	Mature	Mature
Grain shape	Very angular	Angular	Subangular to subround	Subangular

regression equation. The highly indurated character of all Hardyson rock-types precludes grain-size analysis by more conventional means. Thin sections with abundant sutured grain boundaries were excluded from the size analyses. The moment measures, determined graphically according to the methods of Folk (1961), are the statistical parameters used to summarize the frequency distributions.

Grain shape, a complex function of original shape, internal anisotropy, composition, directional properties, and depositional history, was determined on quartz and feldspar in the Hardyston by estimation and comparison with the roundness scale of Powers (1953). Results are summarized in table 2. Roundness ranges from very angular to subrounded, but most sand-size grains are subangular to angular. Roundness sorting is fair to good, and there is no apparent variation in roundness with composition. Shapes (sphericity) do vary with composition, however, because feldspar grains commonly are bounded by cleavage surfaces.

Textural maturity (Folk, 1961) is a function of degree of winnowing (clay content), rounding, and sorting. Hardyston rock types range from immature (greater than 5 percent clay and fine mica) to mature (less than 5 percent clay, well sorted but not well rounded). The Hardyston, however, is most typically submature (less than 5 percent clay, but grains poorly sorted and poorly rounded).

The textural properties described above tend to vary with stratigraphic position. In general, detrital grains are coarser and more poorly sorted in the lower part of the unit. This is only a generality however, and cannot be used to determine the stratigraphic position ("upper" or "lower") of an isolated outcrop.

STRATIGRAPHIC RELATIONS

The Hardyston Quartzite unconformably overlies Precambrian crystalline rocks. The contact is seldom exposed, and a mapper usually must infer its presence and location from float.

Locally, in this area, the Hardyston-Precambrian contact is marked by a deposit of light-green cryptocrystalline material that has been called pinite (Miller and others, 1939; Virgin, 1956), a fine mixture of muscovite with some serpentine or chlorite mineral and iron oxides. Although this material has been interpreted as a metamorphosed remnant of a residual soil developed on underlying gneiss (Miller and others, 1939; Virgin, 1956), more probably it is simply a mylonite or shear zone at the Hardyston-Precambrian interface.

The contact with the superjacent Leithsville Formation is transitional through a series of silty and shaly beds. This contact is also very poorly exposed and commonly faulted. As with the Precambrian contact, the geologist is forced to muster all of his regional biases and intuition in order to locate it with precision.

THICKNESS

There is much uncertainty concerning the thickness of the Hardyston in eastern Pennsylvania and western New Jersey; outcrops are scarce, long sections are nonexistent, and contacts are commonly faulted.

In the Delaware Valley the Hardyston is reported to be as much as 200 feet thick (Johnson and Willard, 1957). Miller (1939) reported a range of 0–200 feet in the Lehigh Valley. Mapping and construction of geologic cross-sections by Drake (1967), Drake and others (1967) and by the writer (unpub. data, Nazareth quadrangle) support a thickness of no more than 100 feet in these quadrangles. A section 45 feet thick was measured at low water in the Delaware River in the Riegelsville quadrangle (see Drake, this volume).

In the many areas where the Hardyston is reported as missing from its stratigraphic position, faulting is undoubtedly responsible. There is no unequivocal field evidence to support erosion or nondeposition in these cases.

Regionally, the Hardyston thickens considerably southwestward along the Reading Prong. Buckwalter (1963) reports a thickness of 600–800 feet in the Womelsdorf quadrangle, at the extreme southwest end of the prong.

AGE

No fossils of conclusive chronologic significance have been found in what is currently mapped as Hardyston in eastern Pennsylvania. The only "fossil" in that area is *Scolithus linearis* (fig. 4C), found at a few localities in the Lehigh Valley. The taxonomic history of *Scolithus* is so diverse and its paleontologic significance so obscure that any interpretations based on it are questionable indeed. There are even opinions to the effect that what has been called *Scolithus* in some areas, including the type area at Chickies Rock, near Columbia, Pa., is not biogenic at all, but tectonic!

Olenellus thompsoni has been found in considerable numbers in calcareous beds of upper Hardyston at three localities in the Raritan quadrangle (fig. 1), New Jersey (Weller, 1903; Bayley and others, 1914). On this basis the Hardyston in eastern Pennsylvania is considered to be Early Cambrian in age.

CORRELATION

In eastern Pennsylvania and adjacent New Jersey the Hardyston directly overlies the Precambrian igneous and metamorphic complex (table 3). It would appear, therefore, to correlate physically with the Hellam Conglomerate Member of the Chickies Quartzite of southeastern Pennsylvania, with the Weverton Sandstone and Loudoun Formation of south-central Pennsylvania and adjacent Maryland and Virginia, and with the Poughquag Quartzite of southeastern New York. All these units are basal clastics in roughly equivalent stratigraphic sections.

In contrast to this apparent physical correlation is the location of the lowest *Olenellus* horizon within the stratigraphic section at the above localities. This horizon is in the upper part of the Hardyston and Poughquag in New Jersey and New York, respectively, but southwestward to central Pennsylvania, Maryland, and northern Virginia, it is progressively 1,200 to 2,600 feet stratigraphically higher, in the Antietam Quartzite. Paleontologically, therefore, the upper part of the Hardyston would appear to be correlative with the Antietam.

What is the significance of this apparent paradox? Is there any more compelling reason to consider the Hardyston necessarily coeval with the Antietam rather than with the Chickies and Weverton? What is the significance of the wandering *Olenellus?* How does the Hardyston fit into the regional stratigraphic picture of the middle Appalachians? An answer to these questions must be based on considerations of the probable origin of these units in terms of Early Cambrian stratigraphy, geography, and tectonics.

ORIGIN OF THE HARDYSTON QUARTZITE

The Cambrian stratigraphic sequences and lithostratigraphic correlations in the middle Appalachians are, for the most part, rather well known (table 3). However, strong and ever-mounting evidence of very complicated structural relations and tectonic history of these rocks greatly hampers attempts to produce an integrated system- and process-oriented sedimentation picture.

Recent work in the Great Valley by the U. S. Geological Survey (Drake, 1967a, b; Drake and others, 1967; J. M. Aaron unpub. data) and the Pennsylvania Geological Survey (Field Conference of Pennsylvania Geologists, 1966) has shown convincingly that much of the Cambrian and Ordovician sequence between the Susquehanna and Delaware Rivers is regionally inverted and allochthonous. These rocks are involved in large-scale nappes and are displaced westward, probably

TABLE 3.   CORRELATION OF CAMBRIAN FORMATIONS IN THE EASTON AREA, PA.–N. J., AND NEIGHBORING AREAS

Age	Cumberland Valley, Pa. (Field Conf. Pa. Geologists, 1966) Autochthonous	Reading area, Pa. (Geyer and others, 1963) Allochthonous	Easton area, Pa.–N. J. (Drake, 1967a) Allochthonous	Eastern New York (Broughton and others, 1962) Allochthonous
Late Cambrian	Conococheague Group 2,150 feet	Conococheague Group 2,000 feet	Allentown Dolomite 1,700 feet	Stockbridge Group 4,000 feet
Middle Cambrian	Elbrook Formation 3,000 feet	Buffalo Springs Formation 1,000 feet	Leithsville Formation 1,000 feet	
Middle Cambrian	Waynesboro Formation 1,000 feet	Leithsville Formation 1,000 feet	Leithsville Formation 1,000 feet	
	Tomstown Formation 1,000–2,000 feet			
Early Cambrian	Antietam Sandstone 500–800 feet	Hardyston Formation 250–600 feet	Hardyston Quartzite 100 feet	Poughquag Orthoquartzite 250 feet
Early Cambrian	Harpers Formation 2,750 feet			
	Weverton Sandstone 1,250 feet			
Precambrian	Volcanic rock	Gneiss	Gneiss	Gneiss

by as much as 35 miles. In contrast, rocks of similar age and gross character in the Cumberland Valley section of the Great Valley, west of South Mountain and southwest from the Susquehanna River, are autochthonous (Field Conference of Pennsylvania Geologists, 1966). Thus tectonic events have juxtaposed initially widely separated parts of the depositional basin, telescoping the original facies relations and obscuring the depositional strike. Whatever relation, if any, exists between the ancient predeformation facies boundaries and the present regional strike of the rocks is totally unknown. Thus, it is extremely difficult at this time to frame the origin of the Hardyston Quartzite in very meaningful paleogeographic terms.

Consideration of the physical and mineralogic properties of the Hardyston (table 4) does not necessarily lead to a unique solution regarding the origin of the detrital material and its environment of deposition. Ancient, complex, poorly exposed units of this sort, including its relatives throughout the Appalachians, do not have entirely convenient analogies among modern sediments. Modern sediments abound with cases in which the application of the interpretative principles that are conventionally applied to ancient rocks would produce conclusions that are diametrically opposite to what we know to be true by direct observation, i.e., interpretations of wind, water current, climate, source, etc. It is much easier, obviously, to understand sediment properties, types, and dispersal patterns when most of the controlling variables are reasonably well known. The reverse path, from sediment properties to the controlling variables, is infinitely more difficult, correspondingly more

TABLE 4. SUMMARY OF PROPERTIES OF HARDYSTON QUARTZITE

Composition: Mineralogy is very similar to subjacent Precambrian rocks, which are probable ultimate source. Quartz and microcline are principal minerals. Cement is both chemical and detrital. Faunal remains are absent to scarce and restricted. Little argillaceous material except in upper part. Authigenic limonite, pyrite, feldspar, and quartz.

Texture: Sorting is poor to moderate in lower part, moderate to good in upper. Rounding is fair to poor. Sand sizes predominate in all mineral species. Highly variable from outcrop to outcrop and from bed to bed.

Sedimentary structures: Crossbedded; exposures too poor to judge abundance reliably; no data on orientation. *Scolithus*(?) burrows in well washed sands. Beds tend to be lenticular and discontinuous, with erosional scour.

Internal organization: Irregularly asymmetric, with grain size decreasing stratigraphically upward. Conglomerates common in lower part.

Contacts and associated lithologies: Nonconformable basal contact. Upper contact transitional into silty shales and carbonates.

speculative, and less reliable. Some reasonable generalizations, however, can be made about the Hardyston, and there are some possible implications which must await further work.

The detrital mineralogy of the Hardyston rather strongly indicates that the ultimate source of the material was a high grade metamorphic and igneous terrane essentially similar to the Precambrian rocks currently exposed in the Reading Prong. The textures of the Hardyston suggest that the source of the detritus was quite local, because even the detrital grains that should have undergone fairly rapid physical modification are little altered, i.e., little rounded and nonspherical. Indeed many crystal faces and forms are still preserved.

All the tangible evidence and the absence of any evidence to the contrary indicate that Hardyston detritus was transported and deposited in water. That at least the upper part of the Hardyston is marine is indicated by the very well-washed, well-sorted sands, the possible glauconite, and the presence of marine fossils in calcareous beds. The poorly-sorted, poorly-washed, laterally and vertically variable, unfossiliferous arkosic sandstones and conglomerates of the lower Hardyston may well be alluvial, however. The paleogeographic picture thus envisioned is one of an Early Cambrian sea gradually inundating the low-lying flank of a shield of Precambrian rocks, which were thinly veneered by a broad apron of alluvial deposits. The encroaching sea reworked the upper part of the alluvial material. Subsequent deposits (upper part of Hardyston) were better sorted, better washed, finer (as the shoreline became more distant), and less variable.

The presence of relatively fresh feldspars in the Hardyston raises some interesting questions about the climate and relief in the source area. It may indicate that the relief was great enough to allow streams to breach the mantle of weathered rock (if there was any), thereby exposing fresh rock to vigorous stream action. Although the lack of strongly altered feldspars of the same initial composition as the fresh would argue against this, there is the possibility that the altered material was virtually destroyed in transport. One clue as to the relief on the basement complex is found in the Blue Ridge of northern Virginia, near Luray, where the Catoctin Formation of Reed (1955) (upper Precambrian or Lower Cambrian volcanic rocks) rests on a granitic basement having a relief of as much as 1,000 feet. Presumably the geomorphic history of the basement complex in northern Virginia was not too significantly different from that of the basement in eastern Pennsylvania, but this is questionable. Another possibility is that the local climate simply was too dry or too cold to support chemical weathering; or perhaps climatic variations were seasonal and extreme, and whatever clays and micromicas formed in a given wet season were spalled away, along with fresh rock, by physical extremes in a thermally variable dry season, and washed away by the next round of torrential rains. If the latter were true, however, claystones ought to be more common throughout the Hardyston rather than virtually confined to the transitional zone with the overlying Leithsville. Or perhaps the clays were transported out of the immediate depositional area of the upper Hardyston by marine currents and were deposited farther from the shore.

Although the location of the source and the sediment dispersal patterns from it are more obscure than its petrologic nature, in all probability the Hardyston detritus came from a westerly or north-westerly direction, specifically a more extensively exposed Canadian Shield. Indeed, some paleogeographic reconstructions of the Early Cambrian (for example Lochman, 1956) place the shield well into eastern Pennsylvania even after partial marine submergence. Moreover, the Precambrian rocks of the Reading Prong itself probably were once part of the shield as the rocks of the prong are quite similar to those of the Grenville province of Canada and New York, and the Adirondacks, as noted by Engle (1956).

Studies of crossbedding in the Weverton Sandstone (Whitaker, 1955) and the Chickies Quartzite (A. Hohl, unpub. data, 1964), both at least partial equivalents of the Hardyston, indicate a westward source for these sediments. Both studies show paleocurrent azimuths trending east and southeast. Unfortunately neither has much to conclude about the probable origin of these units. Both are unfossiliferous and in both the variability of crossbedding orientations is rather low (75 percent of Whitaker's Weverton orientations are within 30° either side of due east), suggestive of an alluvial origin (Potter and Pettijohn, 1963). The plausible possible marine environments (beaches, bars, tidal basins, shallow-water marine) ought to produce either bimodal or more variable orientations (Pettijohn and others, 1965).

Other than the deposits of possible alluvial origin, the character, distribution, and age of the lowest Cambrian clastic rocks in eastern and central North America clearly record a westward-moving marine transgression. The basal Cambrian sequence from southeastern Pennsylvania to southeastern New York is Early Cambrian (Howell, 1956; Fisher, 1956). In the Nittany Valley of central Pennsylvania, the oldest known Cambrian is the Waynesboro Formation of Swartz (1948) of Middle Cambrian age, which is believed to have a westward source (Rodgers, 1956). Along the southeastern flank of the Adirondack Precambrian mass and throughout the eastern and northern interior of the United States the earliest Paleozoic clastic rocks are Late Cambrian (Fisher, 1956; Lochman-Balk, 1956; Bell and others, 1956). Thus, the ubiquitous Cambrian clastic rocks basal to the Paleozoic in the eastern United States are time transgressive and decrease in age and thickness northward and westward. The Upper Cambrian clastic rocks of the central and eastern interior are viewed broadly as facies equivalents of the coeval shales and carbonate rocks of the eastern states, then far removed from shorelines and sources of coarse clastic sediment.

On a local scale, the Hardyston and Leithsville should not necessarily be considered in strictly "layer-cake" terms. They may indeed be partly contemporaneous facies equivalents, being no more than different parts of one and the same transgressive carbonate-orthoquartzite system. At any given Hardyston locality the superjacent Leithsville would, of course, be younger, with the contemporaneous Hardyston-equivalent Leithsville lying somewhat to the east. The transitional contact between these units and the presence of calcareous beds and scattered interbedded carbonate rocks in the upper part of the Hardyston are consistent with this interpretation.

SUMMARY AND CONCLUSIONS

Mineralogy and texture indicate that the Hardyston Quartzite is of essentially local derivation with both alluvial and marine aspects. The alluvial material is in the lower part of the unit and is poorly sorted, poorly rounded, feldspathic, conglomeratic, highly variable both laterally and vertically, and unfossiliferous. It originated as an alluvial mantle on the seaward flank of a Precambrian terrane of probable moderate relief. The upper part of the Hardyston is dominantly of shallow marine origin and is moderately to well-sorted, better rounded, cleaner, finer, less feldspathic, locally calcareous, contains *Olenellus*, and grades up into the overlying Leithsville Formation (dirty dolomite). The upper Hardyston and the Leithsville (along with the Antietam Quartzite-Tomstown

Formation) may be partly contemporaneous facies equivalents in a transgressive carbonate-orthoquartzite sedimentary system. Thus the Hardyston may be genetically related to both the Antietam (marine transgression) and the Weverton-Chickies (alluvium on Precambrian terrane).

Undoubtedly these inferences and speculations are simplified answers to old and complex questions. Although the outcrops are sparse and the data commonly are ambiguous, it is hoped, at the very least, that these problems will be pursued further until the pivotal evidence is found and critically evaluated.

REFERENCES CITED

Adams, W. L., 1964, Diagenetic aspects of Lower Morrowan, Pennsylvanian sandstones, north-western Oklahoma: Am. Assoc. Petroleum Geologists Bull., v. 48, p. 1568–1580.

Bayley, W. S., Salisbury, R. D., and Kümmel, H. B., 1914, Description of the Raritan quadrangle [New Jersey]: U. S. Geol. Survey Geol. Atlas, Folio 191, 32 p.

Bell, W. C., Berg, R. R., and Nelson, C. A., 1956, Croixan type area-upper Mississippi Valley, in Rodgers, J., ed., El Sistema Cámbrico, su paleogeografía y el problema de su base—symposium: Internat. Geol. Cong., 1956, Mexico City, tomo 2, parte 2, p. 415–446.

Broughton, J. G., and others, 1962, Geologic map of New York, 1961: New York State Mus. and Sci. Service, Geol. Survey, Map and Chart Ser. no. 5, 5 map sheets, scale 1:250,000, and text.

Buckwalter, T. V., 1963, Precambrian rocks and Hardyston Formation, in Geyer, A. R., and others, Geology and mineral resources of the Womelsdorf quadrangle: Pennsylvania Geol. Survey, 4th ser., Atlas A 177c (Bull. A 177c), 96 p.

Cozzens, A. B., 1931, Rates of wear of common minerals: Washington Univ. Studies, Science and Technology, no. 5, p. 71–80.

Deere, W. A., Howie, R. A., and Zussman, J., 1966, An introduction to the rock forming minerals: London, Longman's, 528 p.

Drake, A. A., Jr., 1967a, Geologic map of the Easton quadrangle, Pennsylvania–New Jersey: U. S. Geol. Survey Geol. Quad. Map GQ-594.

Drake, A. A., Jr., 1967b, Geologic map of the Bloomsbury quadrangle, New Jersey: U. S. Geol. Survey Geol. Quad. Map GQ-595.

Drake, A. A., Jr., McLaughlin, D. B., and Davis, R. E., 1961, Geology of the Frenchtown quadrangle, New Jersey–Pennsylvania: U. S. Geol. Survey Geol. Quad. Map GQ-133.

———— 1967, Geologic map of the Riegelsville quadrangle, Pennsylvania–New Jersey: U. S. Geol. Survey Geol. Quad. Map GQ-593.

Engle, A. E. J., 1956, Apropos the Grenville, in Thomson, J. E., ed., The Grenville problem: Royal Soc. Canada Spec. Pub., no. 1, p. 74–96.

Field Conference of Pennsylvania Geologists, 31st, Harrisburg, 1966, Comparative tectonics and stratigraphy of the Cumberland and Lebanon Valleys, by D. B. MacLachlan and S. I. Root: Harrisburg, Pa., Pennsylvania Geol. Survey, 90 p.

Fisher, D. W., 1956, The Cambrian System of New York State, in Rodgers, J., ed., El Sistema Cámbrico, su paleogeografía y el problema de su base—symposium Internat. Geol. Cong., 1956, Mexico City, tomo 2, parte 2, p. 321–352.

Folk, R. L., 1961, Petrology of sedimentary rocks: Austin, Texas, Hemphill's Book Store, 154 p.

Friedman, G. M., 1958, Determination of sieve size distribution from thin section data for sedimentary petrological studies: Jour. Geology, v. 66, p. 394–416.

Friedman, G. M., 1962, Comparison of moment measures for sieving and thin-section data in sedimentary petrological studies: Jour. Sed. Petrology, v. 32, p. 15–25.

Friese, F. W., 1931, Untersuchungen von Mineralen auf Abnutzbarkeit bie Verfrachtung im Wasser: Mineralog. petrog. Mitt., v. 41, p. 1–7.

Geyer, A. R., and others, 1963, Geology and mineral resources of the Womelsdorf quadrangle: Pennsylvania Geol. Survey, 4th ser., Atlas A 177c (Bull. A 177c), 96 p.

Goddard, E. N., Chm., and others, 1948, Rock-color chart: Washington, Nat'l. Research Council (repub. by Geol. Soc. America, 1951), 6 p.

Howell, B. F., 1956, The base of the Cambrian in New England and the Central Appalachians, in Rodgers, J., ed., El Sistema Cámbrico, su paleogeografía y el problema de su base—symposium: Internat. Geol. Cong., 1956, Mexico City, tomo 2, parte 2, p. 315–320.

Johnson, M. E., and Willard, Bradford, 1957, Delaware Valley Paleozoics [N. J.–Pa.], in Geol. Soc. America, Guidebook for field trips, Field Trip no. 4, Ann. mtg., 1957, Atlantic City, N. J., p. 125–131.

Kümmel, H. B., and Weller, Stuart, 1901, Paleozoic limestones of the Kittatinny Valley, New Jersey: Geol. Soc. America Bull., v. 12, p. 147–161.

Lochman, Christina, 1956, Stratigraphy, paleontology, and paleogeography of the Elliptocephala asaphoides strata in Cambridge and Hoosick quadrangles, New York: Geol. Soc. America Bull., v. 67, p. 1331–1396.

Lochman-Balk, Christina, 1956, The Cambrian of the middle central interior states of the United States, in Rodgers, J., ed., El Sistema Cámbrico, su paleogeografía y el problema de su base—symposium: Internat. Geol. Cong., 1956, Mexico City, tomo 2, parte 2, p. 447–482.

Miller, B. L., Fraser, D. M., and Miller, R. L., 1939, Northampton County, Pennsylvania: Pennsylvania Geol. Survey, 4th ser., Bull. C-48, 496 p.

Miller, B. L., and others, 1941, Lehigh County, Pennsylvania: Pennsylvania Geol. Survey, 4th ser., Bull. C-39, 492 p.

Pettijohn, F. J., 1957, Sedimentary rocks: New York, Harper and Bros., 718 p.

Pettijohn, F. J., Potter, P. E., and Siever, Raymond, 1965, Geology of sand and sandstone — Indiana Geol. Survey and Indiana Univ. Dept. Geology, Conf. 1965: Bloomington, Inc., Indiana Geol. Survey, 207 p.

Potter, P. E., and Pettijohn, F. J., 1963, Paleocurrents and basin analysis: New York, Academic Press, 296 p.

Powers, M. C., 1953, A new roundness scale for sedimentary particles: Jour. Sed. Petrology, v. 23, p. 117–119.

Prime, F., 1883, The Potsdam Sandstone, in Lesley, J. P., Geology of Lehigh and Northampton Counties [Pa.]: Pa. Geol. Survey, 2nd, D3, pt. 1, p. 205–214.

Reed, J. C., Jr., 1955, Catoctin Formation near Luray, Virginia: Geol. Soc. America Bull., v. 66, p. 871–896.

Rodgers, John, 1956, The known Cambrian deposits of the southern and central Appalachian Mountains, in Rodgers, J., ed., El Sistema Cámbrico, su paleoge-ografía y el problema de su base — symposium: Internat. Geol. Cong., 1956, Mexico City, tomo 2, parte 2, p. 353–384.

Rogers, H. D., 1838, Second annual report on the (1st) geological exploration of the State of Pennsylvania: Harrisburg, 93 p.

—— 1858, The geology of Pennsylvania; a government survey: Philadelphia, 2 v.: 586 and 1046 p.

Swartz, F. M., 1948, Trenton and sub-Trenton of outcrop areas in New York, Pennsylvania, and Maryland: Am. Assoc. Petroleum Geologists Bull., v. 32, p. 1493–1595.

Van Ingen, G., 1902, The Potsdam sandstone of the Lake Champlain basin; notes on field work 1901: N. Y. State Mus., Bull. 69, p. 1176–1227.

Virgin, W. W., Jr., 1956, Pinite schist in eastern Pennsylvania: Pennsylvania Acad. Sci. Proc., v. 30, p. 150–156.

Weller, Stuart, 1903, The Paleozoic faunas: New Jersey Geol. Survey Rept. on Paleontology, v. 3, 462 p.

Whitaker, J. C., 1955, Direction of current flow in some Lower Cambrian clastics in Maryland: Geol. Soc. America Bull., v. 66, p. 763–766.

Willard, Bradford, and others, 1959, Geology and mineral resources of Bucks County, Pennsylvania: Pennsylvania Geol. Survey, 4th ser., Bull. C 9, 243 p.

Wolff, J. E., and Brooks, A. H., 1898, The age of the Franklin White Limestone of Sussex County, New Jersey: U. S. Geol. Survey 18th Ann. Rept., pt. 2, p. 431–457.

THE PRECAMBRIAN GEOLOGY
OF THE CENTRAL AND NORTHEASTERN
PARTS OF THE NEW JERSEY HIGHLANDS

BENNETT L. SMITH
Department of Geology, Rutgers University, New Brunswick, New Jersey 08903

INTRODUCTION

The area generally referred to as the New Jersey Highlands is a belt about 20 miles wide crossing the northern part of the state in a northeast-southwest direction, underlain mostly by Precambrian rocks. The general geologic setting is shown in figure 1. Southeast of the Highlands are Triassic rocks of the Newark group. To the northwest are Paleozoic sedimentary rocks of the Valley and Ridge province. Paleozoic rocks also occupy certain long narrow valleys within the Highlands belt. One such valley, wider and more continuous than most, nearly bisects the Highlands into two blocks which in this report will be referred to as the northern and southern blocks.

The Precambrian rocks extend northeasterly into the Hudson Highlands of New York state and a few miles into Connecticut. They extend southwesterly to the vicinity of Reading, Pennsylvania, and the term Reading Prong has been used by some authors (e.g. Lobeck, 1932) with the geomorphic implication that the Highlands belt represents an extension of the New England Uplands. The term Reading Prong as used in this paper represents a province that is distinctive lithologically and structurally and quite different from the Berkshire and Housatonic Highlands in New England.

PREVIOUS GEOLOGIC WORK

Iron mining, which was begun in pre-Revolutionary days, was a widespread and active industry in the Highlands in the 1800's. The early geologic literature includes many references to the magnetite deposits and Sims (1958, p. 5, 6) in a report on the Dover district, gives a useful summary of this early literature. The first geologic maps of the area which made use of formation names appeared when Spencer (1908) proposed the names Pochuck, Losee and Byram in his report on the Franklin Furnace quadrangle; these names were widely used at one time, and appear on the current geologic map of the state.

Other published reports in the early part of this century include those of Bayley (1910) which comprises an important inventory of detailed information on individual mining properties, Fenner (1914) with a paper on the origin of gneisses in the Highlands, Berkey and Rice (1919) who described the geology of the West Point quadrangle in the Bear Mountain area of New York state, Bayley (1941) with a report on the Delaware Water Gap and Easton quadrangles, and Lowe (1950) who documented the petrology and structure of the Bear Mountain area in the Hudson Highlands of New York and established the term Storm King granite for the major granitic rock type of that area.

By the time of these later reports, iron mining had long since declined because of competition from other sources but during World War II, under the general direction of A. F. Buddington, the U. S. Geological Survey began a study of most of the principal iron ore districts in the New Jersey Highlands as part of its northeastern United States iron ore program (Sims, 1958, p. 6). The resulting publications included those of Hotz (1953) concerning the Ringwood, N. J. and contiguous Sterling Lake, N. Y. areas, Sims and Leonard (1952) describing iron deposits in the Andover mining district, and Sims (1953) with a comprehensive study of the Dover area. More recently Buddington and Baker

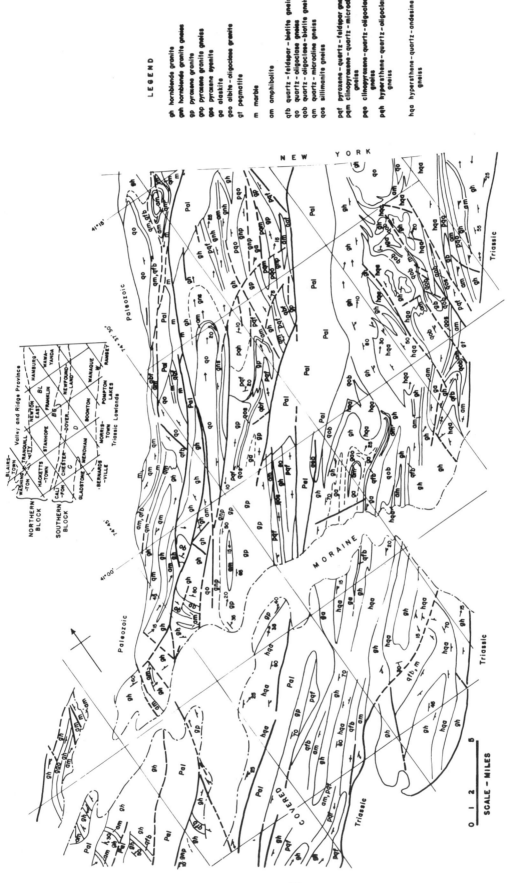

LEGEND

gh   hornblende granite
gnh hornblende granite gneiss
gp   pyroxene granite
gpp pyroxene granite gneiss
gps pyroxene syenite
ga   alaskite
gao albite - oligoclase granite
gt   pegmatite

m   marble
am   amphibolite

qfb quartz - feldspar - biotite gneiss
qo   quartz - oligoclase gneiss
qob quartz - oligoclase - biotite gneiss
qm   quartz - microcline gneiss
qos sillimanite gneiss

pqf pyroxene - quartz - feldspar gneiss
pqm clinopyroxene - quartz - microcline gneiss
pqo clinopyroxene - quartz - oligoclase gneiss
pqh hypersthene - quartz - oligoclase gneiss

hqa hypersthene - quartz - andesine gneiss

Figure 1. Precambrian geology of the central and northeastern parts of the New Jersey Highlands. The small inset map gives the names of the seven and one-half minute quadrangles for the area of the larger map; it also gives the locations of places mentioned in the text: Chester (C), Dover (D), Bowling Green Mountain (BG), Beaver Lake (BL) and Wanaque reservoir (W). Note also the identification of the "northern" and "southern" blocks. On the larger map "Pal" represents Paleozoic rocks. Attitudes of foliation and lineation are given by conventional symbols.

36

(1961) published a geological map of the Franklin and part of the Hamburg quadrangles.

Meanwhile geologists of the New Jersey Zinc Company had mapped a narrow belt along the northwestern fringe of the Highlands in New Jersey and New York. Marble is the host rock for the zinc ores mined at Sterling Hill near Ogdensburg and formerly mined at Franklin and it is in the belt mapped by the zinc company that most of the marble of the Highlands is found. A report of their work, which had involved mapping in great detail, was published in 1956 (Hague and others, 1956; see also Baum, 1957).

A field trip guidebook published in connection with the 1957 Atlantic City meetings of the Geological Society of America (Dorf, 1957) includes papers by Baum, Buddington, Sampson and Smith. Still more recent papers on the Precambrian of the New York state equivalent of the Highlands belt include those by Dodd (1965) and Offield (1967). Buddington's (1966) article on the magnetite deposits of the Adirondacks and New Jersey contains considerable information on lithology and a discussion of the origin of the granite as well as that of the iron ores.

In the past decade, A. A. Drake Jr. and others of the U. S. Geological Survey have mapped quadrangles along the Delaware River (Drake and others, 1961; Drake 1967a, 1967b; Davis and others, 1967). Their maps and accompanying cross sections show the Precambrian rocks of that area resting upon Paleozoic as part of a large nappe structure (see Drake and Epstein, 1967, p. 6).

SCOPE OF THIS PAPER

In 1955 this author began mapping those parts of the Highlands that had not been covered in detail by other modern workers and this led to a re-examination of the area as a whole. The purpose of this paper is to present some of the general results of this regional study. The map (fig. 1) is a quite simplified compilation of the Precambrian geology of 22 seven and one-half minute quadrangles. Most of the field work was concentrated in areas not covered by other recent maps but enough was done in all areas to permit at least tentative reconciliations of map legends and the gathering of petrologic data. A more detailed report on the petrology is in preparation and studies are continuing of the chemical petrology of individual rock units as well as the structural relationships between Precambrian and Paleozoic rocks.

ACKNOWLEDGMENTS

These studies were supported in part by grants from the Research Council of Rutgers University. Grateful acknowledgment is made of the assistance of T. A. Vogel who provided certain analyses and participated in numerous helpful discussions in field and laboratory and in criticism of the manuscript. The author is also indebted to graduate students at Rutgers for their assistance and interest in the project, particularly D. Chapman, R. Goodspeed, E. Johnson, J. Kline, D. Parrillo and J. Vecchioli.

GENERAL GEOLOGY

Table 1 lists the Precambrian rock units used in this paper and the table is intended as an elaboration of the legend of figure 1. As pointed out by certain other authors (for example Sims, 1958, p. 11) "formation" names such as Byram, Pochuck and Losee were defined too broadly and too subjectively to be useful by modern standards and these terms have been abandoned by nearly everyone in favor of names conveying mineralogic information.

DESCRIPTION OF ROCK TYPES

HYPERSTHENE-QUARTZ-ANDESINE GNEISS

Representatives of this group occur throughout a large part of the southern block of the New Jersey Highlands. There is a narrow belt on the southern fringe of the northern block which is a probable equivalent. The rock type is prominent on the map of Sims (1953, 1958) who termed it quartz diorite; he noted (1958, p. 28) that the quartz diorite and related facies are ". . . similar mineralogically and structurally to the noritic members of the charnockitic series of igneous rocks." Sims described the mineral composition in the Dover area as ranging from that of a hypersthene-quartz diorite to hypersthene diorite. He noted also (1958, p. 29) that the rocks within all of the bodies of quartz diorite in his map area

". . . are characterised by a conspicuous compositional layering produced by alternating light and dark bands; typically the dark layers are much thinner than the light layers, and they tend to be lenticular and streaky. Boundaries between light dark layers usually are transitional. At many places the material constituting the light layers cuts across the dark layers. The dark layers are composed of the same minerals as the

TABLE 1. SUMMARY OF MAP UNITS USED IN FIGURE 1

Granite group	Mainly mesoperthite-quartz-hornblende granite (gh) and granite gneiss (gnh), the latter consisting typically of microcline, albite-oligoclase, quartz and hornblende. In a northeast-southwest trending central belt of the Highlands, pyroxene for the most part takes the place of hornblende so that the dominant rock types are pyroxene granite (gp), pyroxene syenite (gps) and gneiss (gnp). There are substantial areas of alaskite (ga), minor local areas of albite-oligoclase granite and alaskite (gao) and pegmatite, widespread in minor amounts but locally abundant (gt).
Marble	The marble unit of figure 1, symbol (m), locally includes skarn, and calc-silicate rock.
Amphibolite	Amphibolite and pyroxene amphibolite (am). Only the larger masses can be shown in figure 1. There are widespread small layers of amphibolite contained in each of the other map units, notably in those listed below.
Quartz-feldspar-biotite gneisses	Because of the limitations imposed by the scale of the map of figure 1, it is necessary to employ in part of the area a general category (symbol qfb) for a rather heterogeneous group of layered gneisses. Locally it is possible to indicate by a letter symbol that one of the following types predominates:

qo: quartz-oligoclase gneiss
qob: quartz-oligoclase-biotite gneiss
qm: quartz-microcline-biotite gneiss
qos: sillimanite bearing quartz-oligoclase gneiss

Pyroxene-quartz-feldspar gneiss	Mostly clinopyroxene-quartz-oligoclase gneiss and clinopyroxene-quartz-microline gneiss. In figure 1, this map unit includes both rock types because on this scale it is not feasible to draw boundaries between them. The general map symbol (pqf) is used for much of figure 1 but where it is possible to indicate that one or the other of these two major types predominates, the more specific symbols (pqo) and (pqm) are shown on the map. Hypersthene-quartz-oligoclase gneiss (pqh) forms a unit of substantial extent in northern Franklin quadrangle.
Hypersthene-quartz-andesine gneiss	Layered rocks composed mostly of andesine, quartz, hypersthene and variable but generally small amounts of clinopyroxene, hornblende and, locally biotite; symbol (hqa).

light layers but contain higher proportions of the mafic minerals; commonly they have the composition of pyroxene amphibolite.

"The weathered surfaces characteristically are brown, streaked with black, and typically tend to be smooth and rounded. The fresh rock is greenish gray, dark gray, or brownish gray and usually has a greasy appearance. These distinctive greenish and brownish shades are caused principally by the coloring of the feldspars, not the mafic minerals; accordingly the rocks are darker than would be expected from their mineral composition . . ."

Typical modal analyses for this rock type from various parts of the Highlands are given in table 2 showing that for the most part it is made up of between 60 and 70 percent plagioclase, 10 to 20 percent quartz, together with orthopyroxene, clinopyroxene and hornblende. The darker layers (see table 4) ordinarily contain little or no quartz and the plagioclase is more calcic. In the lighter layers the anorthite content of the plagioclase tends to cluster around 33 to 35 but ranges up to An_{49}. By contrast the plagioclase of the other dominant gneisses of the Highlands tends to be in the oligoclase range.

Sims' description of this unit (1958, p. 28–30), quoted in small part above, applies without qualification to occurrences southeast of the area he mapped in the quadrangles of Mendham, Bernardsville, Boonton and Pompton Lakes. To the northeast however, the compositional layering becomes less prominent and it is not clear in fact how much of it is due to inclusions or interbeds of amphibolite (depending upon the view one takes as to the origin of the unit). Along the shores of the northern part of Wanaque Reservoir and in the eastern part of Wanaque quadrangle, both light and dark layers tend to become much thicker and the boundaries in some cases are quite sharp, in other cases quite diffuse. There are also considerable masses of relatively homogeneous aspect. North of Wanaque quadrangle, representatives of the map unit (as defined in this paper) are apparently included in Hotz' (1953) rather heterogeneous map unit termed quartz-oligoclase gneiss.

Southwest of the Dover area, in the quadrangles of Chester and Califon, bedrock exposures are sparse and figure 1 is subject to revision in that area; the hypersthene-quartz-andesine gneiss appears to be represented however in part by biotite-garnet-quartz-andesine gneiss.

Recent detailed work in the southwestern part of the Wanaque quadrangle by T. A. Vogel (personal communication) has demonstrated that the unit as a whole is notably uniform in mineral composition across outcrop widths of a mile or more. There seems to be no evidence of large scale differentiation in a direction normal to the layering.

In thin sections the texture is xenomorphic

TABLE 2. MODAL ANALYSES (VOLUME PERCENT) OF HYPERSTHENE-QUARTZ-ANDESINE GNEISS

	A	B	C	D	E	F	G	H	J	K	L
Plagioclase	68	66	68	66½	68	59	64	70	64	64	48
K-feldspar	tr	1	1	tr	½	tr	½			2½	
Quartz	15	9	22	21	13	30	½	6	17½	3	37
Orthopyroxene	11	12	7	4½	10	9	9	15½	5½	14	
Clinopyroxene	1	5	tr	tr		1½	3	1	7½	6½	
Hornblende	1	5		7				4	3½	8	
Biotite	½		½	tr	8		2½				7
Opaques	2		1	1	tr	2	5				tr
Apatite	½		½	tr	tr						tr
Sphene				tr							
Garnet											6
Other	tr	2			tr	2		2	2	3	tr
An content of Plagioclase	32½		34	34	29	30		44	35	41	43

Note: Analyses are by B. L. Smith unless noted.
A: Average of seven analyses by Sims (1958, p. 29) from Dover district.
B: Southwestern portion of Wanaque quadrangle and adjacent part of Newfoundland quadrangle, average of three analyses, Vogel, Smith and Goodspeed, 1968.
C: Boonton quadrangle, average of four analyses (T. A. Vogel, personal communication).
D: Wanaque quadrangle, east of Wanaque reservoir, average of four analyses.
E: Pompton Lakes and Boonton quadrangles, east of the belt mapped by Sims (1958), average of three analyses.
F: Northwestern part of Chester, southeastern part of Stanhope quadrangles, in the "northern block" of the Highlands; average of four analyses.
G: From Offield (1967 p. 20), average of four analyses of "hornblende-feldspar gneiss" from "east block."
H, J, K: Analyses by T. A. Vogel, personal communication, central part of Wanaque quadrangle.
L: Average of three analyses from Chester quadrangle.

granular with some tendency for plagioclase and quartz to be interlocking. The plagioclase is commonly antiperthitic; the nature and origin of this texture in the Wanaque area is discussed by Vogel, Smith and Goodspeed (1968).

Potassium feldspar is relatively scarce as free grains. The orthopyroxene is a moderately pleochroic hypersthene; hornblende is green; the clinopyroxene is a pale to rather strongly colored green mineral that is probably diopsidic. Biotite occurs locally and is more abundant approaching granite contacts. Apatite is a ubiquitous accessory. Sphene is found locally in small amounts but is by no means as abundant as in the pyroxene quartz-feldspar gneiss described below.

Pyroxene-Quartz-Feldspar Gneiss

Most of the rocks assigned to this map unit are found in the southeastern part of the "northern block" of the Highlands but the unit is represented as well in Ramsey quadrangle, in the Dover area, and in the quadrangles of Chester and Califon. As defined for the purposes of figure 1, the unit consists of a rather heterogeneous group of layered quartzofeldspathic gneisses having in common the fact that the dominant feldspar is oligoclase or microcline.

In outcrop appearance, the rocks assigned to this group range from nearly white weathering, light or greenish-gray on the one hand to dark brown weathering, dark greenish-gray on the other hand. There is a general field resemblance to the quartz-oligoclase gneisses described below and the group may in large part be the equivalent of those rocks. In places, particularly along the eastern margin of the "northern block" in Wawayanda and Newfoundland quadrangles and in the structurally lower part of a major synform in the central part of Franklin quadrangle, there is some field resemblance to the hypersthene-quartz-andesine gneisses.

Modal analyses given in table 3 illustrate the range of composition of the two main mineral assemblages. The clinopyroxene is a rather strongly colored green variety, probably diopsidic. In general hypersthene is rare but in the northern part of Franklin quadrangle the map unit as shown in figure 1 includes a substantial unit mapped by Buddington and Baker as hypersthene-quartz-oligoclase gneiss. Sphene is a common accessory. Deep red-brown biotite occurs in traces in many sections. Scapolite takes the place of plagioclase in some. Graphite occurs in some layers, ordinarily finely disseminated.

TABLE 3.   MODAL ANALYSES (APPROXIMATE VOLUME PERCENT) OF
PYROXENE-QUARTZ-FELDSPAR GNEISS

	A	B	C	D	E	F	G	H	J	K
Plagioclase	59	51	51–72	62	67	67	55	4	22	40
K-feldspar		1	0–8	2			3	57	41	25
Quartz		33	4–37	22	12	15	2	33	31	20
Hornblende	13				½		1			12
Clinopyroxene	16	11	1–23	12	18	15	36	3	5	2
Sphene		1	0–2	½	2		1	1	tr	
Apatite		tr	0–2	tr	1					
Opaque	11	3	1–4	1	tr	3	5		1	1
Other	1	tr	0–1	tr	tr			2		
An content of plagioclase	18	15	24–27	26	22	25	12			

Note: Analyses are by B. L. Smith unless otherwise noted.
A: Mendham quadrangle, sample 6-6; contains about 1 percent of chlorite.
B: Ramsey quadrangle, crest of Ramapo Mountain on Skyline Drive, sample 8-51.
C and D: Range and average of five analyses from Wawayanda quadrangle.
E: Stanhope quadrangle, average of two analyses.
F: Califon quadrangle, average of two analyses.
G: Average of two analyses given by Sims (1958, p. 15); trace of biotite in one.
H and J: Two analyses from Wawayanda quadrangle.
K: Quoted by Offield (1967, p. 14) from Buddington and Baker ms. Franklin quadrangle.

Plagioclase is ordinarily only slightly to not at all antiperthitic. Microcline however tends to be perthitic and there are examples of rocks composed of mesoperthite, quartz and small amounts of clinopyroxene which one might rather readily classify in thin section with the granites were it not for the layered field appearance. In the present work the tendency has been to include such layered rocks with the units other than granite but more study is needed concerning the affiliations of these microcline-mesoperthite gneisses. The matter is complicated by the fact that there are sheets of typical hornblende and pyroxene granite and granite gneiss interlayered with the rocks assigned to this pyroxene-quartz-feldspar group.

AMPHIBOLITE

Amphibolite and pyroxene amphibolite are widespread minor rock types in the New Jersey Highlands but only a few layers are thick enough to show on the map of figure 1. Even so, the tendency has been to exaggerate the width of narrow amphibolite bands on the map because they help to delineate the regional structure. Furthermore, in some of the bands which are shown, pegmatite may make up nearly 50 percent of the rock unit at least locally and in other cases plagioclase-quartz-(biotite) gneiss or other rock types may make up 30 percent or more of the unit.

Modal analysis data are given in table 4. Orthopyroxene is almost invariably present in the amphibolite associated with the hypersthene-quartz-

andesine gneiss. Elsewhere, although thin section study has not been done in as much detail, it appears that orthopyroxene is less common but it is found in amphibolite in the eastern part of Wawayanda quadrangle and at least locally elsewhere. Clinopyroxene is far more abundant than orthopyroxene in the amphibolites of Chester and Califon quadrangles. Apatite is a ubiquitous accessory throughout; sphene occurs in traces to as much as four percent in some of the amphibolites in Wawayanda, eastern Franklin and Califon quadrangles.

QUARTZ-FELDSPAR-BIOTITE GNEISS

In the map of figure 1, this is a somewhat heterogenous group of layered gneisses which have in common (1) biotite or hornblende rather than pyroxene as the predominant dark mineral and (2) oligoclase or perthitic microcline as the dominant feldspar. This assemblage is found mainly within five miles of either the northwestern or southeastern edges. It might be noted that, in a general way, the more anhydrous Precambrian rocks occur mainly in a northeast-southwest central zone of the Highlands.

The following rock types can be identified more specifically by letter symbols in parts of the map area.

1. Quartz-oligoclase gneiss (qo) is represented mainly in the northwestern part of Franklin and adjacent Hamburg quadrangle, in the core of the structure called the Beaver Lake anticline by Bud-

TABLE 4. MODAL ANALYSES (APPROXIMATE VOLUME PERCENT) OF AMPHIBOLITES

	A	B	C	D	E	F	G	H	J	K	L
Plagioclase	52½	51	53½	35–69	48	62	78	21–57	40	55	65
K-feldspar	tr			0–tr							
Quartz				0–tr							
Orthopyroxene	11	11		0–16	2	14	9	0–7	2		
Clinopyroxene	11½	17	6	10–56	28	6	9	0–26	6		11
Hornblende	20	13	37½	0–33	19	13	2	17–59	41	40	12
Biotite			1	tr–1½	tr	4	2	0–8	2	tr	9
Garnet			1								
Opaques		8		tr–6	2	½	tr			tr	2
Other	4½							0–10	9		
Apatite		tr	1	tr–1	tr	tr				tr	tr
Sphene			tr	tr–4	1					tr	
Zircon			½	0–tr	tr						
An content of plagioclase	36	38	–	35–53	42	41	35	–	–	38	35

Note: Analyses are by B. L. Smith unless otherwise noted.

A: Average of 7 analyses given by Sims (1958, p. 16). His analysis no. 5 is excluded because it contains 10 percent of orthoclase and the description indicates it is also atypical in other ways.

B: No. 253, Ramapo Mountain, Ramsey quadrangle.

C: Average of two analyses from Wawayanda quadrangle from Johnson (1968).

D and E: Range and average of seven analyses in Chester and Califon quadrangles.

F and G: YN3 and 3A, Wawayanda quadrangle; biotite is deep red plagioclase slightly antiperthitic.

H and J: Range and average of 7 analyses given by Offield (1967, p. 18), including some from New Jersey.

K: Average of two analyses from Blairstown quadrangle.

L: Pyroxene amphibolite from Blairstown quadrangle.

dington (1956). He described the gneiss (1957, p. 80) as composed of "about 30–35 percent quartz although locally about 24 percent, 60–70 percent plagioclase, a little accessory biotite with usually some secondary chlorite and epidote, and accessory magnetite, ilmenite and apatite. Locally there is accessory garnet . . ."

In figure 1 this unit also includes areas mapped by Hague and others (1956, p. 448) as "Losee gneiss;" it was described as a "light-colored, medium- to coarse-grained gneiss with a notably small content of dark minerals. Locally the rock contains accessory hornblende or biotite but in general the only dark mineral is a chlorite that gives the fresh rock a greenish color . . ." (see table 5).

2. Quartz-oligoclase-biotite gneiss (qob) is quite similar to the preceeding type except that biotite is megascopically prominent and the gneiss is ordinarily associated with at least some relatively potassic rocks which are notably scarce in the Beaver Lake anticline. Biotite is mostly brownish to greenish in thin section. Shreds of alteration products may make up one or two percent or more of thin sections and in some cases it is clear that these are the result of alteration of pyroxene. Apatite is a common accessory; sphene is not abundant. There are layers of plagioclase-quartz gneiss containing 5 to 15 percent of hornblende instead of biotite

which could be shown separately in some areas on a larger map scale.

3. Quartz-microcline gneiss (qm) is widely associated with the type of gneiss described immediately above but also makes up substantial and

TABLE 5. MODAL ANALYSES (APPROXIMATE VOLUME PERCENT) OF QUARTZ-OLIGOCLASE GNEISS

	A	B	C	D
Plagioclase	64	59 *	65	60
K-feldspar	½		½	3
Quartz	25½	36	32	25
Pyroxene				5
Hornblende	1	2½		2
Biotite	6	1½		1
Opaques				
Other	3	1	2½	6
An content of plagioclase	11	–	–	–

* Microantiperthitic at least in part.

A: Average of 6 analyses given by Sims (1958, p. 23). The An content of the plagioclase ranges from 4 to 19 and the range of content of plagioclase, quartz and biotite was 50–73, 18½–29 and 0–12 respectively. Plagioclase is apparently antiperthitic at least in part and the potassic feldspar is microperthite.

B: Average of 12 sections of Losee gneiss, Losee pond belt; quoted by Offield (1967, p. 25) from Buddington and Baker manuscript.

C: Average of two sections quoted by Offield (1967, p. 25) from Buddington and Baker ms.

D: Average of 9 analyses of quartz-plagioclase gneiss given by Offield (1967, p. 24).

relatively homogeneous bodies, notably in Newton East and Tranquillity quadrangles and elsewhere along the northern fringe of the Highlands. The microcline in some areas is quite perthitic and in some such cases the rock is apparently identical mineralogically with mesoperthite-quartz-biotite granite.

In field appearance, these rocks are ordinarily well layered, pinkish in color, weathering pale gray to whitish. Biotite is generally a prominent megascopic accessory. Sillimanite occurs locally. Narrow concordant pegmatite layers appear to be more abundant in this unit than elsewhere.

The average of 10 modal analyses of microcline gneiss in the Franklin-Sterling area is given by Hague (1956, p. 444) as 41.3 percent microcline and mesoperthite with 1.8 percent contained albite, 37.5 quartz, 14.0 plagioclase (average An_{10}), 3.9 percent biotite and chlorite and small amounts of magnetite, muscovite and other accessories.

4. Quartz-oligoclase-sillimanite gneiss is quite distinctive because it is well-layered and contains small amounts of graphite, sulphides, garnet and sillimanite. It has been identified in the following localities. (1) One narrow belt, less than 100 ft in outcrop width, has been traced from a point just east of the middle of Wanaque Reservoir southerly to about the southern boundary of Wanaque quadrangle. (2) In the northwestern portion of Dover quadrangle, in the Bowling Green Mountain area, another belt has been traced northwesterly from Route 15 into the extreme southern part of Franklin quadrangle. (3) In Franklin quadrangle, a third similar layer has been traced in a southwesterly direction across the area identified on the U. S. Geological Survey topographic maps as Mt. Paul. (4) Similar graphitic and sillimanite bearing gneiss has been found in the northwestern portion of the Califon quadrangle and east of the community of Chester but, because of poor exposures, little is known of the regional extent.

Elsewhere in the Highlands, the quartzo-feldspathic biotite gneisses and the clinopyroxene gneisses contain graphite, garnet or sillimanite locally.

MARBLE

Nearly all of the marble is found within three miles of the northern boundary of the New Jersey Highlands in Newton East, Franklin, and Hamburg quadrangles. It has been studied in great detail by geologists of the New Jersey Zinc Company (Hague and others, 1956, p. 437–441). They describe two main layers of marble, the "Franklin band" and the

much thinner but otherwise similar "Wildcat band," separated by the "Cork Hill gneiss." The predominant rock of the Franklin Band is described (Hague and others, p. 438) as "white to gray, coarse to locally fine grain crystalline marble, which varies in magnesium content from very low to an almost pure dolomite." In the vicinity of the Sterling Mine, the marble contains minor disseminated graphite, tremolite, mica, pyrite, pyrrhotite, quartz, talc, and a number of other sulphides, silicates, and oxides. (Hague and others, p. 438). The marble contains blocks and discontinuous bands of silicic rock consisting of various types of gneiss, quartzite, and sphene-bearing pyroxene-scapolite rock.

Elsewhere in the New Jersey Highlands, marble is a rare rock type. In the north central part of Franklin Quadrangle, Buddington (1957, p. 82) interprets certain layers of epidotic quartz-scapolite gneisses as representing metamorphosed siliceous carbonate rocks "with some introduction of material." At the northern fringe of the Highlands in Blairstown quadrangle there is a small area of marble with particularly abundant calc-silicates.

In the southern block of the Highlands only five occurrences of carbonate or metacarbonate rock are known. One of these consists of a layer 100 feet in outcrop width or less which has been traced intermittently for about three miles in the central part of Wanaque quadrangle. The others are all much too local to show in figure 1.

In addition to the economic importance of the marble as a host rock for zinc deposits, the marble is quarried for crushed stone and other purposes near Sterling Hill in Franklin quadrangle and at Limecrest, three miles east of Newton in Newton East Quadrangle.

Incidentally, granitic rocks are being quarried for aggregate in four widely separated locations in the Highlands but the last iron mining operation was suspended several years ago.

GRANITE AND RELATED ROCK TYPES

Typical granite in the New Jersey Highlands is a medium grained, pale pink rock weathering tan to light gray, consisting ordinarily of about 25 percent quartz, 40 to 60 percent microcline-microperthite, 10 to 25 percent plagioclase and 5 to 12 percent of hornblende or clinopyroxene or both.

In most of the granite the microperthite consists of about equal quantities of microcline host and plagioclase guest and the term mesoperthite as employed by Buddington for such perthite (1957, 1961) is used in this report. This rock ordinarily displays faint to obscure foliation in the field but a more or less prominent lineation emphasized by clots of mafic minerals.

In other parts of the granite masses the microcline is only slightly or not at all perthitic and there is a correspondingly much higher percentage of "free" plagioclase. Such rocks may display prominent gneissic foliation in the field. Intermediate cases are common in which some of the perthite in a given thin section is of the mesoperthite type while elsewhere the two feldspar phases are in separate grains. The writer agrees with Sims (1958, p. 34) and Buddington (1957, p. 12) that these rocks represent a deformed facies of the granites. In the Beaver Lake anticline in Franklin and Hamburg quadrangles, Buddington (1956) described prominent lineation in the axial region of the structure which gives way to foliation on the margins of the fold structure.

Table 6 gives a summary of 27 modal analyses of hornblende and pyroxene granite that have been made by others and 100 new modal analyses from other parts of the New Jersey Highlands. Zircon and apatite are ubiquitous accessories; sphene occurs in some of the granite of the northwestern block.

Pyroxene is rare to lacking in the southeastern part of the Highlands. Pyroxene predominates in the central portion while both are found toward the northwestern fringe although hornblende predominates.

In comparing the New Jersey Highlands to other areas, note can be made of some features that are rare or lacking. There are no porphyries. Mortar structure is rare to lacking and strain effects in general are notably rare except approaching fault zones. Migmatites are rare. Hybrid rock types and xenoliths, even approaching the margins of granite bodies are not common.

OTHER GRANITIC ROCK TYPES

Alaskite differs from the hornblende and pyroxene granite merely in the percentage of dark minerals. It is found mainly in three localities as follows:

1. In the Dover district, Sims (1958, p. 35) noted that the principal occurrences of alaskite were "at and near the contacts of hornblende granite with country rock and along the anticlinal crests of large folds . . ." He reported that the predominant feldspar in some of this alaskite is microantiperthite; the anorthite content of the plagioclase is in the range 15 to 20.

2. In the southwestern portion of Franklin quadrangle Buddington and Baker (1961) mapped a substantial area of pyroxene alaskite. This also is interpreted by this writer as on an anticlinal structure which extends some distance southwesterly

TABLE 6.   MODAL ANALYSES (APPROXIMATE VOLUME PERCENT) OF GRANITES

	A	B	C	D	E	F	G	H	J	K	L	M	N	O	P
Plagioclase	23	6–47	22	1–24	8	tr–24	12	tr–54	30	3–51	16	22	26	36	29
K-feldspar	43	12–54	41	35–79	64	32–70	46	16–54	34	33–66	52	44	46	42	32
Quartz	24	2–37	24½	5–26	16	5–37	29	16–31	23	1–36	21	29	16	8	29
Pyroxene	½	0–6	1	6–11	8					0–6	1	tr	4		tr
Hornblende	6	3–13	10	0–7	2	7–16	12	6–12	10	0–13	9	3	9	5	6
Biotite	1	0–2	tr			0–tr		0–11	3	0–tr	—	tr			½
Apatite		tr–2	½	0–1	tr	0–1	tr			0–tr	—				
Sphene				0–tr						0–tr	tr				
Zircon		0–tr	tr	0–tr	tr	0–tr	tr			0–1	—				
Opaques		tr–5	1	tr–2	1	tr–4	1			tr–3	1	1			
Other	2	0–2	tr							0–1	—		2	4	5
An content of Plagioclase	12 to 19	7–17	10	8–10	1	8–20	12	11–15	12	8–17	12	12 to 18	15	14	26½

A: Average of seven analyses given by Sims (1958, p. 33).

B and C: Range and average of seven analyses from the eastern slope of Ramapo Mountain in Ramsey quadrangle.

D and E: Range and average of four analyses of pyroxene granite from the southeastern block of the Highlands in Califon and Chester quadrangles.

F and G: Range and average of 12 analyses of hornblende granite from the southeastern block of the Highlands in Chester, Gladstone and Califon quadrangles.

H and J: Range and average of 5 analyses in Boonton and Pompton Lakes quadrangles in the area southeast of that mapped by Sims.

K and L: Range and average of 10 analyses from the northwestern block of the Highlands in the quadranglesof Washington, Hackettstown, Tranquillity and Stanhope.

M: Average of 9 analyses cited by Offield (1967, p. 28).

N: Average of 6 analyses of "Byram gneiss" lacking pyroxene given by Hague and others (1956).

O: Average of 5 analyses of "Byram gneiss" carrying pyroxene given by Hague (1956).

P: Average of 62 samples, Wanaque quadrangle (Goodspeed, 1967, quoted by Vogel, Smith and Goodspeed 1968, p. 1698).

beyond the area mapped by Buddington and Baker.

3. In Wawayanda quadrangle there are extensive masses of pyroxene alaskite and in much of this the plagioclase content is quite low so that the rock is essentially a microcline-quartz alaskite.

In Wanaque quadrangle there are two relatively local masses that are mapped as biotite granite, made up of about 60 percent mesoperthite, about 30 percent quartz with the rest largely biotite. There is some question as to whether or not some of the microcline gneiss along the northern fringe of the Highlands in Tranquillity and Newton East quadrangles should be assigned to the granite group; other comment on this point has been made above in the discussion of pyroxene-quartz-feldspar gneiss.

Albite-oligoclase granite, consisting of about 75 percent plagioclase, the rest mostly quartz, has been described by Sims (1958, p. 30) in the Dover area as characterized in part by subhedral laths of plagioclase showing checkerboard twinning. An attempt has been made to show the approximate extent of the larger bodies of this rock type as mapped by Sims in figure 1. Elsewhere in the Highlands this appears to be a relatively rare and local rock type; it tends to be associated with quartz-oligoclase gneiss, from which it may have been derived by local partial regeneration. The sodic "Losee gneiss" was regarded by the New Jersey Zinc Company geologists as a magmatic intrusion (Hague and others, 1956, p. 458; Baum, 1957, p. 106) but Buddington and Baker classified the similar Losee gneiss of the Beaver Lake anticline with "rocks of uncertain origin." In figure 1 the Losee gneiss of these authors has been grouped with "quartz-oligoclase gneiss."

STRUCTURAL GEOLOGY

The gneisses have been steeply, more or less isoclinally folded, with folds tending to be overturned to the northwest. The various units, including the granites, are notably concordant. There is pervasive mineral lineation, which also tends to be concordant from one rock unit to another. Where fold structures have been delineated, lineations plunge parallel (or at least subparallel) to fold axes. Megascopically the lineations consist of clots of mafic minerals, laths of segregated minerals (with remarkable linear continuity in some of the amphibolite) as well as elongate mineral grains.

There is no clear evidence of more than one time of folding in the gneisses. Minor folds, on outcrop scale, are relatively rare. It would appear that folding took place by relatively uniform flow.

A striking structural feature of the Highlands is the way the rocks have been sliced by many longitudinal faults; these are interpreted for the most part as high angle wrench faults but there is evidence that some are reverse faults. Mylonite zones up to a hundred feet or more in width are associated with the boundary between Precambrian and Triassic rocks in Califon, Gladstone, Chester, Pompton Lakes and Ramsey quadrangles.

There are two other fault sets. One is represented by two clusters of faults striking almost due north-south; one cluster is in Wanaque quadrangle and the eastern part of Newfoundland, the other is in the Northwestern part of the map of figure 1, in the quadrangles of Hackettstown and Tranquillity. Trap dikes, of probable Triassic or Jurassic age, are found occasionally cutting the gneisses and there seems to be some tendency for these dikes to be concentrated in the vicinity of the north-south fault sets. The third set of faults strikes almost due east-west.

STRUCTURAL RELATIONSHIP BETWEEN PRECAMBRIAN AND PALEOZOIC ROCKS

On recently published maps, Drake and others (Drake, 1967a, 1967b, Davis and others 1967, Drake and others, 1961) have interpreted the Precambrian rocks along the Delaware River in southeastern New Jersey and adjacent Pennsylvania as part of a large nappe structure with the Precambrian rocks resting upon Paleozoic. In the valleys occupied by Paleozoic rocks, they interpret the exposed sequence for the most part as being inverted. Arguments that the entire Reading Prong Precambrian might be allochthonous have been reviewed by Isachsen (1964).

These concepts present an important incentive to re-study a large part of the Paleozoic rocks of New Jersey. The matter is beyond the main scope of the present paper but the following observations are offered: (1) There appears to be a normal sedimentary succession in the Paleozoic rocks along the northern flank of the New Jersey Highlands, extending through Wallkill Valley of the Valley and Ridge Provence to the Silurian of Kittatinny Mountain. (2) Relatively flat lying quartzite of Hardyston type is found in contact with or apparently resting upon Precambrian rocks in many locations such as in Franklin (Smith and Baum, 1957, p. 50) at the southwest end of Mohawk Lake in Newton East Quadrangle, in Hackettstown, Tranquillity, and Blairstown quadrangles, as shown in large part of the Raritan Folio (Bayley and others, 1914) and elsewhere (see, for example, Kümmel and Weller, 1901). (3) Offield's (1967) map of an extensive area in New York state immediately adjacent to New Jersey, seems to indicate a normal "right-side-up" sequence for the Paleozoic rocks and he interprets the Precambrian rocks as rooted and true basement—(Offield, 1967, p. 68–70), (4) there seems to be a complete lack of evidence in the central and northeastern parts of the Highlands of any cases in which mining or drilling penetrated from Precambrian into Paleozoic rocks.

For the present, this writer suggests that the Precambrian rocks may be rooted in northeastern New Jersey whilst the Precambrian rocks in the Pennsylvania part of the Reading Prong are allochthonous. The quadrangles between the area mapped by Drake and others and the area of figure 2 need further study. In the Frenchtown quadrangle on the Delaware River it was noted (Drake and others, 1961) that augen and flasergneiss phases are common in the granites and that three sets of lineations were distinguished. In the central and northeastern Highlands no more than one direction of mineral lineation has been identified in any locality and augen and flasergneiss are rare to absent, as is mortar structure and strain effects in general except near major faults.

METAMORPHISM AND SUGGESTED INTERPRETATION OF AGE RELATIONSHIPS

In the hypersthene-quartz-andesine belt of Dover, Boonton and Wanaque quadrangles the mineral assemblage is characteristic of the granulite metamorphic facies. Until further studies are made of the chemical petrology of the area, it is not clear whether variations in other mafic mineral assemblages are related to metamorphic environment or to original composition.

As a model to be tested by additional field work (and possibly by certain age dating techniques) the following age sequence is suggested for the Precambrian rocks of figure 1. Hypersthene-quartz-andesine gneiss appears to be at the base of the exposed section (though these rocks may be intrusive). There follows an assemblage of amphibolite, marble and more or less potassic gneisses, though not by any means necessarily in that order. Relatively sodic gneisses appear to be high in the sequence. The author sees no compelling reason to question that the mesoperthite granite is intrusive and igneous.

SUGGESTED FIELD TRIP ITINERARY

ROAD LOG

A field trip is not planned formally in connection with the 1969 Atlantic City meetings but the following route is suggested for anyone who wishes to examine the principal rock types described in this paper. The suggested trip involves a traverse across the northeastern part of the New Jersey Highlands along a route of excellent rock exposures.

0.0 miles. The mileage log begins at the traffic circle on route 23 where route 202 leaves it in a northeasterly direction. At 2.5 miles, bear left at traffic light, following 202. At 3.3 miles turn right at light, still following 202. At mile 3.4 there is a Triassic trap outcrop on the right and at 3.7 a fanglomerate road cut on the right.

At 4.5 miles, turn left (leaving route 202) on Jefferson Avenue, crossing river. At 4.9, turn right on Colfax Avenue (which runs about along the so-called border fault separating Triassic and Precambrian).

At 6.9 turn left on Skyline Drive (dangerous corner!), climbing the east slope of Ramapo Mountain. Roadside outcrops are mostly typical mesoperthite-hornblende granite.

At 8.3, a convenient stop at a large outcrop on the right in typical hornblende granite with strong lineation and obscure foliation.

From 9.3 (pipeline crossing) to 9.7 (view westerly overlooking Wanaque Reservoir) there is almost continuous outcrop of rather heterogeneous quartz-feldspar gneisses; in part these are clinopyroxene-quartz-oligoclase gneiss, in part biotite bearing, in small part albite-oligoclase granite. Farther down the hill there are outcrops in the amphibolite belt shown in figure 1 but exposures are poor along the road.

At 10.9 to 11.2 large outcrop of hypersthene-quartz-andesine gneiss. At 11.8, bear slightly left at a somewhat unexpected stop sign. At 12.0 turn left at stop sign. At 12.4 large outcrop of biotite bearing hypersthene-quartz-andesine gneiss.

At 13.7 sharp right turn to cross causeway through middle of Wanaque Reservoir. On the west side of the causeway there is a long outcrop in hypersthene-quartz-andesine gneiss. Retrace route to mileage 13.7, at which point the road log is resumed. Continue south.

At 17.5 (3.8 miles southerly from the turn which was made to cross the reservoir), turn right on route 511 (Union Avenue). From 18.1 to about 19.5 there are outcrops of quartz-feldspar-biotite gneiss along with minor amphibolite and other gneisses.

At 19.8, turn left on Main Street and at 20.1 turn right at a rather obscure turn on Mathews Avenue. On the latter, bear right up the hill and at 20.9 there is a road cut on the right in typical sillimanite bearing quartz-oligoclase gneiss containing garnet, graphite and biotite. Continue southerly on Mathews Avenue and at 21.2 turn right on route 23.

At 24.5 there are good outcrops on the right of hypersthene-quartz-andesine gneiss and again at 24.7 to 24.9 (at the turn to Forest Hill Park) where the compositional layering of this rock type is best shown. From 26.3 to 26.8 there are new rock cuts in hornblende granite. At 27.6 it is suggested that one might park and walk ahead 100 yards to examine interesting structures in Paleozoic rocks.

The traverse across the Paleozoic valley is complete at about 31.6. As this is being written, this part of the highway is under reconstruction to a divided four lane road; mileages from this point are therefore approximate. The next two miles of new rock cuts will be in a relatively heterogeneous assemblage of clinopyroxene-quartz-oligoclase and clinopyroxene-quartz-microcline gneisses with some interlayered amphibolite and granite.

From about 35.8 to about 37.8 the new highway cuts will be mostly in quartz-oligoclase gneiss in the core of the Beaver Lake anticline. At about 38.3, the highway intersects route 517 and the road log is discontinued at that point. It is suggested that reference be made to the 1957 field guide for viewing locations in the Franklin area.

REFERENCES CITED

Baum, J. L., 1957, Precambrian geology and structure of the Franklin-Sterling area, New Jersey: in Dorf, editor, Guidebook for field trips, Atlantic City Meeting, Geol. Soc. America.

Bayley, W. S., 1910, Iron mines and mining in New Jersey: New Jersey Geol. Survey, Final Report Ser. v. 7.

——— 1941, Precambrian geology and mineral resources of the Delaware Water Gap and Easton quadrangles, New Jersey and Pennsylvania: U. S. Geol. Survey Bull. 920.

Bayley, W. S., Salisbury, R. D., and Kümmel, H. B., 1914, U. S. Geol. Survey Geol. Atlas, Folio 191.

Berkey, C. P. and Rice, Marion, 1919, Geology of the West Point quadrangle, New York: N. Y. State Museum Bull. 225–226.

Buddington, A. F., 1956, Beaver Lake anticline, N. J.: in The Grenville problem, Thompson ed. Roy. Soc. Canada, Spec. Pub. No. 1.

——— 1957, Magnetite iron ore deposits of the New Jersey Highlands: in Dorf, editor, Guidebook for field trips, Atlantic City meetings, Geol. Soc. America

——— 1963, Isograds and the role of H$_2$O in metamorphic facies of orthogneisses of the northwest Adirondack area, New York: Geol. Soc. America Bull., v. 74, p. 1155–1182.

——— 1966, the Precambrian magnetite deposits of New York and New Jersey: Econ. Geology v. 61, no. 3, p. 484–510.

Buddington, A. F. and Baker, D. F., 1961, Geology of the Franklin and part of the Hamburg quadrangles, New Jersey: U. S. Geol. Survey Misc. Geol. Inv. Map I–346.

Davis, R., Drake, A. A. Jr., and Epstein, J. B., 1967, Geologic map of the Bangor quadrangle, Pennsylvania-New Jersey: U. S. Geol. Survey Quad. Map GQ-665.

Dodd, R. T., Jr., 1965, Precambrian geology of the Popolopen Lake quadrangle, Southeastern New York: Map and chart ser. no. 6, N. Y. State Museum and Science Service.

Dorf, Erling, 1957 (Editor), Guidebook for field trips, Atlantic City meetings, Geol. Soc. America.

Drake, A. A. Jr. and Epstein, J. B., 1967, The Martinsburg formation (middle and upper Ordovician), in the Delaware Valley, Pennsylvania-New Jersey: U. S. Geol. Survey Bull. 1244-H.

Drake, A. A. Jr., McLaughlin, D. B., and Davis, R. E., 1961, Geology of the Frenchtown quadrangle, New Jersey-Pennsylvania: U. S. Geol. Survey Quad. Map GQ-133 (with descriptive text).

Drake, A. A. Jr., 1967a, Geology of the Easton quadrangle, New Jersey-Pennsylvania: U. S. Geol. Survey Quad. Map GQ-594.

——— 1967b, Geologic Map of the Bloomsbury quadrangle, New Jersey, U. S. Geol. Survey Quad. Map GQ-595.

Fenner, C. N., 1914, The mode of formation of certain gneisses in the highlands of New Jersey: Jour. Geology, v. 22, p. 594–612; 694–702.

Goodspeed, R. M., 1967, An investigation of the coexisting feldspars from the Precambrian plutonic rocks in the Wanaque area, New Jersey: unpubl. Ph.D. dissertation, Rutgers Univ.

Hague, J. M., Baum, J. L., Herrmann, L. A., and Pickering, R. J., 1956, Geology and structure of the Franklin-Sterling area, New Jersey: Geol. Soc. America, Bull., v. 67, p. 435–474.

Hotz, P. E., 1953, Magnetite deposits of the Sterling Lake, N. Y. Ringwood, N. J. area: U. S. Geol. Survey Bull. 982-F, p. 153–244.

Isachsen, Y. W., 1964, Extent and configuration of the Precambrian in northeastern United States: Trans. N. Y. Acad. Sc., Ser. II, v. 26, No. 7, p. 812–829.

Johnson, E. L., 1968, Precambrian geology of parts of Passaic County and Sussex County, New Jersey, and infrared absorption studies of biotite: unpubl. Ph.D. dissertation, Rutgers University.

Kümmel, H. B. and Weller, S., 1901, Geological Map of the Green Pond Mountain region, Morris and Passaic Counties: New Jersey Geol. Survey Ann. Rept. of The State Geologist.

Lobeck, A. K., 1932, Atlas of American Geology: The Geographical Press, Columbia Univ., N. Y.

Lowe, K. E., 1950, Storm King granite at Bear Mountain, N. Y.: Geol. Soc. America Bull., v. 61, p. 137–190.

Offield, T. W., 1967, Bedrock geology of the Goshen-Greenwood Lake area, N. Y.: Map and chart series no. 9, N. Y. State Museum and Science Service.

Sampson, E., 1957, The zinc-manganese deposits of the

Franklin-Sterling region: *in* Dorf, editor, Guidebook for field trips, Atlantic City meeting, Geol. Soc. America.

Sims, P. K., 1953, Geology of the Dover magnetite district, Morris County, N. J.: U. S. Geol. Survey Bull. 982-G, p. 245–305.

———— 1958, Geology and magnetite deposits of the Dover district, Morris County, New Jersey: U. S. Geol. Survey Prof. Paper 287.

Sims, P. K., and Leonard, B. F., 1952, Geology of the Andover mining district, Sussex County, N. J.: New Jersey Dept. Cons. and Devel. Bull. 62.

Smith, B. L., 1957, Summary of the Precambrian geology of The New Jersey Highlands: *in* Dorf, editor, Guidebook for field trips, Atlantic City meeting, Geol. Soc. America.

Smith, B. L., and Baum, J. L., 1957, Precambrian of the New Jersey Highlands (Road log of second day of field trip): *in* Dorf, editor, Guidebook for field trips, Atlantic City meeting, Geol. Soc. America.

Turner, F. J., 1968, Metamorphic petrology: McGraw-Hill, New York.

Vogel, T. A., Smith, B. L. and Goodspeed, R. M., 1968. The origin of antiperthites from some charnockitic rocks in the New Jersey Precambrian: Amer. Mineralogist, v. 53, nos. 9 and 10, p. 1696–1708.

TOPOGRAPHIC AND GEOLOGIC MAPPING IN NEW JERSEY

KEMBLE WIDMER [1]
Bureau of Geology and Topography
New Jersey Department of Conservation and Economic Development, Trenton, New Jersey 08625

Geographers have a saying that people attract people. As the population of an area increases, it seems to logically follow that mapping is required in greater and greater detail. Today New Jersey is one of the few States that has complete topographic mapping coverage with relatively up to date maps at scales of 1 mile to 1 inch (1:63,360) — the State Atlas sheets, and 2,000 feet to 1 inch (1:24,000) — U. S. Geological Survey 7½ minute quadrangle maps.

The first map showing the entire State of New Jersey was published by Nicholas Vischero in 1654. It was used by the Duke of York in 1664 to describe the original boundaries of the Province. Inaccuracies in the Vischero map resulted in a boundary dispute between New York and New Jersey more than a Century later.

To assist the Boundary Commission, a new map of New Jersey was drawn in 1769 by Lt. Bernard Ratzer, an officer in the Royal Artillery and English Army topographer. Lt. Ratzer prepared the first "geologic (?)" map of New Jersey for Lord Howe in the fall of 1776. This map, published by William Faden in December of 1777, entitled *The Province of New Jersey* was prepared for British Intelligence operations. It divided the Province into East and West — commonly called *The Jerseys.* It showed every forge, furnace, and iron mine in the Province of New Jersey at that time. The road network was quite complete in areas where the British had control of the Province or where there were numerous Torries. The road network in the Highlands — the area of the most abundant forges and iron mines, and in the Great Valley, was left out apparently because no reliable information was available.

In 1822, the State Legislature recognized the need for more accurate maps and authorized the loan of $1,000 to Thomas Gordon to enable him to attain additional surveys for a "street" map. This map, compiled from earlier surveys, was a great advance in cartography and was the only authoritative map between its publication in 1828 and 1860.

A geologic map of New Jersey was engraved to accompany a report of the Geology of New Jersey by Henry D. Rogers in January, 1839. The Rogers geologic map, on a scale of approximately 6 miles to 1 inch, was probably based on the Gordon map; as like the Gordon map, it includes parts of the adjoining States. Rogers map is hand colored, shows the serpentine of Hoboken, Tertiary marls, bog irons in the Coastal Plain, the greensand marl beds (most of the upper Cretaceous and early Tertiary) and "lower beds of greensand formation" which today is called the Raritan Formation. Some of the overlying clay formations are also included in the "lower beds." In the Rogers maps, the Triassic is shown as "middle secondary red sandstone." The area includes the "trap rocks" and several of the areas of underlying Cambro-Ordovician limestones and border conglomerates, both of which were shown as "calcareous conglomerates of middle secondary." The Highlands are described as gneisses. The northern part of the State is divided into eight numbered formations from "white sandstone formation I" (Hardyston) through "limestone formation II" and "altered limestone formation II" (Jacksonburg) through slates, white sandstone, red sandstone, and "limestone formation VIII," which includes the Devonian limestones and sandstones adjacent to the upper Delaware River. This geologic map,

[1] New Jersey State Geologist.

with its geologic cross sections of the State, was apparently copied in several of the popular geographies of this period.

The next geologic map of New Jersey was a much more extensive affair published in 1868 and again in 1878 (the present Bureau of Geology has two sets so labeled) by George H. Cook and John C. Smock, using a base map prepared by G. M. Hopkins. Cook was then State Geologist and Smock was Assistant Geologist. The map was hand colored, but required four separate maps, each on a scale of 2 miles to 1 inch. The assembled geologic map would thus be nearly 7 feet high and about 4 feet wide. Map no. 1 included the Azoic and Paleozoic. Map no. 2 was of the Triassic, Map no. 3 of the Cretaceous, and Map no. 4 of the Tertiary and Recent. Iron mines and other operations of economic geology, railroads, highways and settlements are shown in a great deal of detail. This series of geologic maps to accompany the early Cook reports was completed with three hachured maps, printed from steel engravings, of iron mines and a map entitled, *Zinc Mines of Sussex County.*

In 1877 the New Jersey Geological Survey found that its geologic work was hampered by inadequate topographic maps and work was begun on the first series of Atlas Sheets which eventually covered the entire State in seventeen overlapping maps (Sheet 1 through 17) on a scale of 1 mile to 1 inch (1:63,360). New Jersey was the first State in the Union to undertake and complete such a program of comprehensive map coverage. Assistance was rendered by the U. S. Coast and Geodetic Survey and by the then relatively new U. S. Geological Survey. The latter organization took over the topographic mapping operation from the State in 1884 when the mapping was about half completed. These maps were published as State Atlas Sheets by the New Jersey Geological Survey.

Upon completion of the topographic series in 1887, whose manuscript maps were drawn to a scale of 3 inches to 1 mile (1:21,120), a base map of the State on a scale of 5 miles to 1 inch (1:316,800), was completed. This map served for many years as a base for special maps showing culture, geology, forests, watersheds, railroads, and other geographic needs of *The New Jersey State Atlas.* In 1903 the overlapping series of Atlas Sheets, numbered 1 through 17, were changed to an edge-match series, and the sheets were changed to numbers 21 through 37, by which they are identified today.

In 1898 the rapid growth of several of the New Jersey cities indicated a need for a larger scale series of topographic maps. The New Jersey Geological Survey began the preparation of 2,000 foot to the inch maps, similar to the present U. S. Geological Survey quadrangle maps. They were, however, designed so that six would cover the area of each of the State Atlas Sheets. Twenty four of these maps were prepared and published, and for a while revised, but this effort was being duplicated by the U. S. Geological Survey mapping which was published as 15 minute quadrangles. The U. S. Geological Survey instituted a 7½ minute quadrangle series on a 2 inches to 1 mile scale (1:31,680) in the 1930's.

In the period from 1902 to 1910 cooperative work between the U. S. Geological Survey and the State Geological Survey resulted in the publication of five geologic folio reports on a scale of 1:62,500. In addition, a New York City folio report included a small part of New Jersey. These geologic folios contained the first detailed mapping of the geology of local areas.

It should be noted also that during the period from 1918 to 1926 the entire State was covered by a series of soils bulletins. Each of the soils bulletins contained a soils map covering the area of one Atlas Sheet. Due to the irregular shape of the State, certain of these reports also covered adjacent quadrangles where there was a limited area of New Jersey. Thus, Atlas Sheet 27, covering the area northwest of Trenton, was incorporated with Atlas Sheet 24 to the north, while that part of the area between Trenton and Camden was incorporated with the sheet to the south. These maps were based on a 3-foot soil profile and were one of the very early efforts to map the agricultural soils. The bulletins still contain much useful information. These soils maps, known as Linwood Lee maps after one of the compilers who became prominant in soils mapping because of this effort, are of interest to geologists because they show a remarkable parallel to the underlying geology. This is not evident in the more recent maps because so many other soil formation factors are taken into consideration. In the 1950's, Rutgers University published a series of Engineering Research Bulletins which provided a map and an evaluation of the engineering soils, as distinct from the agricultural soils, for each of the twenty one New Jersey counties. Complete coverage of New Jersey by a modern set of soils maps combining many of the characteristics of both of these earlier soils surveys is under way by the U. S. Soil Conservation Service.

In 1915 the base map for the entire State, developed in 1887, was found to be too small for easy use. It was, therefore, revised and reprinted on a scale of 1:250,000. The geologic (Sheet 40) and county and municipality map (Sheet 39) at this scale are still available. The county and municipality map was overprinted in 1963 to show the areas covered by the State Atlas Sheets (17) and the U. S. Geological Survey 7½ minute quadrangle maps at a scale of 1:24,000. This latter series, completed in the 1950's, was compiled from Army Map Service and other sources which collectively covered the entire State just prior to World War II and consists of 172 maps.

The present geologic map of New Jersey was first prepared by J. V. Lewis and H. B. Kümmel in 1912. It was revised and reprinted in 1931 by Kümmel and in 1950 by M. E. Johnson. It will be revised again when present stocks are exhausted in the next few years. All revisions since 1912 have been minor because of the large scale of the map and the excellent work done in its original preparation.

Both the U. S. Geological Survey and the New Jersey Bureau of Geology and Topography are preparing geologic maps to cover selected areas in the Federal 1:24,000 quadrangle series. The U.S.G.S. has published sixteen geologic maps in New Jersey in *The Geologic Quadrangle Maps of the United States Series*. The New Jersey Survey has completed mapping in five quadrangles and parts of three others, but has yet to publish any of these quadrangle maps. The publication by the New Jersey Bureau of Geology and Topography of geologic overlay maps to cover the State Atlas Sheets is being considered and much work has been completed.

PRECAMBRIAN AND LOWER PALEOZOIC GEOLOGY OF THE DELAWARE VALLEY, NEW JERSEY–PENNSYLVANIA[1]

AVERY ALA DRAKE, JR.
U. S. Geological Survey, Washington, D. C. 20242

INTRODUCTION

Precambrian rocks crop out in the Delaware Valley in an upland area known as the Reading Prong, a prominent physiographic and geologic province that extends from east of the Hudson River in New York to west of Reading, Pennsylvania (fig. 1). Locally, this upland is better known as the New Jersey Highlands or the Reading and Durham Hills. Sedimentary rocks of Cambrian and Ordovician age crop out both in intermontane valleys within the Prong and in the Great Appalachian Valley to the north. The area is mostly unglaciated, the rocks are deeply weathered, and there is a general appalling lack of exposure. Outcrops are restricted to areas of natural drainage and artificial exposures and most contacts are based, in large part, on float mapping. Most useful information has been gained along the Delaware River which has a sinuous course, but flows generally across the regional strike.

The Precambrian and lower Paleozoic rocks are bounded on the south by the Triassic border fault and on the north by Silurian and younger rocks that are set off by the Taconic unconformity that later served as the detachment surface for the Blue Mountain decollement (fig. 2).

The regional geology of the Delaware Valley is known mostly from reconnaissance mapping in New Jersey (Bayley and others, 1914; Bayley, 1941) and county studies in Pennsylvania (Miller and others, 1939, 1941; Willard and others, 1959). More recently, the U. S. Geological Survey has been carrying out a program of detailed mapping within the Delaware Valley as well as in adjacent areas. The mapping has been carried across the entire Reading Prong and Great Valley into the Valley and Ridge (Davis and others, 1967; Drake, 1965, 1967a, 1967b, 1969; Drake and Epstein, 1967; Drake and others, 1960, 1961, 1965, 1967, 1969) and is currently being extended to the southwest. In addition to this geologic work, regional aeromagnetic surveys have been completed for the area underlain by Precambrian and Triassic rocks in Pennsylvania (Bromery and Griscom, 1967) and for the Precambrian outcrop belt in New Jersey (Henderson and others, 1966). Gravity surveys applicable to this area have also been completed to the southwest (R. W. Bromery, *in* U. S. Geol. Survey, 1964, 1966). This work has shown that the major geologic problem of the area is the geometric and tectonic relation of the Precambrian rocks of the Reading Prong to the sedimentary rocks of the Great Valley and the smaller intermontane valleys within the Prong. A synthesis of all available data suggests that the Precambrian rocks are a far-traveled allochthonous mass.

It is a pleasure to thank J. M. Aaron, D. C. Alvord, R. E. Davis, and I. B. Epstein of the U. S. Geological Survey, my co-workers in the Delaware Valley, for the interchange of data and thoughtful discussions, V. E. Zadnik, now of the U. S. Department of Defense, and K. E. Seifert now of Iowa State University for careful work in the field, P. K. Sims, Director of the Minnesota Geological Survey, for an introduction to the Precambrian rocks of the Prong, and G. H. Wood, Jr. of the U. S. Geological Survey and the late V. E. Gwinn for many discussions

[1] Publication authorized by the Director, U. S. Geological Survey.

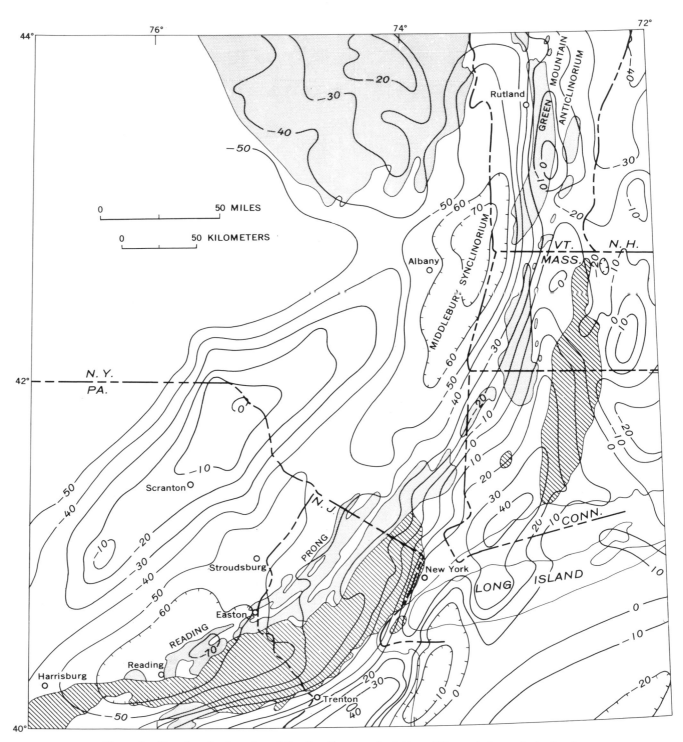

Figure 1. Map showing location of Reading Prong and relation of Precambrian rocks to simple Bouguer gravity anomalies in the northeastern United States. Map modified from Drake (1969). Precambrian rocks, stippled; Triassic rocks, cross-hatched; Paleozoic rocks and Coastal Plain deposits, unpatterned. Contour interval 10 milligals.

which helped to crystallize many of the ideas presented herein.

STRATIGRAPHY AND PETROLOGY

PRECAMBRIAN ROCKS

The Precambrian rocks that crop out in the Delaware Valley are chiefly high-grade quartzo-feldspathic metasedimentary and metavolcanic rocks that are interlayered with smaller amounts of amphibolite and marble and are plutonized and migmatized by sodic granitic rocks, hornblende granite, and alaskite. All these rocks contain variable amounts of magnetite and there are magnetite deposits throughout the Reading Prong. Petrologically, the rocks are not like those exposed in the Blue Ridge to the south or along the Green Mountain axis to the north. They are, however, much like those of the Adirondacks and Grenville province of Canada and New York, so much so, that they were called Grennoble by Engle (1956). A sequence of lower grade interbedded metasedimentary and metavolcanic rocks is believed to overlie the older high-grade rocks.

Traditionally, the Precambrian rocks were divided into four principal units: Franklin Limestone, Pochuck Gneiss (mafic rocks), Losee Gneiss (feldspar dominantly plagioclase), and Byram Gneiss (feldspar dominantly K-feldspar). Bayley (1941) recognized a fifth group of rocks, the Pickering Gneiss, near the Delaware River, and Fraser (in Miller and others, 1939, 1941) recognized a unit called the Moravian Heights Formation in eastern Pennsylvania. The Pochuck, Losee, and Byram were thought to be intrusive and the other units to be metasedimentary and metavolcanic rocks. These units were useful in reconnaissance studies but include rocks of different compositions and origins. A. F. Buddington and his students (Baker, 1956; Buddington and Baker, 1961; Dodd, 1962; Hotz, 1952; Sims, 1958; Sims and Leonard, 1952) were the first to abandon them in favor of more precise lithologic units. These geologists through their studies in northeastern New Jersey and southeastern New York have set the modern standard for the Reading Prong. Similar units have been mapped in the Franklin, New Jersey, area by geologists of the New Jersey Zinc Company (Hague and others, 1956) and in the Reading area of eastern Pennsylvania (Buckwalter, 1959, 1962) as well as in the Delaware Valley. The units from seven areas within the Precambrian outcrop belt are compared and correlated in table 1.

OLDER METASEDIMENTARY-METAVOLCANIC SEQUENCE

Traditionally, all the quartzo-feldspathic gneisses were considered to be igneous rocks that had intruded the Franklin Limestone and minor associated metasedimentary rocks, the dark gneisses being the oldest, the sodic gneisses intermediate in age, and the K-feldspar gneisses the youngest. Fenner (1914) was the first to recognize the abundance of paragneiss and migmatite in the Prong, and all subsequent workers have shown that such rocks occur in far greater proportions than igneous rocks. Judging from my own experience, however, one tends to overmap the igneous units. Thus, the older Precambrian rocks consist of an unknown but thick sequence of paragneiss, migmatite, and marble. The principal rock types are described briefly below.

Amphibolite

Amphibolite, as mapped in the Delaware Valley, includes both hornblende and hornblende-pyroxene amphibolite. These rocks were mapped together because of their similarity of appearance. They can, however, be distinguished in thin section. Layers of this rock crop out throughout the area, and are abundant in all other areas within the Reading Prong. Interlayers of amphibolite are also present in nearly all the quartzo-feldspathic gneisses and granitic rocks.

The amphibolites are dark gray (N3) to grayish black (N2), equigranular to inequigranular rocks that have a crystalloblastic texture. In most exposures the unit has a fair to good foliation marked by the subparallel alinement of tabular minerals or by alternating mafic and felsic layers. Some amphibolite, however, is massive. Andesine and variable quantities of hornblende, clinopyroxene, and orthopyroxene are the essential minerals of the rock. Common accessory minerals include biotite which is fairly abundant in some specimens, magnetite, apatite, and sphene, and some specimens contain epidote, scapolite, and relic patches of calcite. Plagioclase forms 40 to 65 percent of the rock and combined hornblende-pyroxene 30 to 55 percent. Hornblende constitutes 10 to 80 percent of the combined hornblende-pyroxene total, and in most specimens over 60 percent of the combined total is hornblende. Orthopyroxene is rare in the amphibolites of this area, and most specimens contain less than 5 percent of hypersthene. Selected modes of amphibolites are given in table 2.

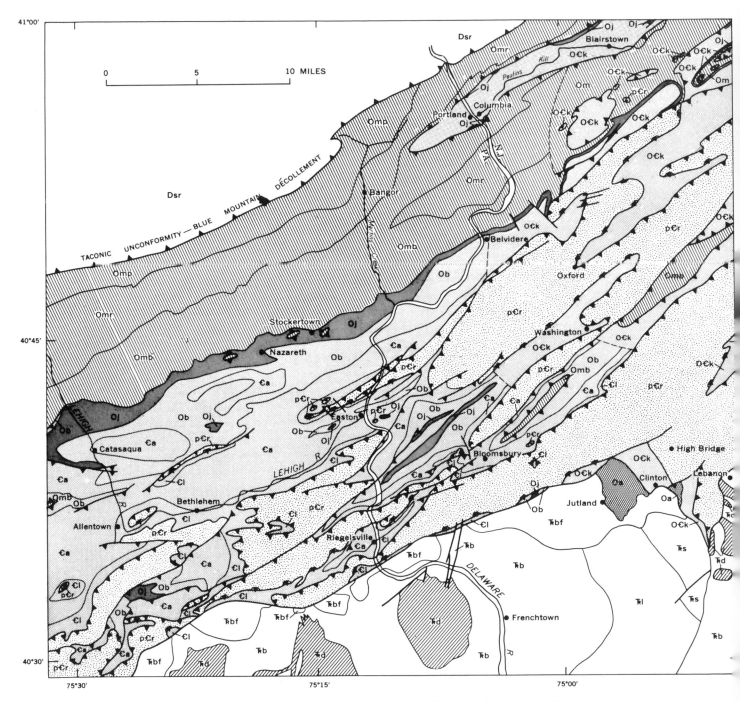

Figure 2.   Generalized geologic map of the Delaware Valley and adjacent areas, New Jersey–Pennsylvania.   Modified from Bayley and others (1914), Drake (1967a, b), Drake and others (1961, 1967, 1969), and Gray and others (1960).

Dashed line between units indicates arbitrary cutoff

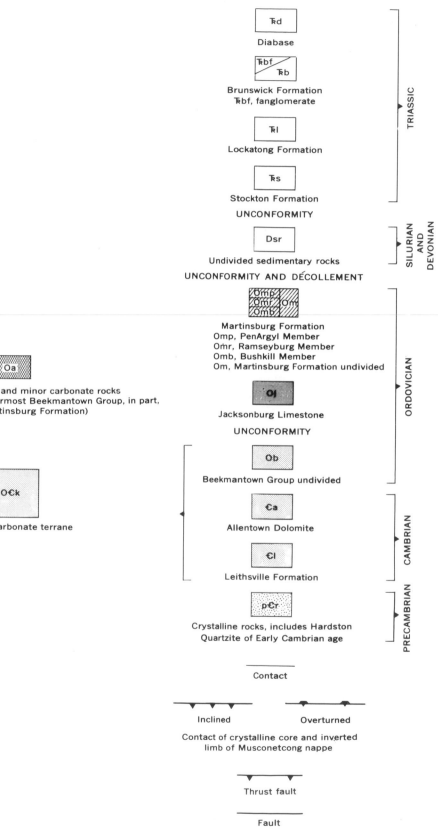

Diabase

Brunswick Formation
Ṟbf, fanglomerate

Lockatong Formation

Stockton Formation

UNCONFORMITY

Undivided sedimentary rocks

UNCONFORMITY AND DÉCOLLEMENT

Martinsburg Formation
Omp, PenArgyl Member
Omr, Ramseyburg Member
Omb, Bushkill Member
Om, Martinsburg Formation undivided

Jacksonburg Limestone

UNCONFORMITY

Beekmantown Group undivided

Allentown Dolomite

Leithsville Formation

Crystalline rocks, includes Hardston
Quartzite of Early Cambrian age

TRIASSIC

SILURIAN AND DEVONIAN

ORDOVICIAN

CAMBRIAN

PRECAMBRIAN

Allochthonous pelitic, psammitic, and minor carbonate rocks
(includes age equivalents of uppermost Beekmantown Group, in part,
Jacksonburg Limestone, and Martinsburg Formation)

Kittatinny carbonate terrane

Contact

Inclined   Overturned

Contact of crystalline core and inverted
limb of Musconetcong nappe

Thrust fault

Fault

55

TABLE 1. COMPARISON, PROBABLE CORRELATION, AND INTERPRETED ORIGIN OF MAJOR PRECAMBRIAN ROCK UNITS IN SEVEN READING PRONG AREAS (AFTER DRAKE, 1969)

Reading area, Pennsylvania (Buckwalter, 1959, 1962)	Delaware Valley, N. J.-Pa. (Drake, 1967a, b; Drake and others, 1961, 1967 and 1969)	Dover district, New Jersey (Sims, 1958)	Franklin area, New Jersey (Hague and others, 1956)	Edison area, New Jersey (Buddington and Baker, 1961)	Sterling Lake, N. Y.-Ringwood, N. J. area (Hotz, 1952)	Popolopen Lake quadrangle, New York (Dodd, 1962)
Granitic gneiss. Mostly intrusive	Hornblende granite, microperthite alaskite, and biotite granite. Mostly intrusive.	Hornblende granite, alaskite, and related rocks. Intrusive.	Byram gneiss. Intrusive.	Hornblende granite and alaskite. Intrusive.	Hornblende granite and related facies. Intrusive.	Hornblende granite, leucogranite, and biotite granite. Intrusive.
....	Microantiperthite alaskite. Intrusive.	Microantiperthite granite. Intrusive.	....	Pyroxene alaskite, pyroxene granite, and pyroxene syenite. Intrusive.	....	....
....	Albite-oligoclase granite and related albite pegmatite. Rheomorphic, probably anatectic, with attendant granitization.	Albite-oligoclase granite and albite-quartz pegmatite. Anatectic.	Granodiorite gneiss. Intrusive.	Albite alaskite. Uncertain origin.	Quartz-oligoclase gneiss (part). Intrusive.	Plagioclase-quartz leucogneiss (part). Probably metavolcanic.
Quartz diorite gneiss. Migmatitic variant of granitic gneiss. (Correlation doubtful.)	Quartz diorite. Plutonic metamorphism of sodic tuff and mafic volcanic rock.	Quartz diorite. Intrusive	....	Hypersthene-quartz-oligoclase gneiss. Metasedimentary.	Quartz-oligoclase gneiss (part). Intrusive.	Hypersthene-quartz-oligoclase gneiss. Metavolcanic.
....	Oligoclase-quartz gneiss. Metavolcanic.	Oligoclase-quartz-biotite gneiss. Metasedimentary.	Losee gneiss and oligoclase gneiss. Intrusive.	Quartz-oligoclase gneiss. Uncertain origin.	Quartz-oligoclase gneiss (part). Intrusive.	Plagioclase-quartz leucogneiss (part). Probably metavolcanic.
Graphitic gneiss and quartz-biotite-feldspar gneiss. Metasedimentary.	Quartzo-feldspathic gneiss, including biotite-quartz-plagioclase gneiss, sillimanite-bearing gneiss, and potassic feldspar gneiss. Metasedimentary.	Biotite-quartz-feldspar gneiss. Metasedimentary.	Microcline gneiss, garnet gneiss, biotite gneiss, and graphitic gneiss. Metasedimentary.	Biotite-quartz-plagioclase gneiss, and quartz-microcline granulite. Uncertain origin. Quartz-potassium feldspar gneiss, quartz-plagioclase gneiss, epidote-scapolite-quartz gneiss, and garnetiferous gneiss. Metasedimentary.	Garnetiferous quartz-biotite gneiss and quartzite. Metasedimentary.	Rusty and nonrusty biotite-quartz-feldspar gneiss. Metasedimentary.
Marble and lime silicate gneiss. Metasedimentary.	Marble and its alteration products, including skarn. Metasedimentary.	Skarn and related rocks. Metasedimentary.	Marble. Metasedimentary.	Marble. Metasedimentary.	Marble and skarn. Metasedimentary.	Marble and pyroxene. Metasedimentary.
Hornblende gneiss. Metasedimentary and metaigneous (?).	Amphibolite. Metaigneous, metavolcanic, and metasedimentary.	Amphibolite. Metasedimentary and metaigneous.	Hornblende gneiss. Metasedimentary, metaintrusive, and metavolcanic.	Amphibolite. Uncertain origin.	Pyroxene amphibolite and amphibolite. Metasedimentary.	Amphibolite, pyroxene-hornblende-plagioclase gneiss, and biotite-hornblende-quartz-feldspar gneiss. Metavolcanic.
....	Pyroxene gneiss. Metasedimentary.	....	Pyroxene gneiss and related rocks. Metasedimentary.	Pyroxene gneiss. Metasedimentary.	....	....
Migmatite and closely interbedded gneiss. Injection gneiss.	Migmatite: microperthite alaskite neosome in amphibolite paleosome, albite-oligoclase granite neosome in amphibolite paleosome.	Hornblende granite and alaskite, and albite-oligoclase granite containing more than 20 percent metasedimentary rocks. Injection gneiss.	....	....	....	....

TABLE 2. MODES [1] (VOLUME PERCENT) OF AMPHIBOLITE FROM THE DELAWARE VALLEY

	1	2	3	4	5
Plagioclase	31.0	53.2	62.0	61.0	46.0
Hornblende	21.0	30.7	28.0	2.5	28.0
Clinopyroxene	15.5		4.0	34.0	9.5
Orthopyroxene				1.0	4.0
Biotite	19.0	1.0			11.0
Magnetite	1.5	9.5	4.0	2.0	0.5
Epidote	4.2	5.4	1.5		0.5
Apatite	tr	tr	1.0	.5	0.5
Chlorite	tr	tr	1.5		
Sphene	tr	tr			
Calcite	1.0		1.0		
Scapolite	5.7				
Total	100.0	100.0	100.0	100.0	100.0

[1] Based on 2,000 points counted by the Chayes method.
  1. Musconetcong Mountain, Frenchtown quadrangle, from near a small magnetite deposit.
  2. Scotts Mountain, Belvidere quadrangle.
  3. Pohatcong Mountain, Bloomsbury quadrangle.
  4. Morgan Hill, Easton quadrangle.
  5. Steeleys Hill, Riegelsville quadrangle.

The unit is migmatitic at many places, and such occurrences have been mapped separately in the Delaware Valley. The division of these units is inherently somewhat subjective, but an attempt was made to separate those rocks that contain more than 10 percent granitic material.

Amphibolite migmatite

Amphibolite migmatite as mapped in the Delaware Valley includes all hybrid rocks that consist of amphibolite paleosome (host) and a granitic neosome (guest). Included in this grouping are true migmatites, veined gneisses, and permeation gneisses. This lumping is necessitated by the generally poor exposure and the impossibility of differentiating the different types on the basis of float. Two mixed rock units have been recognized, the neosome of one type (type 1) is microperthite alaskite whereas that of the other (type 2) is albite-oligoclase granite. Migmatite having a sodic granite neosome is especially common on Musconetcong and Pohatcong Mountains in the Bloomsbury quadrangle where sodic rocks are especially abundant. All transitions occur between unaltered amphibolite and hornblende granite and between amphibolite and albite-oligoclase granite. The relations between the paleosome and neosome in both types of rock are similar, but it seems as if true migmatite is not very abundant in the type 2 unit which consists mostly of veined gneiss and permeation gneiss.

True migmatite consists of alternating, regular, well-defined layers of amphibolite and microperth-

ite alaskite or albite-oligoclase granite. The mafic layers in general are unaltered, but in some specimens they contain biotite, quartz, and some microperthite. The andesine also contains antiperthitic intergrowths. The felsic layers have the same composition as microperthite alaskite or albite-oligoclase granite.

Other phases of the unit contain short discontinuous layers of granitic material that generally parallel the foliation of the amphibolite and lenses, veins, and knots of the neosome throughout the mafic host. These rocks can best be described as veined gneisses. In general, mineralogic relations are as described for the true migmatites.

Still other varieties of the rock contain tiny veins and patches of quartz and feldspar and porphyroblasts of potassic or sodic feldspar generally oriented at angles to the amphibolite foliation.

Type 1 migmatites contain many grains of patch microantiperthite, presumably a result of the interaction of granitic fluids with the feldspar of the amphibolites. These varieties can best be described as permeation gneisses.

Pyroxene gneiss

Pyroxene gneiss is not common in the Delaware Valley, but sizable bodies have been mapped on Rattlesnake Hill in the Riegelsville quadrangle and Scotts and Pohatcong Mountains in the Bloomsbury and Belvidere quadrangles. Much of the rock is a medium-grained, greenish-gray (5GY6/10) to light-grayish green (5G8/1), well-layered to nearly massive granoblastic gneiss composed principally of diopsidic pyroxene and plagioclase and lesser quantities of hornblende, quartz, biotite, epidote, sphene, and magnetite; pyroxene forms from about 10 to 40 percent of the unit and plagioclase 50 to 85 percent. Most specimens of the unit contain relic patches of marble.

Another type of pyroxene gneiss is clinopyroxene-microperthite-quartz gneiss. This rock is light-brownish-gray (5YR6/1), pale-brown-weathering (5YR5/2), green spotted, medium- to medium-coarse-grained, and inequigranular. It consists primarily of microperthite, green augitic clinopyroxene, and sparse quartz. Minor minerals include magnetite, chlorite, epidote, sphene, and zircon. Microperthites are of patch replacement, string, and wormy plume types and appear to be largely intergrowths of albite in triclinic potassic feldspar, but microcline microperthite is present. Patch and string perthites are about equally abundant and plume types are minor. Dull-green augitic clinopyroxene forms 20 to 30 percent of the rock. Much of the pyroxene is strongly altered to chlorite, epidote, and magnetite and contacts with the feldspars are ragged and whispy. Magnetite occurs in small crystals and as clots along cleavage directions in the pyroxene and in larger clots and strings of grains spatially associated with chloritized pyroxene. Quartz is only sparsely present and occurs in tiny veins and interstitially to the feldspar and pyroxene.

The rock is poorly foliated, and, in general, has a granoblastic texture. However megacrysts of both pyroxene and microperthite are present. The unit is easy to recognize because of its spots of green pyroxene. Selected modes of pyroxene gneiss are given in table 3.

TABLE 3.   MODES [1] (VOLUME PERCENT) OF PYROXENE GNEISS
FROM THE DELAWARE VALLEY

	1	2	3	4	5
Plagioclase	55.5	51.5	—	83.5	75.0
K-feldspar	—	—	67.5	—	—
Pyroxene	38.0	38.5	27.5	11.5	15.5
Hornblende	—	3.5	—	—	6.5
Biotite	0.5	1.5	—	—	—
Sphene	2.0	0.5	0.5	2.5	—
Quartz	1.5	2.0	3.5	—	0.5
Epidote	1.5	0.5	0.5	—	—
Magnetite	0.5	1.5	1.5	2.0	2.5
Apatite	0.5	0.5	—	0.5	—
Total	100.0	100.0	100.0	100.0	100.0

[1] Based on 2,000 points counted by the Chayes method.
   1. Rattlesnake Hill, Riegelsville quadrangle.
   2. Pohatcong Mountain, Bloomsbury quadrangle.
   3. Scotts Mountain, Easton quadrangle.
   4. Pohatcong Mountain, Bloomsbury quadrangle.
   5. Scotts Mountain, Belvidere quadrangle.

Marble

Marble is most abundant in the Delaware Valley in Chestnut Hill, Marble Mountain and Scotts Mountain in the Easton and Belvidere quadrangles. Small bodies crop out elsewhere, and a major layer was cut by diamond drilling on Rattlesnake Hill in the Riegelsville quadrangle (Epstein and others, 1967). This rock, its mineralogy, and metamorphism has been the most intensely studied geologic feature of the area. For detailed accounts of these studies, the interested reader is referred to Peck (1905), Miller and others (1939), Bayley (1941), and Montgomery (1955, 1957).

Unaltered marble, which is not abundant in the Delaware Valley, is a medium- to coarse-grained, white to gray, rarely pink, rock. Dolomite marble is by far the most abundant variety, but calcite marble layers are locally present. The marble at most places contains small flakes of graphite that are crudely alined giving the rock a poor foliation. The rock is holocrystalline and dominantly massive, but at most places it is strongly deformed and has a good shear foliation.

Most of the marble has been altered to serpentine rock, talc rock, tremolite rock, or more rarely skarn. These principal alteration types are end points, and virtually any mineralogic combination can be found.

Serpentine rock is largely pure serpentine but is in part micaceous. It is a pale green (5G7/2) to moderate green (5G5/6), massive to schistose rock. Micaceous varieties contain material of a vermiculite-like phlogopite. Lenses and irregular masses of crypto-crystalline silica are common in the serpentine, as are veins of white calcite and pink dolomite. Parts of the rock have been further altered to talcose material and other parts contain relic diopsidic pyroxene and tremolite.

Tremolite rock is white (N9) to very light gray (N8), medium- to coarse-grained and shiny, and consists of close-packed intergrowths of bladed prisms and short crystals of tremolite. Coarse crystals of dolomite and (or) calcite are present in many specimens, as is serpentine which appears to be an alteration product.

The most sheared and altered parts of the marble have been coverted to a roughly foliated rock that principally contains talc, altered phlogopite, and a sparse amount of serpentine. Less pure specimens consist of a felty mass of tremolite needles that contains "floating" grains of diopsidic pyroxene. The tremolite shows all stages of alteration to talc, and the pyroxene is serpentined.

Skarn is sparingly present in the Delaware Valley at places where the marble has been silicated. The skarn is a dusky green (5G3/2) to greenish black (5G2/1), coarse-grained, inequigranular, very heterogeneous rock that has a good foliation. Diopsidic pyroxene and hornblende are the principal constituents, and lesser amounts of biotite, sphene, magnetite, andesine, relic calcite, and very sparse quartz are also present. Pyroxene and hornblende account for over 90 percent of the minerals, and pyroxene commonly forms 60 percent of this total. Hornblende apparently formed later than the pyroxene, and forms replacement textures with it. Some thin layers of skarn contain andesine, and verge on pyroxene amphibolite in composition.

Marble is intruded at many places by pegmatites related to the microperthite alaskite, and by tourmaline-bearing albite-quartz pegmatites. Scattered grains of pyrite, chalcopyrite, galena, sphalerite, molybdenite, and high-thorium uraninite and its secondary products are present at these places. Montgomery (1957) describes the uraninite occurrences.

Quartzo-feldspathic gneiss

Quartzo-feldspathic gneiss is abundant throughout the Delaware Valley and is particularly abundant on Scotts Mountain in the Easton, Bloomsbury, and Belvidere quadrangles. Three end member types have been mapped and are thought to represent slightly different original compositions. All these rocks have been more or less altered by permeating fluids from veins and lenses of microperthite alaskite.

Biotite-quartz-plagioclase gneiss. Biotite-quartz-plagioclase gneiss is a highly variable unit in both composition and texture, but is characterized by the conspicuous presence of biotite and by a prominent compositional layering. Most phases of the unit are very light gray (N8), medium dark gray (N4), or pale yellowish brown (10YR6/2) and weather to shades of reddish brown (10R5/4-10R3/4).

Compositional varieties of the unit include: biotite-quartz-plagioclase gneiss, biotite-magnetite-quartz-plagioclase gneiss, and garnetiferous biotite magnetite-quartz-plagioclase gneiss. Strongly altered hornblende and pyrite are sparsely present in some specimens of the rock.

Most typical phases of the unit are medium fine grained to medium grained, essentially equigranular, and granoblastic. Quartz exceeds plagioclase (largely oligoclase) and biotite forms 5 to 15 percent of the rock. Much of the plagioclase is antiperthitic, and scattered grains of microperthite are common in many specimens. Magnetite is concentrated along planar directions, and seems to have formed, in part at least, at the expense of biotite which has altered to an iron-poor chlorite. Garnet is light reddish brown and occurs in poikiloblastic clusters that are spatially related to areas of antiperthitic intergrowths. Hornblende is sparingly present near contacts in layers of biotite-quartz-plagioclase gneiss that are grossly interlayered with amphibolite. The hornblende is always strongly altered, largely to chlorite, and, in places, intergrowths of chlorite and magnetite retain the hornblende crystal form.

Biotite-quartz-plagioclase gneiss can be differentiated from potassic feldspar gneiss by its general lack of pink potassic feldspar and by its conspicuous biotite content and well-layered structure. It differs from oligoclase-quartz gneiss in its variable composition, well-layered nature, and to a certain extent, color. Selected modes of the unit are given in table 4.

TABLE 4. MODES [1] (VOLUME PERCENT) OF BIOTITE-QUARTZ-PLAGIOCLASE GNEISS FROM THE DELAWARE VALLEY

	1	2	3	4	5
Plagioclase	50.5	43.0	42.5	40.0	57.5
K-feldspar	3.0	1.5	1.5	–	–
Quartz	14.0	37.5	43.5	48.5	27.5
Biotite	21.5	6.5	10.5	10.0	11.5
Garnet	10.0	6.5	–	–	0.5
Magnetite	0.5	2.5	0.5	0.5	2.5
Pyrite	0.5	0.5	–	–	–
Chlorite	–	0.5	–	0.5	0.5
Graphite	–	1.5	tr	0.5	–
Total	100.0	100.0	100.0	100.0	100.0

[1] Based on 2,000 points counted by the Chayes method.
  1. Marble Mountain, Bangor quadrangle.
  2. Chestnut Hill, Easton quadrangle.
  3. Scotts Mountain, Bloomsbury quadrangle.
  4. Morgan Hill, Easton quadrangle.
  5. Musconetcong Mountain, Frenchtown quadrangle.

Potassic feldspar gneiss. Bodies of quartzo-feldspathic gneiss characterized by microcline and (or) microperthite are common on Morgan Hill, Chestnut Hill, Marble Mountain, Scotts Mountain, and Pohatcong Mountain. Potassic feldspar gneiss is grayish pink (5R8/2), pinkish gray (5YR8/1), light gray (N7), or light greenish gray (5G8/1) and weathers very light gray (N8) to grayish orange pink (10R8/2). The rock is fine grained to medium fine grained, equigranular, equidimensional, and has a xenoblastic to granoblastic fabric and a very poor to fair foliation.

Most layers have rather uniform composition, but some layers contain oligoclase in amounts up to 40 percent of the total feldspar. Quartz, microcline and (or) microperthite constitute about 92 percent of the rock. The remainder consists of variable amounts of oligoclase, biotite, chlorite, magnetite, zircon, sphene, apatite, epidote, clinopyroxene, hornblende, and garnet. Aside from oligoclase, biotite and magnetite are the most common of these. Magnetite and biotite occur together, but not in equal amounts.

Microperthite occurs in most specimens, but nonperthitic microcline occurs alone in some. The perthites in most specimens are of the string exsolution type. True patch perthites and worm-like plume perthites are also present, and some grains of string perthite have albite rims. A few specimens, however, contain only patch perthites of the replacement type and many grains are almost completely replaced by albite. Quartz occurs principally as granoblastic grains, but is also present as small inclusions within the feldspar. Biotite forms up to 5 percent of some rocks, and in most specimens contains interleaved chlorite as an alteration product. Magnetite occurs in scattered grains, but more commonly is oriented along planar directions. Some specimens have a little relic hornblende that has largely altered to chlorite and magnetite which preserve the crystal habit.

Layers of oligoclase-quartz gneiss occur within the unit, and layers containing both potassic and sodic feldspar border these layers. Layers of feldspathic metaquartzite also occur within the unit. The mineralogy of these layers is similar to that of the normal rock, except that quartz accounts for 60 to 80 percent of the total composition.

Potassic feldspar gneiss can be distinguished from oligoclase-quartz gneiss and biotite-quartz-plagioclase gneiss by the presence of abundant pink potassic feldspar. It differs from the potassic feldspar-bearing granites in grain size and fabric.

Some phases of the unit are quite granitic in appearance, having many of the aspects of a metamorphosed aplite. The most granitic appearing rock forms small sheets and occurs as veins, lenses, and blotches in rock of obvious metasedimentary origin spatially removed from bodies of hornblende granite or microperthite alaskite. It is possible that the unit actually has an intrusive origin, but its heterogeneous nature (table 5) and rapid variation from impure quartzite to granitic-appearing gneiss suggests an ultimate metasedimentary origin. The most granitic rock (table 6) is chemically unlike typical known granites; note especially the K$_2$O–

TABLE 5. MODES [1] (VOLUME PERCENT) OF POTASSIC FELDSPAR GNEISS FROM THE DELAWARE VALLEY

	1	2	3	4	5
K-feldspar	50.5	26.5	56.0	11.5	52.5
Quartz	45.5	65.0	36.0	78.5	41.0
Plagioclase	1.0	–	1.5	–	0.5
Biotite	0.5	3.5	4.5	0.5	1.0
Magnetite	2.0	0.5	0.5	tr	3.5
Sericite	0.5	0.5	–	4.0	1.5
Garnet	–	4.0	1.5	1.0	–
Pyrite	–	–	–	2.0	–
Zircon	tr	tr	tr	–	tr
Apatite	tr	tr	tr	0.5	–
Total	100.0	100.0	100.0	100.0	100.0

[1] Based on 2,000 points counted by the Chayes method.
1. Pohatcong Mountain, Bloomsbury quadrangle.
2. Chestnut Hill, Easton quadrangle.
3. Scotts Mountain, Bloomsbury quadrangle.
4. Morgan Hill, Easton quadrangle.
5. Scotts Mountain, Belvidere quadrangle.

TABLE 6. CHEMICAL ANALYSES AND CIPW NORMS OF POTASSIC FELDSPAR GNEISS FROM DELAWARE VALLEY COMPARED WITH THOSE OF AVERAGE GRANITES

	1	2	3
Chemical analyses (weight percent)			
SiO$_2$	71.0	72.08	73.86
Al$_2$O$_3$	13.0	13.86	13.75
Fe$_2$O$_3$	3.0	0.86	0.78
FeO	1.3	1.67	1.13
MgO	0.32	0.52	0.26
CaO	0.07	1.33	0.72
Na$_2$O	0.95	3.08	3.51
K$_2$O	9.3	5.46	5.13
H$_2$O–	0.04	–	–
H$_2$O+	0.74	0.53	0.47
TiO$_2$	0.33	0.37	0.20
P$_2$O$_5$	0.06	0.18	0.14
MnO	0.03	0.06	0.05
Total	100.1	100.0	100.0
CIPW Norms			
Quartz	29.3	29.2	32.2
Orthoclase	55.6	32.2	30.0
Albite	8.4	26.2	29.3
Anorthite		5.6	2.8
Corundum	1.0	–	1.4
MgSiO$_3$	–	1.3	0.6
FeSiO$_3$	0.8	1.7	1.1
Magnetite	3.2	1.4	1.2
Apatite	–	0.4	0.3
Ilmenite	0.6	0.8	0.5
Hematite	0.8	–	–
Total	99.7	99.6	99.4

1. SE. abutment of dam, Ingersol Reservoir, Scotts Mountain, Bloomsbury quadrangle. Rapid rock analysis by Paul Elmore, Samuel Botts, Gillison Chloe, and H. Smith, U. S. Geological Survey.
2. Average of 72 analyses of calc-alkali granites (Nockolds, 1954).
3. Average of 48 analyses of alkali granites (Nockolds, 1954).

Na$_2$O ratios. Field relations suggest that the granitic phases may have been generated within the unit by partial anatexis. The presence of microperthite confirms that temperatures were high enough for such a process. Such rock could also result from potash metasomatism, but there are no data to support such a process. In any case, the bulk of the unit seems to be a metasedimentary rock.

Sillimanite-bearing gneiss. Rocks characterized by the presence of sillimanite are restricted to Morgan Hill and Chestnut Hill in the Easton quadrangle. The unit is strongly migmatitic, and as mapped, includes all quartzo-feldspathic rock that contains recognizable sillimanite-bearing lenses or layers. It is thought to be a more aluminous phase of the quartzo-feldspathic gneiss. Sillimanite-bearing gneiss is light gray (N7) to light greenish gray (5GY8/1), more rarely pinkish gray (5YR8/1), pale-yellowish-brown (10YR6/2) to yellowish-gray (5Y7/2) weathering, strongly foliated and lineated, medium grained, inequigranular and heterogeneous. Quartz and lesser amounts of microcline-microperthite, monoclinic potassic feldspar, and strongly sericitized oligoclase(?) are the principal minerals. Potassic feldspar is dominant in most layers; however, other layers contain only oligoclase. Other minerals include sillimanite, magnetite largely altered to hematite, and sparse biotite, ilmenite, zircon, and apatite. Sillimanite is largely altered to light greenish aggregates of clay and mica minerals but the original crystal habit is preserved.

All exposures of sillimanite-bearing gneiss are strongly deformed cataclastically as they occur within the Morgan Hill shear zone (Drake, 1967a) and areas of strong shearing on Chestnut Hill. In thin section, all types of mechanical deformation are present: mylonite zones, mortar structure, and sutured quartz. This deformation has made mapping especially difficult as all sheared quartzofeldspathic units have much the same appearance in outcrop. The study of numerous thin sections is necessary to adequately delineate rock units in these areas. Typical modes of sillimanite-bearing gneiss are given in table 7.

TABLE 7. MODES [1] (VOLUME PERCENT) OF SILLIMANITE-BEARING GNEISS FROM THE DELAWARE VALLEY

	1	2	3	4	5
Quartz	55.5	49.5	52.5	53.0	47.5
K-feldspar	6.0	17.0	38.0	25.5	35.5
Plagioclase	—	—	1.5	2.5	6.5
Sillimanite	29.5	23.5	7.5	10.5	4.5
Sericite	7.5	5.0	—	7.0	3.5
Biotite	—	—	—	—	—
Magnetite	1.5	—	0.5	1.5	—
Pyrite	—	5.0	—	tr	2.5
Zircon	tr	tr	tr	tr	tr
Apatite	—	tr	—	tr	—
Total	100.0	100.0	100.0	100.0	100.0

[1] Based on 2,000 points counted by the Chayes method.
1. East end Chestnut Hill, Easton quadrangle.
2. Morgan Hill, Easton quadrangle.
3. Morgan Hill, Easton quadrangle.
4. West end Morgan Hill, Easton quadrangle.
5. West part Chestnut Hill, Easton quadrangle.

Oligoclase-quartz gneiss

Oligoclase-quartz gneiss crops out throughout the Delaware Valley, but is especially common on Musconetcong Mountain in the Bloomsbury quad-

rangle and on Scotts Mountain in the Bloomsbury and Belvidere quadrangles. The unit contains sparse to many interlayers of amphibolite (fig. 3). The rock is light greenish gray (5G8/1) to grayish green (5G5/2), medium fine grained to medium coarse grained, equigranular, essentially equidimensional and moderately to well layered. Most phases are massive, have a very poor foliation and are nearly granulitic, but the unit has a good foliation where biotite is abundant.

Oligoclase and quartz are the essential minerals of oligoclase-quartz gneiss and varietal minerals include biotite, magnetite, and more rarely hornblende and augitic clinopyroxene. The gneiss is leucocratic, and rarely contains more than 5 percent of mafic minerals. Oligoclase is the most abundant mineral, and in most specimens exceeds quartz by two or three times. In many specimens, the oligoclase is antiperthitic. Microperthite is rare, occurring only near contacts with potassic feldspar gneiss. Quartz occurs interstitially as grains within the oligoclase, and as porphyroblasts; all types are strained. Biotite is commonly oriented in both planar and linear directions. Biotite-rich specimens are commonly magnetite poor, and conversely, as biotite appears to alter to iron-poor chlorite and magnetite. Hornblende and pyroxene are always strongly altered to chlorite, epidote, and magnetite.

Some specimens of the unit appear to be considerably coarsened by metasomatic enlargement. The plagioclase in these specimens is all strongly antiperthitic, microcline patch replacement perthites being the most common type. Many grains form sea and island textures, with well-twinned relics of host oligoclase surrounded by guest microcline. Apparently, late sodic solutions were active in these rocks as microantiperthite grains have replacement rims of albite on their borders. This phase of the oligoclase-quartz gneiss has much the appearance of albite-oligoclase granite or microantiperthite granite.

Oligoclase-quartz gneiss grades imperceptably into albite-oligoclase granite, microantiperthite alaskite, albite pegmatite, and type 2 migmatite and commonly contains blotches and veins of albite-oligoclase granite (figs. 3B–D and 4A). End members are relatively easy to recognize but a certain bulk of rock is difficult to pigeonhole without thin section study. Typical modes are given in table 8, and chemical analyses and norms are presented in table 9. It can be seen, that oligoclase-quartz gneiss is very similar to definite intrusive rocks in composition, and some workers believe

Figure 3. Oligoclase-quartz gneiss and albite-oligoclase granite. Musconetcong Mountain, Bloomsbury quadrangle. A, Oligoclase-quartz gneiss (light colored) and amphibolite (dark colored). Sequence of layers in center of photograph verges on type 2 migmatite. Strong fracture set dips steeply southeast (right) and is tectonically related to cleavage in Paleozoic carbonate rocks. Cut along U. S. Interstate 78. B, Oligoclase-quartz gneiss grading into albite-oligoclase granite (above knife). Cut on U. S. Interstate 78. C, Blotches of albite-oligoclase granite (lower right and under knife) in oligoclase-quartz gneiss. Foliation trends from lower left to upper right. Cut along U. S. Interstate 78. D, Large vein and several small veins of albite-oligoclase granite in oligoclase-quartz gneiss. Veins cross foliation in the gneiss. Quarry just off old Highway 22.

A   B

Figure 4. Albite syenopegmatite. Cut along U. S. Interstate 78, Musconetcong Mountain, Bloomsbury quadrangle. A, Relic body of oligoclase-quartz gneiss within albite pegmatite. Note gradation of small sodic amphiboles into the very large crystals in the pegmatite. B, Pegmatite crosscuts regional foliation which dips into hill. Fine-grained areas are relic oligoclase-quartz gneiss. Dark blotches are crystals and masses of crystals of sodic hornblende.

that the unit is a metamorphosed granitic rock (table 1). There is no question, but that some sodic gneiss have been partly mobile and both Sims (1958) and I (Drake, 1967a, 1967b; Drake and others, 1961, 1967, 1969) believe that in places, the unit has been completely mobilized to form

TABLE 8. MODES [1] (VOLUME PERCENT) OF OLIGOCLASE-QUARTZ GNEISS FROM THE DELAWARE VALLEY

	1	2	3	4	5
Plagioclase (including micro-antiperthite)	62.5	72.0	74.5	44.0	77.0
Quartz	17.5	25.5	22.5	37.5	29.5
Biotite	12.5	.5	—	—	2.5
Hornblende	6.5	1.5	—	1.5	.5
Clinopyroxene	—	—	2.5	—	—
Chlorite	—	—	.5	4.0	—
Magnetite	.5	—	tr	2.5	—
Apatite	.5	—	—	tr	tr
Zircon	tr	tr	tr	tr	tr
Epidote	—	—	—	.5	.5
Total	100.0	100.0	100.0	100.0	100.0

[1] Based on 2,000 points counted by the Chayes method.
1. Elephant Rock, Easton quadrangle.
2. Musconetcong Mountain, Riegelsville quadrangle.
3. Musconetcong Mountain, Bloomsbury quadrangle.
4. Pohatcong Mountain, Bloomsbury quadrangle.
5. Scotts Mountain, Belvidere quadrangle.

albite-oligoclase granite. The bulk of the unit apparently was not mobile and its marked layering (fig. 3B), interlayering with definite metasedimentary and (or) metavolcanic rocks (fig. 3A), fabric, and texture strongly support a non-intrusive origin. Graywacke has been suggested as a possible source for the rock but the $Na_2O - K_2O$ ratio seems far too high to allow such a parentage; sodic tuff or keratophyre seems more likely (table 9).

Origin of the Older Sequence

The older metamorphic sequence originally consisted of interbedded limestone and dolomite (marble, pyroxene gneiss, some amphibolite), pelitic and psammitic sedimentary rocks (quartzo-feldspathic gneisses) and pyroclastic and (or) volcanic sedimentary rocks and related mafic flows and possibly intrusive sheets (oligoclase-quartz gneiss and some amphibolite). Amphibolite is thought to be of at least two different origins, as some contains relic carbonate whereas some is interlayered with probable metavolcanic rocks. Some amphibolite may be metaintrusive, but evidence for this is lacking in the Delaware Valley.

The thickness of the metasedimentary-metavolcanic sequence is not known because of the lack of marker horizons, poor exposure, and extreme struc-

TABLE 9. CHEMICAL ANALYSES FROM C. I. P. W. NORMS OF OLIGOCLASE-QUARTZ GNEISS, ALBITE-OLIGOCLASE GRANITE, AND ALBITE PEGMATITE FROM THE DELAWARE VALLEY COMPARED WITH SIMILAR ROCKS FROM ELSEWHERE IN NEW JERSEY AND WITH TWO OTHER IGNEOUS ROCKS

	1	2	3	4	5	6	7	8	9
	Chemical analyses (weight percent)								
SiO_2	68.9	72.4	75.2	55.2	67.96	68.81	77.53	69.30	75.04
Al_2O_3	16.0	13.8	12.7	15.4	16.13	17.65	13.60	16.81	13.39
Fe_2O_3	.9	.5	.57	2.5	.36	.85	0.23	0.28	1.61
FeO	1.9	1.2	.57	7.4	2.04	.36	.16	1.26	0.37
MgO	.57	.97	.43	2.0	2.00	.20	tr	11.08	0.18
CaO	3.9	3.4	2.6	5.0	1.53	.71	.73	3.34	0.40
Na_2O	4.6	5.6	5.6	3.3	6.54	7.77	6.65	6.00	6.36
K_2O	1.6	1.0	1.2	5.2	1.78	1.83	1.20	1.39	0.83
H_2O+	.84	.48	.44	.86	.52	.66	.18	0.50	1.07
H_2O-	–	–	–	–	.08	.14	.15	–	0.24
TiO_2	.30	.38	.20	1.5	.87	.69	.16	0.23	0.10
P_2O_5	.19	.13	.07	.32	.02	.07	.03	0.03	0.08
MnO	.05	.08	.04	.12	.01	tr	tr	tr	0.05
CO_2	.34	.14	.58	.25	–	–	tr	0.15	0.10
Total	100.0	100.1	100.0	99.1	99.84	99.74	100.62	100.37	99.82

CIPW Norms

	1	2	3	4	5	6	7	8	9
Quartz	26.4	28.3	34.1	1.4	15.87	15.72	32.76	19.9	34.9
Orthoclase	9.4	6.1	7.2	31.1	10.56	10.56	7.23	8.3	5.0
Albite	39.8	47.2	47.2	28.8	55.23	64.98	56.07	50.4	54.5
Anorthite	18.6	9.7	6.1	12.0	7.51	3.06	3.66	15.0	1.1
Corundum	–	–	–	–	.63	1.94	–	–	–
$CaSiO_3$	–	2.7	2.8	4.8	–	–	–	.7	–
$MgSiO_3$	1.4	2.3	1.1	5.0	6.85	.50	–	2.7	.4
$FeSiO_3$	2.1	1.3	.1	9.5	–	–	–	1.7	–
Magnetite	1.4	.7	.9	3.7	.46	–	.23	.5	1.2
Rutile	–	–	–	–	–	.32	–	–	–
Apatite	.3	.3	–	.7	.03	.17	.06	–	.3
Ilmenite	.6	.8	.5	2.9	1.67	.76	.30	.5	.2
Hematite	–	–	–	–	–	.80	.24	–	.8
Total	100.0	100.0	100.0	99.9	98.81	98.81	100.26	99.7	100.0

1. Oligoclase-quartz gneiss from nose of Elephant Rock, Raubsville, Pa., Easton quadrangle. Rapid rock analyses by Paul Elmore, Ivan Barlow, Samuel Botts, and Gillison Chloe, U. S. Geological Survey.
2. Oligoclase-quartz gneiss from quarry just east of old Highway 22, about 1¼ miles southeast of West Portal, N. J., Bloomsbury quadrangle. Rapid rock analyses by Paul Elmore, Ivan Barlow, Samuel Botts, and Gillison Chloe, U. S. Geological Survey.
3. Small sheets of albite-oligoclase granite from large cut in U. S. Interstate 78, top of Musconetcong Mountain, N. J., Bloomsbury quadrangle. Same analysts as 2.
4. Albite pegmatite from large cut on U. S. Interstate 78, top of Musconetcong Mountain, N. J., Bloomsbury quadrangle. Same analysts as 2.
5. Oligoclase-quartz-biotite gneiss, Dover District, N. J. (Sims, 1958, p. 31).
6. Albite-oligoclase granite Dover District, N. J. (Sims, 1958, p. 31).
7. Losee diorite gneiss (albite granite) knob northeast of Berkshire Valley, Raritan quadrangle, N. J. (Bayley, 1941, p. 51).
8. Prondhjemite, Trondhjem, Norway (Turner and Verhoogen, 1960, p. 344).
9. Quartz keratophyre, eastern Oregon (Turner and Verhoogen, 1960, p. 262).

tural complexity and it seems that layers of the various lithologies occur at many different stratigraphic positions. In the Franklin-Sterling area, however, geologists of the New Jersey Zinc Company (Hague and others, 1956) have recognized two major layers of marble as well as other easily recognized thin units and by very detailed mapping in this area of relatively good exposure have identified a section more than 8,000 feet thick. Such work has not been possible in the Delaware Valley.

Metamorphic Facies

Mineral assemblages in the quartzo-feldspathic gneiss, quartz-microcline-sillimanite-almandine (plagioclase-biotite), and amphibolite, plagioclase-hornblende-diopside (biotite, hypersthene), could be considered as representing both the sillimanite-almandine subfacies of the almandine amphibolite facies or the hornblende granulite subfacies of the granulite facies of regional metamorphism (Fyfe and others, 1958). They occur together, however, so it is probable that the rocks in the Delaware Valley belong to the hornblende granulite subfacies. The presence of microperthite, microantiperthite, and abundant migmatites are in keeping with this rank of metamorphism. The alteration of plagioclase and mafic minerals to epidote and chlorite in many of the rocks attests to retrograde metamorphism in the greenschist facies. This alteration is not well understood, but seems to be most common near faults and shear zones and in granulated rock near Precambrian-Paleozoic contacts. In addition, biotite has formed at the expense of hornblende in some of the amphibolites and micromica at the expense of sillimanite in some quartzo-feldspathic gneisses probably as the result of the migmatizing fluids which have passed through the rocks.

INTRUSIVE ROCKS

The principal intrusive rocks in the Delaware Valley are granites, alaskites, and related pegmatite characterized by microperthite; alaskite and lesser granites characterized by microantiperthite, and sodic granitic rocks of complex origin but which are probably mobilized oligoclase-quartz gneiss. Minor amounts of charnockitic quartz diorite are included with the sodic granitic rocks, although they are thought to have had an ultrametamorphic origin.

In addition to these rocks which are more or less present throughout the Reading Prong a different type of pluton has been mapped west of the Delaware River, and a mafic dike of probable Precambrian age has been found in the Bloomsbury quadrangle (Drake, 1967b).

Microperthite-bearing rocks

Microperthite-bearing granites are the most abundant intrusive rocks in the Delaware Valley, as they are throughout the Reading Prong. These rocks are the type Byram Gneiss of earlier workers and are equivalent to the Storm King Granite of Lowe (1950). Hornblende granite is by far the most abundant rock and alaskite is next in quantity in this area. Local bodies of biotite granite have been mapped, but such rock is not common.

Hornblende granite. Hornblende granite includes both gneissoid hornblende granite and hornblende granite gneiss as poor exposure made their differentiation impractical. Large bodies of the rock include small bodies and layers of amphibolite, type 1 amphibolite migmatite, and potassic feldspar gneiss as well as small sheets of microperthite alaskite. Hornblende granite is gradational into microperthite alaskite and type 1 migmatite.

The unit consists of light-gray (N7) to pinkish-gray (5YR8/1), medium- to coarse-grained rocks that weather light brown (5YR6/4), moderate brown (5YR4/4), or white (N9). The granite is equigranular and has an interlocking texture, whereas the gneiss is largely inequigranular and has a granoblastic or more rarely flaser or augen texture.

Microperthite, or commonly mesoperthite, quartz, oligoclase, and hornblende are the essential minerals, and common accessories include magnetite, zircon, and apatite. Sparse amounts of epidote and sphene are present in some specimens. Hornblende commonly is altered to chlorite and magnetite, the magnetite forming sieve textures with the chlorite. Oligoclase is altered to felty masses of clay and mica minerals.

Especially deformed specimens of hornblende granite gneiss contain microcline and oligoclase in partial to complete exclusion of microperthite. This suggests that microperthite completely unmixed during metamorphism to form the separate feldspars. Similar gneissic phases have been noted at other places in the Reading Prong (Sims, 1958).

Typical modes of hornblende granite are given in table 10 and chemical analyses and norms in table 11. The close petrochemical similarity of hornblende granite from these widely spaced localities is readily apparent.

TABLE 10. MODES [1] (VOLUME PERCENT) OF HORNBLENDE GRANITE FROM THE DELAWARE VALLEY

	1	2	3	4	5
Microperthite	47.2	53.4	40.3	41.1	—
Microcline	—	—	—	—	41.6
Quartz	28.3	27.3	24.6	21.6	23.4
Oligoclase	14.6	12.8	31.1	22.8	18.0
Hornblende	9.7	5.4	1.6	13.9	15.8
Magnetite	.3	.6	.2	.6	.7
Apatite	tr	tr	tr	tr	.5
Zircon	tr	tr	tr	tr	tr
Epidote	—	—	1.5	—	—
Sphene	tr	tr	tr	tr	tr
Chlorite	—	.5	.7	—	—
Total	100.0	100.0	100.0	100.0	100.0

[1] Based on 2,000 points counted by the Chayes method.
  1. Musconetcong Mountain, Frenchtown quadrangle.
  2. Steelys Hill, Riegelsville quadrangle.
  3. Morgan Hill, Easton quadrangle.
  4. Musconetcong Mountain, Bloomsbury quadrangle.
  5. Hornblende granite gneiss, Musconetcong Mountain, Frenchtown quadrangle.

TABLE 11. CHEMICAL ANALYSES AND CIPW NORMS OF HORNBLENDE GRANITE FROM THE DELAWARE VALLEY COMPARED WITH SIMILAR ROCKS FROM ELSEWHERE IN THE READING PRONG

	1	2	3	4
Chemical analyses (weight percent)				
SiO_2	75.6	74.89	73.35	58.75
Al_2O_3	12.8	12.33	13.07	17.16
Fe_2O_3	1.4	.69	.74	5.18
FeO	.71	1.50	2.29	3.94
MgO	.12	.16	.08	.91
CaO	.83	.91	1.11	.62
Na_2O	3.5	2.95	2.98	5.72
$H_2O +$	.38	.35	.23	.73
$H_2O -$	–	.10	–	.35
TiO_2	.13	.21	.20	.65
P_2O_5	.05	.04	.06	.20
MnO	.04	.03	.03	.10
CO_2	–	–	–	.13
K_2O	4.6	5.38	5.62	5.40
Total	100.2	99.5	99.8	99.8
CIPW Norms				
Quartz	36.0	34.18	30.70	2.28
Orthoclase	27.2	31.60	33.25	31.60
Albite	29.3	25.15	25.15	48.21
Anorthite	4.2	4.17	5.14	1.39
Corundum	.5	.31	.18	1.43
Hypersthene	.3	2.30	3.51	4.15
Magnetite	1.9	.93	1.07	7.66
Ilmenite	.3	.46	.38	1.22
Hematite	.2	–	–	–
Apatite	–	.10	.13	.34
Calcite	–	–	–	.30
Water	–	.45	.23	1.08
Total	99.9	99.74	99.74	99.75

1. Hornblende granite from north end of cut along Highway 611, Bougher Hill, Riegelsville quadrangle, Pa. Rapid rock analyses by Paul Elmore, Ivan Barlow, Samuel Botts, and Gillison Chloe, U. S. Geological Survey.
2. Hornblende granite from Boonton quadrangle, Dover District, N. J. (Sims, 1958, p. 33).
3. Storm King Granite of Lowe, Bear Mountain, N. Y. (Lowe, 1950, p. 146).
4. Biotite granite from Van Nest Tunnel, Oxford, N. J., Washington quadrangle (Bayley and others, 1914, p. 9).

Biotite granite. Small bodies of biotite granite crop out on Morgan Hill, Chestnut Hill, and Marble Mountain in the Easton quadrangle and on Musconetcong Mountain in the Bloomsbury quadrangle. The rock is light gray (N7) to pinkish gray (5YR8/1), white weathering, and highly contaminated by relic layers of biotite-quartz-plagioclase gneiss. It is fine to medium grained and consists of microperthite, quartz, antiperthitic oligoclase, and variable quantities of biotite. Minor amounts of garnet, magnetite, apatite, and zircon are present in some specimens.

The unit occurs in areas where microperthite alaskite is spatially related to layers of biotite-quartz-plagioclase gneiss. The gneiss near the contacts contains porphyroblasts of potassic feldspar

and veins of microperthite alaskite and passes into a migmatitic rock in which both plagioclase and biotite are coarsened, and finally into a biotite granite that contains many streaks and discontinuous layers of gneiss. It appears that the biotite granite is a hybrid rock formed either by the assimilation of biotite-quartz-plagioclase gneiss by alaskite or by the addition of potash-rich fluids to the gneiss.

Typical modes of biotite granite are given in table 12 and chemical analyses and norms are given in

TABLE 12. MODES [1] (VOLUME PERCENT) OF BIOTITE GRANITE FROM THE DELAWARE VALLEY

	1	2	3
Microperthite	50.0	37.5	22.9
Quartz	38.0	31.0	30.8
Plagioclase	3.0	26.5	38.3
Biotite	6.5	4.5	7.8
Magnetite	.5	.5	tr
Hornblende	.5	tr	.2
Chlorite	1.5	tr	–
Garnet	–	–	–
Apatite	tr	tr	tr
Zircon	tr	tr	tr
Sphene	tr	tr	tr
Total	100.0	100.0	100.0

[1] Based on 2,000 points counted by the Chayes method.
1. Musconetcong Mountain, Bloomsbury quadrangle.
2. Morgan Hill, Easton quadrangle.
3. Marble Mountain, Easton quadrangle.

table 11. The highly variable modal composition supports a hybrid origin as does the chemical analysis. Some of the rock from the Van Nest tunnel contains more quartz and is more granitic in aspect than that reported in table 11.

Microperthite alaskite. Microperthite alaskite, including pegmatitic phases, crops out throughout the Precambrian terrane in the Delaware Valley. Small lenticular masses occur in phacolith-like bodies in small fold crests or troughs, and other bodies are sheetlike in form.

The alaskite is a light-gray (N7), white, pinkish-gray (5YR8/1), or light-brownish-gray (5YR6/1) rock that weathers to a dull white. It has a poor to good foliation that is most prominent where it contains some hornblende and a poor to good lineation. It differs mineralogically from the hornblende granite in containing less than 5 percent mafic minerals and from the microantiperthite alaskite and granite in the type of perthite present. Texturally, it is more like the microantiperthite alaskite and granite. The essential minerals are microperthite (primarily regular string perthites), quartz, and oligoclase; oligoclase is less important than in the hornblende granite. Minor amounts of hornblende and magnetite are present in most specimens. Zircon and apatite are the common accessory minerals. Typical modes of microperthite alaskite are given in table 13 and a chemical analysis and norm in table 14.

TABLE 13. MODES [1] (VOLUME PERCENT) OF MICROPERTHITE ALASKITE FROM THE DELAWARE VALLEY

	1	2	3	4	5
Microperthite	54.0	44.2	46.0	47.3	60.0
Quartz	43.5	29.0	32.5	36.4	30.5
Plagioclase	–	25.8	27.5	11.6	8.5
Hornblende	–	.5	.5	4.4	1.0
Magnetite	.5	.5	2.5	.3	tr
Biotite	–	–	1.0	–	–
Apatite	tr	–	tr	tr	tr
Chlorite	2.0	–	–	–	–
Zircon	tr	tr	tr	tr	tr
Sphene	–	–	tr	tr	tr
Total	100.0	100.0	100.0	100.0	100.0

[1] Based on 2,000 points counted by the Chayes method.
1. Musconetcong Mountain, Frenchtown quadrangle.
2. Lower Pohatcong Mountain, Riegelsville quadrangle.
3. Morgan Hill, Easton quadrangle.
4. Pohatcong Mountain, Bloomsbury quadrangle.
5. Scotts Mountain, Belvidere quadrangle.

Lenticular and tabular sheets and veins and irregular blob-like masses of granite pegmatite are fairly common throughout the hornblende granite-alaskite terrane and in the surrounding metamorphic rocks. The pegmatites have much the same composition of the microperthite alaskite, and are probably genetically related to it.

Origin of the microperthite-bearing rocks. All the microperthite-bearing rocks occur in conformable sheets, pods, and phacoliths and have all the features usually ascribed to syntectonic granites of Buddington's (1959) catazone. The hornblende granite and alaskite are thought to have stemmed from the same magma because of their close petrochemical and spatial relations. Practically all geologists who have worked in the Reading Prong have accepted that the bulk of these rocks were intruded into the older metasedimentary-metavolcanic sequence. Some border phases of the hornblende granite probably formed by the granitization of amphibolite as one can walk from amphibolite through migmatite into "dirty" hornblende granite and finally into pristine hornblende granite. Buckwalter (1962) has found similar relations in the Reading, Pa. area. The biotite granite, as was said above is almost certainly hybrid rock formed by the interaction of alaskite and biotite-quartz-plagioclase gneiss.

Sodic granitic rocks

Smaller and less abundant bodies of sodic granitic rocks and related albite pegmatite crop out throughout the Delaware Valley. Similar rocks are present throughout the New Jersey and New York part of the Prong, but have not been described farther to the west in Pennsylvania. These rocks are always spatially related to oligoclase-quartz gneiss.

Albite-oligoclase granite and albite pegmatite. Albite-oligoclase granite is normally alaskitic, but moderate amounts of mafic minerals are present in exposures that contain relic amphibolite layers.

TABLE 14. CHEMICAL ANALYSES AND CIPW NORMS OF MICROPERTHITE ALASKITE AND MICROANTIPERTHITE GRANITE FROM THE DELAWARE VALLEY COMPARED WITH ALASKITE FROM ELSEWHERE IN THE READING PRONG

	1	2	3
Chemical analyses (weight percent)			
SiO_2	76.5	73.0	77.07
Al_2O_3	12.0	12.3	12.61
Fe_2O_3	.75	1.5	.71
FeO	.50	1.2	.73
MgO	.2	.7	tr
CaO	.63	1.6	.87
Na_2O	3.8	5.2	3.43
K_2O	4.4	2.9	4.06
H_2O-	.06	–	.23
H_2O+	.58	1.3	.62
TiO_2	.14	.27	.12
P_2O_5	.32	.06	tr
MnO	–	.07	.09
CO_2	–	.05	tr
Total	99.9	100.2	100.5
CIPW Norms			
Quartz	37.3	28.9	39.12
Orthoclase	26.1	16.2	24.46
Albite	32.0	44.5	28.82
Anorthite	1.4	2.0	4.45
Corundum	.6	–	.92
Diopside	–	2.0	–
Hypersthene	.5	2.6	.79
Magnetite	1.2	2.1	.93
Ilmenite	.3	.6	.15
Apatite	.7	1.3	–
Water	–	–	.85
Total	100.1	100.2	100.49

1. Microperthite alaskite from cut in Bougher Hill along Highway 611 about 0.3 mile south of Coffeetown, Pa., Riegelsville quadrangle. Rapid rock analysis by Paul Elmore, Lowell Artis, and Samuel Botts, U. S. Geological Survey.
2. Microantiperthite granite from bold outcrop along Lommonsons Road about 0.25 mile south of Lommonsons Glen, N. J., Belvidere quadrangle. Analysts same as 1.
3. Alaskite from Hibernia, N. J., Dover District (Bayley and others, 1914, p. 9).

The unit is completely gradational with micro-antiperthite granite and alaskite where present and always with oligoclase-quartz gneiss and contacts are placed on the amount of microantiperthite present in the rock, its texture, and general fabric.

Albite-oligoclase granite, is pale green (5G7/2) to light gray (N7) on fresh surface and dull white on weathered surface. It is a medium- to coarse-grained rock whose texture ranges from gneissoid to granular. Greenish albite and (or) oligoclase and quartz are the essential minerals, and in some specimens the quartz content is so low that the rock is syenitic. Some hornblende, clinopyroxene, or biotite are present near inclusions of amphibolite, and trace amounts of magnetite, zircon, epidote, sphene, and apatite are not uncommon. The mafic minerals are commonly altered to chlorite and magnetite. Microcline patch microantiperthite is sparse to abundant in many specimens, particularly in those from areas near contacts with microantiperthite and alaskite and granite. Still other varieties contain plagioclase of two colors—one reddish and the other typically greenish. The reddish feldspar is more sodic and apparently owes its color to small amounts of included hematite. Quartz occurs in small rounded grains within the plagioclase and larger grains that interlock with plagioclase crystals. Much of the quartz appears to be late and to have filled pre-existing open spaces.

Typical modes of albite-oligoclase granite are given in table 15 and a chemical analysis and norm in table 9. The close petrochemical relation between the albite-oligoclase granite and oligoclase-quartz gneiss is readily apparent.

TABLE 15. MODES [1] (VOLUME PERCENT) OF ALBITE-OLIGOCLASE GRANITE FROM THE DELAWARE VALLEY

	1	2	3	4	5
Plagioclase	52.8	69.8	81.2	61.8	63.1
Quartz	24.6	27.5	16.5	35.2	30.4
Microcline	1.0	1.2	2.1	—	—
Microantiperthite	12.2	—	—	—	—
Hornblende	7.3	1.1	—	2.5	—
Clinopyroxene	—	—	—	—	4.2
Chlorite	—	—	—	—	2.0
Magnetite	2.1	.4	.2	.5	.2
Zircon	tr	tr	—	tr	tr
Sphene	tr	—	—	—	.1
Apatite	tr	tr	tr	—	tr
Total	100.0	100.0	100.0	100.0	100.0

[1] Based on 2,000 points counted by the Chayes method.
1. Musconetcong Mountain, Frenchtown quadrangle.
2. Steelys Hill, Riegelsville quadrangle.
3. Pohatcong Mountain, Bloomsbury quadrangle.
4. Musconetcong Mountain, Bloomsbury quadrangle.
5. Scotts Mountain, Belvidere quadrangle.

Pegmatites composed primarily of albite or oligoclase, quartz, and lesser quantities of microantiperthite are quite common in the Delaware Valley. Most of the pegmatites were far too small to map at 1:24,000, however, several fairly large bodies crop out on Musconetcong, Pohatcong, and Scotts Mountains in the Bloomsbury and Belvidere quadrangles. Most of the pegmatites form small to large sheets or veins parallel to the foliation of the enclosing rock, but many crosscut the structures. One such pegmatite is well exposed in the cut along U. S. Interstate 78 on Musconetcong Mountain in the Bloomsbury quadrangle. Much of the pegmatite is obviously a coarse-grained facies of the albite-oligoclase granite and is gradational into oligoclase-quartz gneiss (fig. 4A). As other phases seem to be the same type of rock, they are also assumed to be related to the albite-oligoclase granite.

Albite pegmatite is light greenish gray (5G8/1) to greenish gray (5G6/1) and coarse to very coarse grained. Quartz is fairly abundant in most exposures but some phases are syenitic (table 9). Hornblende and pyroxene form large to very large subhedral to euhedral crystals or masses of crystals, the pegmatite in the cut on Highway 78 being especially characterized by very large crystals and masses of sodic hornblende (fig. 4B). The large mafic crystals are present only where the pegmatite is within or adjacent to amphibolite or pyroxene gneiss. Some pegmatites also contain magnetite which appears to have been derived from mafic minerals in adjacent mafic rocks. Epidote is also fairly common and is probably a result of the breakdown of calcic plagioclase in the mafic rocks.

Charnockitic quartz diorite. A few thin layers of hypersthene-bearing rock of quartz diorite composition have been recognized in the Delaware Valley. The rock is characterized by alternating light and dark layers and by a distinctive greasy appearance. Light layers are generally thicker and more continuous than dark layers. The mineralogy of the layers is essentially the same, but the dark layers contain somewhat more mafic minerals and have the composition of pyroxene amphibolite. The light layers have the appearance of pyroxene-bearing albite-oligoclase granite.

Quartz diorite is greenish gray (5G6/1), grayish brown (5YR3/2) weathering, equigranular to inequigranular, and xenoblastic. Oligoclase-andesine, hypersthene, augitic clinopyroxene, quartz, and hornblende are the essential minerals. Other minerals include magnetite, biotite, sphene, apatite, and zircon. Chlorite is a common alteration product of pyroxene and hornblende. Hypersthene is pleochroic in faint shades of pink, and is the most altered mineral. It commonly has magnetite grains oriented along cleavages and magnetite-sphene rims. Clinopyroxene is partly intergrown with hypersthene, and both pyroxenes are partly converted to hornblende. Quartz occurs as poikiloblastic inclusions in oligoclase-andesine and in tiny cross cutting veinlets and lenses that surround and embay plagioclase. Typical modes are given in table 16.

TABLE 16. MODES [1] (VOLUME PERCENT) OF CHARNOCKITIC QUARTZ DIORITE FROM THE DELAWARE VALLEY

	1	2	3	4	5
Plagioclase	37.4	73.5	78.2	65.4	54.2
Quartz	27.6	10.4	5.6	29.5	28.1
Hypersthene	8.6	9.2	12.7	7.5	15.8
Clinopyroxene	7.3	5.8	1.4	.8	1.9
Hornblende	1.1	.5	–	1.1	–
Biotite	.5	–	1.8	.5	–
K-feldspar	16.8	–	–	–	–
Magnetite	.7	.6	.2	.2	tr
Apatite	–	–	tr	tr	tr
Chlorite	–	–	tr	–	–
Zircon	–	–	–	tr	tr
Total	100.0	100.0	100.0	100.0	100.0

[1] Based on 2,000 points counted by the Chayes method.
  1. Musconetcong Mountain, Riegelsville quadrangle.
  2. Musconetcong Mountain, Frenchtown quadrangle.
  3. Musconetcong Mountain, Easton quadrangle.
  4. Musconetcong Mountain, Bloomsbury quadrangle.
  5. Scotts Mountain, Belvidere quadrangle.

The quartz diorite has charnockitic affinities and was first recognized in the Reading Prong by Sims (1958, p. 38). The rock is similar to the noritic members of the charnockite series of igneous rocks and has mineral assemblages similar to the hornblende granulite subfacies of metamorphic rock.

Origin of the sodic granitic rocks. The sodic granitic rocks, like those characterized by microperthite, form conformable sheets, pods, and phacoliths and are typical of syntectonic granites of Buddington's (1959) catazone. Much of the rock has been mobile and is definitely intrusive, but some seem to be the result of granitization as many bodies contain abundant ghostlike relics of isoclinally folded amphibolite (fig. 5) and contacts with the older metamorphic units are transitional and vague. It has been shown (figs. 3A–C) that oligoclase-quartz gneiss grades into albite-oligoclase granite and it is thought, therefore, that the granite is the result of the mobilization of the

Figure 5.   Relic isoclinally folded amphibolite in albite-oligoclase granite. Musconetcong Mountain, Riegelsville quadrangle.

oligoclase-quartz gneiss, the necessary heat for anatexis being supplied by the hornblende granite-alaskite magma.

Albite pegmatite is thought to have had the same origin as the albite-oligoclase granite as much of the pegmatite can be seen to be a coarse phase of the granite. Other pegmatite bodies contain relic bodies of oligoclase-quartz gneiss (fig. 4A). The sparodic mafic crystals and masses in the pegmatite are thought to be reconstituted amphibolite or pyroxene gneiss relics because of their close spatial relation to these metamorphic rocks. It is probably that differential anatexis was involved and that temperatures were high enough to mobilize oligoclase-quartz gneiss but not the mafic rocks.

The origin of charnockitic rocks is a difficult problem throughout the world and, as such rocks are so poorly exposed in the Delaware Valley, little can be added to a solution of the problem. Relations here suggest, however, that the quartz diorite may have an ultrametamorphic or anatectic origin. It has been shown that the albite-oligoclase granite is probably mobilized oligoclase-quartz gneiss, and that in certain areas oligoclase-quartz gneiss and amphibolite are closely interlayered (fig. 3A). It is certainly conceivable that a light and dark layered quartz dioritic rock could be generated by anatexis from such rock. The felsic material would be more mobile which would explain local crosscutting relations. It is even conceivable that a whole body became mobile because of local high temperatures, but no data are available to support this idea.

Microantiperthite alaskite and granite

Microantiperthite-bearing rocks are relatively rare in the Delaware Valley but sizable bodies have been mapped on Musconetcong Mountain in the Frenchtown and Riegelsville quadrangles, Elephant Rock and Pohatcong Mountain in the Easton quadrangle and Scotts Mountain in the Belvidere quadrangle. Other bodies may well be present, but it is difficult to distinguish the unit from microperthite alaskite without thin section study. Microantiperthite alaskite and granite, however, normally weathers to a chalk white and contains glassy, dark-gray, distinctly lenticulated quartz. Another subtle difference is that the mafic minerals are very wispy and strongly altered to chlorite. Microantiperthite alaskite and granite is gradational with oligoclase-quartz gneiss, albite-oligoclase granite and microperthite alaskite.

Microantiperthite, quartz, and oligoclase are the essential minerals, but several specimens contain moderate amounts of microcline. Mafic minerals include hornblende, biotite, magnetite, and augitic clinopyroxene. Pyroxene is rarely present in microperthite-bearing rocks in the Delaware Valley. The mafics invariably occur in discontinuous irregular clots rather than in layers or streaks and are altered to iron-poor chlorite and magnetite. Microcline patch perthite and plume perthite are the most common types, but string mesoperthites are also present. Border replacements are common, as several specimens contain grains of microcline with relic twinned plagioclase in optical continuity. Microcline, where present, appears to fill open spaces. A late stage of soda enrichment is indicated, as microantiperthite and late microcline have albite patches and border rims. Typical modes are given in table 17 and a chemical analysis and norm in table 14.

The texture of the rock ranges from granoblastic to interlocking. It has a poor to fair foliation and a good lineation marked by the lenticulated quartz.

Microantiperthite-bearing rocks seem to have formed where sodic rocks are juxtaposed with microperthite alaskite or hornblende granite. This relation is obvious on Musconetcong Mountain in the Frenchtown and Riegelsville quadrangles (Drake and others, 1961, 1967). It is less obvious elsewhere, but all microantiperthite alaskite con-

TABLE 17. MODES [1] (VOLUME PERCENT) OF MICROANTIPERTHITE ALASKITE AND GRANITE FROM THE DELAWARE VALLEY

	1	2	3	4
Microantiperthite	55.0	42.4	36.5	47.2
Plagioclase	2.7	32.5	14.3	12.8
Microcline	8.0	—	7.6	1.2
Quartz	30.3	20.8	39.4	34.1
Clinopyroxene	—	—	—	2.8
Hornblende	—	3.2	.8	—
Biotite	—	.8	—	—
Chlorite	3.8	1.3	1.2	1.5
Sphene	tr	tr	tr	.4
Magnetite	.2	tr	.2	tr
Apatite	tr	tr	tr	tr
Total	100.0	100.0	100.0	100.0

[1] Based on 2,000 points counted by the Chayes method.
1. Musconetcong Mountain, Frenchtown quadrangle.
2. Elephant Rock, Easton quadrangle.
3. Pohatcong Mountain, Easton quadrangle.
4. Scotts Mountain, Belvidere quadrangle.

tains abundant unmapped sheets and veins of microperthite alaskite. It would seem, therefore, that the rock owes its origin to an interreaction of sodic gneiss or granite with microperthite alaskite magma or fluids therefrom, probably both. This origin, however, is far from obvious in this area of poor exposure and Buddington (*in* Sims, 1958) believes that albite-oligoclase granite in the Dover District formed by the albitization of microantiperthite granite. More work is necessary to completely understand this interesting rock.

Hornblende-clinopyroxene gneiss and clinopyroxene-garnet-quartz granofels

A crudely circular body of hornblende-clinopyroxene gneiss crops out just west of the Delaware River centered on Hexenkopf hill. The eastern part of this body has been mapped (Drake, 1967a; Drake and others, 1967) and its continuation to the west has been studied by J. M. Aaron (oral commun., 1967). Many different types of rock crop out, but all are related.

Most of the rock is medium dark gray (N4) to greenish black (5GY2/1) and consists of hornblende, augitic clinopyroxene, plagioclase (mostly andesine), and quartz and varying amounts of epidote, biotite, sphene, garnet, apatite, magnetite, pyrite, chlorite, and zircon. Perhaps 75 percent of the unit has a fair to good gneissic structure, but other phases have a granofels structure; all phases have a granoblastic texture. In many places, the unit is a veined gneiss, containing veins of both albite pegmatite and microperthite alaskite.

The plagioclase is strongly sericitized and the mafic minerals are altered, in part, to chlorite, magnetite, and sphene. The magnetite commonly has rims of sphene. Some of the hornblende appears to have formed from pyroxene, and biotite appears to form from hornblende that is spatially associated with the granitic veins. The feldspars of the albite pegmatite are essentially unaltered, but are antiperthitic in places. Much of the quartz occurs in areas spatially associated with alaskite veins, but is also intergrown with sericitized plagioclase, hornblende, and pyroxene.

One other distinctive rock type occurs within this unit. It is a heavy, massive, dense, grayish olive (10Y4/2), foliated hornblende-augitic clinopyroxene-epidote gneiss that contains very sparse amounts of biotite, sphene, magnetite, and quartz. No feldspar is apparently present. The hornblende forms at the expense of pyroxene as does at least part of the epidote. The rock has the appearance of a metamorphosed pyroxenite.

Where mapped to date, the body of hornblende-clinopyroxene gneiss has a collar of clinopyroxene-garnet-quartz granofels. This rock is medium gray (N5) to greenish gray (5G6/1), medium-fine grained to medium grained, equigranular, heterogeneous, essentially non-foliated and has a granulitic texture. It is highly siliceous and consists largely of quartz, essentially completely saussuritized plagioclase, and lesser amounts of reddish-brown garnet and greenish-black augitic clinopyroxene. Accessory and varietal minerals include epidote, sphene, hornblende, magnetite, chlorite, and microperthite. The pyroxene is more or less altered to chlorite, epidote, sphene, and magnetite. Garnet occurs in clusters of small grains, spatially associated with pyroxene suggesting that it may have formed by a reaction between feldspar and pyroxene. The small amount of microperthite present in a few specimens occurs in veins with quartz and represents additive microperthite alaskite. The unit can be recognized by its distinctive granulitic texture, high quartz content, and the presence of garnet and clinopyroxene.

The clinopyroxene-garnet-quartz granofels has the fabric, texture, composition, and spatial relations of a hornfels. The hornblende-clinopyroxene gneiss probably is a composite pluton of rock ranging from pyroxenite to mafic diorite in composition, that has been metamorphosed and modified by granitic material. Although it lies athwart the regional tectonic grain, it was deformed with the older metasedimentary-metavolcanic sequence and

clearly predates granite emplacement. The surrounding hornfels has also been retrogressively metamorphosed by the granitic fluids as the garnet and pyroxene are chloritized. It seems, therefore, that this body dates from a period of plutonism older than any recognized heretofore in the Reading Prong.

Lamprophyre

Diabase and other mafic dikes have been mapped at many places within the Reading Prong, most of which are assumed to be of Triassic age. Miller and others (1939, 1941), however, have recognized metadiabase in eastern Pennsylvania which they believe to be Ordovician or post-Ordovician in age. Similar post-Ordovician dike rocks have been described from Franklin, N. J., by Spencer and others (1908). Buckwalter (1959, 1962) has recognized both probable Precambrian and Ordovician metadiabase from the Reading area, Pa. One mafic dike is exposed in the cut of U. S. Interstate 78 that is thought to be of Precambrian age (fig. 6A). The dike is dark gray (N3) to grayish black (N2) and is composed principally of andesine and hornblende and lesser clinopyroxene and magnetite. Apatite is a very abundant accessory mineral. The rock is strongly altered, mafic minerals have altered to chlorite, and plagioclase to saussurite. Some plagioclase crystals have more sodic rims and some contain patches of K-feldspar. The dike contains veins and introduced grains of quartz, sodic oligoclase, and microperthite. Major oxide and minor element analyses of the unit are given in table 18. It can be seen that the rock contains less Na_2O than the camptonites but is more similar to the biotite-augite lamprophyre. The dike contains the high concentrations of barium, strontium, and titanium characteristic of the lamprophyres. Based on its composition the dike is properly a spessartite rather than a camptonite as it does not seem to contain the sodic amphibole barkevikite nor does it contain enough Na_2O, even though it has been enriched in this oxide subsequent to emplacement.

The dike fills a northeast-trending fault in the Precambrian gneisses and has blotches of albite syenopegmatite along its borders (figs. 6A–C). It is very difficult to determine a time relation between the pegmatite and the lamprophyre but it appears that veins of pegmatite cut the dike (fig. 6B) and the pegmatite contains inclusions of lamprophyre (figs. 6B, C). If this is so, the dike predates the albite pegmatite and, hence, is of Precambrian

A

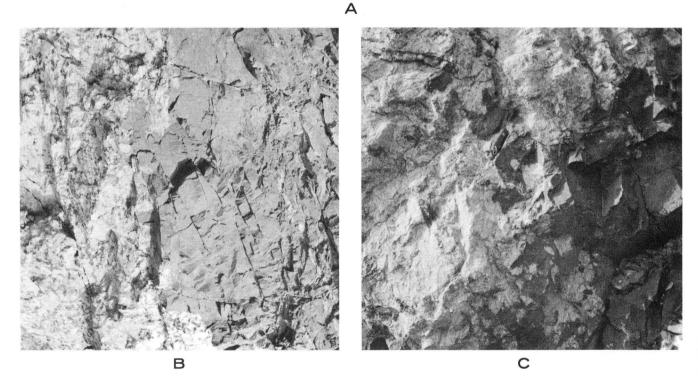

B   C

Figure 6.   Lamprophyre dike cutting Precambrian gneiss in cut on U. S. Interstate 78, Musconetcong Mountain, Bloomsbury quadrangle.   A, Note blotch of albite syenopegmatite on left contact
of dike and thrust fault about halfway up the rock face cut off by the dike.   B, Contact relations
between lamprophyre and pegmatite.   Note vein of pegmatite in lower right and inclusions of
dike rock in the pegmatite.   C, Contact of lamprophyre and pegmatite.   Note soak relations, tiny
veins of pegmatite in the dike, and inclusions of dike rock in the pegmatite.

TABLE 18. MAJOR OXIDE AND MINOR ELEMENT ANALYSES OF LAMPROPHYRES FROM THE DELAWARE VALLEY COMPARED WITH LAMPROPHYRES FROM OTHER AREAS.

	1	2	3	4
Major oxides (weight percent)				
SiO_2	46.9	41.94	47.40	46.37
Al_2O_3	13.0	15.36	17.40	11.98
Fe_2O_3	3.3	3.27	4.65	5.05
FeO	9.1	9.89	5.87	5.16
MgO	3.1	5.01	4.45	8.38
CaO	8.4	9.47	8.30	9.33
Na_2O	3.1	5.15	5.41	2.84
K_2O	2.6	0.19	2.58	4.34
H_2O-	.08	–	–	.71
H_2O+	1.5	3.29	–	2.17
TiO_2	5.0	4.15	3.31	1.80
P_2O_5	2.3	–	.49	1.34
MnO	.21	.25	.14	0.10
BaO	–	–	–	0.31
CO_2	.68	2.47	–	0.24
Total	99.3	100.44	100.0	100.12
Minor elements				
Ba	.07			
Ce	.03			
Co	.003			
Cr	.002			
Cu	.007			
Ga	.0015			
La	.005			
Mo	.0005			
Sc	.0015			
Sr	.07			
V	.02			
Y	.005			
Yb	.0005			
Zr	.03			

1. Lamprophyre dike from cut on U. S. Interstate 78, Musconetcong Mountain, Bloomsbury quadrangle, N. J. Major oxide by X-ray fluorescence supplemented by rapid methods described in U. S. Geol. Survey Bull. 1144-A. Analyses by Paul Elmore, Lowell Artis, and Samuel Botts, U. S. Geological Survey. Minor elements by semiquantitative spectrographic analysis. Results are reported in percent to the nearest number in the series (1, 0.7, 0.5, 0.3, 0.2, 0.15, and 0.1, etc.) which represent approximate midpoint of group data on a geometric scale. The assigned group for semiquantitative results will include the quantitative value about 30 percent of the time. Elements not listed looked for but not found. Analysis by J. L. Harris, U. S. Geological Survey.
2. Camptonite, Campton Falls, N. H. (Turner and Verhoogen, 1960, p. 253).
3. Average camptonite, Tahite (Turner and Verhoogen, 1960, p. 253).
4. Biotite-augite lamprophyre, Spanish Peaks, Colo. (Turner and Verhoogen, 1960, p. 253).

age. The only other lamprophyre dikes reported from New Jersey are post-Martinsburg—pre-Shawangunk in age (Spencer and others, 1908).

No satisfactory origin for lamprophyre has yet been proposed (Turner and Verhoogen, 1960). It is possible that the dike described herein formed by the melting and mobilization of amphibolite by the same sources of heat that mobilized the oligoclase-quartz gneiss in the same locality. If so, the dike would be the extreme case of modification of amphibolite shown so clearly by the large crystals and clots of crystals of hornblende in the albite pegmatite. Some geologists have proposed that the dike represents a "basic behind" produced by metamorphic differentiation within the interlayered oligoclase-quartz gneiss and amphibolite, the pegmatite being the felsic differentiate. This is similar to the proposal presented herein, but some sort of differential anatexis seems more likely than ionic diffusion in the solid state as mineral assemblages clearly show that temperatures were high enough for melting.

Age of the Intrusive Rocks

Data presented heretofore have shown that the hornblende-clinopyroxene gneiss is older than both the microperthite-bearing granites and the sodic granitic rocks and, therefore, presumably older than microantiperthite granite and alaskite and that lamprophyre is slightly older than but intimately associated with albite pegmatite. What then is the relation of the sodic granite to the microperthite granites?

All workers in the Reading Prong have considered the sodic granite to be older than the microperthite granites although conformable contacts have mitigated against direct evidence in most areas. Horz (1952), however, describes one place in the Sterling Lake-Ringwood area where hornblende granite appears to transect sodic granite. In addition, most authors have noted that potassic feldspar replaces the plagioclase in sodic granite where such rock is adjacent to hornblende granite or alaskite.

Tilton and others (1960) report concordant U-Pb ages of 1150 m.y. for zircon from the Canada Hill gneiss in New York, which is the same rock as the oligoclase-quartz gneiss of the Delaware Valley. This can be taken as the age of the original metamorphism of the Reading Prong and fits nicely with its Grenville affinities. Long and Kulp (1962) report that the average K-Ar and Rb-Sr age of mica from sodic rocks in the Prong is about 840 m.y. Part of these rocks appears to be oligoclase-quartz gneiss and the other part to be albite-oligoclase granite. They also report that monazite (Pb-U) from Losee Gneiss at Chester, N. J., is between 720 and 900 m.y. old. These data suggest a major plutonic event at about 840 m.y. ago.

Several isotopic age determinations have been

done on minerals from the microperthite granites and presumably related pegmatites. One zircon age from the Storm King Granite as used by Lowe (1950) of New York suggests that the rock may be more than 900 m.y. old, although K-Ar and Rb-Sr ages of mica from granites of this group average about 840 m.y. (Long and Kulp, 1962). In addition, thorianite has been dated at 800 m.y. (Wells and others, 1933), zircon (lead-alpha method) at 850 m.y. (Long and others, 1959). These minerals were collected from marble quarries along the Delaware River and have a pyrometasomatic origin. The mineralizing fluids stemmed from pegmatites which presumably are related to the microperthite alaskite. It seems, therefore, that the hornblende granite and alaskite are about 840 m.y. old. Hornblende granite and related biotite granite are the oldest microperthite-bearing rocks followed by alaskite and finally by pegmatites that are post-tectonic.

It has been shown that the mica ages of the sodic rock are essentially the same as the ages of various minerals in the microperthite granites (840 m.y.). The major event at the time was certainly metamorphism and the emplacement of the microperthite granite bodies. The mica clocks in the older rocks were reset at that time. This is in keeping with the postulated anatectic origin of the albite-oligoclase granite. The postulated sequence of events during Precambrian time, therefore, is original metamorphism of the older metamorphic sequence at 1150 m.y., emplacement of the hornblende-clinopyroxene gneiss and its hornfels collar, metamorphism and emplacement of the microperthite granites with attendant mobilization of oligoclase-quartz gneiss to form albite-oligoclase granite and probably quartz diorite and post-tectonic emplacement of the residual magmas as granite pegmatite and albite pegmatite and lamprophyre.

YOUNGER METASEDIMENTARY AND METAVOLCANIC ROCKS

Spencer (1908) apparently first recognized the presence of metasedimentary rocks other than marble in the Reading Prong when he described sandstone, siliceous breccia, and indurated carbonaceous shales from near Andover, New Jersey. These rocks were thought to be part of a sedimentary series belonging with marble. Bayley (1914) corroborated Spencer's observations and described quartzite, slaty or schistose rocks having the appearance of sheared volcanic rocks, quartzitic conglomerate, and rocks that are probably

rhyolite from the Delaware Valley. In addition to these rocks, he recognized graphitic mica schist and gneiss and other quartzo-feldspathic gneisses as being metasedimentary. All these rocks comprise his Pickering Gneiss. The bulk of this unit belongs to the now recognized older metasedimentary-metavolcanic sequence that is so widespread throughout the Prong. The remainder, however, is something entirely different and includes rocks of lower metamorphic rank that on the basis of all available evidence is younger than the other metamorphic rocks and probably younger than the plutonic rocks.

Such rocks are relatively abundant in the Delaware Valley along the north border of the Prong (Drake 1967a; Drake and others, 1969) and here include arkose, ferruginous quartzite, quartzite conglomerate, sericite-talc-chlorite schist, sericite-talc-actinolite-chlorite schist, dense purple rocks that are almost certainly metamorphosed rhyolite, greenish-gray rocks that seem to be metatuffs and purple slate.

These metamorphic rocks seem to be interbedded and are spatially related to a major layer of marble, but their exact relation to the marble as well as their intrarelations cannot be determined because of exceedingly complex structure and abominably poor exposure. A prominent conglomerate layer especially well exposed in the Nazareth quadrangle, Pennsylvania (J. M. Aaron, unpub. data) could easily be interpreted as being a basal conglomerate of the sequence.

Many of these rocks appear to have been hydrothermally altered and consist of highly complicated mixtures of minerals. Precise mineralogic studies have not been completed as yet. The following descriptions are based on field and routine microscopic study.

Arkose

Arkose is a grayish-red-purple (5RP4/2) to pale-red-purple (5RP6/2), fine- to medium-grained, dense, massive, granular rock of distinct sedimentary appearance. In places, the rock is strongly sheared, and has yellowish green overcasts due to the presence of epidote on the parting surfaces. The unit is characterized by pink feldspar grains and by a poorly defined bedding. Where sheared, the rock has a prominant foliation and a flaser fabric and might be mistaken for a feldspathic gneiss. Thin section study shows that the rock consists principally of well-rounded microcline, microperthitic monoclinic potassic feldspar, quartz, metamorphic rock fragments, chloritized biotite, sericitized plagioclase, and dolomite that "float" in a matrix of very fine grained clay and mica minerals, feldspar fragments, and mosaic-textured quartz. Some specimens contain sparse grains of hematite, hornblende, and clinopyroxene, the hornblende and clinopyroxene are commonly altered to chlorite and (or) epidote. Hematite also occurs as streaks within the matrix and as coatings on the mineral grains. Feldspar grains and rock fragments in sheared specimens are strongly broken, and what appear to be solid grains in hand specimen are masses of cracked parts of an original grain that retain the original shape. In other specimens the cracked grains are strung out parallel to the foliation and in still other specimens, mortar structure is well developed. Quartz is universally rounded and shows strain shadows.

Ferruginous quartzite

Ferruginous quartzite is a vitreous medium-gray (N5) to medium-dark-gray (N4) foliated to nonfoliated rock that has a splintery fracture. Some specimens, however, are very dusky red purple (5RP2/2), colored by an abundance of hematite. Quartz and hematite are essentially the only minerals in the rock. The quartz occurs in small grains that look like sand particles, in crescent-shaped fragments, and in lenticular mosaic textured masses. Hematite containing sparse admixed magnetite occurs in irregularly shaped masses, as discrete grains, as fillings in cracked quartz grains and in intergrain spaces, and as dusty coatings of quartz grains. The hematite almost certainly was introduced into the rock after it was deformed, and is more abundant in strongly deformed specimens. Fissure filling was the most important process but was not exclusively controlled by the rock foliation as hematite-rich masses commonly form 60–75 degree angles with the planar structure.

Quartzite conglomerate

Quartzite conglomerate differs from the ferruginous quartzite in that it contains pebbles of vein quartz, quartzite, and dark greenish-gray metamorphic rock. These pebbles range to as much as about 2 inches in diameter. Most of the conglomerate examined is strongly deformed and has a flaser, verging on augen, fabric. The rock could easily be mistaken for an augen gneiss, but thin section study definitely shows that the pebbles themselves are sheared and that they are not eyes in a matrix of mylonite. In some specimens the pebble fragments are strung out into a crude

foliation. The metamorphic rock pebbles have in every case been rotated as shown by their relic foliation. Hematite content and distribution is as described for the ferruginous quartzite.

Magnesium-rich schist and metatuff

Several layers of yellowish-gray (5Y7/2) to light- to very-light-gray (N8) schistose rock crop out on Marble and Scotts Mountains in the Easton and Belvidere quadrangles. These rocks are quite complex but appear to be composed of varying proportions of clay mineral-sericite intergrowths, talc, chlorite, quartz, feldspar, phlogopite, actinolite, and more rarely serpentine. Little can be determined from thin section study as the rocks consist largely of felty masses of clay minerals, mica, and altered talc. In some sections phlogopite, quartz, actinolite and magnetite "float" in the felty matrix. Tourmaline is almost universally present in the schists, presumably as a hydrothermal product. These rocks all seem to be magnesium-rich sedimentary rocks related to the marble.

Metarhyolite and purple slate

Still other layers of nearly massive, very dusky red (10R2/2), very fine grained rock look very much like metamorphosed rhyolite. The majority of the rock is now a mass of chlorite, clay mineral-mica aggregate, and very fine grained quartz. Hematite and rarer magnetite occur within this aggregate of flakey minerals as descrete grains and coatings. Most of the grains occur within wavy veins or layers that strongly suggest flow structure.

Some layers associated with the metarhyolite are fine grained and so closely cleaved they are actually slates. During mapping, these rocks were thought to be more strongly sheared parts of the metarhyolite, but a chemical analysis (table 19) shows that the rock contains far too much alumina to be a rhyolite. They are very similar to rocks described by Reed (1955) from the northern Blue Ridge which are thought to be volcanic rock-derived metasaprolite. The association of the purple slates with metarhyolite here suggests a similar origin.

Origin and Age

The relation of these various rock types is obscure, but the general aspect is that of an interbedded sequence of volcanic and sedimentary rocks that have been metamorphosed, closely folded, and strongly sheared. The rocks have also been altered, probably hydrothermally, at some

TABLE 19. CHEMICAL ANALYSIS (WEIGHT PERCENT) OF PURPLE SLATE FROM THE DELAWARE VALLEY COMPARED WITH SIMILAR ROCK FROM THE NORTHERN BLUE RIDGE

	1	2	3
SiO_2	42.7	47.3	44.2
Al_2O_3	20.2	15.7	16.6
Fe_2O_3	15.1	15.3	17.2
FeO	.64	2.1	2.4
MgO	3.7	2.2	3.0
CaO	.16	1.5	1.5
Na_2O	.03	.08	.06
K_2O	7.9	7.3	6.8
H_2O-	.28	2.9	3.6
H_2O+	3.1	–	–
TiO_2	4.6	3.6	3.4
P_2O_5	.19	.72	.11
MnO	.02	.04	.07
CO_2	.99	.05	.05
Total	99.6	99.8	100.0

1. Purple slate from top of Marble Mountain, Easton quadrangle. Rapid rock analysis by Paul Elmore, Samuel Botts, J. Glenn, Gillison Chloe, L. Artis, H. Smith, and D. Taylor, U. S. Geological Survey.
2, 3. Purple slate from northern Blue Ridge (Reed, 1955, p. 892).

time. Presumably this alteration was accomplished by fluids related to the emplacement of granite pegmatite.

The spatial relation of these rocks with marble suggests a genetic relation as well and would imply that this marble was part of an interbedded sequence. Why then are these clastic and volcanic rocks not found in the other major marble belts in the Prong? The answer may be in the fact that marble occurs at many stratigraphic horizons. In this case, the clastic and volcanic rocks would be associated with only the youngest marble which crops out in the Delaware Valley along the north border of the Prong. On the other hand, there is no real evidence that the marble and interbedded metamorphic rocks are part of the same depositional sequence, or for that matter that they are in sedimentary rather than tectonic contact. It seems logical to correlate the metarhyolite of the Delaware Valley sequence with that which crops out at Pottersville Falls, New Jersey (Bayley and others, 1914), and to point out that both occurrences are near the Precambrian-Paleozoic interface and, therefore at the top of the Precambrian pile.

The rocks in the metasedimentary-metavolcanic sequence have been strongly sheared and cataclastically deformed, but are not at the high metamorphic rank of the older metamorphic rocks, nor have they been so homogenized by recrystallization as to completely lose their sedimentary or

volcanic parenthood. Mineral assemblages in the magnesium-rich schists (talc-actinolite-chlorite-quartz, sericite-talc-chlorite-quartz) are suggestive of the quartz-albite-epidote-biotite subfacies of the greenschist facies, whereas, those in the purple slate (quartz-muscovite-chlorite) and arkose (sericite-chlorite) suggest the quartz-albite-muscovite-chlorite subfacies of the greenschist facies. The marble, originally mostly dolomite, contains the assemblage dolomite-calcite-talc-tremolite-serpentine-diopside-phlogopite. This is essentially the same assemblage that occurs in marble that is both overlain and underlain by gneisses at hornblende granulite rank. These minerals result from thermal and hydrothermal rather than regional metamorphism. The marble here is intruded only by granite pegmatite and not by any of the other granitic rocks. The other metasedimentary and metavolcanic rocks are hydrothermally altered but are neither granitized nor migmatized as are all other such rocks throughout the Reading Prong. Some type of retrogressive metamorphic event can not be postulated as these greenschist rocks are in direct contact with sillimanite-bearing rocks.

If it is accepted that this sequence of metasedimentary and metavolcanic rocks is indeed younger than all other Precambrian rocks as is implied by their character and metamorphic grade, it becomes difficult to rationalize the available radiometric age data, as the marble is the host rock for minerals that have been dated as 800 to 850 m.y. old as has been discussed in the section on intrusive rocks. If these ages are even approximately correct, the marble and associated rocks must predate a major period of high grade regional metamorphism during which syntectonic plutons were emplaced. This would seem to be highly unlikely. This dilemma can be avoided, however, if the marble belongs to the older metamorphic sequence rather than the younger. The magnesium schists may also belong to the older sequence, but this is doubted as one would expect to find mineral assemblages characteristic of the higher grades if these rocks had been through a major period of high grade metamorphism. It would seem, therefore, that the younger metamorphic sequence consists of clastic sedimentary rocks, magnesium-rich schists, rhyolite and lesser tuff and volcanic rock-derived saprolite. Though the problem is not solved, as yet, it is at least defined.

LOWER PALEOZOIC ROCKS

In the Delaware Valley sedimentary rocks of Cambrian and Ordovician age flank the crystalline rocks of the Reading Prong on the northwest, crop out within intermontane valleys, and are sparingly present between the Precambrian and Triassic rocks to the southeast. These rocks were deposited in the Paleozoic miogeosyncline and include, from southeast to northwest and oldest to youngest, orthoquartzite-carbonate shelf deposits (Lower Cambrian Hardyston Quartzite through Lower Ordovician Beekmantown Group), deeper water neritic carbonate deposits (Middle Ordovician Jacksonburg Limestone), and graywacke-shale flysch deposits (Middle and Upper Ordovician Martinsburg Formation). More or less similar rocks are present throughout the Great Valley of the Middle Appalachians, but distinct differences occur in the various tectonic units. Correlations and reported thicknesses are given in table 20. In my opinion, the reported thicknesses of the Martinsburg Shale in New Jersey and the Snake Hill Shale in New York are grossly underestimated. These rocks are unconformably overlain to the northwest by molasse deposits of Silurian age, and these and younger rocks will be described by Epstein (this volume) and Sevon (this volume).

At many places in the Appalachians, a Precambrian core or welt separates miogeosynclinal rocks from eugeosynclinal rocks. This is not the case, however, in the Delaware Valley, where autochthonous rocks of eugeosynclinal origin have not been recognized and the same rock units crop out on both sides of the Precambrian highlands.

In 1946, Stose recognized varicolored pelites and other exotic rock types within the Martinsburg outcrop belt between the Susquehanna and Lehigh Rivers in Pennsylvania and attributed their presence to a far-traveled Hamburg klippe. Rocks similar to those in the Hamburg area crop out in the Delaware Valley near Clinton and Jutland, New Jersey. It has been long recognized (Bayley and others, 1914) that these rocks are older than the Martinsburg and are as old as some of the rocks upon which they lie. In my opinion, they must be allochthonous. These rocks are too poorly understood at present to allow direct correlation with the normal Great Valley sequence, therefore, they are not shown on table 20.

Post-Precambrian and pre-Mesozoic intrusive rocks have been mapped at several places in the Reading Prong and Great Valley of New Jersey and Pennsylvania (Buckwalter, 1959, 1962; Miller and others, 1939, 1941; and Spencer and others, 1908). Most of these rocks are either diabase or alkali-rich intrusive bodies, but pegmatites have been recognized. Many of these rocks are clearly post-Ordovician, but some are of early Paleozoic age. The only definite Lower Paleozoic intrusive rocks in the Delaware Valley are pegmatites that cut the Hardyston Quartzite.

The metamorphic grade of the Cambrian and Ordovician rocks in the Great Valley of Pennsylvania, New Jersey, and New York progressively increases northeast from essentially unmetamorphosed rocks near Harrisburg to probable lower greenschist facies at the Delaware River, to biotite zone at the Hudson River, to sillimanite zone where the Reading Prong plunges out in New York. This progression is largely based on the mineralogy of the Martinsburg Formation and equivalents, because carbonate rocks are not sensitive indicators of low-grade regional metamorphism. The metamorphic grade at the northeast end of the Prong is well established (Balk, 1936), as is the relatively unmetamorphosed character of the rock at the southwest end. The grade in the Delaware Valley is a subject of controversy and will be discussed in more detail in the description of the Martinsburg Formation.

CAMBRIAN SYSTEM

The Cambrian System in the Delaware Valley is represented by the Hardyston Quartzite, Leithsville Formation, and Allentown Dolomite. Howell, Roberts, and Willard (1950) divided the Allentown into the Limeport and Allentown (in the restricted sense) Formations. Detailed mapping and bed by bed petrographic study of these rocks throughout eight 7½-minute quadrangles has shown that this division is not valid, and it has not been used in our work.

Hardyston Quartzite

The basal Cambrian clastic unit in the Delaware Valley is the Hardyston Quartzite which unconformably overlies the Precambrian rocks. It grades upward and probably laterally into the Leithsville Formation, the contact being placed at the first recognizable dolomite bed. Although largely a resistant unit, exposures are rare as the formation is thin and at most places is masked by the heavy float of Precambrian rocks. Nearly always, however, it forms a slight topographic bench that can be recognized on the ground, but is seldom pronounced enough to be seen on topographic maps or aerial photographs. In the Delaware Valley, the unit probably does not exceed 100 feet in thickness, 45 feet was measured at low water in the Delaware River south of Riegelsville, N. J. (table 21).

TABLE 20. CORRELATION OF STRATIGRAPHIC UNITS IN THE GREAT VALLEY OF PENNSYLVANIA, NEW JERSEY, AND NEW YORK

Age	Cumberland Valley, Pa., Autochthonous Sequence (Field Conf. Pa. Geol., 1966; MacLachlan, 1967)	Lebanon Valley, Pa., Lebanon Valley Nappe System (Field Conf. Pa. Geol., 1966, MacLachlan, 1967)	Delaware Valley, Pa.–N. J. Musconetcong Nappe (Drake and others, 1961; Drake, 1965; Drake and Epstein, 1967; Drake, 1969)		Kittatinny Valley, N. J. (Lewis and Kümmel, 1940)	Eastern New York (Broughton and others, 1962)
Upper Ordovician	Martinsburg Formation 1,000 feet	Martinsburg Formation Probably 3,000–5,000 feet	Martinsburg Formation	Pen Argyl Member 3,000–6,000 feet	Martinsburg Shale 3,000 feet	Snake Hill Shale 3,000 feet
				Ramseyburg Member 2,800 feet		
Middle and Upper Ordovician				Bushkill Member 4,000 feet		
Middle Ordovician	Chambersburg Formation 750 feet	Hershey Formation 0–1,000 feet	Jacksonburg Limestone	Cement rock facies 300–1,000 feet	Jacksonburg Limestone 135–150 feet	Balmville Limestone 100 feet Holzwasser (1926)
		Myerstown Formation 200 feet		Cement limestone facies 200–400 feet		
	St. Paul Group 1,000 feet	Annville Limestone 120 feet				
Lower Ordovician	Beekmantown Group — Pinesburg Station Dolomite 450 feet	Beekmantown Group — Ontelaunee Formation 600–800 feet				
	Rockdale Run Formation 2,500 feet	Epler Formation 800–1,300 feet	Beekmantown Group	Eppler Formation 800 feet		
	Stonehenge Limestone 775 feet	Rickenbach Dolomite		Rickenbach Dolomite 635 feet		
	Stoufferstown Formation 260 feet	Stonehenge Limestone Rickenbach-Stonehenge interval is 800–1,500 feet				
Upper Cambrian	Conococheague Group 2,150 feet	Conococheague Group 2,000 feet	Allentown Dolomite 1,700 feet		Kittatinny Limestone 2,500–3,000 feet	Stockbridge Group 4,000 feet
Middle Cambrian	Elbrook Formation 3,000 feet	Buffalo Springs Formation 1,000 feet				
	Waynesboro Formation 1,000 feet	Leithsville Formation 1,000 feet	Leithsville Formation 1,000 feet			
	Tomstown Formation 1,000–2,000 feet					
Lower Cambrian and Cambrian(?)	Chilhowee Group 4,500–4,800 feet	Hardyston Quartzite 400 feet	Hardyston Quartzite 100 feet		Hardyston Quartzite 5–200 feet	Poughquag Orthoquartzite 250 feet
Precambrian	Volcanic rock	Gneiss	Gneiss		Gneiss	Gneiss

TABLE 21. MEASURED SECTION OF HARDYSTON QUARTZITE AT LOW-WATER IN THE
DELAWARE RIVER, 0.55 MILE SOUTH OF RIEGELSVILLE, N. J.

	Thickness (feet)
Leithsville Formation (part):	
5. Phyllite, covered	120
Incomplete thickness of Leithsville Formation	120
Hardyston Quartzite:	
4. Quartzite, arkosic, locally finely conglomeratic, poorly sorted, pelitic, medium- to dark-greenish-gray. Becomes progressively less arkosic and less coarsely conglomeratic upwards although it never becomes as clean as unit 3. Differs from unit 2 in that it is not so coarsely conglomeratic and has a slightly brownish color tone	14.0
3. Quartzite, fine-grained, nonconglomeratic, vitreous. Consists almost entirely of quartz in mainly rounded grains, is very clean. Unit seems to lense out to the southeast	6.0
2. Quartz-pebble conglomerate and arkose, dark-greenish-gray. Beds vary erratically in degree of coarseness, as well as in feldspar content. Average attitude: N.55°E., 50°SE. (inverted)	25.0
Total thickness of Hardyston Quartzite	45.0
Precambrian rock (part):	
1. Biotite-bearing oligoclase-quartz gneiss, sheared. Section begins at prominent 8–10-foot cliff	80.0
Incomplete thickness of Precambrian rock	80.0

The formation consists of arkose, feldspathic sandstone, orthoquartzite, quartz-pebble conglomerate, silty shale, and jasper. Arkose and conglomerate (fig. 7) are more abundant near the base of the formation, but these coarse rocks are neither basal nor restricted to the lower part of the unit and no intraformational stratigraphic units can be mapped. Jasper, though rarely present in the Delaware Valley, is abundant to the west in parts of Lehigh County, Pa. (Miller and others, 1941). Current work (Nov. 1968) has shown that this material is also very abundant at the southwest termination of Rattlesnake Hill in the Hellertown quadrangle, Pa.

The formation is the subject of a special study by J. M. Aaron (this volume) and the reader is referred to it for an interpretation of its petrology and origin.

Leithsville Formation

The lowermost carbonate unit in the Delaware Valley is the Leithsville Formation which consists of interbedded light-medium-gray to dark-gray, fine- to coarse-grained dolomite and calcictic dolomite, light-gray to tan phyllite and calcareous phyllite, and very thin beds and stringers of quartz and dolomite sand. Most of the formation approaches pure dolomite in composition. Typical chemical analyses are given in table 22. Fairly thick beds of orthoquartzite are present locally and some workers have mistaken such rocks for the Hardyston Quartzite but microscopic examination

clearly shows well-rounded grains of quartz floating in a largely dolomite cement, not at all like the Hardyston. Other beds contain enough quartz sand or silt so as to remain coherent subsequent to leaching. The formation is characteristically rhythmically bedded, a cycle starting with the deposition of a thin (usually very thin) bed of

Figure 7. Arkose and quartz-pebble conglomerate of Hardyston Quartzite. Note relations of conglomerate to arkose and tiny pegmatite veins (thin white blotches) in lower right of photograph. Morgan Hill, Easton quadrangle, Pennsylvania

TABLE 22.   TYPICAL CHEMICAL ANALYSES OF LEITHSVILLE
FORMATION

	1	2	3	4	5
CaCO$_3$	48.0	55.3	52.16	53.43	51.60
MgCO$_3$	37.3	42.1	43.28	41.43	42.60
SiO$_2$	7.9	.5	1.90	4.77	3.30
Al$_2$O$_3$	—	—	—	—	1.75
Fe$_2$O$_3$	—	—	—	—	.52
R$_2$O$_3$	6.4	2.3	.58	2.20	—
Insol	—	—	—	—	—
P	tr	—	.006	—	—
S	.1	.2	.02	—	—
Total	99.7	100.4	97.95	101.83	99.77

1. East end of Durham Furnace quarry, Durham, Pa., Riegelsville quadrangle (Willard and others, 1959, p. 203).
2. Quarry near Cooks Creek, 1½ miles east of Springtown, Pa., Hellertown quadrangle (Willard and others, 1959, p. 203).
3. Average composition of Leithsville Formation from 108 railroad cars shipped from quarry at Island Park, Pa., Easton quadrangle (Miller, 1934, p. 603).
4. Bethlehem Steel Co. quarry, Redington, Pa., Nazareth quadrangle (Miller, 1934, p. 603).
5. New Jersey Zinc Co. quarry, East Allentown, Pa., Allentown East quadrangle (Miller, 1934, p. 524).

quartz and (or) dolomite sand followed by sandy phyllite, calcareous phyllite, platybedded planar laminated dolomite, and finally by thick-bedded, commonly planar-laminated dolomite (fig. 8A). Lenticular and bedded dark-gray chert is abundant at some places, but is almost completely absent at others. Very large oolites or pisolites have been recognized in the upper part of the unit but are rare and are entirely different from the abundant oolites in the overlying Allentown Dolomite. Edgewise conglomerate (dolorudite) is sparingly present in the upper part of the unit. Mud cracks, ripple marks, graded beds, "fucoids," and possible load casts are common features of the formation. No stromatolites as such have been recognized in the formation, but some beds in the upper part are arcuate, lacking only the fine, conformable, internal laminations of the stromatolites. I have always taken a conservative position on these structures, but the late V. E. Gwinn (oral commun., 1965) believed that they were, without question, stromatolites.

The Leithsville Formation is generally poorly exposed as are its correlates elsewhere in the Appalachians and no satisfactory section is known. The unit is estimated to be about 1,000 feet thick. It is doubtful that it is much thinner, but structural complications could conceal a greater thickness. It is gradational with the underlying Hardyston Quartzite (see J. M. Aaron, this volume) and with the superjacent Allentown Dolomite through an

interval 100–200 feet thick, the contact is placed at the base of the lowest Allentown cycle. In mapping, the contact is placed at the lowest recognized stromatolite colony or oolite bed. In the area mapped in New Jersey, the formation thins and disappears to the northeast so that it is absent from both the valley of Pohatcong Creek and from the Great Valley north of Scotts Mountain. This is thought to be a result of tectonism, not erosion or nondeposition.

I have considered the formation to be of Middle(?) Cambrian age as it is gradational with the Early Cambrian Hardyston Quartzite and the Late Cambrian Allentown Dolomite. There may be some overlap into both the Early and Late Cambrian. There has been no published description of fossils collected from the unit, although F. J. Markewicz (oral commun., 1961) has found specimens of the exterior opercula, and impressions of interior and internal surfaces of what is thought to be the mollusk *Hyolithellus micans*, no body shell was found. These specimens were collected 8 to 10 feet above the Hardyston Quartzite in a quarry in Long Valley near Califon, N. J., and are thought to be of Early Cambrian age, which is in agreement with Aaron's (this volume) ideas as to the Hardyston-Leithsville relation. However, it still seems more logical to consider the bulk of the unit Middle(?) Cambrian.

The sedimentary features to the Leithsville clearly show that it is a shallow-water shelf deposit, probably largely deposited in the subtidal zone. The possible stromatolites and planar-laminated beds suggest that the upper part might have been deposited in the intertidal zone and (or) supratidal zone.

Allentown Dolomite

The Allentown Dolomite is a light-medium-gray or light-olive-gray to dark-medium-gray, fine- to medium-grained rock that weathers characteristically light and dark gray in alternating beds (fig. 8B). In the Delaware Valley, the rocks are completely recrystallized and dolomitized, and chemically approach pure dolomite. Chemical and minor element analyses are given in tables 23 and 24. Bedding varies considerably and more or less regularly from poorly laminated massive beds averaging 3 feet thick to platy or shaly beds less than an inch in thickness. The thicker beds are, in general, fine-grained featureless dolomite or dolarenite, whereas the thin beds are dolorudite or well-sorted oolite. Oolite is very abundant, particularly in the lower and middle part of the

A   **B**

Figure 8. Typical exposures of Leithsville Formation and Allentown Dolomite. A, Leithsville Formation. Orthoquartzite-phyllite cycle is less than 3 inches thick here, although it cannot be seen, it lies between the platy beds and the lower thick bed. Cooks Creek area, Hellertown quadrangle, Pennsylvania. B, Allentown Dolomite. Note the characteristic light- and dark-gray-weathering, variation in bedding characteristics, and numerous different types of stromatolites. Cut at 13th Street Exit, U. S. Interstate 78, Easton, Pa. Photograph by J. M. Aaron, U. S. Geological Survey.

formation. Oolite also forms well-sorted beds and lenses as much as 2 feet thick and, together with more abundant quartz sand, is part of the poorly sorted matrix in the dolorudite beds.

Several types of stromatolites (figs. 9A–C) occur throughout the formation, but are much more abundant in the lower and middle parts. Detailed mapping shows that specific types are not confined to particular stratigraphic intervals and apparently reflect local environmental conditions. The stromatolites, therefore, have very little diagnostic stratigraphic value. Stromatolite zones are invariably topped by desiccation dolomites (fig. 9D).

Quartz sand is ubiquitous throughout the dolarenite and dolorudite beds, but concentrations in lenses, stringers, or beds of orthoquartzite are common and are diagnostic of the formation. According to Zadnik (1960) potassic feldspar is also present in small quantities where quartz is abundant. The greater amounts of Al, Na, K, Ga, and B in samples richest in SiO_2 (tables 23 and 24) suggest that these elements are contained in the feldspar.

Local disconformities, ripple marks, mud cracks, crossbeds, graded beds, load casts, and other sedi-

mentary features are common in the formation. Lenses and irregular beds of dark-gray chert begin to appear in the upper part of the formation at about the same horizon that stromatolites disappear.

The Allentown Dolomite was deposited in cycles as shown by its alternation in color, bedding, and other megascopic features and in textures determined by microscopic study. Zadnik (1960) recognized a succession of six different types of dolomite which he visualized as forming an ideal cycle representing general shallowing conditions of deposition from subtidal to supratidal. An ideal cycle from bottom to top, is as follows: textureless dololutite, dolarenite, oolitic dolarenite, dolorudite formed in part by burrowing and in part by wave action (my interpretation, not Zadnik's), cryptozoon dolomite, and desiccation dolorudite. Individual cycles range from 5 to about 30 feet in thickness. The formation is about 1,700 feet thick in the Delaware Valley; 1,610 feet was measured at the reference section at Carpentersville, N. J., where the basal part is covered. The upper contact is transitional with the Rickenbach Dolomite, the transition being marked by the dis-

TABLE 23.   MINOR ELEMENT ANALYSES OF CARBONATE ROCKS FROM THE DELAWARE VALLEY [1]

	1	2	3	4	5	6
Na	.07	.15	.07	.07	.07	.15
K	.07	.3	.7	.03	.3	.3
Ti	.007	.03	.003	.007	.03	.07
Mn	.007	.007	.03	.007	.007	.015
B	.001	.001	.001	.001	.001	.001
Ba	.0003	.0007	.0003	.0003	.0007	.003
Cr	.0003	.0003	.0003	.0003	.0003	.0003
Cu	.00015	.0007	.00015	.0003	.0007	.0007
Ga	0	.00015	0	0	0	.00015
Pb	.0003	.0003	.0003	.0003	.0003	.0003
Sr	.0007	.0007	.0007	.0003	.0003	.007
Zn	.0015	.0015	.0007	.0015	.0015	.003

[1] Analyses by K. V. Hazel, U. S. Geological Survey. Figures are reported to the nearest number in the series 7, 3, 1, 5, 0.7, 0.3, 0.15, etc., in percent. These numbers represent midpoints of group data on a geometric scale. Comparisons of this type of semiquantitative results with data obtained by quantitative methods show that the assigned group includes the quantitative value about 60 percent of the time. Symbols used: 0, looked for but not found; < with number, below number shown here. Elements not shown were looked for but not found.

1. Lower part of Leithsville Formation from quarry on Highway 611, Raubsville, Pa., Easton quadrangle.
2. Near base of Allentown Dolomite from quarry 0.5 mile north of Riegelsville, N. J., Riegelsville quadrangle.
3. Lower part of Allentown Dolomite from Carpentersville section, N. J., Easton quadrangle.
4. Top of lower 1/3 of Allentown Dolomite from Carpentersville section, N. J., Easton quadrangle.
5. Uppermost Allentown Dolomite from Carpentersville section, N. J., Easton quadrangle.
6. Lowermost Rickenbach Dolomite from Carpentersville section, N. J., Easton quadrangle.

TABLE 24.   TYPICAL CHEMICAL ANALYSES OF ALLENTOWN DOLOMITE

	1	2	3	4	5	6	7
$CaCO_3$	50.9	51.0	47.8	51.4	53.77	47.66	51.73
$MgCO_3$	40.8	32.8	38.28	43.6	42.92	42.92	45.35
SiO_2	5.2	3.7	8.6	2.6	1.68	6.94	1.20
Al_2O_3	1.2	.45	2.1	.59	–	1.95	0.65
Fe_2O_3	.55	.51	.51	.38	–	1.05	0.66
FeO	.06	.29	.40	.31	–	–	–
R_2O_3	–	–	–	–	1.10	–	–
Na_2O	.06	.06	.07	.04	–	–	–
K_2O	.88	.33	1.5	.44	–	–	–
H_2O	.32	.30	.27	.16	–	–	–
TiO_2	.07	.04	.12	.04	–	–	–
P_2O_5	.02	.02	.03	.03	–	–	–
MnO	.01	.02	.02	.01	–	–	–
Total	100.1	99.5	99.4	99.6	99.5	99.9	99.6

1. Collected about 40 feet below upper contact, Carpentersville section, N. J., Easton quadrangle. Rapid rock analysis by P. S. D. Elmore, S. D. Botts, I. H. Barlow, and Gillison Chloe, U. S. Geological Survey.
2. Collected at top of lower 1/3 of formation, Carpentersville section, N. J., Easton quadrangle. Same analysts as 1.
3. Collected from lower part of formation, Carpentersville section, N. J., Easton quadrangle. Same analysts as 1.
4. Collected near base of formation, quarry 0.5 mile north of Riegelsville, N. J., Riegelsville quadrangle. Same analysts as 1.
5. Upper part of formation from quarry on Bushkill Creek just below Lafayette College stadium, Easton, Pa., Easton quadrangle (Miller, 1934, p. 603).
6. Lower part of formation from quarry on west side of Saucon Creek, 0.8 mile southwest of Hellertown, Pa., City Hall, Hellertown quadrangle (Miller, 1934, p. 603).
7. Hardners quarry on Little Lehigh Creek, Allentown, Pa., Allentown East quadrangle (Miller, 1943, p. 524).

Figure 9. Organo-sedimentary structures in Allentown Dolomite. Cut at 13th Street Exit, U. S. Interstate 78, Easton, Pa. A, Large bolster-type stromatolite. Dark lower bed is oolitic dolarenite. Rock above colony is dolorudite having a matrix of quartz sand and oolite. B, Colony of "*Cryptozoon fieldii.*" Dark bed below colony is oolitic dolarenite and bed below that is fine-grained dolarenite. Colony is overlain by typical desiccation dolorudite. C, Digitate-type colony of "*Anomalophycus compactus.*" Best seen just to right of knife. D, Typical bed of desiccation dolorudite (1). Bed below is bolster-type stromatolite (2), and bed below that is a colony of "*Cryptozoon fieldii*" (3). Note lenses of oolite (4) in upper rudite bed.

appearance of the very shallow water stromatolites and desiccation dolorudite, the appearance of dark-gray chert and medium-coarse to coarse-crystalline, light-gray dolomite and shaly dolomite, and an increase in calcium carbonate content (tables 24 and 25). In sections, the contact is placed at the top of the highest typical Allentown cycle. In mapping it is placed below the lowest bed of medium-coarse to coarse-crystalline, thin-bedded, calcareous dolomite, or above the highest stromatolitc colony, oolite, desiccation dolorudite, or orthoquartzite bed.

The lower part of the Allentown Dolomite at Carpentersville, N. J., was dated as early Late Cambrian (Dresbachian) by Weller (1900) who collected the trilobite *Welleraspis jerseyensis* in a now abandoned quarry. Three succeeding generations of geologists have attempted to reestablish this locality to no avail. Dresbachian fossils have also been collected at Peapack, N. J., and Limeport, Pa. (Howell, 1945, 1957). The Limeport locality is definitely in a different tectonic unit, a fault block within the Triassic basin; the tectonic position of the Peapack locality is not known. Late Late Cambrian (Trempealeauian) fossils have been collected from the uppermost Allentown at Newton, N. J., Andover, N. J., Blairstown, N. J., and Portland, Pa. (Weller, 1903; Howell, 1945, 1957). The

localities at Blairstown and Portland are in a different tectonic unit (A. A. Drake, Jr., unpub. data), whereas the locality at Andover is probably in the Musconetcong nappe; the tectonic position of the Newton locality is not known. Middle Late Cambrian (Franconian) faunas have not been found. This relation led Howell, Roberts, and Willard (1950) to divide the Allentown Dolomite into Limeport and Allentown (sensu stricto) Formations and to postulate a disconformity between them embracing all of Franconian time. Much of the above postulate was rationalized on the supposed stratigraphic distribution of different stromatolite types, an assumption now known to be in error. The Franconian disconformity, if present, would have to occur right at the place where the typical Allentown cycles are best developed, a seemingly unlikely happening. In any case, no lithologic features have been recognized that would allow a systematic mapping of a subdivided Allentown Dolomite.

The Allentown-Rickenbach contact is taken as the systemic boundary between the Cambrian and Ordovician for convenience. It is highly unlikely that it is, however, for in the Cumberland Valley sequence in Maryland the boundary is somewhere in a 400–500 foot interval beneath the Beekmantown-Conococheague contact, the lithic break not being the systemic boundary (Sando, 1957).

TABLE 25.　TYPICAL CHEMICAL ANALYSES OF DOLOMITE OF BEEKMANTOWN GROUP.[1]

	1	2	3	4	5
$CaCO_3$	52.6	51.36	51.558	48.630	51.7
$MgCO_3$	35.5	39.72	35.216	40.410	37.1
SiO_2	7.5	4.70	10.750	9.240	8.91
Al_2O_3	1.6	2.68	.140	—	2.31
Fe_2O_3	.41	.92	—	—	—
FeO	.22	—	—	—	—
Na_2O	.07	—	—	—	—
K_2O	1.2	—	—	—	—
H_2O	.38	—	—	—	—
TiO_2	.10	—	—	—	—
P_2O_5	.03	—	—	—	—
MnO	.02	—	—	—	—
S	—	.026	—	.005	—
P	—	.012	.018	.012	—
Total	99.6	99.768	97.682	98.297	100.02

[1] Formation of samples from literature cannot be determined with certainty.
1. Lowermost Rickenbach Dolomite from Carpentersville section, N. J., Easton quadrangle. Rapid rock analyses by P. S. D. Elmore, S. B. Botts, I. H. Barlow, and Gillison Chloe, U. S. Geological Survey.
2. Quarry just north of confluence of Coplay Creek and Lehigh River, Hokendauqua, Pa., Catasauqua quadrangle, average of 183 carloads (Miller, 1934, p. 524).
3. Probably Epler Formation from quarry at Jordon Creek bridge of Reading Railroad, Cementon quadrangle, Pa. (Miller, 1934, p. 524).
4. Quarry 1½ miles northeast of Trexlertown, Pa., Allentown West quadrangle (Miller, 1934, p. 524).
5. Probably Rickenbach Dolomite from outcrops at Veberroth pit of New Jersey Zinc Company's Friedensville mine, Pa., Allentown East quadrangle (Miller, 1934, p. 524).

The position of the boundary here will probably never be determined.

The sedimentary features described above suggest that the bulk of the Allentown was probably deposited in the shallow subtidal to supratidal zone. The disappearance upward of the most shallow water facies suggests that the transition to Beekmantown deposition accompanied slight deepening of water and that these rocks were largely deposited in the subtidal zone.

ORDOVICIAN SYSTEM

The Ordovician System in the Delaware Valley is represented by the Beekmantown Group, the Jacksonburg Limestone, the Martinsburg Formation, and by unnamed pelitic, psammitic and lesser carbonate rocks thought to be allochthonous to the Musconetcong nappe which is itself allochthonous. Rocks deposited in Chazy time are largely lacking in the area, being represented only in the extreme uppermost Beekmantown and in the allochthonous rocks which were deposited elsewhere.

Beekmantown Group

The upper part of the Lower Cambrian-Lower Ordovician carbonate sequence in the Delaware and Lehigh Valleys was named Coplay by Wherry (1909). Miller (1934) substituted Beekmantown recognizing the similarity of these rocks to the Beekmantown of New York and most subsequent workers have followed this usage. Willard (1958) pointed out that the uppermost beds of the Beekmantown were of Chazy age and therefore, the name Coplay should be reinstated. I agree completely with Willard's reading of the rocks, but because of its wide usage, I have retained the name Beekmantown. So far as is known, other workers have as well.

Detailed mapping has shown that the Beekmantown in the Delaware Valley consists of two mappable lithologies, dolomite and interbedded dolomite and limestone, that can be correlated with the Rickenbach Dolomite and Epler Formation of Hobson's (1963) type Beekmantown Group in Berks County, Pennsylvania. His lower and upper formations, the Stonehenge and Ontelaunee, are not present in the Delaware Valley. The Stonehenge disappears along strike to the northeast by facies change (limestone to dolomite) somewhere between the Lehigh and Schuylkill Rivers. The Ontelaunee is absent in the Delaware Valley because of the pronounced unconformity at the top of the Beekmantown. This formation probably is present in the Lehigh Valley, but J. M. Aaron (oral commun., 1965) reports that it is absent as far west as the Nazareth quadrangle, Pa.

R. B. Hoy (written commun., 1958) reports that there are limestones in the lower 70 feet of the Beekmantown in the Saucon Valley, Pa., and he correlates these rocks with the Stonehenge Formation. Callahan (1968), however, refers the entire lower part of the Beekmantown in the Saucon Valley to the Rickenbach as it is visualized in the Delaware Valley. The question immediately arises, is the Beekmantown sequence in the Saucon Valley different from that in the nearby Delaware and Lehigh Valleys, and if it is, is it in a different tectonic unit? This is of great importance as the Beekmantown in the Saucon Valley is the host for the great Friedensville, Pa., zinc deposit. Hopefully, current (1968) work will shed some light on this problem.

Rickenbach Dolomite. The Rickenbach Dolomite, at its reference section at Carpentersville, N. J., includes about 635 feet of various types of dolomite. Typical chemical analyses are given in table 25 and a minor element analysis in table 23. The lower part consists of thin- to thick-bedded, generally massive, light-medium-gray to medium-dark-gray, medium-fine to coarsely crystalline dolomite. The coarsely crystalline beds are calcareous, and both bedded and nodular cherts are common. This part of the formation is characterized by thick-bedded, medium to coarsely crystalline dolomite known informally as "sparkle rock" (fig. 10A), sedimentary breccia (fig. 10B), and by a patchy ruditic texture that is well shown on weathered surfaces but barely visible in freshly broken rock. The clasts in these rudites consist of tiny crystals of dolomite and, more rarely, of calcite and are enclosed in a matrix of more coarsely crystalline dolomite. Sedimentary breccia at this stratigraphic level is the ore horizon at the Friedensville zinc mine.

The upper 200–250 feet of the formation is light-gray to medium-gray, very fine to finely crystalline dolomite containing interbeds of chert. Many beds have a "dirty" appearance because of an abundance of quartz sand and silt. This part of the formation differs from the lower part in being more finely crystalline, generally thin bedded, and laminated.

The Rickenbach grades up into the Epler Formation, the contact being placed at the base of the lowest limestone bed in both sections and in

Figure 10. Typical outcrops of lower part of Rickenbach Dolomite. A, Inverted bedding dips moderately to the southeast (right). Cut along Pennsylvania Railroad under bridge 0.28 mile north of Foul Rift, N. J. (Belvidere quadrangle). B, Sedimentary breccia. Inverted bedding dips southeast (right), poor cleavage dips more gently southeast. Cut along unnumbered road 0.25 mile southeast of Belvidere High School, Belvidere, N. J. (Belvidere quadrangle).

mapping. No fossils have been found in the Rickenbach Dolomite in eastern Pennsylvania or New Jersey. Sedimentary features have not been studied in detail, and post-depositional dolomitization makes a determination of depositional environment difficult but the formation was probably largely deposited in the subtidal zone.

Epler Formation. The Epler Formation consists of an interbedded sequence of dolomite and limestone. The limestones in the lower part of the formation are light to medium gray, thin bedded, averaging less than 1 foot in thickness, and very finely crystalline to cryptocrystalline. Many beds are so finely laminated that they resemble shale. Limestones in the upper part of the formation contain large amounts of calcarenite and, in places, fine fossil hash. These limestones characteristically have fine quartz sand or silt partings which in many places are the only means of determining bedding because of the extreme flowage of the limestone.

Many of the Epler limestones contain sufficient calcium carbonate to have been, or to be potential, sources of stone for the manufacture of portland cement (table 26). Unfortunately, there is no con-

sistency within any given area (compare analyses 1 and 2, table 26) because of dolomitization and the degree to which it has progressed. Much of the limestone has been dolomitized, and in large

TABLE 26. TYPICAL CHEMICAL ANALYSES OF LIMESTONES FROM THE EPLER FORMATION

	1	2	3	4	5
$CaCO_3$	92.57	75.50	82.44	86.036	70.750
$MgCO_3$	5.10	6.07	10.26	4.594	15.256
SiO_2	1.94	12.81	4.74	8.900	11.070
Al_2O_3	–	–	–	.065	.860
Fe_2O_3	–	–	–	–	–
R_2O_3	1.22	5.41	2.30	–	–
P	–	–	–	.016	.019
S	–	–	.026	–	–
Total	100.83	99.79	100.00	99.111	97.955

1. Probable upper part of Epler limestone from quarry ¼ mile north of Catasauqua, Pa., Catasauqua quadrangle (Miller, 1934, p. 591).
2. Same as 1.
3. Average of 10 samples from near Hecktown, Pa., Nazareth quadrangle (Miller, 1934, p. 603).
4. Probable lower part of Epler limestone from quarry at Jordon Creek bridge of Reading Railroad, Cementon quadrangle, Pa. (Miller, 1934, p. 524). From same quarry as dolomite analysis 3, table 25.
5. Lower part of Epler limestone from quarry at Guth's Station, Pa., Cementon quadrangle (Miller, 1934, p. 524).

exposures all phases of the process, from incipient to complete, can be seen. The abundant chemical data presented by Miller (1934) and Miller and others (1939, 1941) clearly reflects the dolomitization process.

Dolomite beds are largely fine crystalline to cryptocrystalline, more rarely medium crystalline, medium light gray to dark medium gray, and commonly laminated or mottled. Chemically, the dolomite differs little from the Rickenbach Dolomite (table 25). Nodular or bedded chert is common throughout the formation, but a zone very rich in chert is near the base, and is a useful criteria in mapping the Rickenbach-Epler contact. Beds of orthoquartzite have also been found in the Epler in the Pohatcong Valley of New Jersey near the Edison limestone quarries (Bloomsbury quadrangle) and along the Delaware River south of Belvidere, N. J. (Belvidere quadrangle). The stratigraphic position of the orthoquartzite is uncertain, but as it seems to be associated with calcarentic limestone, it is probably in the upper part of the formation.

At its reference section at Carpentersville, N. J., the formation is about 800 feet thick. In the Delaware Valley, it is disconformably overlain by the Jacksonburg Limestone. A different Epler lithology is present at each place the upper contact has been observed, making the possible presence of the Ontelaunee Formation difficult to determine. Epler lithologies, however, appear to underlie the Jacksonburg as far west as Nazareth, Pa.

A few fossils have been collected from the Epler in eastern Pennsylvania. Willard (1958) has summarized these and suggests that although the bulk of the formation is of Early Ordovician age, the uppermost part is Chazyan. There is only one reported fossil locality in the Epler of New Jersey (Weller, 1903). It is just south of Columbia, N. J., in the Paulins Kill Valley. The rock here is in a different tectonic unit than that of the Great Valley outcrop belt (A. A. Drake, Jr., unpub. data). The rock here is Early Ordovician and closely underlies the Jacksonburg Limestone.

The Epler was probably deposited intertidally, but the appearance of calcarenite and fossil hash in limestones in the upper part suggest at least local transitions to the shallow neritic zone. Many of these rocks are completely indistinguishable from the cement limestone facies of the Jacksonburg Limestone. The presence of dark dolomites above the limestone suggest a return to intertidal conditions prior to the emergence that produced the unconformity at the top of the Beekmantown.

Jacksonburg Limestone

High-calcium and argillaceous limestones of Middle Ordovician age have long been the source rock of the portland cement industry in the Delaware and Lehigh Valleys and have been referred to collectively as the Jacksonburg Limestone by R. L. Miller (1937a). The formation consists of a lower unit of high-calcium medium- to coarse-grained, largely well-bedded calcarenite and crystalline limestone, that has been termed the cement limestone facies. In the Delaware Valley, the cement limestone facies has a basal unit of calcirudite containing clasts of Beekmantown-type dolomite and chert as much as 50 cm in diameter. The matrix of this calcirudite is largely typical calcarenite. The rudite apparently disappears to the west of Nazareth, Pa. (Sherwood, 1964). Typical chemical analyses are given in table 27.

TABLE 27. TYPICAL CHEMICAL ANALYSES OF THE CEMENT LIMESTONE FACIES OF THE JACKSONBURG LIMESTONE

	1	2	3	4
$CaCO_3$	85.25	81.50	90.0	86.2
$MgCO_3$	5.12	3.35	3.2	4.6
SiO_2	7.00	9.56	4.7	5.8
R_2O_3	2.80	—	1.9	3.0
Al_2O_3	—	3.48	—	—
Fe_2O_3	—	1.28	—	—
Total	100.17	99.17	99.8	99.6

1. Drill hole sample, Lone Star Cement Corporation, Nazareth, Pa., Nazareth quadrangle (Miller and others, 1939, p. 362).
2. Whitehall Cement Company, Cementon, Pa., Cementon quadrangle (Miller and others, 1941, p. 390).
3. Penn Allen Cement Co., Nazareth, Pa., Nazareth quadrangle (O'Neill, 1964, p. 34).
4. Bath Portland Cement Co., Bath, Pa., Catasauqua quadrangle (O'Neill, 1964, p. 34).

The upper unit of the Jacksonburg Limestone is a dark-gray to almost black, fine-grained, thin-bedded argillaceous limestone (calcilutite) that has been termed the cement rock facies. This rock is a natural portland cement and was the basis for the establishment of that industry in this area. This unit also contains two intervals of crystalline limestone, much like the cement limestone facies. These rocks can be seen in the quarries along the Delaware River, but could not be mapped because of poor exposures. Sherwood (1964) mapped the units in the Lehigh Valley. Typical chemical analyses are given in table 28.

Structural complications and poor exposure make it difficult to make a valid estimate of the thick-

TABLE 28. TYPICAL CHEMICAL ANALYSES OF THE CEMENT
ROCK FACIES OF THE JACKSONBURG LIMESTONE

	1	2	3	4	5
$CaCO_3$	73.38	54.62	71.84	75.28	67.80
$MgCO_3$	4.55	4.18	5.50	4.37	6.19
SiO_2	13.96	29.68	15.70	14.44	18.15
R_2O_3	6.76	–	6.84	5.91	–
Al_2O_3	–	6.63	–	–	6.08
Fe_2O_3	–	2.23	–	–	1.78
Total	98.65	97.34	99.88	100.00	100.00

1. Lower cement rock facies from drill hole sample, Lone Star Cement Corporation, Nazareth, Pa., Nazareth quadrangle (Miller and others, 1939, p. 362).
2. Upper cement rock facies from small quarry north of Lone Star Cement Corporation, Nazareth, Pa., Nazareth quadrangle (Miller and others, 1939, p. 362).
3. Composite analysis of 560-foot drill hole, old Coplay Cement Company quarry, Coplay, Pa., Catasauqua quadrangle (Miller and others, 1941, p. 386).
4. Alpha Portland Cement Company, Phillipsburg, N. J., Easton quadrangle (Kümmel, 1901, p. 22).
5. Edison Portland Cement Company, Stewartsville, N. J., Bloomsbury quadrangle (Kümmel, 1901, p. 44).

ness of the Jacksonburg throughout the area although it appears that the cement limestone facies ranges from 200–400 feet thick and the cement rock facies from 300–1,000 feet thick. The formation in this area is in conformable and gradational contact with the overlying Bushkill Member of the Martinsburg Formation. The transition is one of upward decrease in calcium carbonate content. This relation is well shown in core samples taken across the contact (Sherwood, 1964). The contact is easily placed in mapping as cleavage surfaces of Martinsburg become color banded, reflecting bedding, upon weathering whereas those of the Jacksonburg do not. In addition, slate beds only rarely effervesce weakly in dilute hydrochloric acid, whereas the cement rock effervesces slowly and evenly.

The Jacksonburg is abundantly fossiliferous at localities in the Great Valley of easternmost Pennsylvania and New Jersey where the rock is in the upper limb and brow of the Musconetcong nappe. West of Nazareth, Pa., and in the intermontane valleys few fossils are found because of the extreme deformation in the inverted limb of the nappe; strongly smeared bryozoans are most common. Fossil localities and assemblages have been well described by R. L. Miller (1937a), Miller and others (1939, 1941), and Weller (1903). Weller believed that the lowermost Jacksonburg at the type locality at Jacksonburg, N. J., is of Black River (middle Middle Ordovician age), whereas R. L. Miller (1937a) felt that those beds are of the same

age as the Rockland Formation of Canada (late Middle Ordovician) age. In any case, rocks this old are restricted to the Paulins Kill Valley of New Jersey and its extension into Pennsylvania, which is completely fault bounded and is believed to be in a tectonic unit different from the one exposed in the Great Valley and Reading Prong. Elsewhere, the formation ranges from Hull Formation of Canada (late early Trenton) through Sherman Fall Formation of Kay (1929) (middle Trenton) in age, the cement rock being entirely Sherman Fall in age (Miller, R. L., 1937a).

In the Clinton-Jutland, N. J., area only a few thin slivers of Jacksonburg-type rock lie along the contact of a clastic sequence thought to be allochthonous to the Musconetcong nappe. The only rock I have seen here is calcarenite and calcirudite referrable to the cement limestone facies. A fault obviously has to be drawn somewhere, but there is no modern published mapping. Barnett (1965) compared conodonts from these rocks with those collected from several localities in the Great Valley and Reading Prong, and found two different conodont faunas. Conodonts from the normal Jacksonburg outcrop belts belong to the North American midcontinent province, whereas conodonts from the Clinton-Jutland area are northern European in aspect. The relative ages of these two assemblages cannot be determined at this time but these different aspects suggest that the slivers of Jacksonburg in the Clinton-Jutland area may belong with the allochthonous sequence.

The fossiliferous calcirudite, calcarenite, and calcilutite of the Jacksonburg Limestone reflect a deepening environment of deposition from probable shallow neritic shelf of the Epler Formation through deeper neritic shelf to geosynclinal basin. The basal calcirudite is the first sedimentary heralding of the tectonism which culminated in the Taconic orogeny.

Martinsburg Formation

Pelitic and psammitic rocks of Middle and Late Ordovician age in the Great Valley and Reading Prong have long been recognized to belong to the Martinsburg Formation. Throughout the years, there has been considerable controversy regarding the intraformation subdivision, the principal question being whether the formation consisted of three members (Behre, 1927, 1933) or two members, the upper member being Behre's lower member repeated by folding (Stose, 1930; Willard and Cleaves, 1939). Recently completed detailed mapping (Drake and Epstein, 1967) confirmed the tri-partite subdivision and established formal members, much as those visualized by Behre (1927, 1933).

Bushkill Member. The lower part of the Martinsburg Formation has been named the Bushkill Member of Middle and Late Ordovician age (Drake and Epstein, 1967). It consists of thin-bedded (beds mostly less than half an inch thick) dark-gray to dark-medium-gray claystone slate containing very thin interbeds of quartzose and graywacke siltstone and carbonaceous slate. These rock types typically occur in cycles forming ribbon slate. Slaty cleavage, marked by a strong directional orientation of micaceous minerals is the dominant planar element upon which bedding, usually seen as color bands, is readily apparent at most places (fig. 11A).

Dolomite and dolomite sandstone beds occur in the lower 250 feet of the unit (fig. 11B), although other dolomites previously thought to be a part of the Martinsburg (Miller, R. L., 1937b) have recently been shown to be clearly Allentown Dolomite and allochthonous (Aldrich, 1967).

The slate consists largely of quartz and lesser amounts of micromica and chlorite. Minor amounts of calcite are present in the member and some siltstone and slate will effervesce weakly in dilute hydrochloric acid. Typical chemical analyses are given in table 29. Most of the minerals are clay size, although as much as 25 percent of the quartz is silt size.

The Bushkill Member is about 4,000 feet thick. It is overlain conformably and transitionally by the Ramseyburg Member, the contact being placed beneath the lowest prominent graywacke bed, which in most places is more than 1 foot thick.

The Bushkill Member previously has been dated as ranging from middle or late Trenton (late Mid-

TABLE 29. TYPICAL CHEMICAL ANALYSES OF SLATE FROM THE MARTINSBURG FORMATION

	1	2	3	4	5	6	7
SiO_2	59.1	60.8	64.7	64.6	62.0	55.56	57.17
Al_2O_3	13.8	14.6	15.2	15.8	17.2	20.57	14.04
Fe_2O_3	2.8	1.6	2.2	2.1	1.5	6.18	8.45
FeO	3.0	4.6	3.8	4.1	4.4	–	–
MgO	2.8	3.8	3.0	3.2	3.0	2.76	2.66
CaO	3.4	2.5	.36	.10	.64	3.27	3.79
Na_2O	1.1	1.0	1.5	1.0	1.1	1.12	1.53
K_2O	3.5	3.4	3.7	3.8	4.2	2.95	2.83
H_2O+	2.9	3.6	3.6	3.8	3.7	n.d.	3.18
H_2O-	.48	.21	.38	.38	.30	n.d.	–
TiO_2	.66	.68	.70	.73	.82	.66	.67
P_2O_5	.17	.15	.15	.16	.18	n.d.	–
MnO	.11	.12	.11	.09	.07	n.d.	.10
CO_2	2.9	2.3	.15	.07	.42	n.d.	5.54
S as SO_3	3.6	–	–	–	–	–	–
Total S	–	–	–	–	–	1.84	2.73
Total C	–	–	–	–	–	n.d.	2.17
Total	100.32	99.36	99.55	99.93	99.53		104.86

1. Lowermost Bushkill Member containing noticeable calcite and pyrite. Cut along U. S. Rte. 46, 0.8 mile south of Manunka Chunk, N. J., Belvidere quadrangle (Drake and others, 1965).
2. Bushkill Member about 800 feet stratigraphically above 1. Cut along unnumbered road in Pennsylvania, about 1.4 airline miles from Manunka Chunk, N. J., Belvidere quadrangle (Drake and others, 1965). Outcrop shown in figure 11A.
3. Upper part of Bushkill Member from outcrops along abandoned railroad just above U. S. Rte. 46, 0.4 mile north of Manunka Chunk, N. J., Belvidere quadrangle (Drake and others, 1965).
4. Bushkill Member (stratigraphic position not known) from outcrops at Trout Hatchery about 2.4 miles southwest of Asbury, N. J., Bloomsbury quadrangle (Drake and others, 1965).
5. Slate-run in lower part of Ramseyburg Member from outcrop along Mount Vernon Road, about 1.2 miles northwest of its junction with N. J. Rte. 94, Portland quadrangle (Drake and others, 1965).
6. Exceptionally soft bed from slate run in upper part of the Ramseyburg Member from Consolidated No. 1–Star quarry, East Bangor, Pa., Bangor quadrangle (Behre, 1933, p. 174). Note: n.d. not determined, loss on ignition 9.50 percent.
7. Pen Argyl Member from waste pile, Parsons Brothers quarry, Pen Argyl, Pa., Wind Gap quadrangle (Bates and others, 1947).

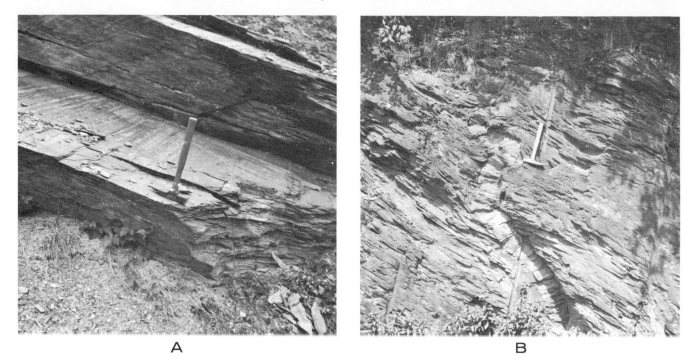

A

B

Figure 11. Bushkill Member of the Martinsburg Formation. Belvidere quadrangle. A, Typical outcrop of ribbon slate. Note laminated bedding which dips gently northwest (left), strong slaty cleavage which dips gently southeast, and well-developed intersection lineation which plunges 10° toward the viewer. B, Dolomite sandstone bed just above contact with Jacksonburg Limestone. Note refraction and fanning of cleavage in dolomite sandstone bed.

dle Ordovician) to Eden (early Late Ordovician) on indirect evidence (Drake and Epstein, 1967). Aldrich (1967) has recently discovered the graptolites *Glyptograptus* cf. *G. teretiusculus* (Hisinger) and a variety of *Orthograptus calaratas* Lapworth, from near Seemsville, Pa. (Catasauqua quadrangle). This locality is probably in the lower third of the member, confirming a Middle Ordovician age for these beds.

The great thickness of the member, its well-marked, uniform, rhythmic bedding, its composition, and its coarsening upwards habit are all characteristic of the flysch facies and reflect the increasing Taconic tectonism.

Ramseyburg Member. The middle part of the Martinsburg Formation is called the Ramseyburg Member of Late (?) Ordovician age (Drake and Epstein, 1967). It consists of alternating beds of claystone slate and light-gray to medium-gray, light-brown- to yellowish-brown-weathering, thin- to thick-bedded graywacke and graywacke siltstone (fig. 12A). Slate in the lower part of the member is identical with that of the underlying Bushkill Member. Higher in the unit the slate becomes progressively thicker bedded and terminates in

slate "runs" like those of the overlying Pen Argyl Member. Thin-bedded, ribbon slate, however, is common throughout. Graywacke constitutes 20 to 30 percent of the unit.

Graywacke forms beds that range from less than an inch to more than 4 feet thick, averaging about 1 foot (fig. 12A). The beds tend to be lenticular and of uneven thickness. The rock is rhythmically bedded, graywacke at the base passes up into medium-gray slate, which in turn passes into grayish-black carbonaceous slate. Each bed normally represents one cycle; some cycles, however, are incomplete. Most of the graywacke is graded and many beds are cross-stratified (fig. 12B). Many beds, however, are planar laminated or cross-laminated (fig. 12C), and convoluted bedding is common as are load casts, slump structures, and groove, flute, and furrow casts. Texturally, most of the graywacke is fine grained, but medium-grained sandstone and siltstone are common and fine conglomerate has been noted locally.

The composition of the average graywacke is 35 percent very fine grained detritus, 34 percent quartz, 16 percent rock fragments (mostly shale), 8 percent authigenic calcite, 5 percent feldspar (mostly microcline), and 2 percent accessory min-

Figure 12. Ramseyburg and Pen Argyl Members of the Martinsburg Formation. A, Ramseyburg Member at the type locality. Bedding dips northwest (left), first generation cleavage dips more gently northwest, and second generation cleavage dips southeast. Note the range in thickness of graywacke beds and of the sedimentary cycles. B, Outcrop of Ramseyburg Member showing cross-stratification, graded bedding, and refracted cleavage. Bedding dips southeast (inverted) and cleavage dips gently southeast. Light-colored beds are graywacke and dark-colored beds are slate. Portland quadrangle, New Jersey. C, Outcrop of Ramseyburg Member showing variation in bedding characteristics in graywacke. Note planar- and cross-laminated beds in upper left. Type locality, Portland quadrangle, New Jersey. D, Typical exposure of Pen Argyl Member in southwest wall of Albion Vein Slate Co. quarry, Pen Argyl, Pa., Wind Gap quadrangle. Beds steepen toward bottom of photograph as they swing into the hinge of a recumbent syncline. A, C, and D are photographs by J. B. Epstein, U. S. Geological Survey.

erals. Part of the detritus has been recrystallized to recognizable porphroblastic muscovite and chlorite, chlorite being more abundant. Typical chemical analyses are given in table 29. Most grains are subangular and ragged, although some are subrounded.

The Ramseyburg is about 2,800 feet thick at the Delaware River and appears to thicken toward the southwest. It grades upward into the Pen Argyl Member with a decrease in coarse clastic content. Graywacke beds become progressively thinner and are replaced by cleaner thin beds of quartzose slate or subgraywacke siltstone in the Pen Argyl cycles. The contact is placed at the top of the highest prominent graywacke interval. East of the Delaware River, the member is unconformably overlain by the Shawangunk Conglomerate of Silurian age.

Only brachiopod debris has been found in the Ramseyburg in the Delaware Valley. Fossils collected from graywacke beds to the west (Behre, 1933; Willard, 1943) suggest that at that locality the beds are of Maysville age (middle Late Ordovician). Some of these fossil-bearing beds appear to be on strike with the Ramseyburg outcrop belt of the Delaware Valley, so it is possible that the Ramseyburg is Maysville in age.

The Ramseyburg Member is a classic flysch-type deposit characterized by cyclic alternations of claystone and graywacke (turbidite?). The appearance of abundant coarse clastic material suggests that Taconic tectonism probably peaked at this time.

Pen Argyl Member. The upper thick-bedded claystone slate unit of the Martinsburg has been called the Pen Argyl Member of Late Ordovician age (Drake and Epstein, 1967). It consists of thick-bedded, dark-gray to grayish-black, medium-gray- to yellowish-brown-weathering, evenly bedded claystone slate rhythmically intercalated with thin beds of quartzose slate or subgraywacke and grayish-black to nearly black carbonaceous claystone slate. A typical cycle starts with quartzose slate or subgraywacke, followed by thick-bedded slate, and, finally, by carbonaceous slate. All members of a typical cycle are not always present. Slate beds commonly are 5 feet thick, but some beds are as much as 15 feet thick. Coarse clastic beds rarely exceed a foot in thickness, but some are as much as 3 feet thick. A typical exposure of the unit is shown in figure 12D.

The slate consists primarily of about equal amounts of muscovite and chlorite, and lesser amounts of calcite, quartz, and groundmass detritus.

A typical chemical analysis is given in table 29. The lighter colored slate (quarryman's gray beds) contains more chlorite than muscovite. The dark beds are carbonaceous and owe their color to graphite and contain noticeable amounts of pyrite and only minor chlorite. Most of the coarse clastic beds are subgraywacke, similar to but cleaner than those in the underlying Ramseyburg, but some beds are quartz-rich phases of the slate.

The Pen Argyl has a minimum thickness of 3,000 feet, but may be as much as 6,000 feet thick. It is unconformably overlain by the Shawangunk Conglomerate of Silurian age and is absent east of the Delaware River. Behre (1927, 1933) recognized that it was similarly unconformably overlapped by Silurian rocks west of Kempton, Pa. Reconnaissance supports Behre's interpretation.

No fossils have been found in the Pen Argyl Member, so it cannot be directly dated. If the Ramseyburg Member is, however, of Maysville age, (middle Late Ordovician) then the Pen Argyl is post-Maysville and pre-Silurian in age.

The Pen Argyl Member represents continued flysch sedimentation and probably reflects the post-peak stages of Taconic tectonism. It is not known over how large an area the unit was deposited, and it may have been restricted to eastern Pennsylvania and New Jersey, as the Martinsburg in central Pennsylvania is succeeded directly by molasse deposits (Juniata Formation) of Late Ordovician age. Basin analysis by McBride (1962) suggests that the Martinsburg basin was deepest at the New Jersey–Pennsylvania border. Perhaps Pen Argyl deposition was restricted to this area.

Metamorphism

The metamorphic grade of the lower Paleozoic rocks in the Delaware Valley has been inferred largely on the basis of the mineral assemblages within the Martinsburg Formation. A correct interpretation of this grade is important to correctly assess regional tectonic relations as well as the fundamental problem of the origin of cleavage.

In the Pennsylvania slate belt and adjoining New Jersey, almost all authors, especially Behre (1927, 1933) who has done by far the most work, have considered the slate to be the result of metamorphic recrystallization, the principal metamorphic minerals being sericite (muscovite) and chlorite. Bates (1947), however, believed that much of the mica in the commercial slate (Pen Argyl Member of the Martinsburg Formation) is illite rather than sericite, although both sericite and chlorite are present in lesser quantities. He thought that the illite was a metamorphic mineral. More recently, McBride (1962) reported that shales and slates from the Martinsburg contain both 2M muscovite and chlorite and seem to be metamorphic. He also reported that these minerals have recrystallized in metamorphosed graywacke. Maxwell (1962), on the other hand, believes that the mineralogy of the Martinsburg in the Delaware Valley results from diagenesis rather than metamorphism and provides abundant references that both degraded and well-crystallized illite, as well as chlorite, can form at low temperatures and pressures. At the present time, the stability fields of the polymorphs of muscovite are so poorly known that it is hazardous to equate degree of metamorphism (or lack thereof) with the polymorph present in a rock.

Maxwell (1962) also reports that carbonate rocks that lie between the Martinsburg and the Precambrian show no signs of plastic deformation or recrystallization. This is in gross error. All writers from the time of the Second Pennsylvania Geological Survey (Lesley and others, 1883) have described and illustrated abundant examples

of flowage and recrystallization in these rocks (see Miller and others, 1939, 1941; Gray, 1951, 1952, 1959; Gray and others, 1958; Geyer and others, 1958, 1963; Prouty, 1959; Drake and others, 1960; Sherwood, 1964; MacLachlan, 1967). Recrystallization is common parallel to cleavage, and in less competent beds, flowage parallel to cleavage has produced a tectonic layering. In such rocks bedding is transposed and is seen only in tiny fold hinges marked by thin silt layers (fig. 13). Abundant examples have been illustrated where dolomite has been boudined out into lenses that now float in limestone that has flowed around them (see Miller and others, 1941). In other places, folds are defined by dolomite beds, whereas the limestones have deformed plastically and have only a flow fabric. In areas of extreme deformation, the Jacksonburg Limestone contains veins of columnar quartz and calcite which parallel cleavage and thrust faults. The Martinsburg Formation contains similar veins of quartz. These veins are obviously secondary and probably are of intraformational origin. Their restriction to areas of strongest deformation suggests that the movement of material and recrystallization resulted

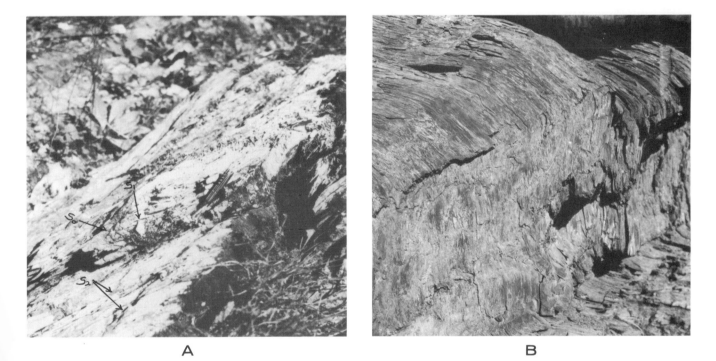

A   B

Figure 13. Transposition of bedding. A, Nearly completely transposed bedding (S$_0$) in keel of recumbent syncline. Strong planar feature is cleavage (S$_1$). Leithsville Formation, Musconetcong valley, New Jersey, Riegelsville quadrangle. Steeper partings are poorly developed second generation cleavage. B, Completely transposed bedding in cement rock facies of Jacksonburg Limestone near Stewartsville, N. J., Bloomsbury quadrangle. Flow cleavage is recumbently folded with the development of a poor, wide-spaced slip cleavage. Tiny fold hinges in bedding can be seen in extreme left center of photograph. Drill hole scar is about 2 inches wide.

from metamorphism rather than during diagenesis as postulated by Maxwell (1962), as plentiful open space in other areas is not filled.

Behre (1927, 1933) reports that mica and chlorite in the Martinsburg lie in the plane of the cleavage (*ab* plane), the long dimension being parallel to the direction of tectonic transport (*a*); the maximum ratio of longest to shortest dimension being 15:1. He also reports that quartz and rutile grains lie in the plane of the grain (*ac* plane = cross joints) and that the long axes of both minerals are parallel to the cleavage in *a* and perpendicular to *b*, the axis of rotation, the maximum axial ratio of quartz grains being 5:1. These features were attributed to secondary growth during metamorphism. Maxwell (1962), however, believes that the spindled quartz grains are purely clastic and that they were oriented by mechanical rotation. It would seem doubtful that the elongated quartz was deposited as such, and if it were, it would be more reasonable for it to have been rotated parallel to *b* rather than *a*. One cannot escape the conviction that these minerals grew under directed stress in a low-grade metamorphic environment.

All the above data support the contention that the lower Paleozoic rocks in the Delaware Valley have undergone low-grade regional metamorphism rather than only diagenetic changes.*

Allochthonous pelitic, psammitic, and minor carbonate rocks

In 1946, Stose recognized varicolored pelites and other exotic rock types that contain fossils ranging from Chazy through Trenton age within the Martinsburg outcrop belt between the Susquehanna and Lehigh Rivers in Pennsylvania and attributed their presence to a far-traveled Hamburg klippe. The faunal evidence shows that at least some of these rocks are older than Trenton, and therefore, older than the oldest Martinsburg. It is now thought that these rocks occur in several klippen that were emplaced during Martinsburg deposition, following a currently popular geologic panacea. Basaltic volcanic rocks crop out near Jonestown, Pa., within the allochthonous pelite and are also probably allochthonous. The source

of these klippen is not known, but it may well be the Cocalico Shale or Wissahickon Formation which crop out south of the Triassic basin.

Rocks similar to those in the Hamburg klippen crop out between the Precambrian rocks of the Reading Prong and the Triassic basin near Clinton and Jutland, N. J. These rocks contain Normanskill (Middle Ordovician) graptolites and were recognized by Bayley and others (1914) as being older than Martinsburg. More recent work by Harry Dodge and R. B. Neuman (*in* U. S. Geological Survey, 1964, p. A83) has established that graptolites of both Early (Deepkill) and Middle (Normanskill) Ordovician age are present in these rocks, which overlie Jacksonburg Limestone of Middle Ordovician (Trenton) age. Here again, older rocks overlie younger rocks and can be presumed allochthonous.

Similar rocks also crop out in the Peapack Valley of New Jersey. The Jacksonburg Limestone is absent in this area and the probable allochthonous rocks lie on what is thought to be carbonate rocks of the Beekmantown Group (Minard, 1959). *Sowerbyella* and rafinesquinoid (earliest known is high in Black River), plectorthid, and dalmanellid brachiopods were collected by Minard (1959) and identified by G. A. Cooper, U. S. National Museum. Cooper states (quoted by Minard, 1959) that these fossils indicate late Middle to early Late Ordovician age. It would seem, therefore, that the allochthonous rocks comprise a sequence ranging from at least Chazy to Trenton or Eden in age, that is, the time equivalent of uppermost Beekmantown through Bushkill Member of the Martinsburg Formation.

Neither detailed maps nor descriptions of the rocks of this sequence have been published. A general correlation has been made with Stose's (1946) Hamburg klippe Taconic sequence (Field Conference of Pennsylvania Geologists, 1956). The sequence, including covered intervals, is estimated to be about 4,200 feet thick in the Jutland, N. J., area and is thought to consist of, from the top down, interbedded dark laminated shale and calcareous fine sandstone or siltstone containing Normanskill graptolites; variegated shale with interbeds of siltstone, limestone with thin shale partings, limestone conglomerate, and chert with shale partings (shale unit contains Deepkill graptolites); variegated shale and interbeds of graded dolomite, siltstone, chert, orthoquartzite, and quartz-pebble conglomerate; and dark interbedded shale and siltstone (could this be the upper unit

---

* J. B. Epstein (written commun., 1969) has recently determined that the mica in a treated sample of slate from the Pen Argyl Member of the Martinsburg Formation from Chapman, Pa., is the 2M polymorph. The presence of this polymorph along with porphyroblastic chlorite and spindled quartz strengthens the argument for metamorphism rather than diagenetic changes.

repeated?).* The dolomite-bearing variegated shale unit (fig. 14) is well exposed in cuts along U. S. Interstate 78 west of Clinton, N. J., and in nearby shale pits. These rocks have features suggestive of shallow water deposition although McBride (1962) believes that similar rocks in Pennsylvania were deposited in deep water.

Dark shale and sandstone crop out just south of the intersection of N. J. Rte. 31 and U. S. Interstate 78 near Clinton, N. J. These rocks are rhythmically interbedded and have sedimentary structures indicative of flysch deposits. They do not, however, have either the appearance, lithology, or tectonic style so characteristic of the Martinsburg of the main outcrop belt. In addition, they differ chemically in that they contain enough manganese so that local concentrations have been exploited commercially (Thurston, 1951). Manganese deposits are unknown in the Martinsburg Formation. These flysch-type units, especially the graptolite-bearing rocks, would normally be assumed to have been deposited in deep water.

In an earlier paper (Drake, 1969), I suggested that these rocks may have been emplaced by submarine gravity slides during Martinsburg deposition in the manner visualized for the klippen near Hamburg and Harrisburg, Pa., and in the Taconic region of New York and New England (Zen, 1967). This seems highly unlikely, however, as reconnaissance suggests that no Martinsburg rocks are present and that the allochthonous sequence everywhere is in contact with carbonate or Precambrian rocks precluding a mechanism of basinward sliding during geosynclinal deposition. An interesting possibility, however, is that the near-shore types, variegated and carbonate units, of the allochthonous sequence may have been emplaced in the flysch-type units of the allochthonous sequence in this manner prior to the transport of the entire sequence to its present position. The lack of knowledge as to the distribution and geometry of the rock types precludes any further specula-

tion. In any case, what appears to be largely clastic geosynclinal rocks are in contact with younger carbonate shelf deposits suggesting that there was indeed an eastern basin of deposition in the Middle Appalachians like that of the Northern Appalachians. Even if these clastic rocks are found to be near-shore deposits, they ultimately had to stem from this eastern basin.

Similar rocks occupy a comparable tectonic position in the Lebanon Valley of Pennsylvania (Geyer and others, 1958, 1963; Gray and others, 1958) and near Limeport, Pa. All these rocks may have been emplaced as thrust sheets subsequent to nappe emplacement as suggested by D. B. MacLachlan (oral commun., 1968).

The abundant red claystones of the sequence are relatively high in ferrous iron. Under conditions of middle-grade metamorphism this iron would be expected to convert to magnetite. Such rock, therefore, should produce strong magnetic anomalies and may be the source of such features in the more metamorphic terrane beneath the Triassic rocks and Coastal Plain deposits to the southeast and northeast.

Post-Precambrian Intrusive Rocks

Post-Precambrian pegmatite was recognized in the Reading Prong by Fraser (1937). The pegmatite forms small sheets along bedding and tiny veins within the Hardyston Quartzite near the contact with microperthite alaskite. The pegmatites are composed primarily of quartz and microperthite and contain magnetite, apatite, and needles of tourmaline as accessory minerals. The microperthite is strongly altered to micromica and small amounts of epidote and chlorite suggest the former presence of hornblende. Pegmatite sheets are as much as 6 inches thick but vary greatly within small distances. Veins are thin, several inches long, and as pointed out by Fraser (1937), apparently fill gashes.

The small pegmatites within the Hardyston are the only clearly lower Paleozoic intrusive rocks in the area, and a preferred interpretation is that they were remobilized from the microperthite alaskite by tectonism during nappe emplacement. It should be pointed out, however, that other late pegmatites that cut only Precambrian rocks also are tourmaline bearing, and could conceivably be of early Paleozoic age. Klemic and others (1959) report that zircon from a rare-earth deposit at the Scrub Oak mine near Dover, N. J., yielded ages of 550 million years (U/Pb, Th/Pb), 545 m.y. (lead/

---

* Since this was written, a field conference in the Taconic region of Vermont, New York, and Massachusetts suggests the following comparisons with rocks of the Taconic sequence: the Normanskill-age dark shale and sandstone with the Pawlet Formation and the next lower unit with the Indian River Slate and Poultney Slate (interestingly enough Poultney-like rocks here are manganiferous as they are in the Taconics); the next lower unit contains Hatch Hill-types of rocks, but the bulk of it could be either a repetition of the supposed overlying rocks or be in fact the West Castleton and Bull Formations of Cambrian age (detailed work will be necessary to answer this question).

Figure 14. Interbedded graded dolomite and laminated red and green claystone in cut on U. S. Interstate 78 west of Clinton, N. J. Light-colored beds are dolomite and dark-colored beds are claystone. Beds strike about north and dip east. Note peculiar fold style which suggests transport from west (right) to east (left), and that the beds are inverted, and perhaps refolded. Photograph by J. P. Minard, U. S. Geological Survey.

alpha), and 600 m.y. (lead/alpha). These ages date the deposit as Cambrian and suggest that there indeed may have been igneous activity during the Cambrian, the problem being one of recognition.

STRUCTURAL GEOLOGY

The Precambrian and lower Paleozoic rocks in the Delaware Valley are involved in two major Appalachian structural elements, the Reading Prong and Great Valley. The Reading Prong traditionally has been considered a first order anticlinorium cored with crystalline rocks that is part of a mountain system that also includes the Blue Ridge-South Mountain anticlinorium to the south and the Green Mountain and Long Range anticlinoria to the north. The Great Valley is a major feature over the entire length of the Appalachians and is generally visualized as a synclinorium. The rocks within the core of the Reading Prong were deformed plastically during the Precambrian, and with the rocks of the Great Valley have been through the Taconic, Acadian, and Appalachian orogenies, and were again deformed during the Triassic. Their structure and current position is the cumulative result of this long and complicated tectonic history.

INTERNAL STRUCTURES IN PRECAMBRIAN ROCKS

In the Delaware Valley, Precambrian rocks occur in six major units: Musconetcong Mountain-Rattlesnake Hill, Lower Pohatcong Mountain-Bougher Hill-Steely's Hill, Pohatcong Mountain, Morgan Hill-Elephant Rock massif, Scotts Mountain, and Chestnut Hill-Marble Mountain. Lithologic boundaries and crystallization foliation trend northeast and dip southeast, although local variations are common. This regional trend is the result of northeast-trending, mostly isoclinal folds in foliation. Structures of Paleozoic age have been superposed on the rocks, but they result from cataclastic rather than plastic deformation.

CRYSTALLIZATION FOLIATION AND LAYERING

The Precambrian rocks in the Delaware Valley have a fair to excellent foliation marked by parallel compositional layers (fig. 3A), lenses, or streaks, or by the arrangement of platy or tabular minerals (fig. 15). Most geologists who have worked in the Reading Prong think that the compositional layering is relic bedding and have found that layering and foliation are parallel. In many exposures in the Delaware Valley foliation and layering are not quite parallel. It seems to me, therefore, that although foliation largely parallels bedding, it probably also parallels cleavage and is, in fact, transposed bedding. If this is so, the lithologic units are the limbs of sheared-out isoclinal folds. This seems much more likely when one considers the evidence gained in modern structural studies shows that regional foliation is almost always cleavage and is additionally supported by the lens-like distributions of the rock units. Foliation in the granitic rocks parallels that in the metamorphic gneisses and is probably flow structure formed dur-

Figure 15. Crystallization foliation in hornblende-bearing albite-oligoclase granite in cut on U. S. Interstate 78, Musconetcong Mountain, Bloomsbury quadrangle, N. J. Note parallelism of foliation with relic amphibolite layers in lower part of photograph.

ing syntectonic emplacement, as it seems best developed near contacts with the gneisses.

LINEATION

Lineation is poorly to well developed throughout the Precambrian rocks in the Delaware Valley and is commonly marked by mineral alinements or streaks, crinkles, rods, or fluted surfaces. Less common lineations include grooves, ridges, boudinage, and minor fold axes (so far as is known, virtually all b axes). A few minor fold axes in the Delaware Valley may be a lineations, but they are sparse and are probably related to intense deformation and b perpendicular to b' type of tectonics.

Lineation has been thoroughly studied throughout the Precambrian highlands of New York, New Jersey, and Pennsylvania as it has been long recognized that magnetite ore bodies plunge parallel to the linear elements in the enclosing rocks. Regionally, plunges are gentle to moderate to the northeast or east-northeast. In the Delaware Valley, the statistical maximum of lineation in the Musconetcong Mountain-Rattlesnake Hill massif is 17°, N.68°E., in Lower Pohatcong Mountain 25°, N.69°E., in Pohatcong Mountain 20°–48°, N.59°E., in the Morgan Hill-Elephant Rock massif 30°, N.68°E., in Chestnut Hill-Marble Mountain 16°, N.58°E., and in Scotts Mountain lineation trends between N.50°–60°E. (or S.55°–56°W.) but plunges range from 63°–17°NE. to 49°–19°SW. reflecting a much more complex tectonic history.

JOINTS

Joints are well developed in the Precambrian rocks and statistical studies show that they can be geometrically related to folds and to lineation as cross, longitudinal, and diagonal joints. The nice geometrical fit is evidence to support the interpretation that these joints were formed during the tectonism that produced the folds.

Cross joints are most abundant, are best developed, and can be seen in most outcrops and more or less parallel the theoretical ac plane. The joints are generally smoothly planar features that at places are filled by pegmatite, and in other places are slickensided or grooved.

Longitudinal joints are less abundant than cross joints, but are present in many outcrops, and are about parallel to the lineation. Diagonal joints are much less well developed, and are difficult to recognize through the blurring overprint of later deformation.

FOLDS

The distribution of lithologic units and the attitudes of foliation indicate that the Precambrian rocks are complexly folded throughout the Delaware Valley. The appalling lack and poor quality of exposures south of the Wisconsin terminal moraine has made detailed structural studies impossible, and in much of the area, fold axes could neither be located nor defined.

Folds range in magnitude from a few inches in wave length and amplitude to as much as 7 miles long parallel to the axis and a mile in width. All are in foliation which is probably transposed bedding. Because the stratigraphic relations are not known, the terms antiforms and synforms are preferred to anticlines and synclines for most of these folds. The folds range from upright and open to isoclinal overturned or, more rarely, isoclinal recumbent; isoclinal folds overturned to the northwest are most common. Most folds plunge northeast essentially parallel to the regional lineation, although southwest-plunging folds have been mapped in the Delaware Valley. All folds seem to have limited vertical extent. The maximum amplitude recorded in the Delaware Valley is 2,600 feet. Most of the folding was disharmonic and was accomplished largely by flexural slip and attendant flowage of the less competent rock. The most competent units, primarily amphibolite (or migmatite), have broken and have been pulled apart in sites of strongest deformation.

Some cross folds have been mapped (Drake, 1967a, b; Drake and others, 1961, 1967). These folds are all minor and almost certainly also formed during the major folding.

No direct evidence has so far been found for more than one episode of folding during Precambrian time. All major and minor structures can be related to one period of plastic folding in the Precambrian or to post-Precambrian deformation. It should be pointed out, however, that the plutonic rocks (840 million years old) are generally thought to have been syntectonically emplaced into a suite of rocks that dates from 1,150 m.y. ago (Long and Kulp, 1962). If this is so, at least some of these sedimentary and volcanic rocks were metamorphosed and therefore probably folded at that time. Two lines of vague and inconclusive evidence bear on this problem. In the Delaware Valley, a sheet of metasomatic albite-oligoclase granite contains several skialiths of isoclinally folded amphibolite (fig. 5). Obviously, the amphibolite was folded prior to its replacement, but the time of deformation is uncertain. The lineation in the amphibolite is within the regional variation, so nothing conclusive can be determined. In Scotts Mountain in the Easton, Bloomsbury, and Belvidere quadrangles (Drake, 1967a, b; Drake and others, 1969) several layers of metasedimentary and metavolcanic rock are spatially related so that they could define several upright synforms and antiforms having vertical axial surfaces. To make this interpretation, however, would require that an upper contact of a unit in one fold be a lower contact of the same unit in an adjoining fold, thus defining a refolded recumbent fold; because of poor exposures, this interpretation was not published (Drake, 1967a, b; Drake and others, 1969). However, it should be pointed out that unusually steep dips and plunges and odd plunge directions (see section on lineation above) add supporting evidence. Although there is good direct evidence for only one period of Precambrian plastic deformation, I personally believe that this deformation obliterated most evidence of any earlier deformation. It would be interesting to see if further evidence of an earlier deformation could be found by detailed structural mapping in the Precambrian outcrop areas north of the Wisconsin terminal moraine.

FAULTS

Faults are difficult to recognize within a crystalline terrane, and where the rocks are so poorly exposed as in the Delaware Valley it is almost impossible to find them. Most faults mapped to date cut both Precambrian and lower Paleozoic rocks, a fact which, in most cases is the only reason they were recognized. Faults cutting only Precambrian rocks were mapped only in a large cut on Musconetcong Mountain along U. S. Interstate 78 in the Bloomsbury quadrangle (Drake, 1967b). At this locality, there are two sets of cross faults and a thrust fault. One set of cross faults strikes northeast and dips moderately southeast. Grooves, slickensides, and fault steps show that movement was largely horizontal and right lateral, displacement being only a few feet. These faults are of Precambrian age, as one is filled by a Precambrian lamprophyre dike (fig. 6A). The other set of faults strikes northwest and dips moderately southwest. Grooves, slickensides, and fault steps show that movement on these faults was largely horizontal and left lateral, displacement being only a few feet. These faults cut the northeast-striking set, but cannot be definitely dated as Precambrian. Faults with similar attitudes can be definitely related to the post-Precambrian deformations.

A nearly flat lying thrust fault (fig. 16) is exposed

Figure 16. Thrust fault within the Precambrian rocks in cut on U. S. Interstate 78, Musconetcong Mountain, Bloomsbury quadrangle, New Jersey.

only along the U. S. Interstate 78 cut. It was not mapped (Drake, 1967b) as it was mechanically impossible to show it, and the cut was not shown on the base map. The thrust could not be traced. It cuts the Precambrian lamprophyre dike and the northeast-striking cross faults but cannot be dated more closely. It seems likely that it is probably of early Paleozoic rather than Precambrian age.

POST-PRECAMBRIAN DEFORMATION

The principal structural problem in the Delaware Valley is the relation of the Precambrian rocks of the Reading Prong to the lower Paleozoic rocks of the Great Valley and the smaller intra-Prong Valleys. This relation and the structures entirely within the Cambrian and Ordovician rocks are part and parcel of the same problem and, therefore, will be treated together.

HISTORICAL SUMMARY

Early geologists in Pennsylvania (Rogers, 1858; Lesley and others, 1883; Miller, 1925) believed that the Precambrian rocks formed anticlinal ridges that were locally overturned and (or) thrust-faulted, and that the intermontane valleys underlain by Paleozoic rocks were synclines and (or) grabenlike blocks downdropped by normal faults. The Pale-

ozoic rocks of the Great Valley were considered to be largely in normal contact with the Precambrian rocks.

In New Jersey, Lewis and Kümmel (1940) recognized that at many places Precambrian rocks were thrust on Paleozoic rocks and believed that both were subsequently folded. In their interpretation, the northern boundaries of the intermontane valleys are high-angle normal faults that formed later than the thrusting and folding, probably during Triassic time.

In New York, all early workers considered the Precambrian rocks of the Hudson Highlands to be autochthonous, although Balk (1936) drew an almost continuous fault around the Highlands. In addition, he drew faults on all sides of several small Precambrian masses along which the Precambrian rocks were interpreted as being upthrust en masse.

In 1935, Stose and Jonas proposed that the Precambrian rocks and Hardyston Quartzite of the mountain area were part of a large, far-traveled thrust sheet and that the inlying limestone valleys were tectonic windows. The Stose-Jonas hypothesis was vigorously denied by B. L. Miller and his coworkers (Miller and Fraser, 1936; Fraser, 1938), whose comments were answered by Stose and Jonas (1939). Shortly before his death, Miller (1944) presented specific evidence that clearly showed that the flat-thrust theory was completely untenable.

Miller's work was unquestioned until 1957, when a systematic geologic study began in the Delaware Valley. Concurrent with this mapping, an aeromagnetic survey was made of the Precambrian outcrop belt in Pennsylvania, and later, gravity studies were made in the southwestern part of the Prong in Pennsylvania and on Pohatcong Mountain in New Jersey. The geologic studies first led to the recognition that thrust faults and overturned folds were more of a factor in the distribution of Precambrian rocks than supposed by Miller and that some of the crystalline ridges were probably klippen (Field Conference of Pennsylvania Geologists, 1961). Preliminary interpretation of the aeromagnetic data (Bromery and others, 1960) suggested that several Precambrian bodies were relatively thin and a few had no magnetic expression whatsoever. These magnetic data combined with the available geologic data suggested that the ridges of Precambrian rocks may indeed be detached parts of a single thrust sheet (A. A. Drake, Jr., and J. B. Epstein, in U. S. Geological Survey, 1962). Later analyses of aeromagnetic and gravity data (R. W. Bromery, in U. S. Geological Survey, 1964) indicated that the crystalline rocks at the west end of the Prong occurred at both the surface and at a deeper level. A still later regional gravity analysis (R. W. Bromery, *in* U. S. Geological Survey, 1966) agreed with the prior interpretation of aeromagnetic and gravity data and confirmed that the Precambrian in the western part of the Prong was thin.

At about the same time of the above work, geologists of the Pennsylvania Geological Survey (Gray and others, 1960; Socolow, 1961; Geyer and others, 1963) recognized that Little South Mountain and other small blocks of Precambrian rocks and Hardyston Quartzite in Pennsylvania were klippen. These interpretations meshed nicely with their finding that the Great Valley was recumbent and inverted.

Isachsen (1964), in a survey paper on the Precambrian in the northeastern United States, interpreted the Reading Prong as being a gigantic klippe composed of both Precambrian and lower Paleozoic rocks. In his interpretation, these rocks were emplaced as an immense allochthonous plum in a Martinsburg pudding, the thrust relation being obscured by later Martinsburg sedimentation.

Many geologists currently working in the Middle Appalachians have accepted an allochthonous interpretation and recently published maps (Geyer and others, 1963; Davis and others, 1967; Drake, 1967a, b; Drake and others, 1967, 1969) show nonrooted Precambrian rocks. Buckwalter (1959, 1962) does not accept the allochthonous interpretation, although his maps show the Precambrian as being almost completely fault bounded. All these studies have culminated in the interpretation that in the Delaware Valley, rocks of the Reading Prong are involved in one grande nappe de recouvrement (A. A. Drake, Jr., *in* U. S. Geological Survey, 1966; J. M. Aaron, *in* U. S. Geological Survey, 1967; Drake, 1969).

REGIONAL GEOPHYSICS

An allochthonous origin is compatible with the regional gravity data (American Geophysical Union and U. S. Geological Survey, 1964). The thrust-faulted but parautochthonous Green Mountain anticlinorium in New England lies on the axis and steep western gradient of the Appalachian gravity high and is bordered on the west by a major gravity low which is coincident with a known basement depression, the Middlebury synclinorium. To the south, the Berkshires continue on the gravity gradient, but in New York, the Prong cuts across the gravity gradient and at the Hudson River is off the gradient entirely. From north-central New Jersey

to Reading, Pa., the Prong lies dead center in the gravity low (fig. 1). To the south, the supposedly autochthonous Blue Ridge-South Mountain anticlinorium lies on the gravity gradient and follows it southwest to the vicinity of Roanoke, Va., where the Blue Ridge becomes obviously allochthonous. From this point southwest, the allochthonous Blue Ridge (Bryant and Reed, 1969) lies in the trough of the gravity low. The comparison of the Prong to the southern Blue Ridge is obvious.

The conclusions of Bromery's regional gravity work have already been mentioned. In addition, he did a gravity survey of Pohatcong Mountain (Bloomsbury quadrangle), New Jersey. No gravity anomaly was associated with the mountain (R. W. Bromery, written commun., 1964). Woollard (1943) had earlier suggested that Musconetcong Mountain in the same quadrangle was a thrust sheet having an average thickness of 500 feet.

Most of the Prong has been mapped aeromagnetically, and regional relations can best be seen on maps by Henderson and others (1966) and Bromery and Griscom (1967). The Precambrian rocks have a characteristic "birds-eye maple" magnetic pattern. In the Delaware Valley and to the west, the Precambrian-Paleozoic interface also has a characteristic pattern. As is shown on figure 17, the Precambrian ridges are bounded on the north by strong negative anomalies and on the south by gradients that drop off into the trench of the negative anomaly north of an adjacent Precambrian ridge. This relation suggests that Precambrian rocks are absent beneath Paleozoic rocks on the north sides of the ridges but are buried beneath Paleozoic rocks on the south sides. The suggested model is a series of southeast-dipping, rootless Precambrian bodies that have nonmagnetic material in front, below, and behind.

Negative anomaly basins also occur at the ends of the Precambrian ridges. If the Precambrian were rooted, these ridge ends would be plunging noses of anticlines, and the magnetic rocks should be expressed for some distance along plunge beneath the cover. This relation is not seen, however, and it seems that the Precambrian rocks spoon out rather than plunge under. This relation is well shown where the Precambrian outcrop ends near Reading, Pa. (Bromery and Griscom, 1967).

Geophysical work also has been done within the Paleozoic outcrop belt that flanks the Prong. Wood and Carter (in press) report that basement lies at a depth greater than 25,000 feet at the Blue Mountain structural front as interpreted from aeromagnetic data. In addition, Wood and Carter

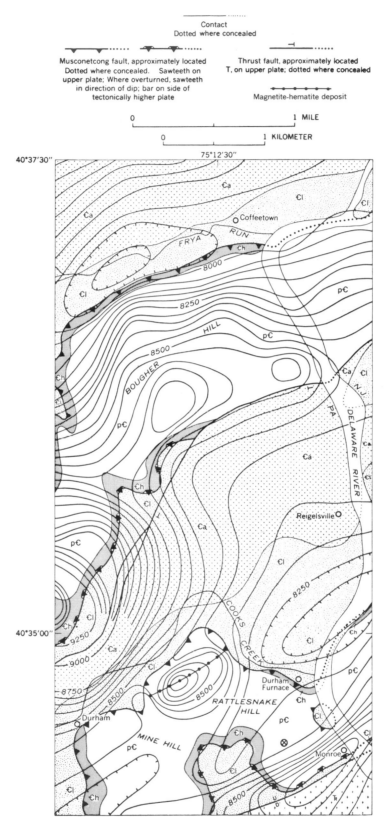

Figure 17. Geologic and aeromagnetic map of part of the Riegelsville quadrangle, Pa.–N. J. (modified from Drake and others, 1967 and Bromery and others, 1960). pͨ, Precambrian rocks; ͨh, Hardyston Quartzite; ͨl, Leithsville Formation; ͨa, Allentown Dolomite; Or, Rickenbach Dolomite; Oe, Epler Formation; Tᴿ, Triassic rocks; ⊗, diamond drill hole.   Contour interval 50 gammas.

(in press) report that oil-company seismic surveys indicate depths to basement between 30,000 and 45,000 feet beneath the Great Valley. All these data are in harmony with the regional gravity data reported above.

THRUST FAULTS ALONG THE PRECAMBRIAN-PALEOZOIC INTERFACE

Thrust faults have been mapped throughout the Delaware Valley and adjacent areas (fig. 2). It will serve no useful purpose here to describe all of them, so I will discuss two specific examples, one from the north border of the Reading Prong, and the other from the south border, to demonstrate tectonic style.

Along the north border of the Prong a thrust has been mapped from near Allentown, Pa., to near Buttzville, N. J., a distance of about 30 miles. Maximum horizontal transport seems to have been about 3 miles although the maximum stratigraphic separation is less than 3,000 feet. Jenny Jump Mountain in New Jersey, a thrust block which is probably related to this thrust, lies to the north and is in contact with rocks as young as the Martinsburg Formation. No other faults have been mapped along the northern Precambrian-Paleozoic interface in New Jersey, although Baum (1967) has recently described thrusting from near Andover, N. J. In Kittatinny Valley there are several klippen of Kittatinny Limestone lying on Martinsburg, as well as two small gneiss klippen (fig. 2). These klippen suggest a minimum horizontal transport of 3 miles and occupy synforms in the folded thrust.

To the northeast in New York, a series of Precambrian klippen lie on Snake Hill Shale between the State line and the Hudson River and are believed to be the remnants of one thrust sheet (Broughton and others, 1962). Isachsen (1964) believes that Stissing Mountain in the Taconic terrane 25 miles north of the Prong probably had the same origin as these klippen. The data presented above, should convince even the most skeptical reader that there has been a good deal of horizontal transport along the north border of the Prong.

Near the south border of the Prong in New Jersey, a thrust fault has been mapped along the front of the Musconetcong-Schooley Mountain mass between Stanhope and Bloomsbury, a distance of about 26.5 miles (Lewis and Kümmel, 1912). Recently, it has been extended another 12 miles into Pennsylvania where this Precambrian mass spoons out beneath Rattlesnake Hill (Drake, 1967a, b; Drake and others, 1967; this report, figs. 2, 18). Tectonic windows have been mapped near West

Figure 18. Contact between Precambrian rocks (above) and Hardyston Quartzite (below) just above thrust fault along north flank of Rattlesnake Hill, Riegelsville quadrangle, Pennsylvania. Brunton lies on contact which dips about 10°–15°SE. (inverted).

Portal, N. J. (Drake, 1967b), and Monroe, Pa. (Drake and others, 1967), and a small klippe sits on carbonate rocks near West Portal (fig. 2), indicating minimum transport of 2 miles. Pohatcong Mountain to the north, however, is completely surrounded by Allentown Dolomite (fig. 2) so that transport had to be much greater. In addition, two water wells on the south flank of this mountain passed from Precambrian rock into Allentown Dolomite. What is almost certainly the same thrust, however, comes to the surface to the south between the Precambrian mass and the Triassic rocks, and the whole Musconetcong-Schooley Mountain ridge is a klippe (Drake and others, 1967). This interpretation has been substantiated recently by diamond drilling (fig. 17) on Rattlesnake Hill in Pennsylvania where inverted Leithsville was cut at a depth of 392 feet. A water well recently completed (1968) near the western extremity of Rattlesnake Hill, passed from Precambrian rock into Leithsville Formation at a depth of about 280 feet. It can be seen that this is an extremely well established thrust, but nowhere in a length of nearly 40 miles is the stratigraphic separation much more than 1,000 feet. This relation will be pursued further, later in this paper.

Two excellent examples of Precambrian rocks thrust onto Paleozoic rocks occur in the tail of Precambrian rocks that extends south from High Bridge, N. J. (fig. 2). One example is along the brook about 0.45 mile east of Annandale, N. J., where amphibolite lies on probable Allentown Dolomite (fig. 19A). The other is a dolomite quarry about half a mile southwest of McPherson, N. J., where Precambrian gneiss clearly overlies dolomite (Leithsville Formation), especially in a cave that has dolomite walls and a gneiss roof (fig. 19B).

Two other water wells in this mountain drilled at Bunnvale, northeast of High Bridge, N. J., passed from Precambrian rock into dolomite at depths of less than 150 feet. According to J. P. Minard (oral commun., 1968) a driller reports that limestone was cut in the bottom of a water well on his property near Chester, N. J., in this southern Precambrian mountain mass.

A study of the geologic map of Pennsylvania (Gray and others, 1960) suggests that other thrusts come to the surface along the south border of the Prong southwest of the Delaware Valley. Evidence is especially compelling in the Reading area (Buckwalter, 1959, 1962) where faults have been mapped almost completely around the Precambrian. These faults faithfully follow the to-pography and suggest at least moderate dips; stream reentrants suggest that at some places the faults dip under the gneiss, not away. In addition, many of the small intermontane valleys that figured so prominently in the Stose-Miller controversy are shown by Buckwalter (1959, 1962) to be underlain by carbonate rocks and in many instances to be fault bound. These areas are, without exception, coextensive with negative magnetic anomalies.

STEEP FAULTS

In New Jersey, steep faults bound the north sides of many of the valleys underlain by Paleozoic rocks. These faults traditionally were considered to be of Triassic age and to be related to the Triassic border fault. Therefore, the valleys were thought to be grabens bounded on the south, in some cases at least, by thrust faults of Paleozoic age and on the north by normal faults of Triassic age. Similar relations were noted in Pennsylvania (Miller and others, 1939, 1941), though there the valleys are not nearly as grabenlike in appearance. It seems strange, however, that these supposed Triassic faults occur without exception between Pre-

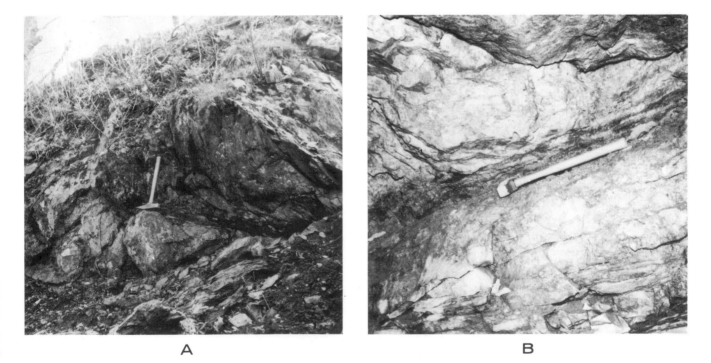

A   B

Figure 19. Contacts of thrust faults. A, Precambrian amphibolite in thrust contact with probable Allentown Dolomite near Annandale, N. J. Hammer sits on fault surface. Note thin zone of cataclastic foliation parallel to thrust and steeper shear fractures. B, Precambrian alaskite in thrust contact with dolomite (Leithsville Formation?) in quarry near McPherson, N. J. Hammer lies on fault surface. Cataclastic foliation is not developed here. The Precambrian rock has the appearance of being brecciated and silicified. The parting surfaces in the dolomite, about parallel to the thrust, are not bedding, but are a crude cleavage.

cambrian and assorted lower Paleozoic rocks on the northwest limbs of overturned folds. How can fate be so consistent? It is even stranger, however, that detailed mapping showed that on ridge noses the "Paleozoic thrust faults" and "Triassic normal faults" came together and that the "Triassic fault" does not displace the older fault. These data seem to indicate north and south bounding faults are actually the same fault.

Some steep faults, however, are indeed separate faults and have been so mapped (Drake, 1967a, b; Drake and others, 1967, 1969). These faults bring rocks as young as Martinsburg Formation into contact with the Precambrian rocks. Such a fault bounds the north side of the Saucon Valley in Pennsylvania and brings rocks as young as Jacksonburg Limestone into contact with Precambrian rocks (Miller and others, 1941). Three holes drilled here by the New Jersey Zinc Company proved conclusively that the Precambrian did not lie in flat thrust contact with the Paleozoic rocks (Miller, 1944). The fault dips about 70°SE. (R. B. Hoy, written commun., 1958). Similar faults are present all through the Prong to the southwest and have been especially noted by Buckwalter (1962). It is not necessary, however, that these faults be normal faults. Minor structures indicate that some are steep thrusts, and sedimentary structures indicate in many places that the rocks in contact with the Precambrian gneisses are overturned. Buckwalter attempted to explain this by faulting out the normal limb of synclines. This is not, however, very convincing mechanically. Strong negative aeromagnetic anomalies are coextensive with such faults and suggest that nonmagnetic rocks are present at depth rather than magnetic basement, which would be the case if the Hardyston were in shallow synclines.

STRUCTURE OF THE GREAT VALLEY

In recent years, a nappe theory has been developed to explain the highly complicated structural relations in the Great Valley of east-central and eastern Pennsylvania and western New Jersey. The grand Alpine-type nappe tectonics were first conceived by Carlyle Gray and his coworkers on the Pennsylvania Geological Survey (Gray, 1951, 1952, 1959; Field Conference of Pennsylvania Geologists, 1954; Gray and others, 1958; Geyer and others, 1958, 1963). Their concept has more recently been elaborated and refined (MacLachlan, 1964; Field Conference of Pennsylvania Geologists, 1966, 1967). This nappe, the Lebanon Valley nappe, includes all carbonate rocks of Cam-

brian and Ordovician age of the Lebanon Valley sequence (table 20) and an indefinite part of the Martinsburg Formation of that valley. The Precambrian rocks of Little South Mountain are involved in this structure (Geyer and others, 1963).

In the Delaware Valley, a nappe has also been recognized (Davis and others, 1967; Drake, 1967a, 1969; Drake and others, 1969). It is believed (Drake, 1969; D. B. MacLachlan, oral commun., 1968) that both areas are on virtually the same structure. The inverted limb crops out in the Lebanon Valley; the brow and normal limb are brought to the surface in the Delaware Valley by the regional northeast plunge. The presence of the nappe is in harmony with the previously cited geophysical evidence which suggests that at least a double, if not a triple thickness of lower Paleozoic rock is present in the Great Valley of eastern Pennsylvania.

In addition to the regional nappe described above, Sherwood (Field Conference of Pennsylvania Geologists, 1961, 1964) in a topical study of the Jacksonburg Limestone, delineated a large recumbent anticline which he called the Northampton nappe. Work by my coworkers and myself has shown that this nappe is the brow and hinge of the regional nappe in the Delaware Valley.

Folds in Bedding

Folds of all sizes and types are related to or are superposed on the nappe. These folds range from crinkles through folds that deflect the outcrop belt such as that in the Jacksonburg at Nazareth, Pa., and in the Martinsburg and younger rocks north of Bangor, Pa. (fig. 2), that can be traced for as much as one mile, to the regional Stone Church syncline (Davis and others, 1967; Drake and others, 1969) which has been traced for at least 16 miles and is defined by the southern outcrop belt of Ramseyburg Member of the Martinsburg Formation shown on figure 2. Some folds formed largely by flexural slip as is shown by striations and slickensides on bedding (fig. 20). These folds are concentric, open, and symmetrical and are limited to the most competent beds in the dolomite units and cement limestone facies of the Jacksonburg Limestone. Most flexure-type folds, however, are flexural-flow folds. These are either concentric or similar and are mostly overturned or recumbent or have been rotated past the horizontal (figs. 21A, B), but some are upright. These folds may have either rounded or peaked hinges and may or may not be thickened in crestal areas (figs. 21C, D). Cleavage commonly is restricted to certain layers, and tends to fan the

Figure 20. Slickensides on bedding produced during flexural-slip folding of Allentown Dolomite, Easton quadrangle, Pennsylvania.

axial surface of the fold (fig. 22A). In some tight hinges (fig. 22B) deformation has reached the passive-slip stage. This type of folding is characteristic of the Cambrian and Lower Ordovician carbonate sequence. Other folds initiated as flexural-flow folds but passed into passive-slip or even passive-flow folds as the rocks become more ductile during continued deformation (figs. 23C, D; 23A). These folds are largely restricted to limestone beds in the Epler Formation and to the Jacksonburg Limestone and characteristically are near isoclinal, have peaked hinges, and are thickened in crestal areas. A special type of fold, called quasi-flexural by Donath and Parker (1964), is common to interbedded limestone and dolomite sequences in the Epler Formation in which dolomite beds deformed by flexural-slip or flexural-flow and limestone beds have accommodated themselves to the form by irregular and contorted flow. Such folds do not necessarily have any asymmetrical relation to each other, or to the regional stress field. These structures have been well illustrated by Miller and others (1939, 1941) and Sherwood (1964). Folds in the cement rock facies of the Jacksonburg Limestone (in the few places where bedding folds can be seen) and in slate of the Martinsburg Formation are of the passive-flow type (fig. 23B). Although the thickest graywacke beds seem to have deformed by flexural-flow, at most places folds in graywacke are of the passive-slip of passive-flow type (figs. 23C, D). In the Delaware Valley these folds are generally isoclinal and recumbent, have peaked hinges, and are notably thickened in crestal areas.

Three principal axial directions can be recognized in the carbonate rocks of the Great Valley in the Delaware Valley: ± horizontal, N.60°E.; 5°, S.44°W.; and 12°, N.80°E. The N.60°E.-folds seem to be first generation as they are most abundant. The S.44°W.-folds seem to be second generation folds and are about parallel to folds in slaty cleavage (see below), and therefore, the N.80°E.-folds could be considered third generation as they are less abundant, but there is no direct evidence. Data on folding and rock fabric are currently (fall 1968) being reevaluated, and it may well be that the N.80°E.-folds are first generation; the N.60°E. are second generation; the S.44°W. are third generation. In the Martinsburg terrane, folds plunge gently between N.48°-65°E. in the Bangor quadrangle and plunge gently N.55°E. in the Portland and Belvidere quadrangles.

Cleavage

At one place or another all rocks from the Precambrian gneisses through the Martinsburg Formation are cleaved. In the gneisses, the Hardyston Quartzite, and the dolomites, cleavage is not everywhere present. Most exposures of Epler and Jacksonburg limestones are cleaved, and cleavage is a regional foliation in the cement rock facies of the Jacksonburg and in the Martinsburg Formation. This cleavage is more or less parallel to the axial planes of macroscopic folds, although it fans slightly in flexural-flow folds (fig. 22A). In the pelitic and semipelitic rocks it is a slaty cleavage. In the carbonate rocks it is a spaced cleavage although in every thin section examined there is at least an incipient parallel growth of micaceous minerals.

I originally believed (Drake and others, 1960) that all this cleavage was related to nappe emplacement and of Taconic age. Maxwell (1962) believed the cleavage to be of Taconic age and Sherwood (1964) believed it to be related to nappe emplacement, the time of emplacement not specified. J. B. Epstein (Field Conference of Pennsylvania Geologists, 1967, and this volume) has shown that the cleavage in the north part of the Martinsburg outcrop belt can be related to folds that deform Silurian and Devonian rocks, and therefore, must have developed during a later Paleozoic orogeny. It would be easy to extrapolate these data to the remainder of the Great Valley and even into the Reading Prong, but at this time, it is dangerous to do so without a careful evaluation.

Figure 21. Fold types. A, Recumbent, concentric, flexural-flow fold in interbedded dolomite and limestone of the Epler Formation, Lehigh Stone Co. quarry, Ormrod, Pa. B, Recumbent concentric and similar folds in interbedded limestone and dolomite of the Epler Formation. Same location as A. C, Tight flexural-flow folds in Leithsville Formation. Note both peaked and rounded hinges reflecting competency of different beds. Riegelsville quadrangle. D, Hinge area of a fold shown on C. Note cleavage development, thickening in more incompetent layers, strain shadow above hinge of competent bed, flow fabric of rock, and luxurious growth of poison ivy.

Figure 22. Fold types. A, Flexure-flow fold in Leithsville Formation. Axial surface is rotated past the horizontal. Note fanning of poorly developed cleavage. Riegelsville quadrangle. B, Very tight small recumbent fold in Leithsville Formation. Deformation was by passive-slip. Riegelsville quadrangle. C, Recumbent, similar, combination flexural-flow and passive-slip fold in limestone of the Epler Formation. Belvidere quadrangle, New Jersey. D, Closeup of C. Note slip along pervasive cleavage.

Figure 23.   Fold types.   A, Recumbent, similar, passive-flow folds in Jacksonburg Limestone. Giant Cement Co. quarry, Egypt, Pa.   B, Hinge of recumbent syncline in Bushkill Member of Martinsburg Formation formed by passive-flow.   A later slip cleavage has been superposed on the fold producing the jumbled relations in the lower part of the photograph.   Belvidere quadrangle.   C, Small fold in Ramseyburg Member of Martinsburg Formation.   Deformation of the graywacke is on border between passive-slip and passive-flow.   Portland quadrangle.   D, Hinge of recumbent anticline in Ramseyburg Member of Martinsburg Formation.   Hammer head is on axial surface.   Graywacke here was deformed by passive-flow.   Belvidere quadrangle.

Slaty cleavage in the interbedded slate and gray-wacke of the Ramseyburg Member is character-istically refracted as it crosses contacts between layers of different competencies (figs. 24A, B). Cleavage in graywacke can be nearly as well de-veloped as in slate, but in the coarser beds it be-comes a distinctly spaced cleavage (fig. 24B). This cleavage seems to have formed largely by pas-sive-slip during folding. This relatively well de-veloped cleavage suggests to me, that the gray-wacke beds were consolidated prior to deforma-tion, contrary to the ideas of Maxwell (1962). If this rock had been unconsolidated, one would ex-pect to find abundant sandstone dikes as illustrated by Maxwell (1962, p. 287). During our study we failed to find any such dikes, although sand-stone beds are commonly sheared up along slip planes.

Slip Cleavage

Slip cleavage is not abundant in the Delaware Valley but it is present in the Martinsburg Forma-tion on both sides of the Stone Church syncline near Ramseyburg, N. J. (fig. 25A), and near Manunka Chunk, N. J. (fig. 25B). It is weakly developed in the Jacksonburg Limestone in the Belvidere quad-rangle, and without question, much of the cleavage mapped in the older carbonate rocks is genetically

related to it. To the west of Bath, Pa., it becomes abundant as the inverted limb of the nappe be-comes exposed (Sherwood, 1964). Where best developed, slip cleavage passes from spaced micro-fractures or crenulations to a perfectly respectable flow cleavage (fig. 25C). The distribution of the slip cleavage about the Stone Church syncline sug-gests a genetic relationship, but an analysis of this problem is not complete as yet. The intense de-formation and attendent slip cleavage at Manunka Chunk has long been spoken of informally as lying along the Manunka Chunk line defining some sort of poorly understood tectonic boundary, that could neither be completely recognized nor mapped. It seems most likely now, that the Stock-ertown fault that is in the subsurface throughout most of the area (Davis and others, 1967; Drake, 1969), a fault of major transport, or more likely an imbricate thereof, comes to the surface here and is the cause of this increased deformation.

Folds in Cleavage

Folds in slaty cleavage are fairly common in the cement rock facies of the Jacksonburg Lime-stone and the Martinsburg Formation in the Dela-ware Valley, and are especially abundant in the southern outcrop belt of the Bushkill Member in the Belvidere and Bangor quadrangles. These

A   B

Figure 24. Cleavage in Ramseyburg graywacke, Martinsburg Formation. A, Refracted cleavage in interbedded slate and graywacke, Bangor quadrangle. B, Typical exposure of passive-slip folded graywacke, Portland quadrangle.

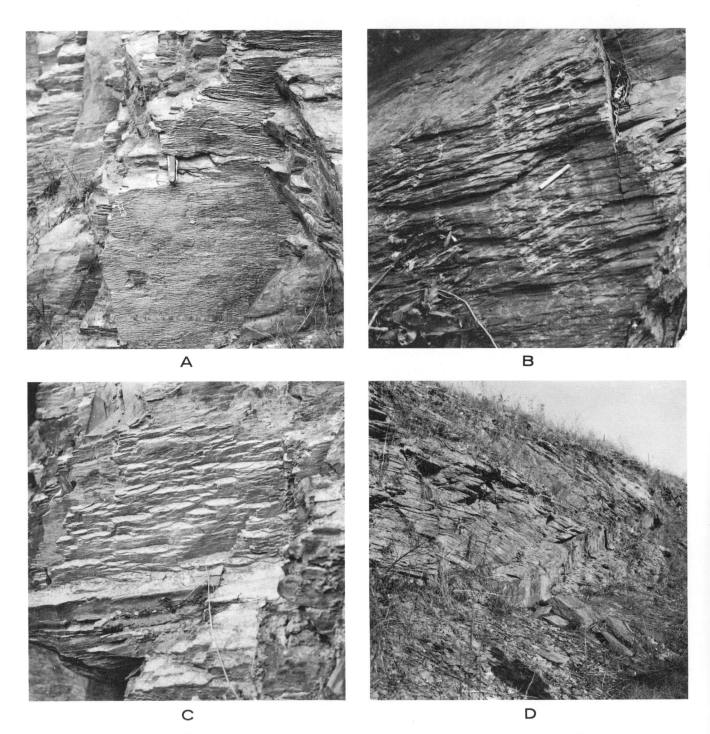

Figure 25.   Slip cleavage and folded slaty cleavage in the Martinsburg Formation.   A, Slip cleavage showing as crenulations on slaty cleavage in Ramseyburg Member near Ramseyburg, N. J.   B, Strongly deformed Bushkill Member near Manunka Chunk, N. J.   Bedding dips steeply right, slaty cleavage gently left, and slip cleavage more steeply left.   C, Flow cleavage, dipping toward viewer, progressively developed from slip cleavage, Ramseyburg area, New Jersey.   Note sheen of new micas on cleavage surface.   Field of view is about 4 feet.   D, Folded slaty cleavage in Bushkill Member, Catasauqua quadrangle.   Note the spaced slip cleavage parallel to the axial surface of the fold.

folds, of course, are far more easily recognized in these rocks, as slaty cleavage is the dominant planar element rather than bedding. There are related folds in bedding in the older carbonate rocks. In this area, these folds range from crinkles to low amplitude kink bands (fig. 25D). To the west, Sherwood (1964) has found that slaty cleavage in the Jacksonburg Limestone has been deformed with bedding into open folds having half wave lengths of one mile. Such large folds have not been recognized in this area, although there is a suggestion that a sizable fold in cleavage may occur within the Stone Church syncline in the Portland and Belvidere quadrangles (Drake and others, 1969). The cleavage arch postulated in the Delaware Water Gap area (Drake and others, 1960) apparently did not form by folding of cleavage, as Epstein and Epstein (this volume) present a much more plausible explanation.

Slip cleavage parallels the axial surfaces of the folds in slaty cleavage and obviously is genetically related. Many folds in cleavage have the under limb sheared out on thrust faults that parallel the slip cleavage forming the cleavage shear zones of Behre (1933).

Fabric elements of the Martinsburg Formation in the Belvidere and Portland quadrangles can be geometrically related to both folds in bedding and to folds in cleavage (fig. 26). Axes of folds in bedding and associated lineations, L_1, average 4° N.55°E. and have clearly associated cross and longitudinal joints. Axes of folds in cleavages and associated lineations, L_2, plunge gently S.38°W. and also have clearly associated cross and longitudinal joints showing that these elements have a regional symmetry, at least over 110 square miles, and are not local aberrant features. The planar element, S_3(?), is not understood at this time.

Faults

Faults of small displacement are commonly seen in good exposures and are certainly present, though undetected, throughout the area. Most faults noted are thrust faults that are more or less parallel to cleavage or to slip cleavage. In the Jacksonburg Limestone and Martinsburg Formation the faults are commonly filled by veins of columnar quartz and (or) calcite, and in places, faults can be traced by following vein quartz float. Most, however, cannot be traced and have not been mapped.

High angle normal faults and tear faults have been mapped where they can be seen in quarries

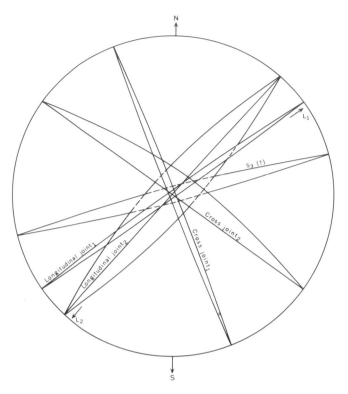

Figure 26. Stereographic projection (lower hemisphere) of some fabric elements of Martinsburg Formation, Portland and Belvidere quadrangles.

or large outcrops, or obviously displace formational contacts. Such faults are certainly more abundant than are shown on the geologic maps.

The most important fault in the Great Valley of this area is the Stockertown fault (Davis and others, 1967). This fault is largely in the subsurface in the Delaware Valley but is exposed in three antiformal windows near Stockertown and Nazareth, Pa. (fig. 2) where it lies along the Jacksonburg-Martinsburg contact. It probably extends all along the Great Valley to the southwest being expressed by the mapped thrusts and antiformal windows at this stratigraphic and tectonic position shown by Gray and others (1960). Several imbricate faults splay from the fault in the Bangor quadrangle (Davis and others, 1967) and even more have been recognized in the adjoining Wind Gap quadrangle (J. B. Epstein, oral commun., 1966). The possible relations near Manunka Chunk, N. J., have been mentioned above. Total displacement on this fault and its imbricate faults is at least several miles (displacement was grossly underestimated by Davis and others, 1967) and it is one of the key structures in the Middle Appalachians. Current analysis suggests that the Fish Hatchery fault in the Musconetcong Valley (Drake, 1967b) may be the same fault. Its place in the regional synthesis will be presented later.

RELATION OF THE PRECAMBRIAN TO THE PALEOZOIC ROCKS

The data presented above point to a regional synthesis which must include both allochthonous Precambrian rocks and a regional nappe in the lower Paleozoic rocks. As Miller (1944) conclusively disproved the Stose's flat-thrust hypothesis, some other relation of Precambrian rock to Paleozoic cover must be sought. Geophysical and geologic relations allow two possible interpretations: (1) several southeast-dipping imbricate thrust sheets bringing the Precambrian rocks out onto the nappe in the valley, or (2) a nappe de recouvrement, that is, a fold nappe in which both the Precambrian and Paleozoic rocks are involved.

The imbricate-thrust hypothesis is attractive in that very large distances of tectonic transport can be avoided. Regional stratigraphic relations, however, make this interpretation implausible, as each Precambrian body has nearly normal stratigraphic relations with the Paleozoic rocks that surround it. Except for the greater transport along the north edge of the Prong, the maximum stratigraphic separation recognized to date is about 1,500 feet. It is difficult to visualize thrust sheets that can be traced for more than 40 miles along strike having such paraconformable relations with autochthonous rocks, and even more difficult when the thrust sheets have similar displacements along normal faults on their south sides.

A nappe de recouvrement interpretation is favored to explain the regional relations. The nappe is visualized as embracing all rocks up to and including at least the Bushkill Member of the Martinsburg Formation, the Precambrian rocks forming the core. The geometric relations seen at the present level of erosion are the result of refolding and faulting of the core and inverted limb of the nappe, so that Precambrian rocks occupy the troughs of synforms and stratigraphically youngest rocks lie in the cores of antiforms. Shearing between the Precambrian and cover rocks produced the stratigraphic separations on the "thrust faults," the apparent normal faults on the south sides of the ridges are actually thrust faults developed after emplacement of the nappe. The greater transport along the north border is the result of the core shearing through the cover and riding out on the normal limb. The klippen of Precambrian and carbonate rocks in the Kittatinny Valley of New Jersey (fig. 2) are a result of this thrusting. Figure

27 is an interpretation of the relations in the Delaware Valley where the structure has been called the Musconetcong nappe.

The brow of the nappe is in the Martinsburg Formation in the Delaware Valley (Davis and others, 1967; Drake and others, 1969), and as it plunges northeast, the inverted limb is exposed in Pennsylvania and the normal limb in New Jersey. In the Delaware Valley, however, there seems to be a plunge culmination from which synforms cored by Precambrian gneiss spill both to the northeast and southwest. The passage from the upper limb to the lower limb around the core is exposed to the southwest in Little South Mountain (Geyer and others, 1963). Geophysical evidence suggests a similar relationship north of Reading, Pa., and to the northeast throughout most of New Jersey. Unfortunately so little is known in detail of New Jersey geology that the relations there are difficult to interpret.

There may be a transition between allochthonous and autochthonous Precambrian in New Jersey similar to that between the southern and northern Blue Ridge in Virginia. If so, it is not readily apparent. It is just as likely that much of the lower Paleozoic rock in eastern New Jersey and adjoining New York is upper limb, for there, Silurian and Devonian rocks occur in infolds, and are unconformable on all rocks from Precambrian through Ordovician. If the Paleozoic rocks are upper limb, the Precambrian rocks would appear to be rooted although in actuality they are floating. This would go a long way toward reconciling the thoughts of those who have worked to the northeast with those who have worked in Pennsylvania, as upper limb infolds should be preserved in the least eroded part of the structure. The Precambrian rocks are much more cataclastically deformed to the southwest, and contain tectonic windows and occur in isolated klippen. This may indicate that the sole of the core is near and that the body of Precambrian rock is relatively thin. The maximum thickness of the Precambrian core is not known. The Blue Ridge thrust sheet is estimated to be as much as 5 km thick (J. C. Reed, Jr., oral commun., 1967), so it is not unreasonable to assume that the nappe core is as much as 2 km thick.

The magnitude of the nappe is tremendous. The Precambrian core crops out over a width about 9 miles in the Delaware Valley, the total width of the structure being about 20 miles. The root zone is unknown, but almost certainly is beneath the Triassic basin, in parts of which, the aeromagnetic reflection of Precambrian rocks can be seen through

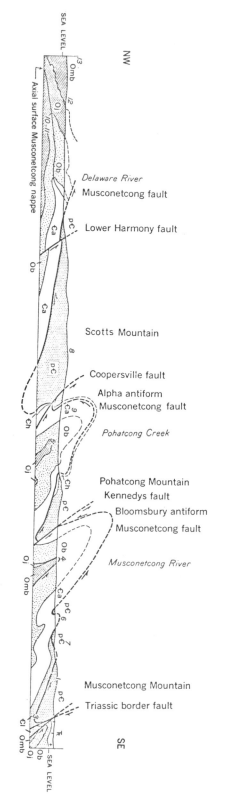

Figure 27. Geologic section across the Reading Prong and part of the Great Valley in the Delaware Valley (modified from Drake, 1969). ꞮꞮ̵b, Triassic rocks; Omb, Martinsburg Formation; Oj, Jacksonburg Limestone; Ob, Beekmantown Group; Ꞓa, Allentown Dolomite; Cl, Leithsville Formation; Ꞓh, Hardyston Quartzite; p-Ꞓ, Precambrian rocks; 5, tectonic position of field trip stop.

the Triassic cover. Finding the root may not be this simple, however, as Woodward (1964) has postulated that a right-lateral wrench fault along which there has been as much as 80 miles movement lies buried beneath the Triassic rocks. Bromery (*in* U. S. Geological Survey, 1966, written commun., 1967) has found good geophysical evidence that such a fault does exist about on the line Doylestown-Birdsboro-Mount Hope, Pennsylvania, and has a displacement about as visualized by Woodward. These data, suggest that the nappe may be laterally displaced from its root zone. It is possible, however, that this fault is the root zone of the nappe as well as a wrench fault as visualized by Reed and others (1969) for the Brevard, and Burchfiel and Livingston (1967) for several Alpine nappes and the Brevard. If this fault is the root zone, in the Delaware Valley the Precambrian rock has been displaced about 30 miles from its root. To the southwest, displacements apparently were about the same (MacLachlan, 1967). No data are available to interpret the relations further to the east in New Jersey and New York.

Origin of the Nappe

It is difficult to formulate an origin for a structural feature as poorly understood as the Musconetcong nappe. It probably developed high on the Appalachian geanticline by the upthrusting of basement into cover rocks. The movement of the Precambrian rocks into the core of the structure was largely brittle, as there is no sign of post-Precambrian plastic deformation in the crystalline rocks. Shear is manifested along Precambrian-Paleozoic interfaces by envelopes of mylonite or cataclasite in the Precambrian rocks 1 foot to several tens of feet thick (fig. 19B). In some places, the crystalline rock has been cataclastically refoliated parallel to the contact, but only for several inches to several feet (figs. 28A, B). Carbonate rocks in the contact zone are commonly shattered, silicified, and granulated; bedding is absent (figs. 19A, B and 28A). Thin-bedded shaly material has been converted to phyllite and in most places is now a mass of sheared-out isoclines.

It is postulated that as the nappe continued to grow under the influence of northwest-directed stress, it was unable to support its own weight and collapsed. Gravity probably played a large part in the northwestward transport of the structure from the geanticlinal high toward the basin. The skeptic might ask why, if this is so, do most of the structural elements now dip southeast? In part this is due to later folding, but a solution borrowed from

Wood (Wood and Carter, in press) seems applicable. He points out that the Newark Triassic basin has subsided several thousand feet subsequent to the Appalachian deformations, probably owing to regional collapse of the geanticline. Such subsidence tilted this part of the Appalachian region to the southeast and thereby rotated the original northwest-sloping surfaces to the southeast.

Much of the northwest transport under the influence of gravity was probably by shearing within the incompetent units, the Martinsburg Formation and Jacksonburg Limestone, and within incompetent parts of other units, such as phyllite intervals in the Leithsville Formation. It is thought that much of this transport has been along the Stockertown fault (see above). If this is true, and if the Stockertown and Fish Hatchery faults are one, the inverted limb of the nappe would contain the Cambrian and Ordovician dolomites, a thinned and partly absent Jacksonburg Limestone, and some Bushkill Member, as this rock crops out in the Musconetcong Valley (Drake, 1967b). Movement probably also took place on several other faults such as the unnamed faults in the Jacksonburg

Limestone in the Easton quadrangle (Drake, 1967a).

In an earlier paper (Drake, 1969) I suggested that the Musconetcong nappe was emplaced atectonically and that penetrative features can be related to later deformation. In the Great Valley, most such features can be related to the same stress that deformed Silurian and Devonian rocks. Within the Prong most of the cleavage in the Cambrian and Ordovician dolomites seems geometrically related to the post-nappe antiforms. The Martinsburg in the Musconetcong Valley has two flow cleavages and structures there seem to be refolded isoclinal folds. The same situation holds for the Pohatcong Valley where bedding and cleavage are largely parallel in the Jacksonburg and are folded together. It seems likely that the first generation cleavage in both places may be related to nappe emplacement. If this is so, the development of penetrative elements during nappe emplacement may have been restricted to the incompetent units as we have been unable to find parallel bedding and cleavage in all parts of the nappe. This problem is yet to be solved.

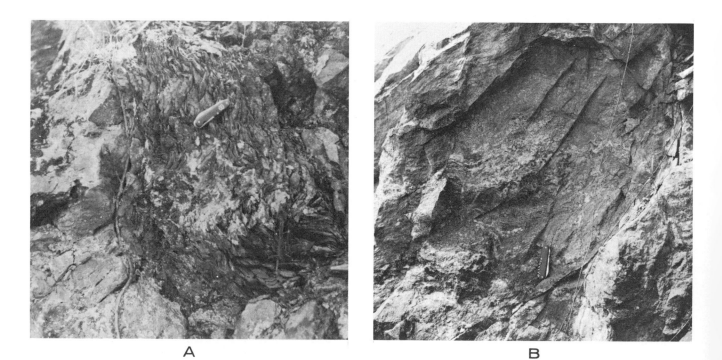

A   B

Figure 28. Cataclastic foliation and slip cleavage in Precambrian rocks. A, Cataclastic foliation in amphibolite along core-cover interface of Musconetcong nappe near Annandale, N. J. Note crinkle folds and slip cleavage allowing amphibolite to accommodate to the synform in the dolomite. B, Cataclastic foliation and slip cleavage in migmatite near core-cover interface of Musconetcong nappe at Rattlesnake Hill, Pennsylvania. Cataclastic foliation is well developed and here is nearly horizontal. Note the crinkling of the foliation and the development of parallel fractures.

Bedding is transposed in incompetent layers within the Leithsville Formation and is cut by a slip cleavage formed during later deformation which seems to parallel the only cleavage in more competent beds that formed during antiformal folding. A similar relation can be seen in figure 29, where the recumbent anticline formed by slip along cleavage; a more poorly developed, relic cleavage approximately parallels bedding in the fold limbs. Of course, shearing and cataclasis in the Precambrian rocks and flowage in the purer limestones can be attributed to nappe emplacement. It seems, therefore, that the Musconetcong nappe is atectonic in gross aspect and that tectonic features related to its emplacement can be found only where they were best preserved from subsequent tectonism. It may well be, however, that the search for relic fabric elements has not been thorough enough.

The recognition that nappes of large transport can be emplaced largely atectonically and with only the lightest metamorphism similar to such major structures in the Alps, Pyrenees, and Caledonides (den Tex, 1963) is of major importance to Appalachian geology. Many of us have a tendency to correlate far transport, tectonite fabric, and at least middle grade metamorphism. The past generation of geologists has used the similarity of fabric in supposed allochthonous and supposed autochthonous units, to deny the presence of the far-traveled structures described by their predecessors. It is hoped that the recognition that a penetrative fabric can be put into rocks subsequent to far transport will allow advances here as there have been in Europe.

Late deformation resulted in the antiforms and synforms now seen. Probably much of this folding is actually the result of imbrication of the Precambrian core of the nappe. The Precambrian rocks contain, in addition to a cataclastic foliation, shear fractures or a slip cleavage (fig. 28) that is best developed and most common near the borders of Precambrian bodies and is geometrically related to the cleavage in the Paleozoic rocks. Joint sets are related to this late deformation. Steep, northwest-striking joints are best developed; most are slickensided and grooved or have fault steps suggesting that there has been movement along them. The Precambrian rocks have not been folded themselves, but have been accommodated in the structures by the relative incompetence of the Paleozoic rocks and by movement on shear fractures and slip cleavage, joints, and other fractures.

Time of Nappe Emplacement

The nappe was probably emplaced during the Taconic orogeny immediately after or during the waning stages of Martinsburg flysch deposition, as no younger rocks are involved. No isotopic dates are available for the Delaware Valley. However, Long (1962) reports that mica in the Hudson River pelite east of the Hudson River in New York dates from 450 m.y. ago, a Taconic date. MacLachlan (1967) believes that the allochthon in the Harrisburg area was emplaced during the Taconic and that later major thrusting and folding probably occurred during the Appalachian orogeny (D. B. MacLachlan, oral commun., 1968). The folding of the nappe core and lower limb cannot be directly dated in the Delaware Valley. J. B. Epstein (Field Conference of Pennsylvania Geologists, 1967; this volume) has recently shown that the cleavage in the Martinsburg formation in the upper limb of the nappe along the Blue Mountain structural front is a result of folding during the Appalachian orogeny, rather than of diagenesis as postulated by Maxwell (1962). Because this cleavage can be traced south and seems to be related to the refolding, it is concluded that the refolding is a result of the Appalachian orogeny.

Determining which features formed during which deformation is exceedingly difficult in this region because all deformation during the Paleozoic was essentially coaxial. Structural trends apparently diverge no more than 20° and in this poorly exposed area present a problem of the greatest magnitude. Perhaps this can be solved by current investigations.

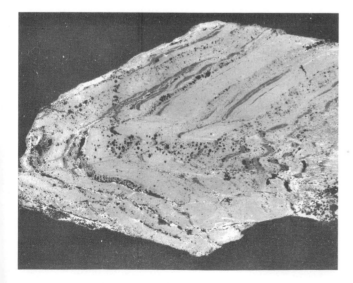

Figure 29. Small recumbent fold in Allentown Dolomite formed by slip along cleavage. Note relic cleavage approximately parallel to bedding in fold limbs. Fold is 5 inches long. Negative print of an acetate peel by J. B. Epstein, U. S. Geological Survey.

ROAD LOG

FIRST DAY

Departure from the Holiday Inn Motel, Warren, N. J., at 8:00 A.M. Stop locations are shown on figure 30.

Mileage

0.00   Leave Holiday Inn parking lot and proceed east on U. S. Rte. 22. Area here is underlain by the cement rock facies of Jacksonburg Limestone. The lack of exposures can serve as an introduction to the glories of working in this part of the Reading Prong and Great Valley.

0.45   Bear left onto U. S. Interstate 78. Hill tops in this area are covered by glacial drift that is older than Wisconsin in age. This is the Jerseyan drift of the New Jersey Geological Survey. In mapping, one soon learns that it is fruitless to attempt to find outcrops on these smooth, featureless hills. The three high ridges seen are from the south, Musconetcong, Pohatcong, and Scotts Mountain and are underlain by Precambrian rocks.

1.95   Pohatcong Creek. Outcrops are inverted cherty dolomite of the Epler Formation. A northwest-trending tear fault having a horizontal displacement of about 1,500 feet lies just south of bridge.

2.35   Spooning out nose of Pohatcong Mountain. Poor badly sheared outcrops are potassic feldspar gneiss that has a cataclastic foliation dipping about 53°S. The mountain is surrounded by Allentown Dolomite, the Leithsville Formation being absent. It is bounded on the north by the Musconetcong fault, the Precambrian-Paleozoic interface of the Musconetcong nappe, and on the south by a steep thrust fault. No gravity anomaly is associated with this mountain and aeromagnetic data show that the Precambrian rocks have a limited vertical extent.

3.55   Musconetcong River, once a famous trout stream, now a product of man's abuse of nature. Outcrops are Allentown Dolomite. The rock is only now beginning to take on the alternating light- and dark-weathering characteristic of the formation.

5.85   Deep cut. Rock is mostly interlayered oligoclase-quartz gneiss and amphibolite and a large cross cutting albite syenopegmatite. Some biotite gneiss has formed here at the expense of the amphibolite.

6.86   Deep cut.

STOP 1. PRECAMBRIAN ROCKS OF MUSCONETCONG MOUNTAIN

45 Minutes

This is a very dangerous locality, be extremely careful of automobile and truck traffic. Most of the rock here is interlayered oligoclase-quartz gneiss and amphibolite, biotite gneiss (an alteration product of amphibolite), and albite syenopegmatite. These rocks are described in the text of this paper. Note also the northwest-striking tear fault, the northeast-striking fault, small thrust faults, and lamprophyre dike. Everyone is invited to have a go at determining the relative age of the pegmatite and lamprophyre. I have tended to change my mind every time I have visited this cut. Layering and foliation in rock here ranges from north-northwest to nearly north in strike and dips moderately to the east. Lineation plunges about down the dip as this locality is about on the axis of a fair-sized northeast-plunging antiform of Precambrian age. Note relation of joints to lineation. This location is about 260 feet vertically above a window in the Precambrian rocks (younger rocks found in nearby railroad tunnel substantiate the window). Note the lack of later deformation in the rocks here as compared with those to be seen later today at Stops 3 and 7.

Mileage

6.85   Continue east on U.S. Interstate 78.

8.50   About on contact of Precambrian hornblende granite and Allentown Dolomite. The high-angle normal fault shown on the geologic map of the Bloomsbury quadrangle (Drake, 1967b) is probably incorrect, as current analyses suggest that this fault is more likely a thrust fault of Triassic age.

9.15   About on contact of lower Paleozoic carbonate rocks and quartzite fanglomerate of Triassic age.

11.10   Roadcut and shale pit.

STOP 2. ALLOCHTHONOUS PELITIC PSAMMITIC, AND MINOR CARBONATE ROCKS (PSEUDO-MARTINSBURG)

30 Minutes

Outcrops of rhythmically interbedded variegated claystone, mudstone, and minor fine conglomerate and silty dolomite. At the time of this writing (fall 1968) these rocks have not been

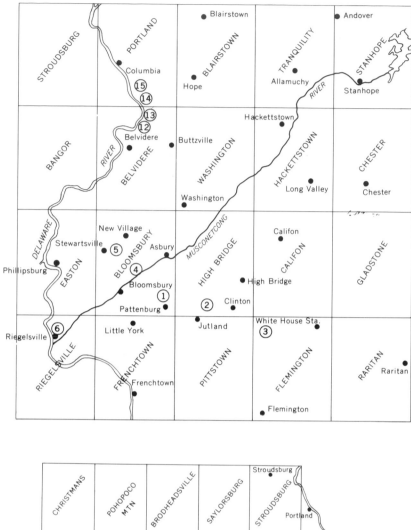

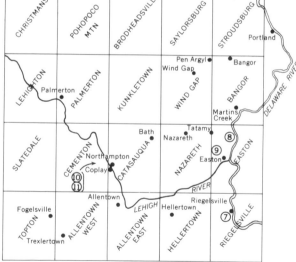

⑨ Stop location

Figure 30. Index maps of parts of eastern Pennsylvania and western New Jersey showing field trip stop locations and quadrangle map coverage.

studied in detail.   Stop is made to demonstrate the difference between these rocks and the Martinsburg Formation within the Reading Prong, Stop 4 today, and Great Valley, Stops 12–15 tomorrow.   The rocks here strike nearly north and dip moderately east and are in fault contact with carbonate rocks to the east and probably with Precambrian rocks to the north.   Small folds plunge 5°–10°, S.28°E.   Some beds have two cleavages, N.32°E., 35°SE., and N.30°E., 62°SE.   The intersections of bedding with the flatter cleavage plunges 6°, S.43°W.   According to F. J. Markewicz (oral commun., 1961) conodonts can be found in shaly bedded red mudstone here and graptolite fragments in the more brownish mudstone.

Mileage

11.10   Continue east on U. S. Interstate 78.

13.05   Clinton, N. J., area is underlain by carbonate rocks of the Beekmantown Group.

14.05   Turn right following N. J. Rte. 31S.

14.95   Outcrops on right are rhythmically interbedded dark "paper" shale and siltstone of the pseudo-Martinsburg.   These rocks would seem to be flysch deposits.   Bedding here is N.25°E., 35°SE., and cleavage is N.3°E., 85°NW., the intersection plunging 4°, N.4°E.   Small folds plunge 5°–10°, N.12°E.   These rocks are in fault contact with carbonate rocks of the Beekmantown Group to the east and north and are covered by Triassic fanglomerate to the west and south.

17.15   LEFT TURN onto Molasses Hill Road.

18.30   RIGHT TURN onto narrow gravel road.

18.45   Quarry in dolomite.

STOP 3.   CONTACT BETWEEN PRECAMBRIAN ROCKS AND LEITHSVILLE FORMATION(?)

30 Minutes

This quarry exposes Precambrian microperthite alaskite thrust on dolomite, Leithsville Formation(?).   At north end of quarry a cave is exposed which has walls of dolomite and a roof of Precambrian rock.   Both rocks are extremely deformed, the Precambrian rock has the aspect of a silicified breccia.   Bedding is obliterated in the carbonate, which near the thrust is dark gray, milled-up dolomite containing veins of white, recrystallized calcite.   At the cave the thrust strikes N.73°E., and dips 29°SE., and contains from 3 to 12 inches of gouge.   Shear planes in the footwall parallel the thrust.   The thrust surface rolls around considerably changing to a northwest strike and southwest dip spooning out toward the south, the rolls plunging ±30°, S.2°E.   Several small faults at about

N.25°W., 90°± cut the footwall block.   Strong slickensides indicate dip slip movement, probably related to the accommodation of the Precambrian rock in the synform.   The locality is just about the southernmost exposure of Precambrian rock in the Reading Prong and shows that thrusting has been every bit as active here as along the north border.   Unfortunately this area has never been mapped in detail.

Mileage

18.45   Turn around—follow gravel road.

18.65   LEFT TURN on Molasses Hill Road.

19.80   RIGHT TURN on N. J. Rte. 31.

22.88   Complicated cloverleaf intersection, FOLLOW Interstate 78 WEST toward Phillipsburg and Easton.

33.15   RIGHT TURN at Bloomsbury exit and follow U. S. Rte. 22W.

33.75   Cross Musconetcong River.   RIGHT TURN immediately on secondary blacktop road.   Outcrops along here are Allentown Dolomite.   Higher slopes and top of hill to left is mantled with pre-Wisconsin drift.

35.70   Outcrop of Epler Formation on right.

35.90   RIGHT TURN onto grounds of Warren County Rod and Gun Club and Trout Hatchery.

35.95   Parking area.

STOP 4.   BUSHKILL MEMBER OF MARTINSBURG FORMATION, FISH HATCHERY FAULT, AND BLOOMSBURY ANTIFORM LUNCH

130 Minutes

The Bushkill Member is extremely deformed in the Musconetcong Valley and is in fault contact with Epler Formation on all sides, the fault closing around both the lensing southwest and northeast extremities.   This fault has been called the Fish Hatchery fault (Drake, 1967b).   In many places in the valley, bedding and first generation slaty cleavage are parallel suggesting isoclinal folding.   Second generation cleavage here is another flow cleavage having progressed beyond typical cleavage.   In the pavement exposures just below the parking area Bushkill Member and Epler Formation are juxtaposed, although the interface cannot be seen.   Fabric elements in the Bushkill seem to be: S_0, N.47°E., 35°SE.; S_1, N.58°W., 63°NE.; S_2, N.45°E., 84°SE.; intersection of S_0 and S_1, 32°S.70°E.; and intersection of S_1 and S_2, 16°N.14°E.   From this exposure walk down hill to river valley and follow path downstream to right for about 1,200 feet to another outcrop of Bushkill Member.   Here S_0 is N.87°E., 55°NW.; S_1 is N.56°E., 82°SE., S_2 is N.28°E., 80°SE., intersection of S_0 and S_1 is

52°, N.62°E., and intersection of S_1 and S_2 is 80°, S.56°E. The Bushkill Member and bounding Fish Hatchery fault everywhere appears to underlie carbonate rocks and is believed to be in the lower limb of the Musconetcong nappe brought to the surface in an antiform. It should be pointed out to those who cannot accept a refolded nappe interpretation for the Reading Prong that the geometry and distribution of this rock require that it be bounded by a folded thrust of some sort. A disbeliever's interpretation, therefore, should be that this rock is in an infolded higher level younger over older thrust sheet, which also involves grand tectonics!

Mileage

35.95 Turn around and leave parking area.

36.0 RIGHT TURN on blacktop road.

36.85 LEFT TURN on Franklin Road.

37.55 South contact of Precambrian rocks of Pohatcong Mountain with carbonate rocks of Musconetcong Valley. This contact is a steep thrust fault, the Kennedys fault, which brings Allentown Dolomite against oligoclase-quartz gneiss.

37.60 Road intersection keep left. Rock in this area, no outcrops naturally, is potassic feldspar gneiss.

38.17 Road intersection, keep left. Layer of amphibolite in this area.

38.40 Intersection with secondary road, keep right. Rock here is potassic feldspar gneiss.

38.60 White boulders in field to left are potassic feldspar gneiss with 1/8- to 1/4-inch coatings of cement dust, souvenirs of the former Edison Cement Company in the valley below. These rocks will effervesce rapidly in hydrochloric acid, prompting a field assistant to report that cement limestone cropped out within the Precambrian rocks.

39.15 LEFT TURN on Good Spring Road. We have just crossed the north boundary of Pohatcong Mountain. Here a slice of Hardyston Quartzite lies along the Musconetcong fault between Precambrian rocks and the Allentown Dolomite.

39.65 TURN RIGHT on Willow Grove Mill Road. Small carbonate-floored valley to left is the result of the erosion of the Precambrian rocks of Pohatcong Mountain and reflects the relative thinness of these allochthonous rocks at this place.

40.30 Pohatcong Creek. The geologic relations are very complex and can only be generalized at the scale of 1:24,000. A tear fault about parallels the creek here (fig. 31); outcrops to right of road are Rickenbach Dolomite and those to the left Epler Formation.

40.45 Park along shoulder of road near narrow dirt road leading off to right.

STOP 5. EDISON CEMENT LIMESTONE QUARRY, CONTACT OF EPLER FORMATION AND JACKSONBURG LIMESTONE, AND ANTIFORMAL HINGE

45 Minutes

Follow dirt road into small quarry in Jacksonburg Limestone. Geologic relations are shown on figure 31. The contact of the Epler Formation and Jacksonburg Limestone is well exposed here, particularly at the southeastern corner of the quarry. Here the Epler is interbedded, somewhat limy, dololutite and dolarenite. Both facies contain patches of coarsely crystalline, dark-gray to black calcite. Bedding in the Epler is N.55°E., 55°SE. (inverted) and is marked by strong slickensides that plunge 45°, S.51°E. Cement limestone facies of the Jacksonburg Limestone consists of both calcarenite and crystalline limestone. Some beds of both lithologies contain pebbles of dolomite and some beds contain abundant bryozoans, brachiopods, and corals. Bedding is N.43°E., 31°SE. (inverted) and is marked by slickensides that plunge 31°, S.47°E. Along the south wall of the quarry the Jacksonburg-Epler contact appears to steepen upward and then flop over to a more gentle southeast dip, defining reverse drag folds (see Drake, 1967b, sections B-B' and C-C') probably related to nappe emplacement. There probably isn't much more than 50 feet of cement limestone present in this area but it is difficult to determine accurately. There are abundant drilling data, but these were not made available to me. There are so many complex structures here that it is difficult to sort them all out. Folds in bedding can be seen when defined by competent beds but in other places bedding and slaty cleavage seem to be about parallel, especially in the argillaceous limestone of the cement rock facies, and are folded together with the development of slip cleavage about parallel to the axial surfaces. An upward closing fold cored with cement rock facies (the youngest unit involved) is exposed about in the center of the west wall of the quarry. It can be best seen from the east side of the quarry (looking across the small lake) as the relations are too jumbled by minor structures to make much sense close up. This antiform, the Alpha antiform is thought to be the principal structure in this valley and to have been formed in the inverted limb of the nappe.

Mileage

40.45 Continue north on Willow Grove Mill Road. Note outcrops of pre-Wisconsin drift in cut banks along road.

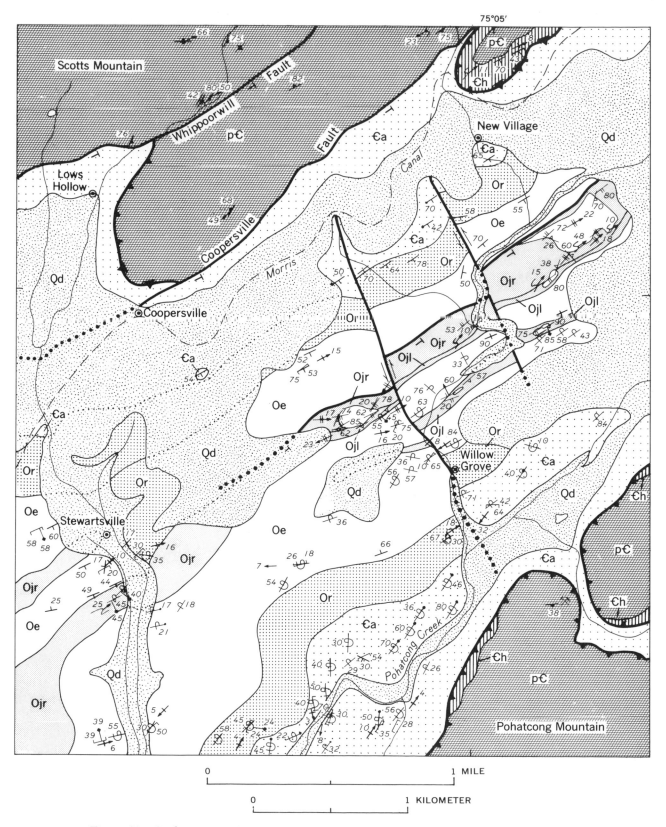

Figure 31. Geologic map of the Edison cement quarry area (modified from Drake, 1967b).

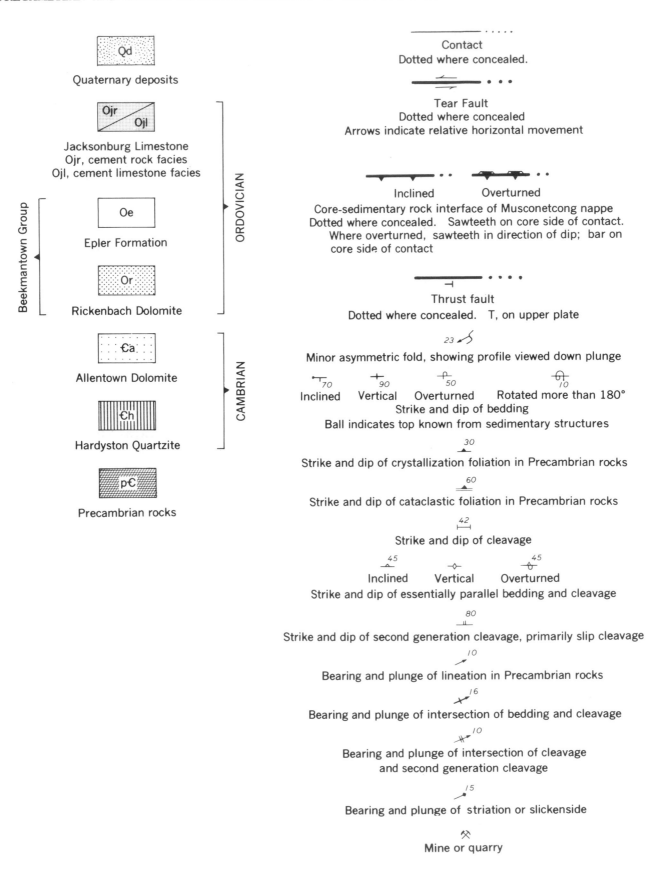

Qd
Quaternary deposits

Ojr / Ojl
Jacksonburg Limestone
Ojr, cement rock facies
Ojl, cement limestone facies

Oe
Epler Formation

Or
Rickenbach Dolomite

Beekmantown Group

ORDOVICIAN

Ɛa
Allentown Dolomite

Ɛh
Hardyston Quartzite

CAMBRIAN

pɛ
Precambrian rocks

Contact
Dotted where concealed.

Tear Fault
Dotted where concealed
Arrows indicate relative horizontal movement

Inclined   Overturned
Core-sedimentary rock interface of Musconetcong nappe
Dotted where concealed.   Sawteeth on core side of contact.
Where overturned, sawteeth in direction of dip; bar on
core side of contact

Thrust fault
Dotted where concealed.   T, on upper plate

23
Minor asymmetric fold, showing profile viewed down plunge

70   90   50   10
Inclined   Vertical   Overturned   Rotated more than 180°
Strike and dip of bedding
Ball indicates top known from sedimentary structures

30
Strike and dip of crystallization foliation in Precambrian rocks

60
Strike and dip of cataclastic foliation in Precambrian rocks

42
Strike and dip of cleavage

45   45
Inclined   Vertical   Overturned
Strike and dip of essentially parallel bedding and cleavage

80
Strike and dip of second generation cleavage, primarily slip cleavage

10
Bearing and plunge of lineation in Precambrian rocks

16
Bearing and plunge of intersection of bedding and cleavage

10
Bearing and plunge of intersection of cleavage
and second generation cleavage

15
Bearing and plunge of striation or slickenside

Mine or quarry

41.50   LEFT TURN on New Village Road.

41.85   Flashing light in Stewartsville, N. J. RIGHT TURN on Main Street.

42.80   LEFT TURN on N. J. Rte. 24, Scotts Mountain on right bounded by a steep thrust fault, the Whippoorwill fault.

45.10   Warren, N. J., intersection with U. S. Rte. 22. Continue west on U. S. Rte. 22.

48.80   RIGHT TURN at exit to Phillipsburg, N. J., and South Main Street. Follow signs to Broad Street and Easton, Pa. Outcrops are Allentown Dolomite.

49.74   Delaware River bridge, entering Easton, Pa. Turn left immediately on Pa. Rte. 611 and follow signs to Rte. 611 South.

50.30   Lehigh River bridge. Keep to left and follow U. S. Rte. 611S. Large outcrop is Allentown Dolomite.

50.70   Hill across river is a klippe of Precambrian gneiss on Allentown Dolomite protruding through outwash deposits of Wisconsin age. Railroad cut has been driven through the klippe. Outcrops to right are Allentown Dolomite.

52.95   Morgan Hill shear zone in Precambrian rocks to right. This shear zone probably formed along the sole of the nappe core. The sheared rocks are in contact with the carbonate rocks along the Whippoorwill fault. Rocks across river are Allentown Dolomite. Note quarries and kilns. This rock was burned locally for agricultural lime.

54.70   Good exposures (rare) of Leithsville Formation in small antiformal valley, note extremely deformed phyllite in outcrops to left of road. This rock has been mapped as Precambrian rock by some geologists because of its extreme deformation.

55.10   Elephant Rock. A thin spooning-out wedge of Precambrian rock is exposed here, bounded on north by Musconetcong fault and on south by Epler Formation which is brought up from lower limb by the post-nappe Raubsville fault.

55.90   Outcrops of inverted Allentown Dolomite.

57.10   Bougher Hill, the southwestern extension of Lower Pohatcong Mountain. This hill is largely underlain by hornblende granite and microperthite alaskite and is bounded on the south by a steep thrust fault which brings Allentown Dolomite into contact with the Precambrian rocks.

58.80   Riegelsville, Pa., turn left at stop light. Town is built on terrace of outwash gravel of Wisconsin age.

59.00   Delaware River bridge, enter New Jersey.

59.10   Riegelsville, N. J., park by hotel near railroad station.

STOP 6. CARBONATE ROCKS AND STRUCTURES OF THE MUSCONETCONG VALLEY

30 Minutes

The upper part of the Leithsville Formation and the superjacent Allentown Dolomite are relatively well exposed in the cut behind the Pennsylvania Railroad Station and in outcrops and quarry exposures to the north. The contact is transitional and is placed at the first stromatolite-bearing bed. These rocks are regionally overturned and are refolded into the Finesville antiform, the trace of the axial surface being about 0.65 mile to the north. Small recumbent folds are common in this area where deformation style can be studied.

Mileage

59.10   Recross Delaware River into Pennsylvania.

59.35   LEFT TURN immediately after crossing canal.

60.00   Rejoin U. S. Rte. 611, continue on south.

60.60   Rattlesnake Hill, park on left near canal locks.

STOP 7. PRECAMBRIAN ROCKS, CONTACT OF HARDYSTON QUARTZITE WITH PRECAMBRIAN ROCKS, RATTLESNAKE HILL WINDOW, TRIASSIC LIMESTONE FANGLOMERATE, AND SATELLITIC FOLDS IN LEITHSVILLE FORMATION

Remainder of afternoon or until participants have had enough.

It is suggested that individuals begin by observing folds in Leithsville Formation north of intersection of Routes 212 and 611 and proceed south along Rte. 611 viewing whatever strikes their fancy. It must be emphasized that this is an extremely dangerous area and the greatest caution must be exercised because of the heavy traffic along Rte. 611.

Rattlesnake Hill and its extension in New Jersey, Musconetcong Mountain, has been established to be allochthonous by tectonic windows, water wells, and drilling. The contact between the carbonate rocks and the Precambrian-Hardyston mass is not exposed (at least not in 1968). The following data were gained in 1964, when geologic relations in this area were best exposed. The Precambrian-Hardyston Quartzite contact is exposed about 400 feet south of the intersection of Routes 611 and 212 where it is due east, 8°S. (inverted), but rolls around in the area to an extreme of N.60°W., 45°SW. (inverted). The contact zone contains 2 to 5 inches of mylonitic rocks cut by fracture cleavage having a shear sense of southeast wall up. Both the Precambrian rock and the Hardyston are cut by a high-angle fault marked by slickensides that plunge vertically down the dip and by steps indicating that the south wall has moved up. The Precambrian rock is mylonized, has a cataclastic foliation, and has a slip cleavage and other shear-type structures more or less parallel to the slip cleavage in the carbonate rocks. It is difficult to determine, but compositional layering near the contact is about N.30°E., 53°NW. Cataclastic foliation is crinkled or folded into tight, nearly isoclinal similar folds; the cataclastic foliation has a general southeast dip the general attitude at the contact being N. 60°E., 58°SE. Axes of folds and crenulations in this foliation average about 17°, N.40°E. Slip cleavage and shear fractures are about parallel to the axial surfaces of these folds and are N.55–80°E., 55–75°SE. The first outcrop north of the contact is Hardyston containing cleavage that is N.70°E., 60°SE. Thirteen feet farther north is an outcrop of strongly sheared, iron-stained Hardyston mylonite that is underlain by rubble breccia, most probably fault breccia. Shear planes are N.60°E, 50°SE. More blocks of mylonitic Hardyston are exposed in the cut bank for another 17½ feet. The first dolomite, probably not in place, is 50 feet farther north. The rock has a strong flow fabric, but no attitudes were taken. Good carbonate rock crops out 40 feet farther north. The first rock exposed is dolomite with a strong flow fabric followed by sheared platy dolomite and phyllite all of which is bleached, leached, and silicified. Slip cleavage, N.65°E., 70°SE., marked by slickensides which plunge 50°E., is the dominant planar element. In places flow cleavage is about parallel to bedding and is N.62°W., 22°SW. The intersection of bedding (and flow cleavage) with slip cleavage is 8°, S.45°W. At the north end of the outcrop area bedding rolls around to a north dip and then back to steeply southeast presumably being involved in folds like those seen north of Rte. 212.

Walking south of contact note strong cataclastic foliation in the Precambrian rock. About 300 feet south of the contact microperthite alaskite passes into what was mapped as migmatite (Drake and others, 1967) but which is now known to be a strongly silicated and injected layer of dolomite marble. About 800 feet south of the contact dolomite of the Rattlesnake Hill window (fig. 17) is exposed. In first exposures, the planar element is N.42°E., 62°NW., and seems to be cleavage. In a small quarry about 100 feet higher than road level the following fabric elements were found: bedding, N.22°W., 13°NE.; flow cleavage, subparallel but flatter; slip cleavage, N.30°E., 70°SE.; intersection of bedding and cleavage, 5°, N.10°E.; intersection of bedding and slip cleavage, 11°, N.34°E.

Strongly deformed mixed Precambrian rocks are exposed south of the window to the village of Monroe. Carbonate rocks exposed here between the Precambrian and Triassic rocks are badly sheared and silicified. Where some sense can be made, it is apparent that bedding (and flow cleavage?) has been transposed and occurs only in small hinges between slip cleavage, reflecting the presence of the Musconetcong fault south of the Precambrian rocks.

The fault along the Triassic border mapped as a high-angle normal fault (Drake and others, 1967) is now thought to be a thrust fault. Kümmel (1898) originally interpreted the border fault in this area as a thrust fault but subsequently changed to a normal fault interpretation. Other geologists have suggested that the rocks are in depositional contact. Details have not as yet been worked out.

Mileage
60.60   TURN around and go north on U. S. Rte. 611 toward Easton.
70.85   Delaware River bridge, enter New Jersey. Take road straight up hill, N. J. Rte. 24E.
71.70   Junction with U. S. Rte. 22. Continue east on Rte. 22.
75.40   TURN RIGHT into Holiday Inn parking lot.

END OF FIRST DAY

SECOND DAY

Departure from Holiday Inn Motel, Warren, N. J., at 8:00 A.M. Tonight will be spent in East Stroudsburg, Pa., so be sure to have all your belongings.

Mileage
75.40   Leave parking lot of Holiday Inn following U. S. Rte. 22 WEST.
79.90   Delaware River bridge, stop and pay toll, cross river, RIGHT TURN on Pa. Rte. 611N. Outcrops along road are Allentown Dolomite.
81.31   C. K. Williams marble quarry.

STOP 8.   PRECAMBRIAN MARBLE, ARKOSE, AND MINERAL COLLECTING

30 Minutes

This quarry has not been worked since 1947 so it is now badly overgrown, however, no trip to this area is complete without a look at the marble. The marble here is quite dolomitic and has largely been altered to magnesium minerals. The main face in the northern part of the quarry exposes tremolite rock mixed with pale-green serpentine. Dark-green serpentine veined by fibrous calcite and tremolite crops out above the central wall. The rock in the southwest wall is a mixture of serpentine and talc. Microperthite alaskite and related pegmatites crop out north of the quarry, above the quarry and in the south wall which is also marked by the Marble Mountain fault.

In the past the following minerals could be found here: calcite, diopside, dolomite, phlogopite, quartz, chrysotile and massive serpentine, talc and tremolite. In addition crystals of sphene, thorianite, and zircon could occasionally be found.

Calcareous arkose can be seen at river level below the Easton Water Works (orange brick building) about 600 feet south of the quarry.

Mileage

81.31 Retrace route along Rte. 611 SOUTH toward Easton.

82.76 Junction with U. S. Rte. 22. Follow Rte. 22W toward Bethlehem and Allentown. Outcrops are Allentown Dolomite.

84.56 RIGHT TURN at 13th Street Exit, Easton. Follow around to left onto 13th Street. Buses discharge passengers on 13th Street, turn around and park on *right* side (facing north) on 13th Street.

STOP 9. ALLENTOWN DOLOMITE

45 Minutes

Highway cut at this location exposes a classic section of Allentown Dolomite containing all the described rock types, several types of stromatolites, all sorts of sedimentary structures, bedding slickensides, and stretched oolites. Rock here is in lower half of formation as beds transitional with Leithsville Formation are exposed in outcrops just east of here. Dip slopes on Chestnut Hill to the north are Leithsville.

Mileage

84.56 Follow signs to U. S. Rte. 22W and continue west toward Bethlehem and Allentown.

85.26 Narrow ridge of Precambrian rock to right spoons out to the west. Outcrops are jumbled up and bleached Leithsville Formation.

91.96 Low hill to left is a klippe of Precambrian rock lying on Rickenbach Dolomite. Hill slightly farther west is also a klippe. Featureless plain in this general area is caused by a mask of pre-Wisconsin drift.

98.16 Lehigh River bridge. Outcrops are Rickenbach Dolomite.

99.16 RIGHT TURN on MacArthur Road Exit, Allentown (Rte. 145), continue north on Rte. 145.

103.93 LEFT TURN on Rte. 329 (Main Street) toward Egypt, Pa.

104.5 Entering Egypt, Pa.

105.1 LEFT TURN on blacktop road at Mary's Cafe.

105.7 LEFT TURN on Lehigh Street next to Ormrod Hotel.

105.75 RIGHT TURN at first opportunity.

105.80 LEFT TURN and enter Ormrod quarry, Whitehall Cement Manufacturing Co.

105.90 Continue straight.

106.10 RIGHT TURN and descend into quarry.

106.40 Park along road.

STOP 10. ORMROD QUARRY, WHITEHALL CEMENT MANUFACTURING CO.

60 Minutes

This stop and Stop 11 are in an area currently under investigation by the U. S. Geological Survey, but have not been studied in detail as of this writing (fall 1968). This stop is included because of its importance in showing the geologic relations in the inverted limb and brow of the nappe in the Great Valley. The outline for these stops was prepared by W. C. Sherwood for a trip we jointly led (Field Conference of Pennsylvania Geologists, 1961).

Regionally, this stop is near the west termination of the Jacksonburg outcrop belt in the Lehigh Valley. Just west of here all geologic features are deflected about 4–5 miles south (see Gray and others, 1960). The reasons for this are not known, but presumably the deflection must be some sort of major fold hinge. The Jacksonburg exposed here is in the lower limb of the nappe and is arched into an antiform, Beekmantown rocks being exposed to both the north and south. The Jacksonburg-Beekmantown contact is exposed in the southwest face of the quarry and here is conformable (compare with Stop 5 of yesterday). Bedding in the Jacksonburg is accentuated by thin bentonites that define tight, high-amplitude folds as shown by repetition of beds, slight divergence of bedding, and at least one fold hinge. Two or possibly three bentonites are exposed near the base of the quarry face directly below the contact and another bed is exposed in the southeast wall of the quarry. This bed contains euhedral crystals of pyrite and according to Sherwood (Field Conference of Pennsylvania Geologists, 1961) can be traced over a large part of Lehigh and Northampton Counties, Pa.

Mileage

106.40	Retrace route to quarry entrance.
106.70	LEFT TURN.
106.90	RIGHT TURN, leave quarry property.
107.0	RIGHT TURN, follow blacktop road.
107.15	LEFT TURN.
107.20	Bear right across small bridge.
107.25	RIGHT TURN.
107.40	Lehigh Crushed Stone Co., park along road.

STOP 11. FOLDS IN BEEKMANTOWN IN LOWER LIMB OF NAPPE; LUNCH

120 Minutes

This stop is about 500 yards east of Stop 10 and is in the inverted limb of the nappe and the south limb of the antiform at Stop 10. The quarry is mainly in rocks of the Beekmantown Group, and the Jacksonburg-Beekmantown contact is exposed at the southwest end of the quarry. Complex folds are exposed in the northeast wall of the quarry and seem to be largely restricted to the contact zone. The folds are largely concentric and seem to have formed by flexural flow as shown by slickensided bedding surfaces and flowage of material into hinge areas. Probably some should be called quasi-flexural folds, however, as the incompetent limestones have adjusted themselves to the form surface by flowage rather than flexure. Tight recumbent folds are exposed in the south wall of the small abandoned quarry northeast of and adjacent to the main quarry. Here the best exposed fold has a thinned lower limb and incipient boudinage.

Mileage

107.40	Turn around and proceed back toward Egypt, Pa.
107.80	Continue straight.
108.70	Stop sign, RIGHT TURN on Rte. 329.
109.30	Bear left, follow Rte. 329.
109.50	Fork in road, bear right, follow Rte. 329.
109.90	Traffic light, continue straight. Escarpment ahead is underlain by Martinsburg Formation.
111.00	Lehigh River bridge, enter Northampton, Pa., continue on Rte. 329.
112.30	Universal Atlas Cement Co. plant and quarry.
115.90	Traffic light, continue on Rte. 329.
116.40	BEAR LEFT, follow Rte. 329.
117.30	Keystone Portland Cement Co. plant and quarry.
118.20	Intersection Rtes. 329 and 45. RIGHT TURN, follow Rte. 45 into Bath, Pa.
118.25	LEFT TURN, follow Rte. 45.
118.30	Traffic light. RIGHT TURN on West Northampton Street (Rte. 45).
118.40	Traffic light, continue straight on Rte. 45.
118.70	Outcrop of Bushkill Member of Martinsburg Formation on left.

119.00	Pass over Martinsburg escarpment.
120.00	Penn-Dixie Co. plant and quarry.
122.10	Another Penn-Dixie Cement Co. plant and quarry.
122.90	Junction Rtes. 45 and 191. Continue straight following Rte. 191N. Enter Nazareth, Pa.
123.10	LEFT TURN on South Broad Street (Rte. 191).
123.60	RIGHT TURN on East Walnut Street.
123.80	Stop sign, continue on East Walnut Street.
126.40	Flashing traffic light, continue straight toward Martins Creek.
126.90	Fork in road, bear LEFT across bridge toward Martins Creek.
127.40	Traffic light, continue toward Martins Creek.
129.90	Stop sign, bear LEFT down hill.
130.30	Outcrops of cement rock facies of Jacksonburg Limestone on right.
131.50	Traffic light in Martins Creek, Pa. Join Pa. Rte. 611, continue straight on Rte. 611.
131.70	Outcrops of cement rock facies of Jacksonburg Limestone in roadcut.
132.55	Road intersection, continue straight toward Belvidere. Scotts Mountain on the horizon is underlain by Precambrian rocks. Gravel pits to right along Delaware River are in outwash deposits of Wisconsin age. Quarries belong to Alpha Portland Cement Co. and are in Jacksonburg Limestone. The journey to the east will be across gravel deposits related to the Wisconsin Glaciation. We will, in general, follow the Jacksonburg-Epler contact. The ridge to left is underlain by Bushkill Member of Martinsburg Formation.
135.65	Break in slope. This is probably the southwestern edge of the Wisconsin terminal moraine which is poorly defined in this area.
136.05	Gravel Hill. Glacial topography much better developed here.
137.25	Delaware River bridge, enter New Jersey. Smaller stream to right is Pequest River. This stream serves as a temporary habitat for trout in their journey from hatchery to frying pan.
137.65	Traffic light in Belvidere, N. J. Continue straight. Gravel of Wisconsin terminal moraine on left.
138.95	Outcrops of Jacksonburg Limestone in roadcuts.
139.25	Junction with Rte. 46, bear left and follow Rte. 46.
139.95	HARD RIGHT TURN onto access road for highway maintenance road.

STOP 12. LOWERMOST PART OF BUSHKILL MEMBER OF MARTINSBURG FORMATION

105 Minutes

This outcrop is rather famous, having been illustrated by many geologists, see for example Maxwell (1962) and de Sitter (1964, p. 276).

This is the southernmost outcrop of Bushkill Member in the Great Valley of New Jersey and is fairly limy (see table 25). Some geologists have considered this rock to be cement rock facies of Jacksonburg Limestone, but it contains far too little calcium carbonate. Probably the most interesting features of this outcrop are the dolomite sandstone beds (figs. 11B, 32) that dip steeply southeast and are thrown into series of essentially concentric folds. The surrounding slate is deformed by passive-flow folds, the slaty cleavage averaging N.32°E., 28°SE. Cleavage within the sandstone is a spaced bedding normal cleavage and fans the folds. The cleavage could, and has been, called fracture cleavage, but in actuality the surfaces are marked by alined minerals. The dolomite sandstone beds were almost certainly competent at the time of deformation as cleavage in the slate is dragged against the sandstone (fig. 32) demonstrating slip along the interface. Bedding in the slate is preserved in strain shadows near synclinal fold hinges (fig. 32). J. B. Epstein (this volume) has built upon this feature to develop a solution to problems in the relation of cleavage in the Martinsburg to the Martinsburg-Shawangunk contact. Maxwell (1962) on the basis of some

direct measurements of dolomite sandstone beds suggests that the section has been flattened by 40 percent, whereas de Sitter (1964), by calculation, suggests that the sandstones have been flattened by 36 percent but that the total section has been flattened by about 70 percent.

Fabric elements in this southern exposure of Martinsburg average: bedding, N.59°E., 65°SE.; slaty cleavage, N.80°E., 14°SE.; very poorly developed slip cleavage N.85°W., 22°NE.; intersection of bedding and slaty cleavage, 11°, N.63°E.; intersection of slaty and slip cleavage, 2°, N.88°E.; and cross joint, N.18°W., 85°SW. Regionally this area is in the upper brow of the nappe.

From this station, it is planned for the participants to walk north about 1.25 miles along the abandoned railroad grade observing the structural features shown in the relatively good exposures of Bushkill Member in these cuts. Especial attention is directed to the increasing development of slip cleavage as one goes north, until it is the strongest cleavage. Broughton (1947) believed that the strong slaty cleavage we have seen here is actually S_2 and that S_1 has been obliterated. This does not seem to be the case, however, and it is hoped that progressive

A   B

Figure 32. Concentrically folded dolomite sandstone bed within passive-flow folded slate of Bushkill Member of Martinsburg Formation. A, Note bedding-normal spaced cleavage in sandstone, drag on the interface, lateral transport of the faulted sandstone along the slaty cleavage, and the bedding traces preserved in the strain shadow to the right of the synclinal hinge. B, Closeup of A. Note quartz veins parallel to cleavage in the dolomite sandstone and preserved bedding in slate in strain shadow to right of synclinal hinge.

northerly development of slip cleavage can be demonstrated. Buses will follow the directions below.

Mileage
139.95 Continue north on Rte. 46.
141.15 Manunka Chunk, N. J. Buses park on shoulder near point where narrow gravel road goes uphill to railroad grade.

STOP 13. MANUNKA CHUNK LINE

TIME—included in Stop 12

This is the area of extreme deformation in the Bushkill Member and in the Martinsburg Formation as a whole. Speculations as to the cause of this deformation have already been presented and the suggestion made that it is related to the cropping out of the largely subsurface Stockertown fault or more probably an imbricate therefrom. The deformation may, on the other hand, be related to structures at depth that have only recently been recognized. Here the slaty cleavage has been folded and has slip cleavage (now a flow cleavage) about parallel to the axial surfaces. Many thrust faults are more or less parallel to the slip cleavage, many of which are filled with veins of columnar quartz. In the strongest fold in cleavage, quartz rolls around the hinge and forms pods in the fold crest. Fabric elements here are: bedding, N.40°W., 16°NE.; slaty cleavage, N.7°W., 34°NE.; slip cleavage, N.16°E., 32°SE.; intersection of bedding and slaty cleavage, 10°, N.12°E.; and intersection of slaty and slip cleavage and axes of folds in cleavage, 10°, N.53°E. Recumbent folds in bedding are hard to find, but there is one small fold in the cut about 150 feet north of the road-railroad intersection. Any suggestions as to other solutions to the problems here will be greatly appreciated. Detailed mapping to the northeast should prove highly enlightening.

141.15 Continue north on Rte. 46.
142.25 Road to Hope and Land of Make Believe (this sign should have pointed the way to Stop 1 yesterday), RIGHT TURN. Buses discharge passengers a short distance up the Hope Road, turn around (will necessitate going some distance east on Hope Road), and park along road just east of railroad overpass.

STOP 14. RAMSEYBURG MEMBER OF MARTINSBURG FORMATION AND SLIP CLEAVAGE

30 Minutes

This is the type locality of the Ramseyburg Member the best exposures of Ramseyburg can be seen in the railroad cut above the road. The long outcrop of interbedded slate and graywacke siltstone along the Hope Road contains the most beautifully developed slip cleavage in the area. Slaty cleavage is largely, but not everywhere, parallel to bedding. The rock tends to break into long pencils or rods bounded by the two cleavages. Fabric elements are: bedding, N.54°E., 44°NW.; slaty cleavage (where not parallel to bedding), N.51°E., 72°NW. Slip cleavage, N.32°E., 53°SE.; intersection of bedding and slaty cleavage, 4°N., 49°E.; and intersection of slaty and slip cleavage, 5°, S.27°W. The slip cleavage here is not a pure fracture phenomenon but has alined micas that can be seen megascopically.

In the railroad cut only bedding and a southeast-dipping cleavage are readily apparent, the cleavage having the attitude of the slip cleavage in cut along Hope Road. In these outcrops slaty cleavage is not apparent, although bedding surfaces are crinkled and the rock breaks into rods as it does where two cleavages are present. Poorly developed slaty cleavage was finally found essentially parallel to bedding in outcrops along the next road north of here and thin section study shows that a poor slaty cleavage is present in these rocks.

Structurally this area is on the southeast limb of the Stone Church syncline.

Mileage
142.25 Continue north on Rte. 46.
145.15 North end of Columbia cut. Buses park along side of road near pizza parlor.

STOP 15. BUSHKILL MEMBER OF MARTINSBURG FORMATION

30 Minutes

This is an unusually fine exposure of the Bushkill Member and the outcrop from which Maxwell (1962) described a sandstone dike. To date (fall 1968) I have been unable to find this dike but hopefully by the time of this trip, it will be located. In our case, participants will have the opportunity to search for such structures and see one of the areas critical to Maxwell's (1962) ideas on the origin of cleavage.

The area is on the northwest limb of the Stone Church syncline and is adjacent to the completely fault bounded Paulins Kill valley (Drake and others, 1969). The carbonate rocks in the valley are anticlinal and are thought to belong to a separate tectonic unit. This analysis is not complete, however, and it may be in error.

Fabric elements at this northern outcrop are: bedding, N.40°E., 58°NW.; slaty cleavage, N.37°E., 54°SE.; slip cleavage, N.40°E., 50°NW.; intersection of bedding and cleavage, 8°,

N.42°E.; and intersection of slaty and slip cleavage 2°, N.36°E. Several fold hinges and thrust faults can be seen in this cut to the south and are typical of the tectonic style of the area.

Mileage

145.15   Continue north of Rte. 46 on alluvial terrace.

145.75   Terrace, about 25 feet high, in glacial deposits on right.

146.65   Exit to Portland—keep straight on U. S. Interstate 80. Exposures of Allentown Dolomite to right.

147.85   Exposures of Allentown Dolomite to right.

150.45   Gravel and till on right and across Delaware River to left probably in kame terraces. This Wisconsin drift covers the Pen Argyl Member of the Martinsburg Formation from here to Delaware Water Gap. Kittatinny Mountain, underlain by the Shawangunk Conglomerate of Silurian age, straight ahead.

154.95   Contact of Shawangunk Conglomerate and Martinsburg Formation covered by talus on right.

159.85   Contact of Shawangunk Conglomerate and Bloomsburg Red Beds on right.

164.85   Kame terrace on right.

170.15   Folds in Bloomsburg Red Beds to right.

175.85   Cross Delaware River into Pennsylvania.

182.15   Toll booth. Exposures of Shawangunk Conglomerate.

188.65   Contact of Shawangunk Conglomerate and Bloomsburg Red Beds on left. Keep west on U. S. Interstate 80.

195.35   Cross flood plain of Delaware River.

202.85   Cut through Port Ewen Shale of Early Devonian age in Godfrey Ridge.
         and Pa. Rte. 447.

210.65   Turn right at exit 52 onto U. S. Rte. 209 north

218.85   Turn left into Holiday Inn Motel.

END OF SECOND DAY AND PRECAMBRIAN-LOWER PALEOZOIC TRIP

REFERENCES CITED

Aldrich, M. J., Jr., 1967, Cambrian dolomite in the Martinsburg Formation, eastern Pennsylvania: Pennsylvania Acad. Sci. Trans., v. 41, p. 133–140.

American Geophysical Union and U. S. Geological Survey, 1964, Bouguer gravity anomaly map of the United States (exclusive of Alaska and Hawaii): Washington, D. C., U. S. Geol. Survey, 2 sheets, scale 1:2,500,000.

Baker, D. R., 1956, Geology of the Edison area, Sussex County, N. J.: U. S. Geol. Survey open-file report, 275 p.

Balk, Robert, 1936, Structural and petrologic studies in Dutchess County, New York: Geol. Soc. America Bull., v. 47, p. 685–774.

Barnett, S. C., III, 1965, Conodonts of the Jacksonburg Limestone (Middle Ordovician) of northwestern New Jersey and eastern Pennsylvania: Micropaleontology, v. 11, no. 1, p. 57–80.

Bates, T. F., 1947, Investigation of the micaceous minerals in slate: Am. Mineralogist, v. 32, p. 625–636.

Bates, T. F., and others, 1947, Properties and new uses of Pennsylvania slate: Pennsylvania State College Mineral Industries Expt. Sta. Bull. 47, 168 p.

Baum, J. L., 1967, Stratigraphy and structure of an anomalous area in the vicinity of Andover, Sussex County, N. J. [abs.]: Geol. Soc. America, Northeast Section Mtg., 1967, Program, p. 14.

Bayley, W. S., 1914, The Precambrian sedimentary rocks in the Highlands of New Jersey: Internat. Geol. Congress 12th, Toronto, 1913, Compte Rendus, p. 325–334.

———— 1941, Pre-Cambrian geology and mineral resources of the Delaware Water Gap and Easton quadrangles, New Jersey and Pennsylvania: U. S. Geol. Survey Bull. 920, 98 p.

Bayley, W. S., Salisbury, R. D., and Kümmel, H. B., 1914, Raritan, N. J.: U. S. Geol. Survey Geol. Atlas, Folio 191.

Behre, C. H., Jr., 1927, Slate in Northampton County, Pa.: Pennsylvania Geol. Survey, 4th ser., Bull. M9, 308 p.

———— 1933, Slate in Pennsylvania: Pennsylvania Geol. Survey, 4th ser., Bull. M 16, 400 p.

Bromery, R. W., and Griscom, Andrew, 1967, Aeromagnetic and generalized geologic map of southeastern Pennsylvania: U. S. Geol. Survey Geophys. Inv. Map GP-577.

Bromery, R. W., and others, 1960, Aeromagnetic map of part of the Riegelsville quadrangle, Bucks and Northampton Counties, Pennsylvania, and Hunterdon and Warren Counties, New Jersey: U. S. Geol. Survey Geophys. Inv. Map GP-236.

Broughton, J. G., 1947, An example of the development of cleavage: Jour. Geology, v. 54, p. 1–18.

Broughton, J. G., and others, 1962, Geologic map of New York—1961: New York State Mus. and Sci. Service Geol. Survey Map and Chart Ser., no. 5, scale 1:250,000.

Bryant, Bruce, and Reed, J. C., Jr., 1969, Structural and metamorphic history of the southern Blue Ridge, *in* Fisher, G. W., and others, eds., Studies in Appalachian geology—the central and southern Appalachians: New York, John Wiley and Sons (in press).

Buckwalter, T. V., 1959, Geology of the Precambrian Rocks and Hardyston Formation of the Boyertown quadrangle: Pennsylvania Geol. Survey, 4th ser., Geol. Atlas 197, 15 p.

———— 1962, The Precambrian geology of the Reading 15-minute quadrangle: Pennsylvania Geol. Survey, 4th ser., Prog. Rept. 161, 49 p.

Buddington, A. F., 1959, Granite emplacement with special references to North America: Geol. Soc. America Bull., v. 70, p. 671–747.

Buddington, A. F., and Baker, D. R., 1961, Geology of the Franklin and part of the Hamburg quadrangles,

New Jersey: U. S. Geol. Survey Misc. Geol. Inv. Map I-346.

Burchfiel, B. C., and Livingston, J. L., 1967, Brevard zone compared to Alpine root zones: Am. Jour. Sci., v. 265, p. 241–256.

Callahan, W. H., 1968, Geology of the Friedensville Zinc mine, Lehigh County, Pennsylvania, in Ridge, J. D., ed., Ore deposits of the U. S., 1933–1967, the Graton-Sales Volume: New York, Am. Inst. Mining, Metall., and Petroleum Engineers, 1880 p.

Davis, R. E., Drake, A. A., Jr., and Epstein, J. B., 1967, Geologic map of the Bangor quadrangle, Pennsylvania–New Jersey: U. S. Geol. Survey Geol. Quad. Map GQ-665.

Dodd, R. T., 1962, Precambrian geology of the Popolopen Lake quadrangle, southeastern New York: unpublished Ph.D. thesis, Princeton Univ., 178 p.

Donath, F. A., and Parker, R. B., 1964, Folds and folding: Geol. Soc. America Bull., v. 75, p. 45–62.

Drake, A. A., Jr., 1965, Carbonate rocks of Cambrian and Ordovician age, Northampton and Bucks Counties, eastern Pennsylvania and Warren and Hunterdon Counties, western New Jersey: U. S. Geol. Survey Bull. 1194-L, 7 p.

———— 1967a, Geologic map of the Easton quadrangle, New Jersey–Pennsylvania: U. S. Geol. Survey Geol. Quad. Map GQ-594.

———— 1967b, Geologic map of the Bloomsbury quadrangle, New Jersey: U. S. Geol. Survey Geol. Quad. Map GQ-595.

———— 1969, Structural geology of the Reading Prong, in Fisher, G. W., and others, eds., Studies in Appalachian Geology—the central and southern Appalachians: New York, John Wiley and Sons (in press).

Drake, A. A., Jr., Davis, R. E., and Alvord, D. C., 1960, Taconic and post-Taconic folds in eastern Pennsylvania and western New Jersey: U. S. Geol. Survey Prof. Paper 400-B, p. B180–B181.

Drake, A. A., Jr., Denny, W. V., and Hamlin, H. P., 1965, Evaluation of the Martinsburg Shale and two younger formations as sources of lightweight aggregate in the Delaware River area, Pennsylvania–New Jersey: U. S. Geol. Survey Prof. Paper 525-D, p. D156–D162.

Drake, A. A., Jr., and Epstein, J. B., 1967, The Martinsburg Formation (Middle and Upper Ordovician) in the Delaware Valley, Pennsylvania–New Jersey: U. S. Geol. Survey Bull. 1244-H, 16 p.

Drake, A. A., Jr., Epstein, J. B., and Aaron, J. M., 1969, Geologic map and sections of parts of the Portland and Belvidere quadrangles, New Jersey–Pennsylvania: U. S. Geol. Survey Misc. Geol. Inv. Map I-552.

Drake, A. A., Jr., McLaughlin, D. B., and Davis, R. E., 1961, Geology of the Frenchtown quadrangle, New Jersey–Pennsylvania: U. S. Geol. Survey Geol. Quad. Map GQ-133.

———— 1967, Geologic map of the Riegelsville quadrangle, Pennsylvania: U. S. Geol. Survey Geol. Quad. Map GQ-593.

Engle, A. E. J., 1956, Apropos the Grenville, in

Thomson, J. E., ed., The Grenville problem: Royal Soc. Canada Spec. Pub., no. 1, p. 74–98.

Epstein, J. B., D'Agostino, J. P., Drake, A. A., Jr., and Lampiris, Nicholas, 1967, Preliminary log of drill hole near Riegelsville, Pa.: U. S. Geol. Survey open-file report, 8 p.

Fenner, C. N., 1914, The mode of formation of certain gneisses in the highlands of New Jersey: Jour. Geology, v. 22, p. 594–612; 694–702.

Field Conference of Pennsylvania Geologists, 20th, Hershey, 1954, Structure and stratigraphy of Lebanon County, by Carlyle Gray, C. E. Prouty, and J. R. Mosely: Harrisburg, Pa., Pennsylvania Geol. Survey, 44 p.

Field Conference of Pennsylvania Geologists, 22d, Trenton, 1956, Guidebook: Trenton, New Jersey Geol. Survey, 62 p.

Field Conference of Pennsylvania Geologists, 26th, Bethlehem, 1961, Structure and stratigraphy of the Reading Hills and Lehigh Valley in Northampton and Lehigh Counties, Pennsylvania. Edited by J. Donald Ryan: Bethlehem, Pa., Lehigh Univ., 82 p.

Field Conference of Pennsylvania Geologists, 31st, Harrisburg, 1966, Comparative tectonics and stratigraphy of the Cumberland and Lebanon Valleys, by D. B. MacLachlan and S. R. Root: Harrisburg, Pa., Pennsylvania Geol. Survey, 90 p.

Field Conference of Pennsylvania Geologists, 32d, East Stroudsburg, 1967, Geology in the region of the Delaware to Lehigh Water Gaps, by J. B. Epstein and A. G. Epstein: Harrisburg, Pa., Pennsylvania Geol. Survey, 89 p.

Fraser, D. M., 1938, Contributions to the geology of the Reading Hills, Pa.: Geol. Soc. America Bull., v. 49, p. 1199–1212.

Fyfe, W. S., Turner, F. J., and Verhoogen, Jean, 1958, Metamorphic reactions and metamorphic facies: Geol. Soc. America Mem. 73, 259 p.

Geyer, A. R., and others, 1958, Geologic map of the Lebanon quadrangle, Pa.: Pennsylvania Geol. Survey, 4th ser., Atlas 167-C.

Geyer, A. R., and others, 1963, Geology of the Womelsdorf quadrangle, Pa.: Pennsylvania Geol. Survey, 4th ser., Atlas 177-C.

Gray, Carlyle, 1951, Preliminary report on certain limestones and dolomites of Berks County, Pa.: Pennsylvania Geol. Survey, 4th ser., Prog. Rept. 136, 85 p.

———— 1952, The high calcium limestone of the Annville belt in Lebanon and Berks Counties, Pa.: Pennsylvania Geol. Survey, 4th ser., Prog. Rept. 140, 17 p.

———— 1959, Nappe structures in Pennsylvania [abs.]: Geol. Soc. America Bull., v. 70, no. 12, p. 1611.

Gray, Carlyle, Geyer, A. R., and McLaughlin, D. B., 1958, Geologic map of the Richland quadrangle, Pa.: Pennsylvania Geol. Survey, 4th ser., Atlas 167-D.

Gray, Carlyle, and others, 1960, Geologic map of Pennsylvania: Pennsylvania Geol. Survey, 4th ser., scale 1:250,000 with sections.

Hague, J. M., and others, 1956, Geology and structure

of the Franklin-Sterling area, New Jersey: Geol. Soc. America Bull., v. 67, p. 455–474.

Henderson, J. R., Andreasen, G. E., and Petty, A. J., 1966, Aeromagnetic map of northern New Jersey and adjacent parts of New York and Pennsylvania: U. S. Survey Geophys. Inv. Map GP-562.

Hobson, J. P., 1963, Stratigraphy of the Beekmantown Group in southeastern Pennsylvania: Pennsylvania Geol. Survey, 4th ser., General Geology Rept., G 37, 331 p.

Holzwasser, Floric, 1926, Geology of Newburgh and vicinity: New York State Mus. Bull. 270, 95 p.

Hotz, P. E., 1952, Magnetite deposits of the Sterling Lake, New York-Ringwood, New Jersey area: U. S. Geol. Survey Bull. 982-F, p. 153–244.

Howell, B. F., 1945, Revision of the Upper Cambrian faunas of New Jersey: Geol. Soc. America Mem. 12, 46 p.

—— 1957, Upper Cambrian fossils from Bucks County, Pennsylvania: Pennsylvania Geol. Survey, 4th ser., Bull. G 28, 39 p.

Howell, B. F., Roberts, Henry, and Willard, Bradford, 1950, Subdivision and dating of the Cambrian of eastern Pennsylvania: Geol. Soc. America Bull., v. 61, p. 1355–1368.

Isachsen, Y. W., 1964, Extent and configuration of the Precambrian in northeastern United States: New York Acad. Sci. Trans., ser. 2, v. 26, no. 7, p. 812–829.

Kay, G. M., 1929, Stratigraphy of the Decorah formation: Jour. Geology, v. 37, no. 7, p. 639–671.

Klemic, Harry, and others, 1959, Radioactive rare-earth deposit at Scrub Oak mine, Morris County, N. J.: U. S. Geol. Survey Bull. 1082-B, p. 29–58.

Kümmel, H. B., 1898, The Newark system or red sandstone belt: New Jersey Geol. Survey Ann. Rept. 1897, p. 25–159.

—— 1901, Report on the cement industry: New Jersey Geol. Survey Ann. Rept. 1900, p. 9–101.

Lesley, J. P., and others, 1883, The geology of Lehigh and Northampton Counties: Pennsylvania Geol. Survey, 2d ser., v. 1, 283 p.

Lewis, J. L., and Kümmel, H. B., 1912, Geologic map of New Jersey: New Jersey Dept. Conserv. and Devel., Atlas Sheet no. 40, revised 1931 by H. B. Kümmel, and 1950 by M. E. Johnson, scale 1:250,000, with sections.

—— 1940, The geology of New Jersey: New Jersey Dept. Conserv. and Devel. Bull. 50, 203 p.

Long, L. E., 1962, Isotopic age study, Dutchess County, N.Y.: Geol. Soc. America Bull., v. 73, p. 997–1006.

Long, L. E., Cobb, J. C., and Kulp, J. L., 1959, Isotopic ages on some igneous and metamorphic rocks in the vicinity of New York City: New York Acad. Sci. Annals, v. 80, p. 1140–1147.

Long, L. E., and Kulp, J. L., 1962, Isotopic age study of the metamorphic history of the Manhattan and Reading Prongs: Geol. Soc. America Bull., v. 73, p. 969–996.

Lowe, K. E., 1950, Storm King Granite at Bear Moun-tain, N. Y.: Geol. Soc. America Bull., v. 61, p. 137–190.

MacLachlan, D. B., 1964, Major nappe in the Susquehanna River (Pennsylvania) region of the Great Valley [abs.]: Geol. Soc. America Spec. Paper 82, p. 126.

—— 1967, Structure and stratigraphy of the limestones and dolomites of Dauphin County, Pennsylvania: Pennsylvania Geol. Survey, 4th ser., Bull. G 44, 168 p.

McBride, E. E., 1962, Flysch and associated beds of the Martinsburg Formation (Ordovician), central Appalachians: Jour. Sed. Petrology, v. 32, p. 39–91.

Maxwell, J. C., 1962, Origin of slaty and fracture cleavage in the Delaware Water Gap area, New Jersey and Pennsylvania, in Petrologic studies—A volume in honor of A. F. Buddington: New York, Geol. Soc. America, p. 281–311.

Miller, B. L., 1925, Mineral resources of the Allentown quadrangle, Pennsylvania: Pennsylvania Geol. Survey, 4th ser., Atlas 206, 195 p.

—— 1934, Limestones of Pennsylvania: Pennsylvania Geol. Survey, 4th ser., Bull. M 20, 729 p.

Miller, B. L., 1944, Specific data on the so called "Reading Overthrust:" Geol. Soc. America Bull., v. 55, p. 254.

Miller, B. L., and Fraser, D. M., 1936, Comment on Highlands near Reading, Pennsylvania; an erosion remnant of a great overthrust sheet, by G. W. Stose and A. I. Jonas: Geol. Soc. America Bull., v. 46, p. 2031–2038, Reply by G. W. Stose and A. I. Jonas, p. 2038–2040.

Miller, B. L., and others, 1939, Northampton County, Pennsylvania: Pennsylvania Geol. Survey, 4th ser., Bull. C 48, 496 p.

Miller, B. L., and others, 1941, Lehigh County, Pennsylvania: Pennsylvania Geol. Survey, 4th ser., Bull. C 39, 492 p.

Miller, R. L., 1937a, Stratigraphy of the Jacksonburg Limestone: Geol. Soc. America Bull., v. 48, p. 1687–1718.

—— 1937b, Martinsburg Limestone, in eastern Pennsylvania: Geol. Soc. America Bull., v. 48, p. 93–112, 2034–2037.

Minard, J. P., 1959, The geology of Peapack-Ralston Valley in north-central New Jersey: unpublished M. S. thesis, Rutgers Univ. (New Brunswick), 103 p.

Montgomery, Arthur, 1955, Paragenesis of the serpentine-talc deposits near Easton, Pennsylvania: Pennsylvania Acad. Sci. Proc., v. 29, p. 203–215.

—— 1957, Three occurrences of high-thorium uraninite near Easton, Pennsylvania: Am. Mineralogist, v. 42, p. 804–820.

Nockolds, S. R., 1954, Average chemical compositions of some igneous rocks: Geol. Soc. America Bull., v. 65, p. 1007–1032.

O'Neill, B. J., Jr., 1964, Atlas of Pennsylvania's mineral resources, Pt. 1, Limestones and dolomites of Pennsylvania: Pennsylvania Geol. Survey, 4th ser., Bull. M 50, 40 p.

Peck, F. B., 1905, The talc deposits of Phillipsburg, N. J., and Easton, Pa.: New Jersey Geol. Survey Ann. Rept. 1904, p. 161–185.

Prouty, C. E., 1959, The Annville, Myerstown, and Hershey Formations of Pennsylvania: Pennsylvania Geol. Survey, 4th ser., Bull. G 31, 47 p.

Reed, J. C., Jr., 1955, Catoctin Formation near Luray, Virginia: Geol. Soc. America Bull., v. 66, no. 7, p. 871–896.

Reed, J. C., Jr., Bryant, Bruce, and Myers, W. B., 1969, The Brevard Zone in the Grandfather Mountain area of North Carolina—a reinterpretation of its movement direction and relation to thrusting, in Fisher, G. W., and others, eds., Studies in Appalachian geology—the central and southern Appalachians: New York, John Wiley and Sons (in press).

Rogers, H. D., 1858, The geology of Pennsylvania, a government survey: Philadelphia, 2 N, 1045 p.

Sando, W. J., 1957, Beekmantown Group (Lower Ordovician) of Maryland: Geol. Soc. America Mem. 68, 161 p.

Sherwood, W. C., 1964, Structure of the Jacksonburg Formation in Northampton and Lehigh Counties, Pa.: Pennsylvania Geol. Survey, 4th ser. General Geology Rept. G 45, 64 p.

Sims, P. K., 1958, Geology and magnetite deposits of Dover district, Morris County, New Jersey: U. S. Geol. Survey Prof. Paper 287, 162 p.

Sims, P. K., and Leonard, B. F., 1952, Geology of the Andover mining district, Sussex County, New Jersey: New Jersey Dept. Conserv. Devel. Bull. 62, 46 p.

Sitter, L. V. de, 1964, Structural geology, 2d ed.: New York, McGraw-Hill, 551 p.

Socolow, A. A., 1961, Geologic interpretation of certain aeromagnetic maps of Lancaster, Berks, and Lebanon Counties, Pennsylvania: Pennsylvania Geol. Survey Inf. Circ. 41, 19 p.

Spencer, A. C., and others, 1908, Franklin Furnace, N. J.: U. S. Geol. Survey Geol. Atlas, Folio 161.

Stose, G. W., 1930, Unconformity at the base of the Silurian in southeastern Pennsylvania: Geol. Soc. America Bull., v. 41, p. 629–657.

——— 1946, The Taconic sequence in Pennsylvania: Am. Jour. Sci., v. 244, no. 10, p. 655–696.

Stose, G. W., and Jonas, A. I., 1935, Highland near Reading, Pa.: An erosion remnant of a great overthrust sheet: Geol. Soc. America Bull., v. 48, p. 757–779.

——— 1939, Discussion of the geology of the Reading Hills, Pa.: Am. Jour. Sci., v. 237, no. 4, p. 281–286.

Tex, E. den, 1963, A commentary on the correlation of metamorphism and deformation in space and time: Geologie Mijnb., v. 42, p. 170–176.

Thurston, W. R., 1951, Geology and mineralogy of the manganese deposit at Clinton Point, N. J.: New Jersey Dept. Conserv. and Econ. Devel., Geol. Ser. Bull. 60, 22 p.

Tilton, G. R., Wetherill, G. W., Davis, G. L., and Bass, M. N., 1960, 1000-million-year-old minerals from the eastern United States and Canada: Jour. Geophys. Research, v. 65, p. 4173–4179.

Turner, F. J., and Verhoogen, John, 1960, Igneous and metamorphic petrology: 2nd ed., New York, N. Y., McGraw-Hill Book Co., 964 p.

U. S. Geological Survey, 1962, Geological Survey Research 1962: U. S. Geol. Survey Prof. Paper 450-A, 257 p.

——— 1964, Geological Survey Research 1964: U. S. Geol. Survey Prof. Paper 501-A, 367 p.

——— 1966, Geological Survey Research 1966: U. S. Geol. Survey Prof. Paper 550-A, 385 p.

——— 1967, Geological Survey Research 1967: U. S. Geol. Survey Prof. Paper 575-A, 377 p.

Weller, Stuart, 1900, Description of Cambrian trilobites from New Jersey: New Jersey Geol. Survey Ann. Rept. for 1899, p. 47–53.

——— 1903, Report on Paleontology, Vol. III, The Paleozoic faunas: New Jersey Geol. Survey, 462 p.

Wells, R. C., Fairchild, J. G., and Ross, C. S., 1933, Thorianite from Easton, Pennsylvania: Am. Jour. Sci., v. 26, p. 45–54.

Wherry, E. T., 1909, The early Paleozoics of the Lehigh Valley district, Pennsylvania: Science, n.s., v. 30, p. 416.

Willard, Bradford, 1943, Ordovician clastic sedimentary rocks in Pennsylvania: Geol. Soc. America Bull., v. 54, p. 1067–1122.

——— 1958, Lower Ordovician, Lehigh and Delaware Valleys: Pennsylvania Acad. Sci. Proc., v. 32, p. 177–183.

Willard, Bradford, and Cleaves, A. B., 1939, Ordovician-Silurian relations in Pennsylvania: Geol. Soc. America Bull., v. 50, p. 1165–1198.

Willard, Bradford, and others, 1959, Geology and mineral resources of Bucks County, Pa.: Pennsylvania Geol. Survey, 4th ser., Bull. C 9, 243 p.

Wood, G. H., Jr., and Carter, M. D., Tectonics of the Anthracite region, Pennsylvania: Pennsylvania Geol. Survey, 4th ser., Bull., in press.

Woodward, H. P., 1964, Central Appalachian tectonics and the deep basin: Am. Assoc. Petroleum Geologists Bull., v. 48, p. 338–356.

Woollard, G. P., 1943, Geologic correlation of aerial gravitational and magnetic studies in New Jersey and vicinity: Geol. Soc. America Bull., v. 54, p. 791–818.

Zadnik, V. E., 1960, Petrography of the Upper Cambrian dolomites of Warren County, N. J.: U. S. Geol. Survey open-file report, 96 p.

Zen, E-an, 1967, Time and space relationships of the Taconic allochthon and autochthon: Geol. Soc. America Spec. Paper 97, 107 p.

GEOLOGY OF THE VALLEY AND RIDGE PROVINCE BETWEEN DELAWARE WATER GAP AND LEHIGH GAP, PENNSYLVANIA[1]

JACK B. EPSTEIN[2] AND ANITA G. EPSTEIN
U. S. Geological Survey, Beltsville, Md., and Washington, D. C.

ABSTRACT

The stratigraphy, structure, glacial geology, geomorphology, and economic geology of Middle Ordovician through Middle Devonian rocks and overlying surficial deposits in the area between the Delaware and Lehigh Rivers, eastern Pennsylvania, are so interrelated that these topics should not be considered separately. Pre-orogenic basin deepening resulted in deposition of thick Middle and Upper Ordovician flysch deposits (Martinsburg). A thick clastic wedge of Silurian rocks (Shawangunk, Bloomsburg) was deposited in a continental to marginal marine zone by rivers flowing from the highlands produced by the Taconic orogeny. The source lands were lowered in time, so that the streams, which initially had steep gradients and low sinuosity during most of Shawangunk time, meandered on a low coastal plain of alluviation in Bloomsburg time. This molasse littoral was followed by a complex series of infratidal to deep neritic carbonates, orthoquartzites, siltstones, and shales, deposited during a transgressive-regressive-transgressive cycle in Late Silurian to early Middle Devonian time.

Because of dissimilarities in gross lithology, the flysch, molasse, and shelf sequences were deformed differently during the later Paleozoic orogeny. Each sequence, a lithotectonic unit, was deformed semi-independently of rocks above and below and is presumably separated from these rocks by décollements. Type and amplitude of folds were controlled by lithic variations within each lithotectonic unit. The most obvious structural features in the area, including the prominent slaty cleavage in the Martinsburg, are assigned to the Appalachian orogeny; the profound effects of the Taconic orogeny are more elusive.

Erosion of the folded heterogeneous stratigraphic units produced the linear ridges and valleys characteristic of the Valley and Ridge province. It is concluded that the location of wind and water gaps that cut through these ridges was structurally controlled. This is contrary to the widely held concept of superposition.

The topography had a pronounced effect on the direction of travel of the Wisconsin glacier, as well as on the manner of its retreat and on the deposits that were formed. Older drift, possibly Illinoian in age, is found southwest of the limit of Wisconsin glaciation.

Economic products in the area are sand and gravel, slate, building stone, and clay. They are the result of processes of sedimentation and, in part, deep weathering; structural conditions have controlled the location of some of them.

The road log, which includes nine stop descriptions, covers most of the major aspects of the geology in the area.

INTRODUCTION

The field conference in the Valley and Ridge province of easternmost Pennsylvania will demonstrate the interrelationship of the stratigraphy,

---

[1] Publication authorized by the Director, U. S. Geological Survey.

[2] Work done in cooperation with the Pennsylvania Geological Survey.

structure, geomorphology, glacial geology, and economic geology of Middle Ordovician through part of the Middle Devonian strata and overlying surficial deposits in the area between the Delaware and Lehigh Rivers (fig. 1). The area was the scene of a field trip that we led in 1967 (Field Conference of Pennsylvania Geologists, 1967), and much of the material herein is modified from that earlier report. Part of the work is still in progress, and some of the conclusions presented in this report should be regarded as tentative.

Data for this field trip have been gathered by (1) nearly completed bedrock and surficial mapping of all or parts of eight 7½-minute quadrangles by J. B. Epstein, U. S. Geological Survey (Stroudsburg, East Stroudsburg, Portland, Saylorsburg, Wind Gap, Kunkletown, Palmerton, and Lehighton—the last five being done in cooperation with the Pennsylvania Geological Survey), and (2) stratigraphic studies of Upper Silurian and Lower Devonian rocks of northeastern Pennsylvania, New Jersey, and southeasternmost New York by Epstein and others (1967). In addition, we have benefited from discussions and exchange of data with several co-workers, including A. A. Drake, Jr., U. S. Geological Survey, and J. D. Glaeser, J. F. Wietrzychowski, and W. D. Sevon of the Pennsylvania Geological Survey. D. W. Kohls supplied the road-log description of The New Jersey Zinc Company operations at Palmerton, Pa. G. F. Smith has included a summary of the clay operations of Universal Atlas Cement at Kunkletown, Pa.

We will examine the probable facies relationships and correlations between rocks in the eastern and western sections of the area and consider the environmental conditions under which these strata were deposited. The rock stratigraphic units have been grouped into four successive lithofacies, each of which has a different style of deformation. Folds produced in these rocks are disharmonic, and it is believed that each rock sequence is set off from sequences above and below by a décollement, or zone of detachment. Movement was northwest into the Appalachian basin, primarily by gravitational sliding, aided by directed tectonic forces. The contact between the Shawangunk Conglomerate of Silurian age and Martinsburg Formation of Ordovician age, believed by most workers to be an angular unconformity, may also be one zone of detachment. The apparent deformational effects of the Middle to Late Ordovician Taconic orogeny may not be as intense in this area as some geologists believe.

Erosion of the folded heterogeneous bedrock in this part of the Appalachians has produced linear ridges and valleys. This topography had a profound effect on the manner of retreat of the Pleistocene glaciers. We shall demonstrate that during the Wisconsin Glaciation a moraine-dammed proglacial lake existed in the Saylorsburg, Pa., area. Other glacial features will be examined, including older, possibly Illinoian, drift in the western part of the area. Knowledge of the underlying bedrock lithology and drift petrography aids in the interpretation of the direction of Pleistocene glacier movement. The evolution of the modern Delaware-Lehigh River drainage system is closely tied to bedrock structure. Detailed mapping has shown that the many wind and water gaps in this area are structurally controlled, placing doubt upon the hypothesis of regional superposition. Finally, an intimate knowledge of all these facets of geology

Figure 1. Panoramic view from the limestone quarries in Cherry Ridge at Bossardsville, Pa., showing some of the variety of geologic features to be seen on the field trip. View from southeast (right) to northwest (left). 1, Blue Mountain underlain by moderately to steeply northwest-dipping Silurian clastic rocks of the Shawangunk Conglomerate and Bloomsburg Red Beds. The Bloomsburg forms a synclinal valley (Popular Valley, 2) and an anticlinal ridge (Kemmererville anticline, 3). The scenic Cherry Valley is choked with Wisconsin kame deposits, 4. The irregular serrated crest of Godfrey Ridge, 5, reflects complex folding of heterogeneous Upper Silurian through Middle Devonian rocks. The folds in Godfrey Ridge (lithotectonic unit 3) are of a smaller scale than those in the underlying Shawangunk and Bloomsburg (lithotectonic unit 2).

yields insight into the origin of the economic deposits of the area.

Figure 2 is an index map of the field-trip area showing trip routes and quadrangle coverage. A generalized geologic map of the field-trip area showing the limit of Wisconsin drift is presented as figure 3. Figure 4 is a physiographic diagram of the field-trip area showing stop locations and the position of the Wisconsin terminal moraine.

On the first day, we will examine a delta deposited in proglacial Lake Sciota and then see representative exposures of the four lithotectonic units in the western part of the field-trip area. These stops will include a slate quarry in the Martinsburg Formation; the Martinsburg-Shawangunk contact and effects of the Taconic and Appalachian orogenies at Lehigh Gap; sedimentary features of Silurian rocks and their interpretation at Lehigh Gap; and tightly folded Upper Silurian-Lower Devonian sedimentary rock saprolites near Kunkletown.

On the second day, we will stay in the Delaware Water Gap area and compare this eastern sequence with the sequence seen on the first day. At Bossardsville, rocks structurally similar to those at

Kunkletown, but unweathered, will be seen. This and other stops will deal with sedimentary structures and interpretation of Wisconsin glacial deposits, interpretation of the environments of deposition of Silurian through Middle Devonian sediments, structure and stratigraphy of Godfrey Ridge, and finally the origin of wind and water gaps from atop Kittatinny Mountain, and a discussion of the overall tectonics of the area.

EARLY HISTORY

The area between the Delaware and Lehigh Water Gaps seen on this field conference has had an interesting geologic history, the subject of this report. The human history of the region is also of interest; some of the Indian lore and legend is given here, most of which is taken from Brodhead (1870), Miller and others (1939), Stokes (n.d.), and Woodward (1944).

The Lenni-Lenape Indians, better known as the Delawares, a branch of the Algonkians, occupied the Delaware River drainage basin from New York to Delaware. Their ancestors, according to legend, had great antiquity, having originated west of the

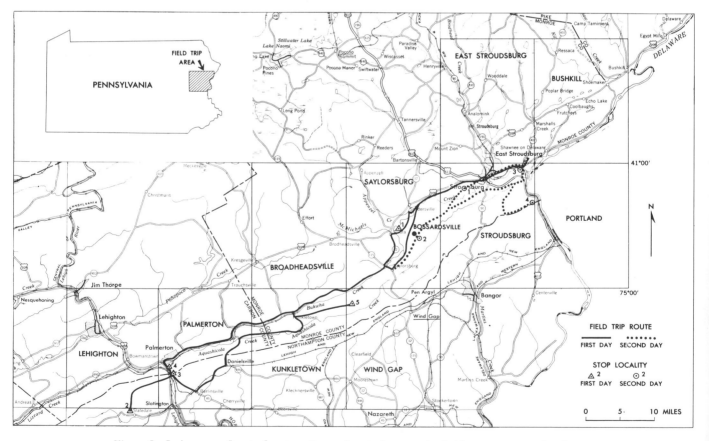

Figure 2. Index map of part of eastern Pennsylvania showing the field-trip route, stop localities, and 7½-minute-quadrangle coverage.

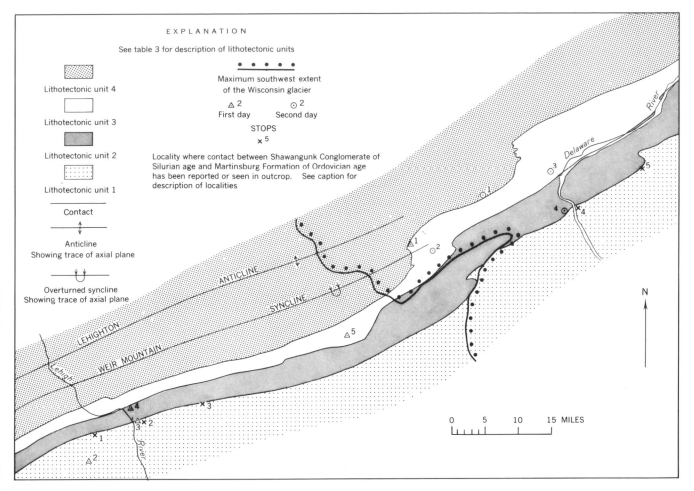

Figure 3.   Generalized geologic map of the Delaware River–Lehigh River area, Pennsylvania–New Jersey.   Modified from Gray and others, 1960.   Martinsburg-Shawangunk contact localities: 1, Northeast extension of the Pennsylvania Turnpike tunnel through Blue Mountain; 2, Lehigh Gap; 3, water tunnel for city of Bethlehem, at Little Gap; 4, Delaware Water Gap; 5, Yards Creek hydroelectric project near Blairstown, N. J.

"Mamoesi Sipi" (Mississippi).   The Wolf clan of the Delawares occupied the area covered by this field conference and had their main village, Minisink, in New Jersey just north of Delaware Water Gap.   The Wolf clan and finally the entire Lenni-Lenape nation were brought into subjugation through war and stratagem by the Iroquois or Five-Nation Confederacy who occupied lands to the north, at about the time of arrival of William Penn in Pennsylvania.   Penn began buying land from the Delawares in 1682.   Most of the land south of the "Endless Mountains" (Blue and Kittatinny Mountains) (specifically the colony of "Penn's Woods,") was deeded to the white man by the warring Indians in the peace treaty of 1736.

The white man perpetrated all manner of trickery against the Indian.   The crowning injustice came with the notorious "Walking Purchase" in 1737.   In 1734, Thomas Penn claimed to have found a copy of a deed dated 1686 from the Delaware chiefs to his father, William Penn, giving title to certain lands north of Blue and Kittatinny Mountains, the area occupied by the Minsi tribe of the Wolf clan.   The tract of land was outlined, in part, by the distance a man could walk in a day and a half.   In 1737, the Delaware Indians, believing the walk would be an ordinary day and a half journey with pauses for rest and food, finally allowed the land to be measured.   Penn, however, hired trained athletes and had them make preliminary trips to find the shortest and fastest routes.   A reward was offered to the runner who covered the greatest distance.   The "walk" began on September 19 near Wrightstown in Bucks County.   By finish time, all runners had fallen exhausted except one who reached Broad Mountain in Carbon County, about 65 miles north of the starting point and about twice as far as the Indians thought a man could

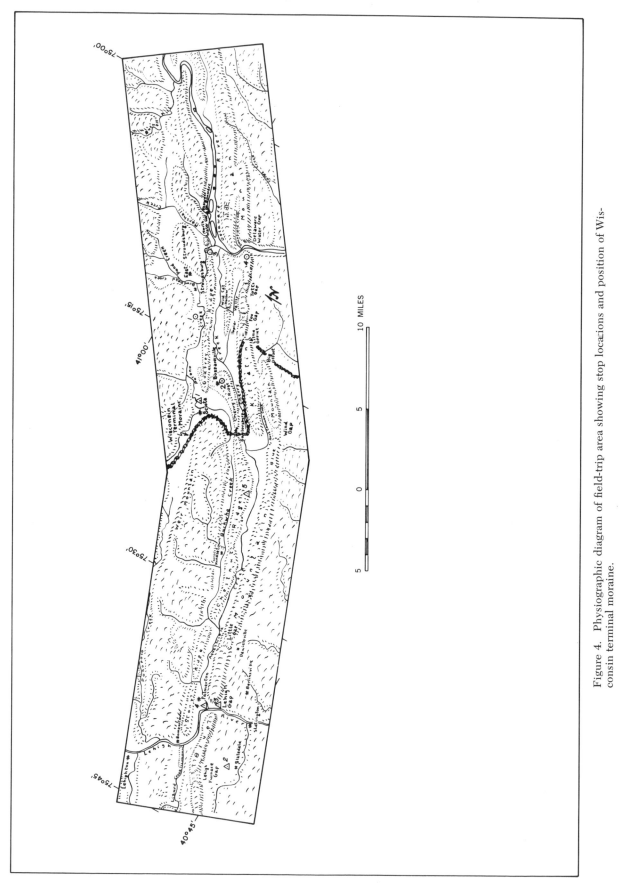

Figure 4. Physiographic diagram of field-trip area showing stop locations and position of Wisconsin terminal moraine.

walk in the time allotted. According to the deed, all land east of the line of walk to the Delaware River would be assigned to Penn, but Penn drew the line in a northeasterly direction, thus acquiring an extra 25 miles of territory.

The Indians felt they were cheated and refused to vacate their lands. They were finally expelled by the Pennsylvania authorities with the help of the Six-Nation Indians. The Minsi tribe migrated to the western part of the State and then to Canada, where their identity with the Lenni-Lenape nation was lost. It is not surprising that many of these Indians joined with the French against the English during the French and Indian War in 1754.

During the French and Indian War, Chief Teedyuscong led the Delawares in many bloody massacres against the white settlers. In 1756, he joined an alliance with the Shawnees, Nanticokes, and Mohicans, and the Governor of Pennsylvania declared war on the Indians. A chain of forts was built from the Delaware to the Susquehanna. Bounties were offered for Indian scalps. Many atrocities were committed on both sides. In 1756, Tom Quick, Sr., was killed and scalped by Indians near Milford, Pa. His son escaped and vowed vengeance. Near the end of his life, Tom Quick, Jr., regarded by his friends as a hero but outlawed by the government, claimed that he had killed 99 Indians and regretted that he did not make it an even hundred.

In 1762 peace was negotiated, Teedyuscong claiming that the reason he went to war was because the Delawares were cheated by the Walking Purchase. He and many other Delawares were murdered by a war party of Six-Nation Indians in 1763. After 1770, the Lenni-Lenape journeyed to Ohio and then finally to Oklahoma where they settled permanently and lost their identity as the Lenni-Lenape nation.

The earliest geological investigation in the area occurred in the middle of the 17th century under Dutch auspices. The Dutch discovered a copper deposit in the Bloomsburg Red Beds on the north side of Kittatinny Mountain about 6 miles northeast of Delaware Water Gap village in Pahaquarry Township, N. J. They believed the Pahaquarry mine to be a New World Kupferschiefer and built a road ("The Old Mine Road") to Esopus (now Kingston), N. Y., a distance of more than 100 miles, to transport the ore. This was the longest and best road in the colonies at the time. Unfortunately, the mine was unprofitable, and attempts at working it have been unsuccessful to the present day. The Dutch were on friendly terms with the Lenni-Lenape Indians of the Minisink valley and their

chief, Wissinoming. The chief had a daughter, the princess Winona. The head of the Dutch exploration group, Hendrich Van Allen, fell in love with the princess, and they often climbed the mountains around Delaware Water Gap together. When the Dutch surrendered New Amsterdam (New York) to the English in 1664, Van Allen was ordered to abandon mining operations. His position was such that he could not take Winona with him to Holland as his wife. When he reported the news to Winona, she bade farewell and ran to the edge of a precipice. He made an attempt to stop her, and in doing so they both fell to their deaths. The part of the cliff overlooking Delaware River where this episode is supposed to have taken place is now known as "Lover's Leap" and is located along the Appalachian trail below stop 4 of the second day of this field trip.

STRATIGRAPHY

The stratigraphic units with which this field trip is concerned are described in tables 1 and 2. Table 1 describes the units in the eastern part of the field-trip area; table 2 deals with the western part. Figures 5 and 6 show the lateral and vertical relationships of these rock units, from northeast to southwest, in the field-trip area.

GEOGRAPHIC AND TECTONIC ENVIRONMENTS OF DEPOSITION

The Middle Ordovician through part of the Middle Devonian rocks with which this field trip is concerned compose virtually one complete flysch—molasse—carbonate-orthoquartzite shelf succession.

From Cambrian through Late Ordovician time, the Appalachian basin in the Delaware Water Gap-Lehigh Gap region gradually deepened and the influx of terrigenous sediment gradually increased. The rock record indicates that the area slowly changed from a supratidal-shallow neritic shelf (Hardyston Quartzite, Leithsville Formation, Allentown Dolomite, and Beekmantown Group: dolomite, limestone, arkose, and orthoquartzite) to a deeper neritic-flysch basin (fossiliferous calcarenite and calcilutite of the Jacksonburg Limestone to thick accumulations of rhythmically bedded graywacke and slate of the Martinsburg Formation). Jacksonburg time marked the onset of flysch-type sedimentation, which culminated in Martinsburg time with rhythmic graded sequences of graywacke turbidites and dark-gray slate. The Martinsburg sediments reflect Taconic orogenic activity, which reached its peak with the emergence of the area sometime in the Late Ordovician.

System	Series	Lithotectonic Unit	Formation	Member	Description	Thickness (feet)	Localities where unit will be seen on field trip and economic use
DEVONIAN	Middle	4	Mahantango Formation		Medium-dark-gray siltstone and silty shale; contains Centerfield coral beds, calcareous siltstone biostromes having abundant horn corals.	about 2,000	
			Marcellus Shale		Dark-gray silty shale. Basal 200 feet consists of interbedded medium-dark-gray calcareous silty shale, silty shale and slightly calcareous shale. Sparsely fossiliferous. Upper contact gradational.	800	Quarried for fill.
			Buttermilk Falls Limestone of Willard (1938)	Upper member	Medium-gray limestone and argillaceous limestone containing nodules and beds of dark-gray chert. Fossiliferous and burrowed. Upper contact probably abrupt.	150	Quarried for road metal.
				Middle member	Medium-gray calcareous argillite containing lenses of light-medium-gray limestone. Fossiliferous.	40	Quarried for road metal.
				Lower member	Medium-dark-gray calcareous siltstone and argillite and argillaceous limestone containing beds, pods, and lenses of dark-gray chert. Fossiliferous; large crinoid columnals abundant in lower half; burrowed.	80	Quarried for road metal.
	Lower		Schoharie Formation		Medium- to medium-dark-gray massive argillaceous calcareous siltstone. Fossiliferous. Burrowed, *Taonurus* in lower half, vertical burrows in upper half. Upper contact gradational.	100	Quarried for road metal.
			Esopus Formation		Medium- to dark-gray silty shale and shaly to finely arenaceous siltstone. Poorly fossiliferous. Burrowed, *Taonurus*. Upper contact gradational.	180	Quarried for road metal. Stop 3, 2d day.
		Oriskany Group	Ridgeley Sandstone		Medium- to light-gray fine- to coarse-grained calcareous sandstone and quartz-pebble conglomerate containing minor beds and lenses of siltstone, arenaceous fine-grained limestone, and dark-gray chert. Fossiliferous. Upper contact abrupt.	14–16	Quarried for sand and road metal. Stop 3, 2d day.
			Shriver Chert		Medium-dark-gray siliceous calcareous shale and siltstone and beds, lenses, and pods of dark-gray chert, and minor calcareous coarse-grained and conglomeratic sandstone. Fossiliferous, burrowed. Upper contact abrupt.	54–85	Quarried for road metal. Stop 3, 2d day.
		Helderberg Group	Port Ewen Shale		Lower 60 feet consist of medium-dark-gray poorly fossiliferous, irregularly laminated calcareous shale and siltstone; upper 90 feet consist of medium-dark-gray fossiliferous, burrowed, irregularly bedded calcareous siltstone and shale. Upper contact gradational.	150	Stop 3, 2d day.
			Minisink Limestone		Dark- to medium-gray fine-grained argillaceous fossiliferous limestone. Upper contact abrupt.	14	Stop 3, 2d day.
			New Scotland Formation	Maskenozha Member	Dark-gray silty calcareous laminated fossiliferous shale containing beds and lenses of medium-gray fine-grained argillaceous fossiliferous limestone. Upper contact abrupt or gradational.	43–48	Stop 3, 2d day.
				Flatbrookville Member	Medium-dark-gray silty and calcareous fossiliferous shale containing beds and lenses of medium-gray fine-grained argillaceous very fossiliferous limestone and dark-gray chert. Upper contact gradational.	20–33	Stop 3, 2d day.

TABLE 1—(Continued)

System	Series	Lithotectonic Unit	Formation	Member	Description	Thickness (feet)	Localities where unit will be seen on field trip and economic use
DEVONIAN	Lower	3 / Helderberg Group	Coeymans Formation	Stormville Member	Lenses of medium-gray fine- to coarse-grained biogenic limestone, fine- to medium-grained arenaceous limestone, fine- to coarse-grained crossbedded and planar-bedded calcareous limonitic sandstone and quartz-pebble conglomerate. Contains nodules and lenses of dark-gray chert. Fossiliferous. Upper contact abrupt.	2–26	Stop 3, 2d day.
				Shawnee Island Member	Nonbiohermal facies, medium-gray fine- to medium-grained argillaceous and arenaceous slightly limonitic irregularly bedded fossiliferous and burrowed limestone. Upper 10–25 feet contain dark-gray chert lenses and nodules. Biohermal facies, medium-light-gray to light-pinkish-gray very coarse grained unbedded to crudely bedded biogenic limestone. Upper contact abrupt.	35–60	Stop 3, 2d day.
				Peters Valley Member	Medium-gray arenaceous limestone to light-medium-gray fine- to coarse-grained pebbly calcareous sandstone. Crossbedded; fossiliferous. Upper contact gradational.	3–9	Stop 3, 2d day.
				Depue Limestone Member	Medium- to dark-gray fine- to medium-grained arenaceous and argillaceous, partly straticulate fossiliferous limestone. Upper contact abrupt or gradational.	13–17	Stop 3, 2d day.
SILURIAN AND DEVONIAN	Upper Silurian and Lower Devonian		Rondout Formation	Mashipacong Member	Medium-dark-gray to light-gray shale, calcareous shale, and very fine grained to medium-grained argillaceous limestone. Mud cracks, cut-and-fill structure. Upper contact abrupt or gradational.	8–11	Stop 3, 2d day.
				Whiteport Dolomite Member	Dark- to medium-dark-gray mud-cracked laminated dolomite. Upper contact abrupt or gradational.	5–9	Formerly quarried for agricultural lime. Stop 3, 2d day.
				Duttonville Member	Dark- to medium-gray calcareous shale and argillaceous limestone. Mud-crack intervals and biostromal limestone beds. Upper contact abrupt.	12–17	Stop 3, 2d day.
SILURIAN	Upper		Decker Formation	Wallpack Center Member	Lenses and beds of quartz-pebble conglomerate, calcareous sandstone and siltstone, argillaceous and arenaceous fine- to coarse-grained limestone, and dolomite. Crossbedded to planar-bedded and flaser-bedded, fossiliferous. Upper contact abrupt.	84	Stop 5, 1st day; stops 2 and 3, 2d day.
			Bossardville Limestone		Lower 30–40 feet dark- to medium-gray, laminated, very fine grained to fine-grained argillaceous limestone locally containing deep mud cracks (as much as 20 feet deep) and the upper 60–70 feet dark-gray very fine grained laminated limestone. Poorly fossiliferous, leperditiid ostracodes. Upper contact gradational.	100	Quarried for bituminous concrete, road metal, and high calcium limestone. Stops 2 and 3, 2d day.
			Poxono Island Formation of White (1882)		Calcareous and dolomitic, light-olive-gray to green laminated fissile to nonfissile shale, olive-green dolomite, sandstone, and siltstone. Upper contact gradational.	about 700	Stop 2, 2d day.

TABLE 1—(Concluded)

System	Series	Lithotectonic Unit	Formation	Member	Description	Thickness (feet)	Localities where unit will be seen on field trip and economic use
SILURIAN	Middle and Upper	2	Bloomsburg Red Beds		Red, green, and gray conglomeratic sandstone, sandstone, siltstone, and shale partly in upward-fining sequences. Fifty percent sandstone, 45 percent shale and siltstone, and 5 percent conglomerate. Crossbedded and laminated, mud cracks; cut-and-fill structure; scattered ferroan dolomite concretions. Fish scales locally. Partly burrowed. Upper contact gradational.	about 1,500	Quarried for fill and road metal. Stop 4, 2d day.
	Lower and Middle		Shawangunk Conglomerate	Upper quartzite-conglomerate member	Medium-gray to medium-dark-gray fine- to coarse-grained planar-bedded to crossbedded, limonitic, pyritic, conglomeratic (quartz and argillite pebbles as much as 2 in. long) evenly to unevenly bedded quartzite containing about 2 percent dark-gray argillite. Rare ferroan dolomite beds and calcite and ferroan dolomite concretions. Upper contact gradational.	816	Stop 4, 2d day.
				Middle quartzite-argillite member	Medium-light-gray to medium-dark-gray and light-olive-gray fine- to coarse-grained, laminated to planar-bedded, and crossbedded, evenly to unevenly bedded, rippled and flaser-bedded, limonitic, pyritic, and graphitic (rare), quartzose sandstone containing burrows and trails and rare ball-and-pillow structure, interbedded with medium-dark-gray to dark-gray laminated, flaser-bedded, evenly to unevenly bedded, burrowed siltstone and shale containing rare fossils (eurypterids and Dipleurozoa (reported) and *Lingula*). Rare beds of collophane (carbonate fluorapatite), siderite, and chlorite nodules and quartz pebbles.	273	Stop 4, 2d day.
				Lower quartzite-conglomerate member	Light-gray to medium-dark-gray and light-olive-gray medium- to coarse-grained crossbedded and planar-bedded, limonitic, pyritic, unevenly to moderately evenly bedded thin- to thick-bedded quartzite, conglomeratic quartzite, and quartz-, chert-, and shale-pebble conglomerate (quartz pebbles as much as 2 in. long). About 7 percent dark-gray irregularly bedded laminated locally mud-cracked argillite. Upper contact gradational.	300	Stop 4, 2d day.
ORDOVICIAN	Middle and Upper	1	Martinsburg Formation	Pen Argyl Member	Dark-gray to grayish-black, thick- to thin-bedded, evenly bedded claystone slate, rhythmically intercalated with beds of quartzose slate or subgraywacke and carbonaceous slate. Upper contact abrupt and unconformable.	3,000–6,000	Quarried for slate, lightweight aggregate, and fill. Stops 2 and 3, 1st day.
				Ramseyburg Member	Medium- to dark-gray claystone slate alternating with beds of light- to medium-gray, thin- to thick-bedded graywacke and graywacke siltstone. Graywacke composes about 20–30 percent of unit. Upper contact with Pen Argyl Member gradational.	about 2,800	Upper part quarried for slate in Bangor-Pen Argyl area.
				Bushkill Member	Dark- to medium-gray thin-bedded claystone slate containing thin interbeds of quartzose and graywacke siltstone and carbonaceous slate. Upper contact gradational.	about 4,000	Formerly quarried for slate.

Silurian through lower Middle Devonian litho-facies (molasse followed by orthoquartzite-carbonate shelf sedimentation) and their inferred environments of deposition are shown in figure 7.

The Silurian clastic sequence in the field-trip area (Shawangunk-Bloomsburg interval) consists of three distinct units: a basal unit of sandstone and conglomerate; a middle unit of interbedded very fine to medium-grained sandstone, siltstone, and minor shale; and an upper redbed unit of fining-upward fine- to coarse-grained sandstone, siltstone, and silty shale. These compose a clastic wedge derived from sourcelands to the southeast that rose during the Taconic orogeny. The major environments of deposition given for the Shawangunk-Bloomsburg interval in this report are in agreement with those of Smith (1967) who made an independent study of this interval from New York to central Pennsylvania.

Initially, uplift was very rapid, judging from the unconformable nature of the Shawangunk-Martinsburg contact and the coarse texture of basal Shawangunk rocks. The lower and middle units of this clastic sequence fine upwards through more than

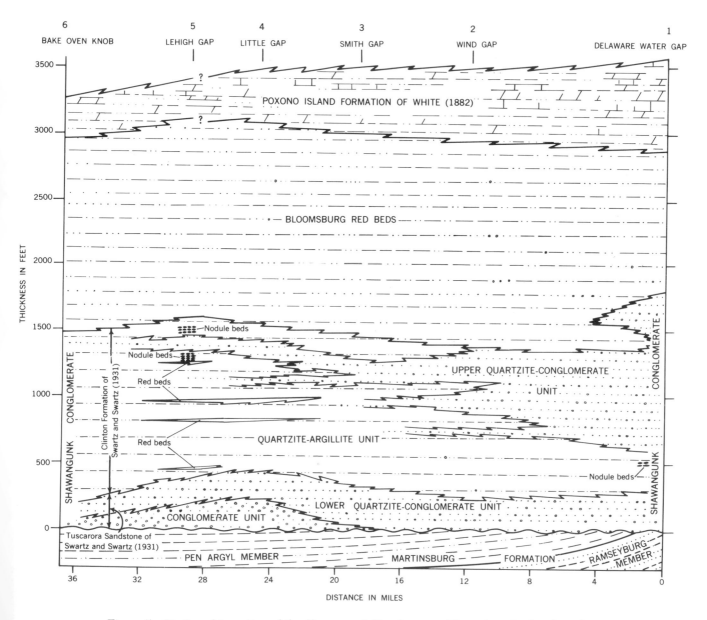

Figure 5. Stratigraphic section of the Shawangunk Conglomerate, Bloomsburg Red Beds, and Poxono Island Formation of White (1882) from Delaware Water Gap to Bake Oven Knob, Pa. See figure 6 for location of sections. Modified from Field Conference of Pennsylvania Geologists (1967).

System	Series	Lithotectonic Unit	Formation	Member	Description	Thickness (feet)	Localities where unit will be seen on field trip and economic use
DEVONIAN	Middle	4	Marcellus Shale		Predominantly dark-gray carbonaceous silty shale. Dark-gray to grayish-black shale in basal 30 feet (Union Springs Member) overlain by 40 feet of calcareous medium-dark-gray to dark-gray shale and shaly fine-grained limestone (Stony Hollow Member of Cooper. 1941).	about 600	Dug for fill. Stop 5, 1st day.
			Buttermilk Falls Limestone of Willard (1938)		Deeply weathered cherty argillaceous limestone. Medium-gray argillaceous limestone and dark-gray chert where unweathered.	40–80	Clay used in white cement. Stop 5, 1st day.
			Palmerton Sandstone of Swartz (1939)		Weathered, very massive, partly conglomeratic, coarse- to very coarse grained sandstone. A few siltstone and fine-grained sandstone beds near base. Rare molds of crinoid columnals and favositids. Medium-gray to medium-dark-gray where unweathered. Upper contact abrupt.	0–about 110	Quarried for sand. Stop 5, 1st day.
	Lower		Schoharie and Esopus Formations, undivided		Predominantly deeply weathered fossiliferous, partly cherty siltstone and fine-grained sandstone. Upper part, in fresh exposure, is dark-gray slightly calcareous siltstone. Upper contact abrupt.	48–about 110	Stop 5, 1st day.
			Oriskany Group — Ridgeley Sandstone		Beds and lenses of weathered crossbedded and planar-bedded to laminated quartz-pebble conglomerate and conglomeratic sandstone, locally containing intervals of abundant brachiopod molds. Upper contact abrupt.	25–45	Quarried for sand. Stop 5, 1st day.
			Oriskany Group — Shriver Chert		Beds and lenses of deeply weathered chert, sandstone, and conglomerate. Brachiopod molds locally abundant. Upper contact abrupt.	25–45	Stop 5, 1st day.
		3	Helderberg Group — New Scotland Formation		Deeply weathered fossiliferous chert and silty and sandy shale. Upper contact abrupt.	0(?)–55	Clay used in white cement. Stop 5, 1st day.
			Helderberg Group — Coeymans Formation	Stormville Member	Weathered very fine grained to conglomeratic, partly fossiliferous crossbedded to planar-bedded sandstone containing interbeds of arenaceous, probably calcareous shale and siltstone. Upper contact abrupt.	0–45	Stop 5, 1st day.
SILURIAN & DEVONIAN	U. Silurian L. Devonian		Andreas Red Beds of Swartz and Swartz (1941)		Red, green, and gray partly pebbly sandstone, siltstone, and shale. Northeasternmost outcrop approximately 1.5 miles west of Bowmanstown, Pa. Overlain either by New Scotland Formation or Shriver Chert.	0–about 50	
SILURIAN	Upper		Decker Formation		Lenses and beds of medium-gray and medium-light-gray shale and siltstone, dark-yellow-orange to greenish-gray very fine grained to conglomeratic slightly calcareous fossiliferous sandstone, and medium-gray partly arenaceous fine- to coarse-grained fossiliferous limestone. Some beds of calcareous dolomite and dolomitic limestone. Upper part has red interbeds.	°	Stop 5, 1st day.

TABLE 2.—(Continued)

System	Series	Lithotectonic Unit	Formation	Member	Description	Thickness (feet)	Localities where unit will be seen on field trip and economic use
SILURIAN	Upper		Bossardville Limestone		°	°	
			Poxono Island Formation of White (1882)		°	about 300–400	
	Middle and Upper	2	Bloomsburg Red Beds		Fining-upward sequences of very fine to coarse-grained sandstone with scattered clay galls, siltstone, silty shale, and shale. Mainly red with fewer green and gray beds than in eastern area. Crossbedded, laminated to thick-bedded; mud-cracked locally, partly burrowed; scattered ferroan dolomite concretions; ball-and-pillow structure near base. Upper contact probably gradational.	about 1,500	Stop 4, 1st day.
	Lower and Middle		Shawangunk Conglomerate	Quartzite-argillite Member	Medium-gray to greenish-gray very fine to medium-grained thin- to thick-bedded quartzite containing a few intervals of dark-grayish-red-purple fine-grained partly silty and shaly hematitic quartzite. Grayish-orange to light-olive-gray shale, silty shale, and siltstone with some grayish-red-purple beds. Medium-gray to greenish-gray very fine to medium-grained, thin- to thick-bedded, laminated to planar-bedded and crossbedded, evenly to unevenly bedded, rippled and flaser-bedded sandstone containing burrows and a few intervals of dark-grayish-red-purple fine-grained partly argillaceous hematitic sandstone interbedded and interlaminated with grayish-orange to light-olive-gray and minor grayish-red-purple, evenly to unevenly laminated, flaser-bedded, burrowed shale, silty shale, and siltstone. Upper half contains scattered beds and lenses of collophane (carbonate fluorapatite), siderite, and chlorite nodules, quartz pebbles, siltstone and shale intraclasts, and *Lingula* fragments. Upper contact gradational.	about 1,225	Stop 4, 1st day.
				Lower quartzite-conglomerate Member	Medium-light-gray and greenish-gray, medium- to very coarse grained partly conglomatic (with quartz pebbles as much as 2 in. long and clay galls as much as 7 in. across); crossbedded to planar-bedded quartzite and very light gray to medium-light-gray and greenish-gray predominantly medium grained quartzite. Upper contact gradational.	200–300	Stop 3, 1st day.
				Conglomerate Member	Greenish-gray to medium-gray crossbedded and planar-bedded medium- to thick-bedded quartz-, chert-, quartzite-, argillite-pebble conglomerate (quartz pebbles as much as 6 in. long), with clay galls up to 8 in. across. Medium-dark-gray medium- to very coarse grained conglomeratic quartzite and a few beds of greenish-gray argillite. Upper contact gradational.	0–225	Stop 3, 1st day.
ORDOVICIAN	Middle and Upper	1	Martinsburg Formation	Pen Argyl Member	°	°	° Stops 2 and 3, 1st day.
				Ramseyburg Member	°	°	°
				Bushkill Member	°	°	°

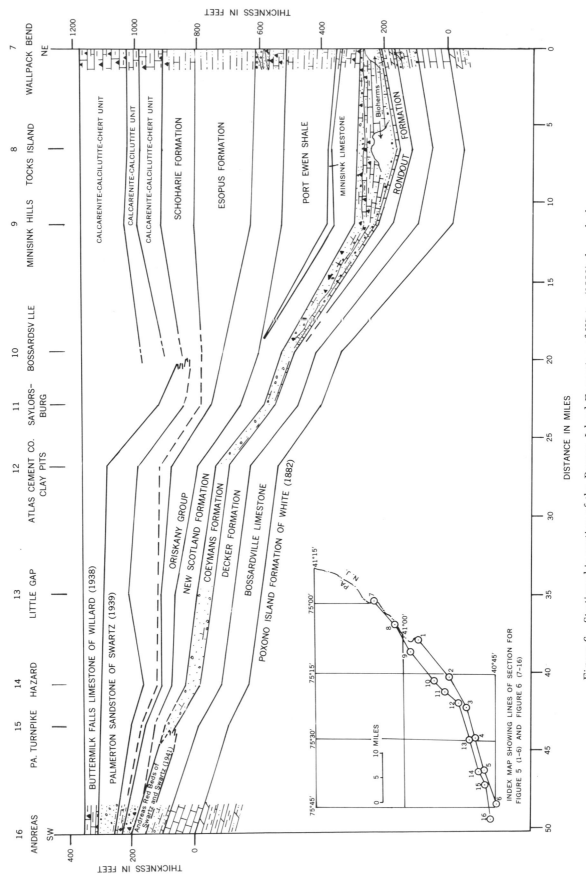

Figure 6.   Stratigraphic section of the Poxono Island Formation of White (1882) through the Buttermilk Falls Limestone of Willard (1938) from Wallpack Bend to Andreas, Pa. Modified from Field Conference of Pennsylvania Geologists (1967).

1,000 feet of rock representing a transgressive phase and lowering of the source area concomitant with basin subsidence. The initial deposits of this sequence, the conglomerate unit of the Shawangunk Conglomerate at Lehigh Gap (possibly also once present southeast of Delaware Water Gap, but removed by erosion; fig. 5) are characterized by rapidly alternating conglomerate beds, medium- to very coarse grained sandstone, and very minor argillite (fig. 8A). The conglomerates, containing pebbles as much as 6 inches long, are indistinctly to planar bedded and the sandstones are planar bedded to crossbedded, indicative of relatively rapid flow (upper lower flow regime to upper flow regime; Simons and Richardson, 1962; Fahnestock and Haushild, 1962), and probably are channel and point bar deposits of streams. Grain size is variable and pebbles are well rounded to subangular, indicative of a fluviatile environment (Sames, 1966). The sandstones are generally immature (high muscovite-chlorite matrix). Paleocurrent trends are unidirectional to the northwest, also indicating a fluvial environment. Bed forms and sedimentary structures indicate deposition by streams which had great competency and steep gradients and are characteristic of streams of low sinuosity (braided). Properties of braided streams have been discussed in many papers; e.g., Doeglas (1962) and Allen (1965a). The lack of deep channel-fill deposits with relief greater than 5 feet also indicates that the streams were not confined to a single channel but occupied many anastomosing channels that shifted position continuously. The nearly complete absence of fine siltstone and shale in the conglomerate unit is also characteristic of braided stream deposits. Braided streams shift so rapidly across the fluvial plain that they are able to remove fine-grained sediment. Some of the shale and silt was incorporated in the coarser sediment as flattened mud balls (fig. 8B).

Glaciofluvial deposits of Pleistocene age in easternmost Pennsylvania are very similar to the conglomerates and sandstones in the Shawangunk. The glacial sediments were undoubtedly deposited by braided streams with high velocity and coarse load. Fahnestock (1963) described similar deposits of the White River, Mount Rainier, Wash. Some of the glacial deposits in the field-trip area are described at stop 1, first day, and alternate stop 1, second day.

The overlying quartzite-conglomerate unit is similar to the conglomerate unit, except that pebbles more than 2 inches long are absent (fig. 8C) and argillites, probably representing overbank or backwater deposits, are more abundant. Mud cracks in at least one argillite bed (fig. 8D) show that these deposits were subject to subaerial exposure. The sedimentation units show rapid superposition, as in the conglomerate unit and also represent deposition by braided streams. The finer grain size suggests that the source highlands were lowered or eroded back at this time.

The long linear outcrop belt of the lower part of the Shawangunk suggests that it was deposited on a coastal plain of alluviation with a linear source to the southeast. The pebbles in the conglomerate unit are as much as 6 inches long near Little Gap, Pa., suggesting that the Fall Line could not have been far to the southeast, perhaps in the area of the Reading Prong (see Yeakel, 1962). The maturity of the pebbles (quartz, chert, and quartzite) and heavy minerals (preliminary studies indicate a preponderance of zircon, tourmaline, and rutile, with rare hornblende) and lack of appreciable feldspar, suggest a sedimentary source. The absence of kaolinite in the clay-size fraction of the Shawangunk excludes a deeply weathered crystalline source. The rocks now exposed in the Prong (Drake, this volume) or rocks in the Great Valley that probably were eroded from the Prong area, apparently could not have been source beds. Could it be that rocks dissimilar to those now found to the southeast were the source for the Shawangunk, emplaced in their position in thrust sheets or nappes but long since eroded?

The fluviatile deposits of the lower Shawangunk grade rapidly up into interbedded shale, siltstone, and sandstone of the quartzite-argillite unit. This unit is generally evenly bedded (fig. 9) and is interpreted to have been deposited in a complex transitional (continental-marine) environment. The following subenvironments are probably represented: tidal flat, barrier bar or beach, estuary, tidal channel and gulley, lagoon, and shallow subtidal (prodelta?). In addition, other environments characteristic of deltas (Shepard and Lankford, 1959; Coleman and Gagliano, 1965; Bernard and LeBlanc, 1965; Donaldson, 1966) may be represented; their differentiation, however, will require more detailed work. The unit is thickest and most complex at Lehigh Gap (stop 4, second day).

Deposits believed to be tidal flat in origin consist of irregularly interlaminated to finely interbedded burrowed shale, siltstone, and sandstone. Flaser bedding (ripple lensing; fig. 10A), a characteristic of sediments reworked by tidal currents (Häntzschel, 1939), is common. Flaser-bedded sands, however, can be deposited in deeper waters

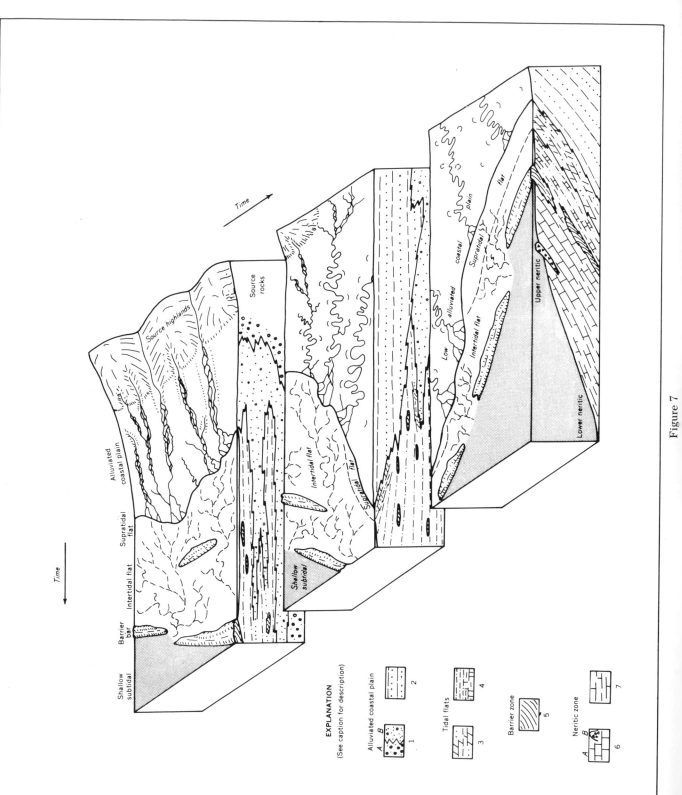

Figure 7

during storms (Reineck, oral commun., 1968), so that some of the beds in the quartzite-argillite unit could be subtidal.

Crossbedded sandstones containing mud clasts may be tidal channel or tidal gulley deposits (van Straaten, 1961). Thicker crossbedded sandstones, as much as 8 feet thick, may be fluviatile deposits that prograded out onto the flats (upper part of the unit, fig. 5). Even-bedded to planar-laminated mature sandstones (orthoquartzites) with primary current lineation (fig. 10B) are believed to be beach or barrier-bar deposits. Some ripple-topped sandstones may be sand-flat deposits (Evans, 1965) (fig. 10C). Rare silty sandstone beds have ball-and-pillow structure (subaqueous soft-rock slump features possibly produced during storms; fig. 10D).

Red burrow-mottled siltstones may have formed as supratidal deposits in the oxidizing zone above mean tide level (McKee, 1957; Nichols, 1962);

these may be the landward equivalents of inter-tidal-flat green beds.

Finely laminated shales and siltstones, some red, may be high tidal flat beds, deposited from suspension (van Straaten, 1961), and in part bound by algae (Evans, 1965). Evans, however, described abundant mud cracks in high tidal-flat laminated sediments. Mud cracks were not observed in the laminated beds of the quartzite-argillite unit, which, therefore, may have been deposited in quiet bodies of water (lagoons) in the tidal flat area (McKee, 1957).

The quartzite-argillite unit thins towards Delaware Water Gap where irregularly bedded siltstone and sandstone contain phosphate nodules (fig. 10A). These beds also contain mud flasers such as have been found in recent tidal channel deposits (Reineck and Singh, 1967). Other sandstone and siltstone beds in this interval, at Delaware Water

---

Figure 7. Generalized block diagram showing sedimentary environments and major lithofacies in northeasternmost Pennsylvania from Silurian through early Middle Devonian time.

Alluviated coastal plain
  1, Streams of high gradient, coarse load, low sinuosity (braided).
    A, Bedforms in upper flow regime (planar beds, antidunes) and upper lower flow regime (dunes); chiefly conglomerates and sandstones; conglomerate unit of Shawangunk Conglomerate.
    B, Bedforms in lower upper flow regime (planar beds) and upper lower flow regime (dunes); chiefly conglomeratic quartzites and quartzites; lower and upper quartzite-conglomerate units of Shawangunk Conglomerate.
  2, Streams of low gradient, medium load and fine flood-plain deposits, high sinuosity (meandering). Bedforms in lower flow regime (dunes and ripples). Sandstones, siltstones, and shales. Bloomsburg Red Beds and Poxono Island Formation of White (1882).

Tidal flats
  3, Supratidal flat, may include tidal creeks. Dolomite, limestone, shale, sandstone. Laminated (algal), massive, mud-cracked, intraclasts, sparse fauna. Quartzite-argillite unit of Shawangunk Conglomerate and Poxono Island, Rondout, and Decker Formations.
  4, Intertidal flat, may include tidal channel and gulley, estuary, lagoon, beach. Shale, siltstone, sandstone (equivalent limestone in areas of low terrigenous influx), minor nodules and oolites of collophane, siderite, and chlorite. Irregularly bedded and laminated, graded, rippled, flaser bedded, cut-and-fill, ball-and-pillow structure, burrowed, restricted fauna (*Lingula*, Dipleurozoa, eurypterids, in noncarbonates; abundant leperditiid ostracodes in carbonates). Quartzite-argillite unit of Shawangunk Conglomerate, Poxono Island Formation, Bossardville Limestone, and Rondout and Coeymans Formations.

Barrier zone
  5, Offshore bar and beach. Conglomerate, sandstone, and siltstone, and equivalent limestone. Foreshore laminations, crossbedding, scouring, abundant wave-tossed shell debris, textural maturity. Quartzite-argillite unit of Shawangunk Conglomerate, Decker and Coeymans Formations, Oriskany Group, Palmerton Sandstone of Swartz (1939).

Neritic zone
  6, A, Upper neritic. Cherty, calcisiltite and calcareous siltstone, unevenly bedded (burrowed), diverse fauna.
    B, Bioherm and biostromal bank. Biolithite and biogenic limestone containing minor terrigenous sediment. Decker, Coeymans, and New Scotland Formations, Minisink Limestone, Port Ewen Shale, Oriskany Group, Schoharie Formation, Buttermilk Falls Limestone of Willard (1938).
  7, Lower neritic. Shaly siltstone, calcareous shaly siltstone, and calcareous shale. Laminated, burrowed, skeletal debris less abundant than in upper neritic. New Scotland Formation, Port Ewen Shale, Esopus and Schoharie Formations.

Figure 8. Sedimentary structures in the lower part of the Shawangunk Conglomerate. A, Planar-bedded conglomerate with well-rounded to subangular quartz, chert, and quartzite pebbles as much as 3 inches long and crossbedded medium- to coarse-grained quartzite; conglomerate unit of Shawangunk Conglomerate, Lehigh Gap (stop 3, first day). B, Shale pebbles (flattened mud balls) more than 7 inches long in quartzite float boulder; lower quartzite-conglomerate unit of Shawangunk Conglomerate, Lehigh Gap. C, Crossbedded conglomeratic quartzite containing quartz pebbles not more than 1 inch long; lower quartzite-conglomerate unit, Delaware Water Gap, N. J. D, Mud-cracked shaly siltstone, probably an overbank or backfill deposit; lower quartzite-conglomerate unit, Delaware Water Gap, N. J. Underside of bed exposed about 50 feet above U. S. Interstate 80 at south entrance of gap.

Gap, may be fluviatile or estuarine in origin. Most estuarine deposits, however, are difficult to distinguish from fluviatile deposits (Land and Hoyt, 1967). Associated with these beds are vertical burrows as much as 10 inches long.

The quartzite-argillite unit is poorly fossiliferous. Eurypterids (Clarke and Ruedemann, 1912) and specimens of Dipleurozoa are believed by Johnson and Fox (1968) to indicate deposition in an intertidal environment. *Arthrophycus,* a burrow, was seen in a loose quartzite block at Dela-

ware Water Gap. *Lingula,* shown by Craig (1952) to favor very shallow warm brackish waters was found in several beds containing collophane nodules at Lehigh Gap.

Parts of the quartzite-argillite unit are iron rich (hematite, goethite, lepidocrosite, iron-rich chlorite, and siderite). Siderite and some chlorite, as well as collophane (calcium fluorapatite) occur as nodules (fig. 11A, B). The iron is believed by Hunter (1960) to have been derived from an extensively weathered source area, transported by

Figure 9.   Even-bedded sandstones, siltstones, and shales; quartzite-argillite unit of Shawangunk Conglomerate, Delaware Water Gap, N. J.

streams in solution or as colloids, and deposited close to shore.

Green pleochroic chlorite occurs as intergranular fillings and replacements, nodules and oolites, fossil replacements, and possibly as pellet replacement. Preliminary X-ray studies indicate that the chlorite may be iron-rich chamosite. Chamosite has been reported in equivalent facies elsewhere (Alling, 1947; Hunter, 1960; and Schoen, 1964).

The thin rim of chlorite on some siderite nodules, the flattening of some nodules over clastic grains, the protrusion of quartz grains from the nodules into surrounding matrix, the occurrence of coarser quartz in the nodular beds, and other soft-rock features (fig. 11B) indicate that the nodules formed as early diagenetic replacements in an agitated environment (oolitic, in part). Shell fragments of *Lingula* suggest that the environment where these nodules formed was in shallow warm waters, possibly brackish, at depths probably less than 60 feet (Craig, 1952). Porrenga (1967a, b) has shown that chamosite is presently found in tropical waters at depths generally less than 60 meters and temperatures greater than 20°C, whereas glauconite forms in deeper and (or) colder waters. Hunter (1960) showed a northeast-trending belt of chamosite, succeeded by a glauconite belt to the north-

west, in the Appalachian basin in Clinton time. The association of siderite with chamosite is considered by Hunter to indicate deposition in quiet waters that were deeper than the agitated oxygenated waters closer to shore, in which hematite was deposited.

Phosphate nodules occur with the chlorite and siderite nodules and with shell fragments of *Lingula*. *Lingula* must have been the main intrabasin source for the phosphate (see caption, fig. 11). Solution and reprecipitation of scattered phosphatic shell material at or just below the water-sediment interface probably produced phosphatized silt and mud locally (Clarke, 1924). (Many of the phosphate nodules have the same soft-rock deformation features described above for the siderite nodules. Most of the phosphate nodules at Lehigh Gap are clearly phosphatized siltstone.) Occasionally, waves and tidal currents churned these muds and also comminutized some of the phosphatic shell material. Many of the *Lingula*-nodule beds occur as heavy lag concentrates in well-sorted sandstones having foreshore laminations (beach deposits; fig. 11A). There can be little doubt that the phosphate nodules are of very shallow water origin (see Bushinski, 1964, and references cited therein).

The association of chlorite, siderite, and phosphate nodules and oolites, all of apparently very early diagenetic or penecontemporaneous origin, requires a unique set of physical, chemical, and biological conditions for their formation. Their limited occurrence in a few scattered thin beds (fig. 5), and the absence of concentrated phosphate in nearby beds (fig. 11A) suggests that the restricted conditions for their deposition occurred during limited periods of time and possibly in limited areas. These nodules may have formed in a subtidal environment or possibly in the reducing environment of lagoons (Sheldon, 1964), and then were reworked and concentrated by storm waves (fig. 11A).

The sediments in the quartzite-argillite unit were carried by northwest-flowing streams and deposited in the transitional environment. Much of the sediment was then reworked by waves, tides, and longshore currents. Small deltas may have formed where the amount of material entering the basin exceeded the removal capacity of ocean waves and currents. The upper quartzite-conglomerate unit, similar to the rocks in the lower Shawangunk, is a fluvial equivalent of the upper part of the quartzite-argillite unit.

Regression probably began during middle to late

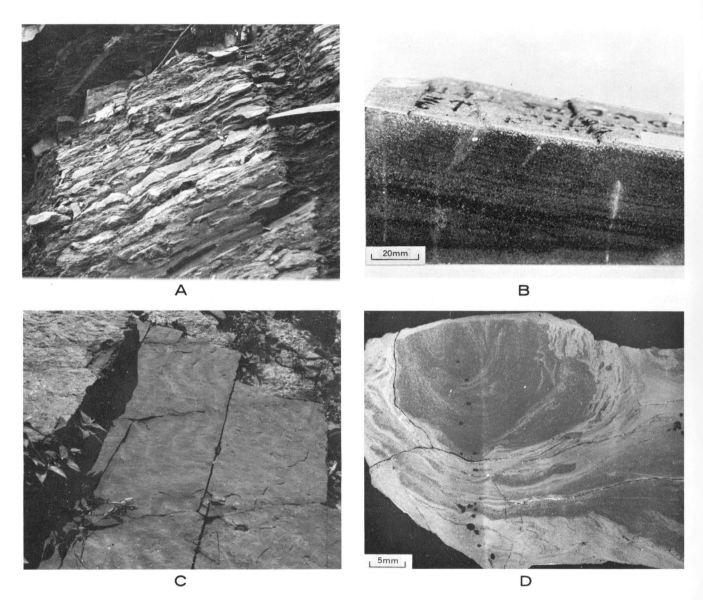

Figure 10.   Sedimentary structures in the quartzite-argillite unit, Delaware Water Gap, N. J. A, Flaser-bedded fine-grained sandstone irregularly interbedded with burrowed muddy silt-stone.   Many of the sands have load cast sole markings.   Black phosphate nodules that weather white, as much as 1½ inches long, are scattered throughout.   The nodules are angular and some have a partial mud rind.   These beds were probably reworked by storm waves while the sedi-ment was still soft.   Hammer head gives scale.   B, Laminated, well-sorted fine-grained sand-stone, believed to be a beach or barrier-bar deposit.   Primary current lineation (not apparent in photograph) is present on the upper bedding surface and parallels the dip of the laminae. Small ridges on bedding surface are not ripples, but offsets of small soft-rock faults (arrow) that paralleled the ancient strand line.   C, Asymmetric ripples in fine-grained sandstone.   Ripples have wavelengths of about 3½ inches and amplitudes of about ½ inch.   D, Convolutions (ball-and-pillow structure) in very fine grained sandstone and shale, due to soft-rock slumping.   Note overturned and refolded flow fold at top.   Negative print of acetate peel.

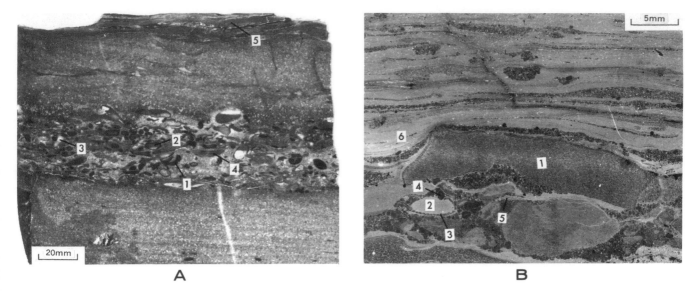

A   **B**

Figure 11. Nodular beds in the quartzite-argillite unit of the Shawangunk Conglomerate, Lehigh Gap, Pa. A, Nodules and oolites of carbonate fluorapatite (1), rimming iron-rich shale pebbles (2), quartz pebbles as much as 8 mm long (3), and shale pebbles (4) that were probably ripped up from substrate similar to the shale at 5. This nodular bed lies between laminated well-sorted medium-grained quartzite. The shale and phosphate nodules may have been deposited in the reducing environment of a lagoon and washed onto a sandstone bar during a storm. Chemical analysis of the shale (5) shows it to contain 8.2 percent iron expressed as Fe_2O_3, and 0.19 percent P_2O_5. Photograph of polished section. B, Nodules of siltstone partly replaced by collophane (1), siderite (2) rimmed with chlorite (3) and penetrated by quartz grain (4) (showing that the nodule was soft at time of deposition), and fragments of *Lingula* (5). Note lapping of irregularly laminated shale and very fine grained sandstone on large siltstone clast (6). Serial sections of the siltstone clasts show that they are extremely irregular in shape, suggesting deposition while the siltstone was wet and plastic. Negative print of acetate peel.

quartzite-argillite time and culminated in a return to fluviatile conditions with the deposition of the Bloomsburg Red Beds.

The Bloomsburg, as seen at Lehigh Gap, stop 4, first day, consists predominantly of red sandstone, siltstone, and shale which occur in poorly to well-defined upward-fining cycles (fig. 12). These cycles are as much as 14 feet thick and consist of, from bottom to top:

(1) Very fine to coarse-grained, large-scale cross-bedded to planar-bedded sandstone that may contain red shale clasts as much as 3 inches long. The basal contact is a sharp erosion surface of very low relief that cuts into underlying finer grained beds. The lower few inches of the sandstone may be gray. The abruptness of the basal contact may be accentuated by bedding slippage described at stop 4, first day.

(2) Finely interbedded and irregularly interlaminated siltstone and very fine grained sandstone containing small-scale ripples. Mud clasts are occasionally found in the sandstone beds.

(3) Shaly siltstone that is extensively burrowed, may be indistinctly mud cracked and contain scattered irregular dark-yellowish-orange concre-

tions averaging about 1 inch in length. These were suggested by N. D. Smith (oral commun., 1968) to be calcareous nodules characteristic of overbank deposits (Bernard and Major, 1963; Moody-Stuart, 1966). X-ray diffraction analysis and staining tech-

Figure 12. Alternating crossbedded sandstone, irregularly laminated sandstone and siltstone, and poorly bedded shaly siltstone in fining-upward cycles, Bloomsburg Red Beds, Lehigh Gap, Pa.

niques show that the nodules in the Bloomsburg are ferroan dolomite.

The cycle described above is idealized and does not represent all rocks in the Bloomsburg.

These fining-upward cycles are generally considered to have been deposited by meandering streams (e.g., Allen, 1965b). The coarse basal sands were deposited in stream channels and point bars through lateral accretion as the stream meandered. The mud clasts were derived from bank caving. Large-scale crossbedding and planar-bedding are indicative of the upper lower and lower upper flow regimes, respectively. These basal beds grade up into laminated sandstone and siltstone containing small-scale crossbedding that indicates decreasing flow regime. These are interpreted to be levee and crevasse-splay deposits (Allen, 1965b). Next in succession are the fine overbank or flood-plain deposits that accumulated by vertical accretion. Burrowing animals obliterated stratification in many of these beds in this low-energy tranquil environment. Mud cracks show that subaerial exposure was common. Evaporation at the surface may have caused the precipitation of calcareous concretions, but, as mentioned previously, the concretions in the

Bloomsburg are ferroan dolomite. It is possible, for reasons mentioned below, that the Bloomsburg rivers were not far above the strandline, and the dolomite concretions may have formed in a high tidal flat environment by a mechanism similar to the one described for the dolomitization of calcium carbonate later in this section.

Fossils are very rare in the Bloomsburg. Fish have been found in the lower part of the Bloomsburg near Delaware Water Gap which Beerbower and Hart (1959) believe lived in a fluvial or lagoonal environment.

Cyclicity in the Bloomsburg is readily explained by vertical and lateral accretion from migrating streams concomitant with basin sinking. Each fining-upward cycle represents superposition of beds of successively lower flow regime.

The lower contact of the Bloomsburg with the quartzite-argillite unit is transitional through 117 feet of red and green rocks, and marks the change from a marginal marine-continental environment to a low alluviated coastal plain. The greenish rocks are well laminated with flaser bedding and primary slump structures (fig. 13). The color is believed to have been derived at the site of de-

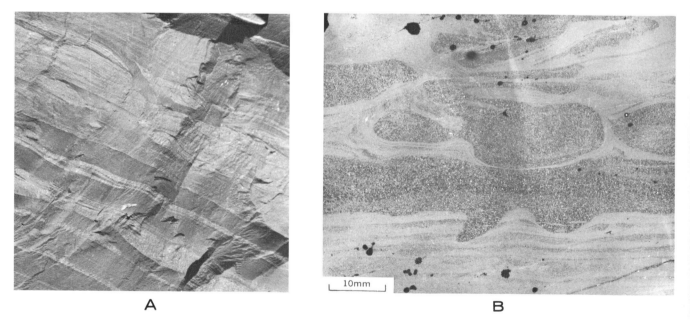

10mm

A   B

Figure 13. Interlaminated greenish-gray siltstone and sandstone, 80 feet above base of Bloomsburg Red Beds, Lehigh Gap (stop 4, first day). A, Interlaminated medium-greenish-gray muddy siltstone (dark) and partly flaser-bedded light-greenish-gray fine-grained sandstone (light). Partly convoluted sandstone near top shown in detail in B. Tip of hammer at top gives scale. B, Disrupted bedding due to slump of incoherent mud and sand. Sand was disrupted during movement and separated by pinching and swelling into irregular discoid lumps as much as 2 inches long and ½ inch thick (sedimentary boudinage). Some of the sand lumps have complex flowage folds and a few are nearly completely surrounded by clay laminae indicating that the mud adhered to the sand as the sand rolled on the mud substrate; the sand was more mobile than the mud. Downward projections in lower half are probably burrows. Negative print of acetate peel.

position (Walker, 1967a, b) and the red-green transition represents the oxidation-reduction zone at mean tide level (Nichols, 1962; McKee, 1957). A similar situation prevails at the top of the Bloomsburg where supratidal-flat deposits of the Poxono Island Formation are interbedded with red beds in a transition interval a few hundred feet thick. Clearly, the land surface in Bloomsburg time was at low altitudes. Encroachment of broad tidal flats during Poxono Island time marked a general transgressive episode that continued with minor regression through part of Early Devonian time.

Following the end of molasse sedimentation in Late Silurian (Bloomsburg) time, until Oriskany time, deposits in this part of the Appalachian basin are of the marine shelf carbonate-orthoquartzite facies. During the first half of this interval, the rate of sedimentation and basin sinking was about equal and the area was maintained near sea level, so that rocks from the Poxono Island through lower Coeymans (fig. 6) were deposited in tidal flats (supra- and intertidal) and barrier bars or biostromal banks (intertidal and shallow subtidal). Repetition of lithofacies characterizes this stratigraphic sequence. These lithofacies are indicative of broad subenvironments within this major shallow marine transitional complex (fig. 7).

Sediments that accumulated in supratidal flats are characterized by: (1) laminations of organo (algal)-sedimentary origin; (2) slightly quartzose dominantly very fine to fine-grained laminated and very thin bedded to massive dolomite, limestone, and interlaminated dolomite and limestone (figs. 14 A–C, 62A, B); (3) very restricted fauna (mainly leperditiid ostracodes); (4) mud cracks (figs. 14A–C, 62A, B). Supratidal sediments occur in the Poxono Island Formation of White (1882) (fig. 14A), Bossardville Limestone (fig. 14B), Decker Formation (fig. 63A), and Rondout Formation (fig. 14C).

The intertidal zone can be subdivided into intertidal flats and barrier bars or beaches. Intertidal flat sediments are characterized by: (1) generally graded and rippled laminae to thin beds of slightly quartzose very fine to medium-grained limestone; (2) cut-and-fill structure, small-scale crossbedding, and intraclasts; (3) abundant ostracodes (mainly leperditiids); (4) scattered beds containing a diverse marine fauna believed to be storm-tossed skeletal debris on a tidal flat behind a barrier bar; (5) rare mud cracks. Barrier-bar deposits of the intertidal zone are distinguished by: (1) quartzose fine- to coarse-grained limestone and calcareous sandstone and conglomerate containing foreshore laminations and large-scale crossbedding, (2) cut-and-fill struc-

ture and intraclasts, (3) abundant skeletal debris of a variety of marine organisms, (4) scattered burrows. Intertidal sediments occur in the Poxono Island Formation, Bossardville Limestone (fig. 15A), Decker Formation (fig. 63), Rondout Formation, and lower Coeymans Formation (fig. 15B).

Sediments that may have accumulated in the shallow subtidal zone of a barrier bar are characterized by: (1) laminated to bedded fine- to coarse-grained quartzose limestone and calcareous siltstone to sandstone containing scattered to abundant burrows, (2) flaser bedding (ripple lenses), (3) abundant and diverse marine fauna. Shallow subtidal sediments occur in the Poxono Island Formation, uppermost Bossardville Limestone, Decker Formation (fig. 63), Rondout Formation, and Coeymans Formation.

From Coeymans to Oriskany time, the area underwent a major transgressive-regressive episode. Northeast of Bossardsville, Pa., the basin steadily deepened from Coeymans through most of Port Ewen time (sinking exceeded sediment accumulation); southwest of Bossardsville, however, basin deepening did not begin until New Scotland time, and the deeper neritic phase was of shorter duration. Regression began in late Port Ewen time and culminated with post-Oriskany emergence.

In northeastern Pennsylvania (from Wallpack Bend to Andreas, fig. 6), the Lower Devonian outcrop belt crosses several Early Devonian depositional zones. In Coeymans time, the area of Andreas, Pa., was probably emergent (Andreas Red Beds of Swartz and Swartz, 1941, in part possibly a nonmarine clastic correlative of the Coeymans). The shoreline lay somewhere near Hazard, Pa., where sediment-bearing streams entered the basin and marine currents spread the quartzose silt, sand, and gravel out into the basin as well as along the shore as far northeast as the New York–New Jersey border (Epstein and others, 1967), as evidenced in the Stormville and Peters Valley Members of the Coeymans Formation (table 1). In middle Coeymans (Shawnee Island) time, the following zones existed: (1) near Kunkletown, Pa. (Atlas Cement Company clay pits, fig. 6; stop 5, first day), the barrier-bar zone (Stormville Member of Coeymans Formation, fig. 16 and table 1: lenses and beds of planar- and crossbedded calcareous quartz siltstone, sandstone, and conglomerate); (2) near Minisink Hills, Pa. (stop 3, second day), the biostromal-bank zone, an area slightly more distant from shore and (or) clastic sediment influx (Shawnee Island Member of Coeymans Formation, fig. 17 and table 1: burrowed biogenic limestone containing as much as

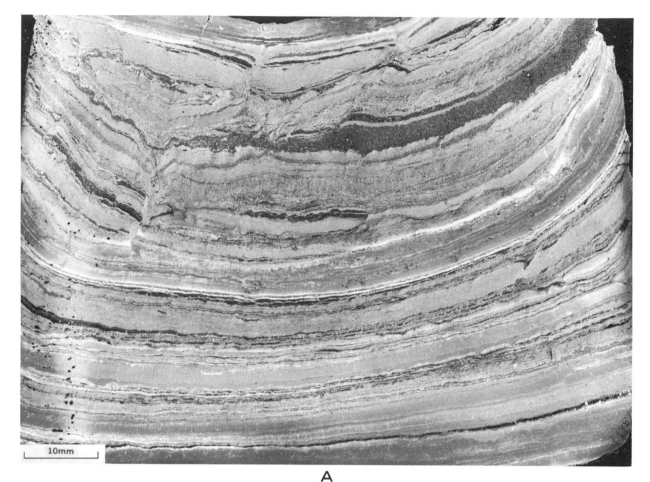

A

B   C

Figure 14. Mud cracks in supratidal deposits. A, Uppermost beds of Poxono Island Formation of White (1882) at Bossardsville, Pa. (stop 2, second day). Curled mud-crack column consists of laminated to very thin bedded slightly quartzose calcite (white to light gray), ferroan dolomite (medium gray), and dolomite (dark gray) mudstone. Note intraclasts of dolomudstone, small healed mudcrack, and thickening of some laminae toward center of mud-crack "microbasin." Cleavage is conspicuous in upper right. Specimen stained with Alizarin Red S and potassium ferricyanide. Negative print of acetate peel. B, Lowermost beds of Bossardville Limestone in creekbed at Shawnee on Delaware, Pa. Compressed mud-crack columns of laminated limestone are approximately 20 feet long. Bedding is near vertical. C, Whiteport Dolomite Member of Rondout Formation in roadcut on north side of Cherry Valley Road (mile 21.2, road log, second day). Mud-crack columns in dolomudstone are about 7 feet long.

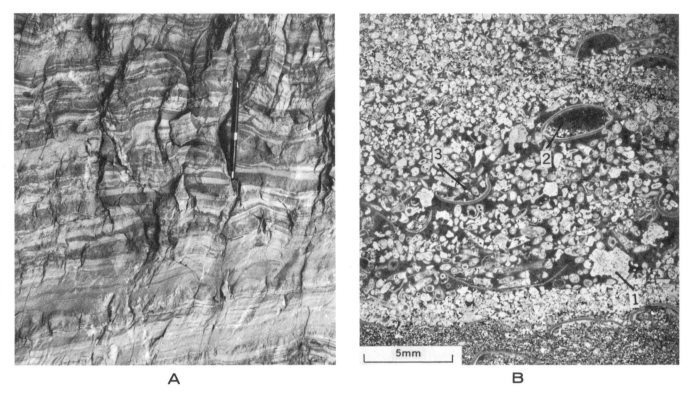

A   B

Figure 15. Intertidal flat deposits. A, Bossardville Limestone, Hamilton Stone Company quarry, Bossardsville, Pa. (stop 2, second day). Graded laminae and beds with fine- to medium-grained limestone containing ostracodes at base grading up into very fine to fine-grained limestone. The basal contact of a graded unit is generally sharp and may contain intraclasts; the top is usually scoured. B, Depue Limestone Member of Coeymans Formation, along southeast bank of Brodhead Creek approximately 0.1 mile above power dam (near stop 3, second day). Negative print of thin section. Calcareous pelletal quartz siltstone (micrite fills interstices) containing ostracodes is abruptly overlain by quartzose ostracode-rich fine- to coarse-grained limestone containing pellets and intraclasts (sparry calcite fills interstices). Note at (1) many large intraclasts from underlying calcareous siltstone, (2) sparry calcite beneath ostracode umbrellas and geopetal structure, and (3) concave upward ostracode carapaces filled with debris.

30 percent clay- to sand-sized quartz); (3) near Tocks Island, N. J., the reef and inter-reef zone, a high energy marine environment where wave activity over the reefs swept clay- to sand-sized clastic and biogenic debris beyond the reefs or into inter-reef areas (biohermal and nonbiohermal facies of the Shawnee Island Member, fig. 17 and table 1: very coarse grained crudely bedded to unbedded biogenic limestone and quartzose burrowed biogenic limestone). In late Coeymans time, the area of outcrop from Lehigh Gap to the Pennsylvania–New Jersey border (figs. 6, 7) became a barrier-bar zone.

A deeper subtidal phase followed Stormville deposition. Rickard (1962) interprets the lithofacies of the New Scotland Formation and Port Ewen Shale as indicative of a deep neritic environment. From New Scotland through most of Port Ewen time, the basin slowly subsided (fossiliferous burrowed quartzose calcilutite and calcisiltite containing beds and lenses of very fossiliferous limestone, both having an abundant benthonic fauna, and beds

Figure 16. Stormville Member (barrier-bar facies) of the Coeymans Formation, intersection of State Highway 90 and Cherry Valley Road (mile 23.5, road log, second day). From bottom to top: laminated to planar-bedded quartzose crinoidal fine- to coarse-grained limestone; crossbedded calcareous sandstone; and planar-bedded calcareous quartz-pebble conglomerate, pebbles as much as 1 inch long (from Epstein and others, 1967).

Figure 17. Biohermal and nonbiohermal facies (reef-inter-reef zone) of Shawnee Island Member of Coeymans Formation overlain by the Stormville Member (barrier-bar zone) of Coeymans Formation; southeast slope of Wallpack Ridge, north of Tocks Island; bedding distorted because of foreshortening in photograph. (From Epstein and others, 1967.)

and lenses of chert, gradually give way to laminated quartzose calcisiltite and calcareous siltstone with a less abundant fauna). Northeast of Bossardsville, sediment influx exceeded basin deepening from late Port Ewen through Oriskany time. Deposition, however, was continuous, and a gradual transition from a deep- to shallow-water neritic shelf environment proceeded from Port Ewen to Oriskany time. The abundance of bioturbation structures and skeletal debris increases from the base of the Port Ewen into the Oriskany (figs. 18, 19). Southwest of Bossardsville, the basin did not become as deep as to the northeast, and the transgressive phase was of shorter duration. The Oriskany Group thins to the southwest and overlies successively older strata as well as successively nearer shore facies southwestward; near Kunkletown, Pa., the Oriskany overlies the New Scotland Formation, whereas farther to the southwest at Andreas, Pa., it overlies the Andreas Red Beds (fig. 6). In late Oriskany time, the strandline continuously shifted to and fro across the field-trip area (barrier-bar deposits consisting of lenses and beds of laminated, planar-bedded, and crossbedded quartzose sandstone and conglomerate abruptly overlie and grade up into burrowed quartzose calcisiltite containing abundant, dominantly spiriferid, brachiopod hash, probably of a shallow subtidal probar apron, fig. 20). The entire area was emergent following Oriskany time.

Deep neritic conditions were re-established throughout most of the field-trip area following post-Oriskany emergence. Shallow neritic deposits of this transgression are absent. During Esopus-Schoharie time the area of greatest subsidence was once again northeast of Bossardsville,

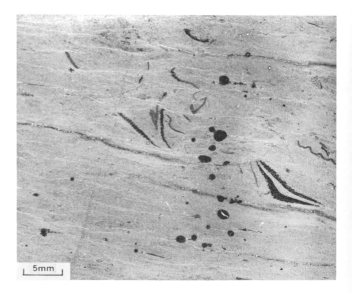

Figure 18. Calcareous siltstone to quartzose calcisiltite containing scattered skeletal debris of predominantly benthonic organisms. Bedding obliterated by extensive burrowing which is not obvious in photograph. Port Ewen Shale, 95 feet above New Scotland–Port Ewen contact, roadcut along U. S. Interstate 80, approximately 0.4 mile southwest of Minisink Hills, Pa. (stop 3, second day). Negative print of acetate peel. Specimen stained with Alizarin Red S (calcite, mainly in shells, dark gray to black). Cleavage dips gently to right in photograph.

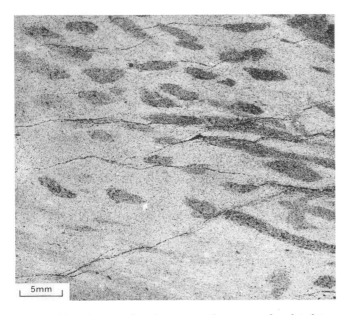

Figure 19.   Burrowed calcareous siltstone and calcisiltite. Shriver Chert, 10 feet above contact with Port Ewen, roadcut along U. S. Interstate 80 (stop 3, second day). Specimen stained with Alizarin Red S (darker gray grains are calcite). Negative print of acetate peel taken about parallel to bedding.

where deeper neritic deposits are thicker. The Esopus and lower part of the Schoharie Formations are interpreted to be deep neritic deposits because: 1) they do not contain abundant skeletal debris, 2) they are relatively thick and widespread, 3) their lithologies are persistent over a wide geographic area (Valley and Ridge province and outliers of New York and New Jersey and Valley and Ridge province of northeastern Pennsylvania, 4) trace fossils change from an abundance of lateral burrows (*Taonurus*, fig. 21; the *Zoophycos* facies of Sielacher, 1967) in the Esopus and lower part of the Schoharie to predominantly long vertical burrows in the upper part of the Schoharie (deep neritic in Esopus-early Schoharie time to shallower neritic through late Schoharie time). The Esopus-Schoharie interval thins southwest. The Palmerton Sandstone of Swartz (1939) overlies the Schoharie southwest of Bossardsville. It is a possible near-shore equivalent (beach or barrier bar?) of the upper Schoharie and lower Buttermilk Falls of the northeastern area (fig. 6). The Palmerton is massive and mysteriously unbedded. Perhaps radiographs, a technique used by Hamblin (1962) for apparently structureless rocks, would show internal sedimentary structures that would give more definite clues as to its origin. Marcellus through Mahantango time records a return of geosynclinal flysch sedimentation (carbonaceous shale with a depauperate fauna, Marcellus Shale, grades upward into rhythmically bedded fossiliferous graded siltstones, Mahantango Formation). The Middle and Upper Devonian formations, Marcellus Shale through the Catskill Formation, repeat, with minor variations, the Middle Ordovician through Upper Silurian flysch-molasse succession.

STRUCTURAL GEOLOGY

Field mapping in rocks of Ordovician to Devonian age in eastern Pennsylvania and northwesternmost New Jersey indicates that rocks of differing lithology and competency have different styles of deformation. Folding is thus disharmonic. Four rock sequences, lithotectonic units, have been recognized. Each sequence has been deformed semi-independently of rocks above and below and presumably is set off from these by décollements (detachments along a basal shearing plane or zone). Type and amplitude of folds are apparently controlled by lithic variations within each lithotectonic unit. The lithotectonic units, their lithologies, thicknesses, and styles of deformation are listed in table 3.

Three décollements, or zones of décollement in relatively incompetent rocks, are believed to separate the four lithotectonic units. The Martinsburg-Shawangunk contact is interpreted to be a zone of detachment between lithotectonic units 1 and 2 and will be seen at Lehigh Gap (stop 3, first day). Thin fault gouge and breccia and development of bedding-plane slickensides with microscarps or steps indicating northwest movement of the overlying Shawangunk Conglomerate, which is corroborated by drag folds in the slickensided zone (fig. 22), will be seen at this stop.

A thick detachment zone probably separates lithotectonic units 2 and 3. The change in style of deformation between the two units takes place in the Poxono Island Formation of White (1882), but considerable northwest movement is indicated by wedging and bedding slip in the Bloomsburg Red Beds (fig. 25). At stop 4, first day, it will be suggested that the net telescoping in the Bloomsburg may amount to many miles. Certainly the amount of shortening in lithotectonic unit 3 in the western part of the field-trip area amounts to several miles. Field evidence does not support an interpretation made by Wietrzychowski (1963) that the folds in lithotectonic unit 3 are related to the extension of the Sweet Arrow fault of Wood and Kehn (1961). The Sweet Arrow fault must terminate several miles west of the Lehigh River.

The movement between lithotectonic units 3 and 4 appears to have occurred within the Marcellus

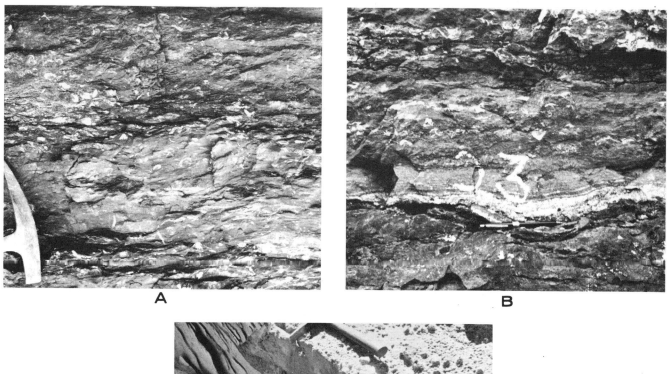

A

B

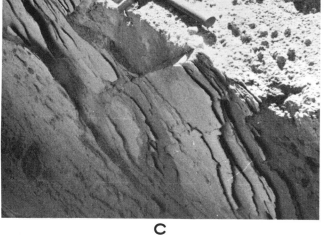

C

Figure 20. Intertidal and shallow subtidal deposits in the Oriskany Group. A, Extensively bur-rowed calcisiltite and calcareous siltstone containing beds, lenses, and pods of dark-gray chert and abundant spiriferid brachiopod debris (shallow subtidal). Fifty feet above Port Ewen–Oriskany contact, roadcut along U. S. Interstate 80 (stop 3, second day). B, Laminated and cross-bedded calcareous quartzose sandstone and siltstone abruptly overlies burrowed quartzose cal-cisiltite and calcareous siltstone containing beds, lenses, and pods of dark-gray chert and abundant spiriferid brachiopods. Chert intraclasts occur in some of the sandstones, suggesting an early diagenetic origin for the chert. Unit 23 shown in photograph is graded from quartz sandstone at base, into quartz siltstone and quartzose calcisiltite at top (intertidal barrier-bar zone and subtidal probar apron[?]). Seventy-five feet above Port Ewen–Oriskany contact (same locality as A). C, Ridgeley Sandstone (upper part of Oriskany Group), Universal Atlas Cement Company clay pits (stop 5, second day). Low-angle foreshore laminations in sandstone (inter-tidal zone of barrier bar). Bedding is vertical, hammer handle parallel to strike of beds.

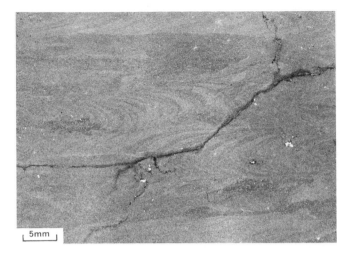

Figure 21. Horizontal burrows, *Taonurus*, in lower half of Schoharie Formation, roadcut on Pa. Rte. 402 near Buttermilk Falls. Negative print of acetate peel.

Shale. This incompetent formation is extensively faulted in the Lehigh Gap area (J. D. Glaeser, oral commun., 1967). In the Stroudsburg area, to the east, the possible zone of movement in the Marcellus Shale is not seen because of thick drift cover. Wedging and bedding slip observed in the Catskill Formation north of the field-trip area indicates that considerable movement may have taken place within lithotectonic unit 4. Glaeser (1967) has documented some of these structures in the lower Catskill Formation. Examples of small-scale internal disharmony within all lithotectonic units are numerous.

Because the detachment zones generally dip to the northwest and may be rootless, northwest movement into the Appalachian basin may have been primarily by gravitational sliding, albeit aided by directed tectonic forces. Because cleavage in the

TABLE 3. LITHOTECTONIC UNITS IN THE DELAWARE RIVER-LEHIGH RIVER AREA, PENNSYLVANIA. MODIFIED FROM FIELD CONFERENCE OF PENNSYLVANIA GEOLOGISTS (1967).

Litho-tectonic unit	Age of lithotectonic unit and stratigraphic sequence (see tables 1 and 2)	Lithologic characteristics	Style of folding	Average size of folds
4	Upper and Middle Devonian Marcellus Shale and younger rocks	More than 10,000 feet of sandstone, conglomerate, siltstone, and shale	Nearly symmetric, concentric, predominantly flexural slip	Two major folds in the area (Weir Mountain syncline and Lehighton anticline—fig. 3); half-wavelengths of more than 5 miles; amplitudes diminish from 4,200 feet near Kunkletown to zero near Bossardsville
3	Middle Devonian to Upper Silurian Buttermilk Falls Limestone of Willard (1938) to upper part of Poxono Island Formation of White (1882)	700 to 1,500 feet of limestone, shale, siltstone, sandstone, and dolomite; heterogeneous stratigraphic units between 3 and 180 feet thick	Asymmetric, concentric, and similar, flexural slip and flow, passive slip and flow. Cascade folds in the east (fig. 23A, B) and flaps (antiformal synclines and synformal anticlines) in the west (fig. 23C)	Wavelengths 1,000 to 1,500 feet; amplitudes about 250 feet in the east increasing to about 1,500 feet in the west
2	Upper to Lower Silurian Lower part of Poxono Island Formation of White (1882) to Shawangunk Conglomerate	3,100 feet of sandstone, siltstone, shale, and conglomerate; coarser toward base of sequence	Asymmetric, concentric; flexural slip with minor passive slip and flow. Extensive bedding slip and wedging in the Bloomsburg Red Beds. Folds tighter in western part of area with development of faulted-out cores of anticlines west of Lehigh River	Wavelengths about 1 mile; amplitudes 1,500–5,000 feet
1	Upper and Middle Ordovician Martinsburg Formation	About 12,000 feet of thick monotonous sequences of slate and graywacke	Asymmetric, similar nearly isoclinal, nearly recumbent in eastern section (fig. 24); mainly passive flow and slip, flexural slip near contact with Shawangunk Conglomerate. Folds probably superimposed on upright limb of large regional nappe	Wavelengths 1,000 to 3,000 feet; amplitudes 400 to 2,000 feet. Small-scale imbricate faults and major thrusts with possible displacements in miles south of field-trip area

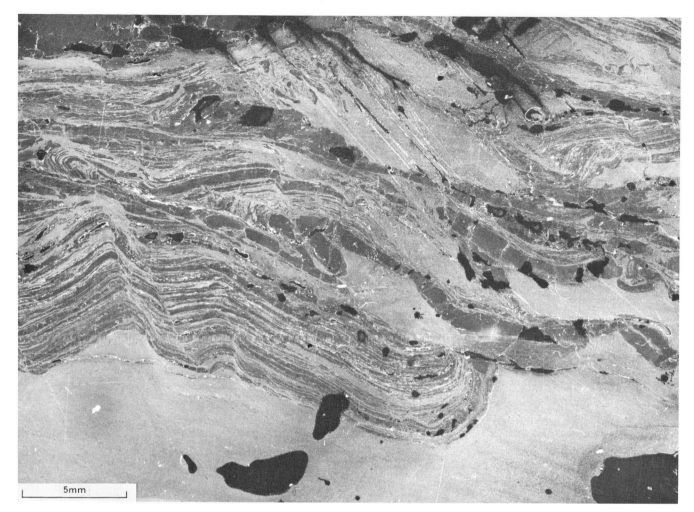

Figure 22.   Negative print of acetate peel from 2–3 inch-thick slickensided bed that underlies a 6 inch-thick fault gouge and breccia at the contact of the Martinsburg Formation of Ordovician age and Shawangunk Conglomerate of Silurian age, Lehigh Gap, Pa. (stop 3, first day).   Northwest is to the left.   The drag folds that are overturned to the northwest and small faults with displacement of overriding beds to the northwest corroborate the movement indicated by bedding slickensides.

Bloomsburg Red Beds is deformed at slippage planes and because the cleavage fans the folds, the sliding must have occurred after folding began and possibly was the result of the folding; i.e., the necessary slope for sliding was developed by folding. Upward shearing of these décollements, as described by Gwinn (1964) in décollements in central and western Pennsylvania, has not been observed in this area.   Possibly these décollements are similar to or continuous with those described to the southwest by Gwinn.

The axial planes of the folds in lithotectonic unit 3 maintain a near-perpendicularity to the form surfaces of the underlying and overlying units. The folds, therefore, may have developed early in the deformational sequence and were later refolded.

Certainly this should be a considered working hypothesis.

The intensity of deformation increases from east to west across the field-trip area: (1) In the east, the structure is generally monoclinal with numerous superimposed folds, many of which are overturned (fig. 26).   In the west, many of the folds are recumbent and isoclinal, and continued tightening has produced faults in lithotectonic units 1 and 2 because of insufficient space in the cores of anticlines (fig. 27).   (2) The Weir Mountain syncline increases in amplitude from a few but near Bossardsville (figs. 3, 26) to more than 4,200 feet west of Kunkletown.   (3) Flow cleavage in the Marcellus Shale is recognizable in the Stroudsburg area, but farther west the cleavage is so well developed that a shaly interval in the Mahantango Formation has been quarried for slate (Behre, 1933, p. 121). Other evidence may be cited. Most notable is the change in width of the Appalachian Mountain section of the Valley and Ridge province. In the east, this section is not more than 5 miles wide, but to the west it rapidly increases to about 40 miles west of the Lehigh River.

Two mechanisms were operative in producing the folds: (1) flexural folding, in which bedding was active and movement was either by slip (flexural slip) or flow (flexural flow), and (2) passive folding, in which movement was along laminar-flow planes (passive flow) (fig. 24) or slip planes (passive slip) (fig. 28), and in which bedding was passive and merely indicates deformation in the direction move-

A

B

C

Figure 23. Folds in lithotectonic unit 3. A, Folds interpreted to be cascades in the eastern part of the field-trip area. Ridgeley Sandstone (massive bed and rocks underlying it) and overlying Esopus Formation (forms convergent cleavage fan in syncline) along Brodhead Creek, 1 mile east of Stroudsburg, Pa. B, Upper cherty limestone member of Buttermilk Falls Limestone of Willard (1938) in roadcut along U. S. Interstate 80, Stroudsburg, Pa. Road parallels strike of axial plane of fold. The variation in plunge is apparent. The folds in lithotectonic unit 3 in the eastern part of the field-trip area are noncylindrical and nonplanar. C, Flap fold in lithotectonic unit 3 in western part of the field-trip area. View looking west across Lehigh River from Bowmanstown, Pa. Palmerton Sandstone of Swartz (1939) lies above dashed line and younger rocks underlie it. The folds are therefore, from left to right, an antiform and a synform. The Northeast Extension of the Pennsylvania Turnpike is seen in middle of photograph where Dyson (1956) first correctly interpreted the structure. Compare with figure 27.

Figure 24.   Recumbent syncline in Pen Argyl Member of Martinsburg Formation, Strunk quarry, ½ mile southeast of North Bangor, Pa.   Axial-plane slaty cleavage dips gently southeast (left).   This is an example of passive flow in lithotectonic unit 1.   Beer bottles in flooded quarry give scale.

ment (see Donath and Parker, 1964).   Flexural-slip folding in the field-trip area is characterized by extensive development of bedding-plane slickensides and by maintenance of constant orthogonal bedding thickness in all parts of the fold, whereas in flexural-flow folding, thickness perpendicular to bedding need not be constant.   Passive folds are essentially similar in form, and constant axial-plane thicknesses (the length between bedding planes measured along cleavage, for example) are generally maintained.   If the planes of movement (the cleavage direction) are macroscopically discon-

Figure 25.   Wedging and bedding slip in the Bloomsburg Red Beds, Lehigh Gap, Pa. (stop 4, first day).   Underside of sandstone bed just above hammer (lower center) is slickensided. Steps in slickensides indicate that the sandstone bed moved downdip to the northwest, not updip as would be expected if the slickensides were due to interbed slip during flexural folding. The northwest movement is corroborated by a small wedge to the left of the hammer and by northwest dragging of the moderately dipping southeast cleavage at the slippage surface. Compare with figure 51

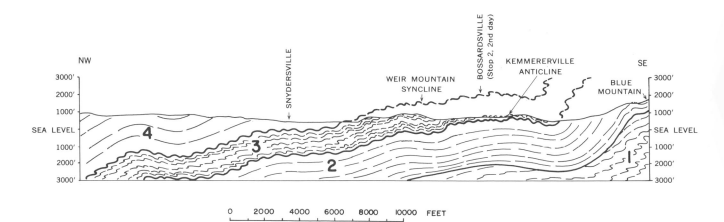

Figure 26.   Geologic section through Bossardsville (stop 2, second day) and Snydersville in the eastern part of the field-trip area, showing disharmonic folding in four lithotectonic units and the angular unconformity between the Martinsburg Formation of lithotectonic unit 1 and the Shawangunk Conglomerate of lithotectonic unit 2.

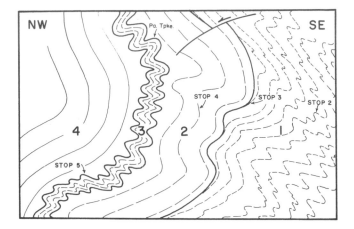

0 2000 4000 6000 FEET

Figure 27. Reconstructed geologic section near Lehigh Gap in the western part of the field-trip area showing disharmonic folding in four lithotectonic units and the angular unconformity between the Martinsburg Formation of lithotectonic unit 1 and the Shawangunk Conglomerate of lithotectonic unit 2. Position of stops 2, 3, 4 and 5 of the first day projected to plane of section. Vertical scale same as horizontal.

tinuous, the folding is by passive slip, but if the movement planes are so closely spaced as to be indistinguishable to the unaided eye, the folding is termed passive flow. In this day when the terminology of cleavages is so confused, it is tempting to call the first type of cleavage "slip cleavage" and the second type "flow cleavage," leaving fracture cleavage, along which there has been no apparent movement, as another end type of a continuous spectrum. Slaty cleavage thus becomes a descriptive term referring to the property, dependent upon the parallelism of platy minerals, whereby a rock can be split into very thin slabs of slate (Martinsburg Formation, stop 2, first day).

Many examples of secondary deformation of cleavage occur in the field-trip area. A second slip cleavage is developed locally in the Martinsburg slates (fig. 29A), as well as in younger rocks (fig. 29B). Occasionally in the Martinsburg, the earlier cleavage is folded by subsequent movement along bedding surfaces (fig. 29C). These deformations of cleavage should not be confused with "refraction" of cleavage through beds of different lithic types (fig. 29D).

THE TACONIC OROGENY IN EASTERN PENNSYLVANIA

The contact between the Martinsburg Formation of Ordovician age and Shawangunk Conglomerate of Silurian age has attracted the attention of geolo-

gists ever since Rogers (1838) recognized that it was an unconformity and later proclaimed that the orogeny was the ". . . most momentous . . . revolution" in North America (Rogers, 1858, p. 785). White (1882) described the contact as unconformable at Lehigh Gap, Pa., and Otisville, N. Y., but Chance (in White, 1882) and Lesley (1883) maintained that the angular relations were due to faulting. Later, Clarke (1921) and Keith (1923), among others, maintained that the angular unconformity seen between Ordovician and Silurian rocks to the northeast is not to be seen in Pennsylvania.

Miller (1926) disagreed. He believed that an angular unconformity is present in Pennsylvania and based his conclusions on the following reasons: (1) the disconformable relations seen in exposures; (2) sericitized slate pebbles in the basal beds of the Shawangunk Conglomerate, apparently derived from the underlying Martinsburg; (3) omission of beds along strike; (4) the Martinsburg Formation is more highly metamorphosed than Devonian shales a few miles away; (5) structures in Ordovician and Cambrian rocks are more complex than those in Devonian and Silurian rocks; (6) the cleavage in the Martinsburg which was formed during the Taconic orogeny is itself deformed into folds and was faulted by the later Appalachian orogeny.

Behre (1924, 1933) argued that the Taconic orogeny produced slaty cleavage, close overturned

Figure 28. Slip cleavage in small passive fold superimposed on vertically dipping laminated shale (white), dolomite (medium gray), and limestone (dark gray), in the Poxono Island Formation of White (1882), along tracks of Central Railroad of New Jersey, ½ mile southeast of Bowmanstown, Pa. Cleavage dips about 50°SE. (toward right in photograph). Negative print of thin section. Top toward upper left.

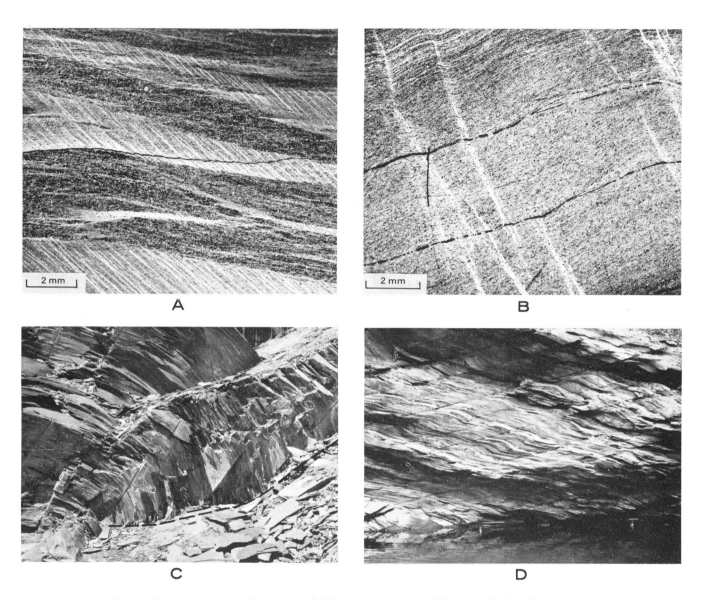

Figure 29. Structures in slates in the field-trip area. A, Nearly horizontal slaty cleavage (S₁) wrinkled by a later moderately dipping slip cleavage (S₂) in Martinsburg Formation, Delaware River Valley. Bedding (S₀) dips slightly steeper than the slaty cleavage. Note that the slip cleavage is absent in the siltier beds. Negative print of thin section. B, Slaty cleavage (S₁) in quartzite-argillite unit of Shawangunk Conglomerate deformed by steeper dipping slip cleavage (S₂). Bedding (S₀), which is slightly steeper than the slaty cleavage, is not apparent in photograph. Negative print of thin section. Outcrop just south of where Aquashicola Creek joins the Lehigh River, ½ mile south of Palmerton, Pa. C, Dragging of slaty cleavage into bedding slip fault (dashed line), Pen Argyl Member of Martinsburg Formation, Albion quarry, Stephans-Jackson Slate Company, ½ mile south of Pen Argyl, Pa. The southeast-dipping cleavage in the quarry has been rotated to the northwest in the area of the photograph. In a narrow zone near the fault the cleavage is overturned to the southeast. A slip cleavage (at arrow) that dips 66°SE. (to left) parallels the axial plane of the drag fold. D, "Refracted" slaty cleavage in Pen Argyl Member of Martinsburg Formation, Strunk quarry, ½ mile southeast of North Bangor, Pa. Bedding dips gently northwest (left) and cleavage dips gently southeast. Cleavage is steeper in silty beds. Beer bottles floating in flooded quarry give scale.

folds, and thrust faults and was more intense than later Appalachian deformation which merely distorted the slaty cleavage. Behre (1927, 1933) divided the Martinsburg into three members, a lower and upper slate separated by a sandstone unit. Stose (1930), however, maintained that the upper slate member of Behre is the lower repeated by folding; hence, the Taconic orogeny must have been intense, for the Shawangunk Conglomerate rests on the lower member of the Martinsburg. Recent work by Drake and Epstein (1967) has reestablished the threefold subdivision of the Martinsburg.

Willard and Cleaves (1939) showed that the angular unconformity extended as far southwest as Susquehanna Gap in Pennsylvania, where the Bald Eagle Conglomerate of Grabau (1909) rests conformably on top of the Martinsburg Formation. Willard (1938) earlier pictured the unconformity at Delaware Water Gap. His cross section through the gap is not in agreement with the information presented in this report and will be discussed at stop 4 of the second day.

Hess (1955) believed that the Taconic orogeny was so intense that it was not the cause of folding of the sediments in the Appalachian geosyncline, but rather the cause of the geosyncline itself. Woodward (1957) maintained that the slate belt of the Martinsburg is the result of the superposition of three periods of folding, each having a different trend. However, there is no field evidence to support Woodward's views.

Recent work in the Delaware Valley by Drake and others (1960) led to the interpretation that the Taconic orogeny was more severe than the Appalachian orogeny in that area, but that far more complex structural terrane was present to the southwest, and that the slaty cleavage in the Martinsburg is Taconic in age. At the same time, Arndt and Wood (1960) concluded that the Appalachian orogeny was by far the stronger. Wood and others (1963, p. 78) suggested that the discordant contact between the Martinsburg and Shawangunk might be largely the result of faulting. Maxwell (1962) concluded that the flow cleavage in the Martinsburg Formation in the Delaware Water Gap area was produced during the Taconic orogeny, but he maintained that the cleavage was the product of only slight stress on pelitic sediments with high porewater pressures. The slate that was produced, there, is not a metamorphic rock, but is rather a product of diagenesis. As a consequence, Maxwell concluded that the Taconic orogeny was minor. Fracture cleavage was produced in the Martinsburg

and younger rocks during the later, more intense, Appalachian orogeny. Maxwell's ideas are stimulating and will be discussed at the end of this section.

The recent interpretation that a large nappe underlies the Great Valley in easternmost Pennsylvania (Drake and others, 1961, 1967; Drake, 1967a, 1967b; Drake and Epstein, 1967; Davis and others, 1967) with no structural counterpart in rocks younger than Ordovician, strongly suggests an intense Taconic orogeny. Nappes have been reported in other parts of the Great Valley (Stose, 1950; Gray, 1954; Field Conference of Pennsylvania Geologists, 1966).

Thus, the relative intensities of the Taconic and Appalachian orogenies in eastern Pennsylvania have not been resolved, and there has been disagreement whether the Ordovician was ever a period of mountain building in eastern Pennsylvania. On the basis of field evidence outlined below, it is concluded that (1) there is an angular unconformity between the Martinsburg Formation of Ordovician age and Shawangunk Conglomerate of Silurian age; (2) the prominent slaty cleavage in the Martinsburg and younger rocks is Appalachian in age; (3) most of the folding seen in the field-trip 1-B area is Appalachian in age; and (4) the effects of the Appalachian orogeny in this small area were far more intense than the observed effects of the Taconic orogeny; however, if the Taconic nappe reported in the Great Valley was atectonitic (no tectonic fabric was developed), then evidence for this nappe would not be present in the field-trip area and the Taconic orogeny may have been extremely intense. Work currently in progress by I. B. Alterman, Columbia University, in the Martinsburg outcrop belt from southeasternmost New York to eastern Pennsylvania, indicates that in several localities an earlier cleavage, previously unreported, which has been nearly obliterated by the later deformation, can be identified. Alterman believes that this less distinct, earlier cleavage developed as an axial plane cleavage in the nappe (written commun., October 1968).

The contact between the Martinsburg Formation and Shawangunk Conglomerate has been seen or reported at five localities in this area (fig. 3). On the basis of observations at these localities and from data gathered during mapping of 40 miles along the Martinsburg-Shawangunk contact, the following generalizations may be made:

1. Slaty cleavage penetrates all pelitic rocks of the Martinsburg Formation except within 100–200 feet of the contact with the Shawangunk Conglom-

erate where, as at Lehigh Gap (stop 3, second day), the conspicuous slaty cleavage disappears and bedding-plane slickensides become prominent (fig. 30).

2. In the Delaware Water Gap area, moderately dipping southeast cleavage in the Martinsburg Formation flattens within 2,000 feet of the Shawangunk Conglomerate and then dips gently to the northwest as the contact is approached. Drake and others (1960) and Maxwell (1962) attributed this arching of the slaty cleavage to refolding during the Appalachian orogeny. This phenomenon is accompanied by the development of bedding-plane slickensides, and in view of what has been said concerning the disharmonic relations between lithotectonic units 1 and 2, it was earlier suggested that the arching of the cleavage may be due to drag as the Shawangunk moved over the Martinsburg on a décollement (Field Conference of Pennsylvania Geologists, 1967). There are, however, several smaller scale examples of cleavage dying out or being deflected as more competent rocks are approached. It is therefore suggested that a pressure-shadow mechanism, described below, may be responsible for the dying out and arching of cleavage shown in figures 32, 48, and 70.

In thin sections of slaty shales and siltstones, cleavage is seen to curve around clastic grains, small lenses of sandstone, or sand-filled burrows. Cleavage is most intensely developed (flattening is greatest) on top and bottom of these more competent clastic bodies (fig. 31). Cleavage is poorly developed or absent in the areas of maximum extension to the sides of the grains in the areas of "pressure shadows."

In many small folds involving interbedded cleaved argillites and uncleaved or poorly cleaved more competent rocks, cleavage diverges around synclinal troughs and is either poorly developed or absent in the pressure-shadow area next to the trough (fig. 32A). If the relationship is true on a larger scale (fig. 32B), it may explain the arching of cleavage at Delaware Water Gap and dying out of cleavage at Lehigh Gap.

3. A second (slip) cleavage cuts the earlier slaty cleavage in the Martinsburg in the southern part of the eastern section of the area. To the west, the slip cleavage appears higher in the Martinsburg, and in the Lehigh Gap area it is found in overlying formations.

4. No fragments of Martinsburg slate have been found in basal Shawangunk beds.

5. At all localities where the contact between the Shawangunk Conglomerate and Martinsburg Formation has been seen, there is a thin clay zone believed to be a fault gouge. At Lehigh Gap and along U. S. Interstate 84 about 4 miles east of Port

A   B

Figure 30.   Disappearance of cleavage in the Martinsburg Formation near contact with overlying Shawangunk Conglomerate at Lehigh Gap, Pa.   A, Well-developed southeast-dipping (to right) cleavage in the Martinsburg, 300 feet stratigraphically below the contact with the Shawangunk.   Bedding dips moderately to the left.   Martinsburg consists of interbedded claystone slate and siltstone and sandstone graywacke.   B, Contact of Martinsburg (below) and Shawangunk Conglomerate at man's hand.   Note lack of cleavage in the Martinsburg.   Bedding dips moderately northwest (left).   Bedding-plane slickensides are extensively developed (see description of stop 3, first day).

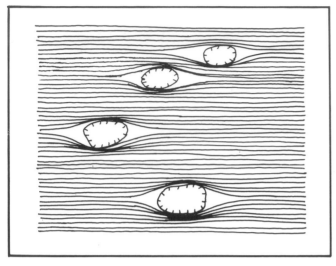

Figure 31. Diagrammatic sketch from thin section showing curving of cleavage around coarse silt-sized quartz grains in silty shale. Quartzite-argillite unit of Shawangunk Conglomerate, Lehigh Gap, Pa. Quartz grains average 0.05–0.06 mm in length.

Jervis, N. Y., there is a thin fault breccia at the contact. Steps on slickensided bedding surfaces and minor internal structures in the fault zone (fig. 22) indicate that the overriding beds in the Shawangunk moved to the northwest.

6. The attitudes of cleavage in the Martinsburg Formation within a few miles of the Shawangunk contact are very similar to attitudes of cleavage in overlying rocks, strongly suggesting that the Martinsburg cleavage was not deformed by the stresses that produced cleavage in the younger rocks and implying that all cleavages were formed during a single period of orogeny (fig. 33). Martinsburg cleavage, however, is folded into cleavage arches farther to the southeast, and a second (slip) cleavage is extensively developed, indicating that deformation increased in intensity in that direction.

7. In a few localities where they have been mapped, fold axes pass from the Shawangunk Conglomerate into the Martinsburg Formation without deflection, showing that the folds are post-Taconic in age. Cleavage in the Martinsburg is parallel to the axial planes of the folds, or fans the folds slightly, showing that the cleavage is also post-Taconic in age.

8. Quartz, chert, and quartzite pebbles in the basal beds of the Shawangunk, in places more than 5 inches long, indicate that the Martinsburg was breached at the time the basal part of the Shawangunk was deposited so that underlying stratigraphic units were exposed and supplying pebbles (possibly chert from the Beekmantown Group, quartz-

ite from the Hardyston Quartzite, and vein quartz from Precambrian rocks). Thus, an unconformity is indicated. The sharp lithologic break at the contact also indicates that rocks of inferred deep-water origin (Martinsburg Formation) are overlain by fluviatile-terrestrial deposits (Shawangunk Conglomerate).

A

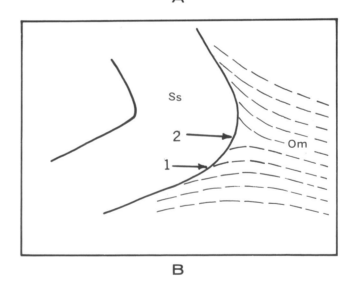

B

Figure 32. Pressure shadows and arching of cleavage in interbedded rocks of different competencies. A, Flattened folds in lower part of Bushkill Member of Martinsburg Formation, along U. S. Route 46, 1.8 miles northeast of Belvidere, N. J. Slaty cleavage in pelite diverges around the syncline in the more competent micaceous fine-grained dolomite and is less well developed at the trough. B, Diagrammatic fold drawn from A showing the structural relations, on a larger scale, of cleavage in the Martinsburg (Om) near the contact with the Shawangunk Conglomerate (Ss) at Delaware Water Gap (1) and Lehigh Gap (2). Compare the arching of cleavage at 1 with figure 70, and the dying out of cleavage at 2 with figure 48.

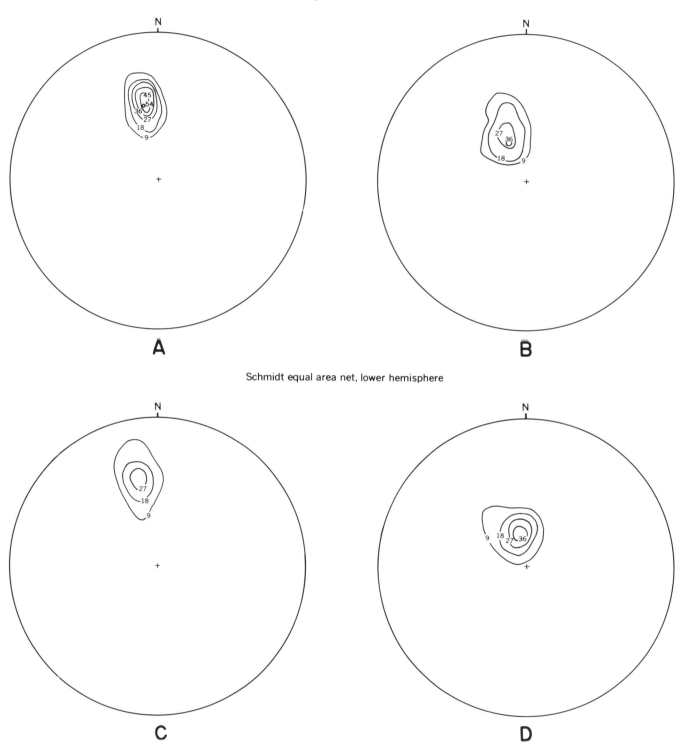

Schmidt equal area net, lower hemisphere

Figure 33.   Equal-area projections of cleavage (S₂) in the Martinsburg Formation and Marcellus Shale and Mahantango Formation in the Palmerton and Stroudsburg quadrangles.   Contour interval 9 percent per 1 percent area.   See figure 2 for location of quadrangles.

A, Mahantango Formation, Palmerton quadrangle; 42 poles; maximum at N.79°E., 42°SE.;
   data from W. D. Sevon, Pennsylvania Geological Survey.
B, Marcellus Shale and Mahantango Formation, Stroudsburg quadrangle; 28 poles; maximum
   at N.66°E., 23°SE.
C, Martinsburg Formation, Palmerton quadrangle; 224 poles; maximum at N.78°E., 50°SE.
D, Martinsburg Formation, Stroudsburg quadrangle; 90 poles; maximum at N.75°E., 20°SE.

9. There is an angular discordance at all localities where the Martinsburg-Shawangunk contact has been seen in outcrop or reported in tunnels. Figure 3 shows these localities, and they are described below:

(1) Pennsylvania Turnpike Extension tunnel through Blue Mountain: Martinsburg is overturned and dips 35°SE.; the Shawangunk is also overturned but dips 45°SE.

(2) Lehigh Gap: Martinsburg dips 46°NW.; Shawangunk dips 36°NW. (Seen at stop 3 of the first day.)

(3) Water tunnel for the city of Bethlehem, Pa., at Little Gap, from unpublished data by B. L. Miller, 1940: Martinsburg strikes N.35°E. and dips 52°SE., overturned; the Shawangunk strikes N.68°E. and dips 55°SE., overturned.

(4) Delaware Water Gap: the contact was formerly exposed and reported by Beerbower (1956) to be unconformable by 3° in strike and 1° in dip. Mapping of graywacke beds in the Martinsburg in the gap area shows that the strike of the Martinsburg differs from that of the Shawangunk by about 15°, and the contact between the Pen Argyl and Ramseyburg Members of the Martinsburg Formation is unconformably overlain by the Shawangunk Conglomerate 2 miles west of Delaware Water Gap (see fig. 70).

(5) Yards Creek hydroelectric pumped storage project near Blairstown, N. J.: the middle (Ramseyburg) member of the Martinsburg Formation is unconformably overlain by the Shawangunk Conglomerate. The cutting out of beds in the Martinsburg is discernible, but the angular relations are not readily measurable.

As mentioned previously, Maxwell (1962) has stimulated widespread interest with his interpretation of a nonmetamorphic origin (high pore pressures and fluid expulsion in water-bearing muds with only slight Taconic deformation) for the slaty cleavage in the Martinsburg Formation in the Delaware Water Gap area. According to Maxwell, Appalachian deformation was brittle, and produced only fracture cleavage in the Martinsburg and younger rocks. Carson (1968), for example, has attributed the progressively better development of slaty cleavage in lower beds of the Martinsburg to increasing overburden with higher water pressures in the original muds. However, some of the field relations presented by Maxwell,

particularly the ages of structures produced during Taconic and Appalachian deformation, do not agree with, or do not readily explain, data gathered during mapping of the area between the Delaware and Lehigh Rivers.

Without going into detail, some of the items that should be considered in the evaluation of Maxwell's ideas are:

(1) The age of the prominent slaty cleavage in the Martinsburg is not Taconic, but younger, for reasons presented above.

(2) Small "sandstone dikes," similar to the ones shown by Maxwell (1962, p. 287) are found in rocks younger than the Martinsburg (i.e., quartzite-argillite unit of the Shawangunk Conglomerate, Poxono Island Formation). These, however, may be sand sheared out along cleavage. If these are sandstone dikes in the post-Ordovician rocks, could it be that cleavage developed in these rocks as envisioned for the Martinsburg by Maxwell and that this "diagenetic" cleavage extends down into "metamorphic" cleavage in the Martinsburg as suggested below?

(3) The arching of cleavage in the Martinsburg as the contact with the overlying Shawangunk Conglomerate is approached at Delaware Water Gap (Maxwell, 1962, p. 285; fig. 70, this report) is perhaps best explained by the pressure shadow mechanism described above (with its implication of a post-Silurian age for the origin of the cleavage) rather than by post-Taconic arching of a cleavage produced during Taconic deformation as suggested by Maxwell.

(4) The occurrence of bedding-plane slickensides deformed by slaty cleavage (fig. 34), indicating that the Martinsburg was competent enough to deform by flexural slip prior to passive deformation, does not support the hypothesis that the cleavage was imposed upon a water-bearing pelite.

(5) The flow cleavage in the Martinsburg is a product of metamorphism because chlorite porphyroblasts were seen in many thin sections. Behre (1933, p. 177) has also shown that many quartz grains have a length to width ratio of as much as 5:1, which he attributes to metamorphism, not to original sedimentary shape. The dominant mica in the Martinsburg is 2M muscovite, as shown by X-ray diffraction analyses. This contradicts the assertion that the mica is "illite" (Maxwell, 1962, p. 298). Incomplete evidence suggests that the original detrital muscovite is both 1M and 2M polymorphs (sample without prominent cleavage at Lehigh Gap near the Shawangunk contact; seen at stop 3, first day) which becomes more ordered at

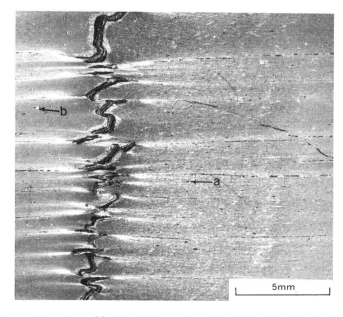

Figure 34.   Bedding plane slickensides (vertical calcite bed) offset by horizontal slaty cleavage in Pen Argyl Member of Martinsburg Formation, showing that flexural slip folding preceded passive folding.   Chlorite porphyroblasts appear gray (a).   Pyrite grains (white) have chlorite and muscovite developed parallel to cleavage in their pressure shadow area (b).   These features are easily discernible in thin section at high magnification, but are not readily apparent in this negative print of a thin section at low magnification.   Sample from dumps of inactive New Peerless quarry, 3,000 feet southeast of North Bangor, Pa.

lower stratigraphic levels, presumably due to metamorphism.

(6) Slaty cleavage is not confined to the Martinsburg, as Maxwell claims.   Many of the shaly formations overlying the Martinsburg have cleavage with preferred mineral alinement (e.g., Port Ewen Shale, Marcellus Shale).   In particular, pelites in the Mahantango Formation have been quarried for slate near Aquashicola, Pa., a fact noted many years ago by Dale (1914, p. 108) and Behre (1933, p. 119).

(7) In some exposures of the Martinsburg, the later slip cleavage has nearly obliterated the earlier slaty cleavage, and a well-defined cleavage banding has formed.   In thin section, the second cleavage has nearly perfect mineral alinement; in hand sample, the rock can be split into thin laminae parallel to the second cleavage.   If the second cleavage had pervaded the entire rock (transposition had been complete), a perfectly respectable slate would have resulted.   In other words, a second slaty cleavage may possibly form as suggested by Broughton (1946, p. 13).

In summary, some of Maxwell's conclusions concerning the origin and relative ages of cleavages are open to question; others deserve additional

consideration.   In the Delaware River to Lehigh River area, the prominent slaty cleavage in the Martinsburg Formation is not Taconic in age, but formed during the latest Paleozoic deformation at the same time cleavage formed in post-Ordovician rocks.   Slip cleavage and fracture cleavage, found in rocks of all ages, may have formed during a later phase of that orogeny, or may have formed at the same time slaty cleavage developed in more pelitic rocks.

GEOMORPHOLOGY

The area of this field trip lies in the Appalachian Mountain and Great Valley sections of the Valley and Ridge physiographic province (fig. 4).   The area has been extensively studied by geomorphologists concerned with the classic problems of Appalachian drainage evolution, peneplanation, origin of wind and water gaps, etc.   Discussion of one of these problems necessitates consideration of the others.   However, on this field trip we shall limit our discussion to one topic, the origin of wind and water gaps.

It has been concluded that gaps in the Stroudsburg area are structurally controlled (Epstein, 1966; reprint provided in back of this report); that is, the gaps are located where folds die out over short distances, where folding is locally more intense, or where resistant rocks dip steeply and have a narrow width of outcrop.   In addition to the gaps discussed in the included reprint, three gaps in Blue Mountain, in the western part of the field-trip area, also show structural control; Little Gap (on the limb of a recumbent, nearly isoclinal fold, fig. 35); Lehigh Gap (two folds die out abruptly to the west, stops 3 and 4, first day); and Lehigh Furance Gap (where a longitudinal fault emerges from the core of an overturned anticline and has been rotated about a vertical axis during refolding; fig. 52, discussed at stop 4, first day).

GLACIAL GEOLOGY

There is evidence of at least two glaciations in the field-trip area.   Deeply weathered till and outwash deposits, possibly Illinoian in age, occur beyond the limit of the fresher till and melt-water deposits of the Wisconsin drift sheet.   The older (Illinoian?) drift will be seen at stop 2, first day.

Glacial sediments of Wisconsin age are composed of varying proportions of gravel, sand, silt, and clay. On the basis of texture, internal structure, bedding and sorting characteristics, and generally well-preserved landforms, the deposits are subdivided into till (ground, end, and terminal moraine) (fig. 36A)

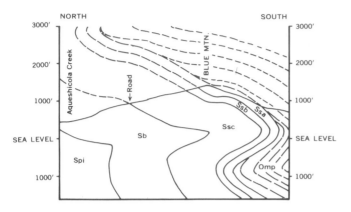

Figure 35. Geologic section through Blue Mountain, 2,000 feet east of Little Gap. Some data from unpublished cross section of water tunnel for city of Bethlehem, Pa., by B. L. Miller, 1940. Horizontal scale same as vertical. Spi, Poxono Island Formation of White (1882); Sb, Bloomsburg Red Beds; Ssc, quartzite-argillite unit of Shawangunk Conglomerate; Ssb, lower quartzite-conglomerate unit of Shawangunk Conglomerate; Ssa, conglomerate unit of Shawangunk Conglomerate; Omp, Pen Argyl Member of Martinsburg Formation.

and stratified drift (kame-terrace, kame, outwash, deltaic, and glacial lake-bottom deposits) (fig. 36B). The deposits and glacial history of the Stroudsburg quadrangle, in the eastern part of the field-trip area, have been described (Epstein, 1969).

Wisconsin deglaciation occurred by northeastward retreat and stagnation. Periods of stagnation are characterized by ice-contact deposits such as kames, whereas end moraines and large ice-margin

deltas are interpreted to reflect periods of normal retreat (i.e., the ice-margin was continuous with the retreating main ice mass). A conspicuous terminal moraine marks the limit of the Wisconsin Glaciation (figs. 3, 4). Because the slope of the land just in front of the terminal moraine was eastward, toward the Delaware River, a basin formed in which lake sediments probably accumulated (Mackin, 1941). The lake deposits, covered by gravel, now appear as a flat plain in the area west of Saylorsburg, in the valleys of Buckwa and Aquashicola Creeks (fig. 4). The pre-Wisconsin drainage divide in these valleys is about 5 miles west of Saylorsburg where the valleys are the narrowest (fig. 37). Melt waters from the northeastward-retreating ice were trapped between the ice front and the terminal moraine, forming a lake, Lake Sciota (fig. 37). Stratified deposits, including an ice-margin delta and rhythmically laminated lakebeds, were laid down in the proglacial lake. At first, the outlet of the lake was in the terminal moraine at Saylorsburg, at an altitude of about 680 feet, to which the delta seen at stop 1, first day, is graded. During the following period of ice stagnation, drainage was through Delaware Water Gap after it was uncovered during deglaciation. Kame terraces were deposited between the stagnant ice and adjacent valley walls and deltas were deposited in locally ponded areas. The kame terraces and deltas were laid down at successively lower altitudes as the ice melted. In Cherry Valley lakebeds and kames were deposited

Figure 36. Unstratified and stratified drift. A, Bouldery till (muddy gravel) with striated boulders as much as 10 feet long. Thirty-foot high exposure along Delaware, Lackawanna, and Western Railroad tracks north of Brodhead Creek, 4,000 feet east of East Stroudsburg State Teachers College, East Stroudsburg, Pa. B, Stratified lenticular sand and gravel in kame deposit in pit 500 feet southeast of Lake Poco and about two miles southwest of Portland, Pa.

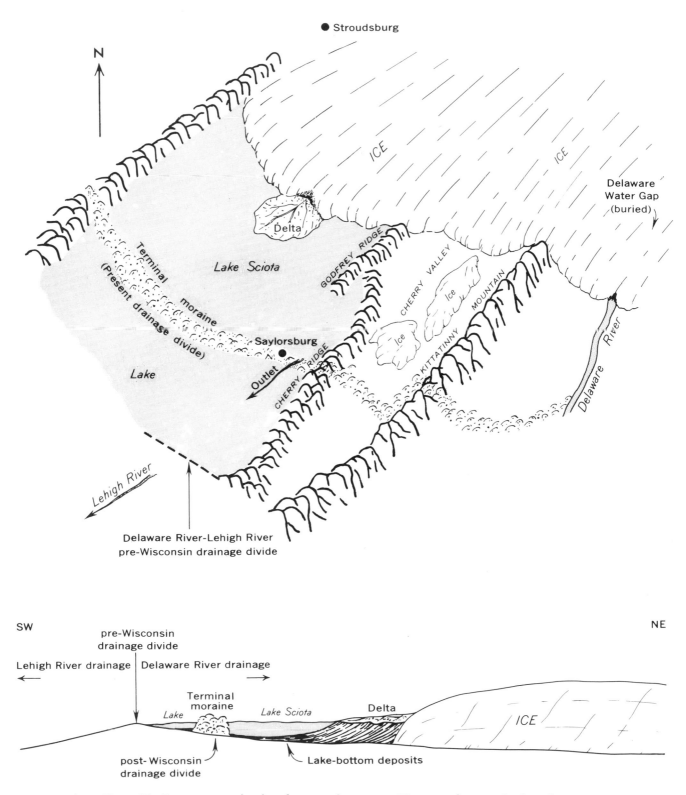

Figure 37. Diagrammatic sketch and section showing pre-Wisconsin drainage divide and position of Wisconsin glacier at the time of deposition of delta in Lake Sciota (stop 1, 1st day). The terminal moraine marks the present Delaware River–Lehigh River drainage divide. The ice front was probably not as straight as depicted, and the isolated stagnant blocks of ice probably lay southwest of the main ice mass.

during periods of stagnation and kame moraines mark active ice-margin positions.

Following deglaciation, alluvial fans, alluvium, talus, and swamp deposits accumulated.

Bedrock topography has been subdued in many places by the drift cover. Examples of drainage modification are numerous. Talus deposits, congelifractates, congeliturbates (fig. 38), rock streams, and block cities are believed to be partly due to frost action in a periglacial environment.

The Wisconsin ice was locally more than 1,300 feet thick, as shown by glacial striae along the crest of Kittatinny Mountain (seen at stop 4, second day).

Pebble-count data in the Stroudsburg area corroborate the south-southwestward direction of ice movement indicated by glacial striae (fig. 39). This conclusion is based on comparison of pebble composition with regional bedrock lithology (Epstein, 1969). The ice moved about S.25°W., obliquely across the outcrop belts of bedrock which strike N.60°E. Pebbles derived from the underlying bedrock in most areas form a large part of the overlying drift, in some areas more than 95 percent. Dilution was very rapid away from the bedrock source. Pebbles from one lithologic belt make up, on the average, only 10 percent of the pebbles in drift 5 miles from their bedrock source.

Grain-size data show that: (1) the stratified drift is more than twice as well sorted as the till; (2) the

Figure 39. Two sets of glacial striae on bedding plane of Bloomsburg Red Beds. One set trends S.16°W. (heading towards observer) and is crossed by a second set trending S.27°W. About ¾ mile south of village of Delaware Water Gap, Pa.

Figure 38. Convolutions probably due to frost action in a periglacial environment (a congeliturbate) involving slate fragments from the Martinsburg Formation. In a 75-foot-long and 3-foot-high roadcut ¼ mile east-southeast of Filetown and about 1½ miles north of Nazareth, Pa.

till is generally a muddy to muddy sandy gravel (classification of Folk, 1954), whereas the coarse stratified drift contains less clay (sandy gravel, gravelly sand, to gravelly muddy sand); and (3) the older till (Illinoian?) is distinctly more clayey (gravelly sandy mud) than the Wisconsin Till.

X-ray diffraction studies (Epstein, 1969) show that quartz, feldspar, illite, and chlorite are present in all silt and clay samples of Wisconsin drift; quartz and feldspar are more abundant in the silt. Hematite was noted in many samples, and vermiculite, swelling chlorite or montmorillonite, degraded chlorite or vermiculite, and kaolinite are present in a few samples. Kaolinite and vermiculite are much more abundant in the older drift and are probably products of prolonged weathering.

Sand, gravel, and peat have proven and potential economic value in the area of Wisconsin Glaciation. Glaciation apparently had a profound effect on the location of residual clay deposits of commercial importance. Clay deposits are present in Lower Devonian limy shale and shaly limestone in Cherry and Chestnut Ridges, Monroe County, Pa. These deposits are southwest of the Wisconsin terminal moraine and are believed to have formed during pre-Wisconsin weathering. No clay deposits are present northeast of the moraine, and that any clay that formed in this area presumably was stripped away during the Wisconsin glacial advance.

ROAD LOG

FIRST DAY

Departure from the Holiday Inn Motel, East Stroudsburg at 8:00 A.M. Field trip will follow route shown on figure 2.

Mileage

0.0   Leave Holiday Inn parking lot and proceed south on U. S. Rte. 209.

0.3   RIGHT turn onto U. S. Interstate 80. On right, roadcuts in the lower cherty limestone member of the Buttermilk Falls Limestone of Willard (1938). On left, Godfrey Ridge, underlain by Upper Silurian and Lower Devonian rocks.

0.9   Roadcut in Wisconsin ground moraine.

2.1   Cross Brodhead Creek.

2.3   Roadcut on left (south) exposes upper cherty limestone member of Buttermilk Falls Limestone on the northwest limb of an anticline. Roadcut parallels axial trace of fold. Note variation in plunge of fold axis; folds in Godfrey Ridge are noncylindrical and nonplanar.

3.8   On left, sand and gravel in Wisconsin terrace, probably a valley train deposit.

4.3   Bear RIGHT to U. S. Rte. 209S.

4.7   Coming out of cloverleaf, ridge ahead, lying at right angles to road, is a linear kame (to be discussed at stop 1, 2nd day).

5.2   Road passes through linear kame and continues south onto silt and very fine sand (lakebeds) and sand and gravel (kame).

6.5   On left, green house in birch grove sits on 1,000-foot-long ridge of stratified sand and gravel, an esker, that leads into a kame delta to the south.

7.1   On left, sand and gravel pit exposes topset and foreset beds of a kame delta (see alternate stop 1, 2nd day).

8.2   Flat plain on left is underlain by gravel, probably valley train deposits, and laminated lake-bottom beds.

8.5   Roadcut through Marcellus Shale.

9.0   The surrounding flat plain is underlain by gravel and glacial lake-bottom beds. Godfrey Ridge to left. Hills on right underlain by Marcellus Shale.

9.5   Leave interstate highway and bear RIGHT to Snydersville exit.

9.7   Stop sign, RIGHT turn.

9.9   Stop sign, LEFT turn.

10.1   Evergreen-capped ridge to right is an esker that leads into the ice-marginal delta of stop 1.

10.5   Buttermilk Falls Limestone and Schoharie Formation exposed in roadcut to left.

11.1   Sand and gravel pit in an ice-marginal delta deposit.

STOP 1. GRAVEL PIT IN ICE-MARGINAL DELTA AND ESKER

The delta and esker at this locality are very well preserved (fig. 40). Note the flat top of the delta underlain by horizontally stratified sand and gravel (fig. 41A), the lobate foreset slope as we pass south of it (fig. 41B), and the typical nodular form of the esker to the northwest. Rhythmically laminated sand, silt, and clay were uncovered in the lakebeds to the south during construction of the overpass for U. S. Rte. 209 in 1962, 2,000 feet to the south (fig. 41C). The topset and foreset beds are well displayed in the gravel pit (fig. 41D). The esker was deposited subglacially or englacially. The hummocky proximal slope of the delta marks the position of the ice at the time of deposition (fig. 40). Kettle holes in the topset plain of the delta were areas where drift collapsed because of melting of buried ice masses. The drift in these areas, where exposed in the pit, lacks stratification.

The delta was deposited in glacial Lake Sciota, an ice-defended, moraine-dammed lake, during Wisconsin time. The outlet in the moraine was at Saylorsburg, 3 miles to the southwest (fig. 37). The level of the lake was at about 680 feet, the approximate altitude of the foreset-topset interface. During the ensuing period of deglaciation, kames, deltas, lake-bottom beds, and terraces were deposited, some of which were seen enroute to this stop.

Note the fresh appearance of the sand and gravel that make up the topset and foreset beds of the delta. Dark-gray cherty limestone pebbles

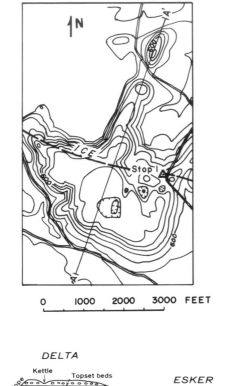

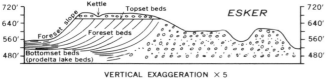

Figure 40. Topographic map and geologic section of esker and delta at stop 1, first day, showing inferred position of ice margin at time of deposition. The level of Lake Sciota in which the delta was deposited was near the topset-foreset interface. Topography from U. S. Geological Survey, Saylorsburg 7½-minute quadrangle, Pennsylvania.

react readily with dilute hydrochloric acid. Compare this with the more distinctly weathered older drift, Illinoian(?), southwest and west of the terminal moraine and at stop 2, in which all limestone has been leached and the chert has weathered to a very light gray.

Also note the similarity of the coarse planar-bedded gravels with similar deposits to be seen in the conglomerate unit of the Shawangunk Conglomerate at stop 3, today. These beds were probably deposited by streams of low sinuosity (braided pattern) in the upper flow regime of Simons and Richardson (1962).

The relatively high velocities necessary to maintain the upper flow regime was supplied by rapid melting of ice in the case of the Pleistocene delta, and interpreted steep slope due to tectonic uplift in the case of the Silurian deposits in the Shawangunk.

Mileage

11.1   Continue on road towards Sciota.

11.3   Road curves around lobate foreset slope of delta. Lake-bottom beds in valley to left.

11.6   RIGHT turn towards Sciota.

12.4   Sciota. Stop sign at road junction. LEFT turn onto U. S. Rte. 209. Follow road to Saylorsburg.

12.5   Marcellus Shale in pit to right.

12.9   Glacial delta of stop 1 to left. Note the flat topset plain, the lobate forest slope, and the valley underlain by lake-bottom beds in front of the delta. Blue Mountain in skyline; Cherry Ridge in middle ground. Note termination of Godfrey Ridge to left.

13.8   Route crosses Wisconsin terminal moraine. Note undulating topography and coarse texture of drift.

14.8   To left, Saylors Lake, a kettle lake.

15.1   Stop sign in Saylorsburg; LEFT turn onto Pa. Rte. 115.

15.2   RIGHT turn onto Kunkletown Road. At this point we are on terminal moraine near the outlet of glacial Lake Sciota, whose waters discharged via Buckwha Creek (fig. 4) into Aquashicola Creek (at Little Gap, 12 miles to the west) and finally into the Lehigh River (5 miles farther west).

15.3   Outlet channel parallels road on left.

15.7   Wisconsin drift in roadcut on right. Note undulatory topography on terminal moraine.

16.5   On right, shale pit in Marcellus Shale showing crudely stratified shale-chip gravel, possibly the result of periglacial solifluction, a congeliturbate.

17.1   Roadcuts in Marcellus Shale on right. For the next 10.7 miles, the road follows the outcrop belt of the Marcellus Shale and Mahantango Formation. On left, Chestnut Ridge, held up by the Ridgeley Sandstone and Palmerton Sandstone of Swartz (1939). Buckwha Creek valley narrows to the west as we approach the pre-Wisconsin Delaware River-Lehigh River drainage divide.

19.1   Outcrop of Centerfield coral bed of the Mahantango Formation in creek on left, a calcareous siltstone biostrome.

19.8   Turkey farm on left.

20.6   Panoramic view, see figure 42.

21.4   On left, roadcut in Mahantango Formation.

21.8   Note the narrow valley of Buckwha Creek on left. This is probably near the pre-Wisconsin drainage divide between the Lehigh and Delaware Rivers.

22.9   Roadcut in steeply-dipping uppermost Marcellus Shale on right.

23.0   Stop sign in Kunkletown; proceed straight ahead. Sand pits to left are in the Palmerton Sandstone of Swartz (1939) and Ridgeley Sandstone. SLOW!

A

B

C

D

Figure 41. Internal sedimentary structures and morphological features in glacial delta near Sciota, Pa. A, View of sand and gravel pit as exposed in 1962. Flat-topped terrace underlain by horizontally stratified topset beds of sand and gravel. Blue Mountain, underlain by the Shawangunk Conglomerate and Bloomsburg Red Beds of Silurian age, in the background; cultivated fields on Cherry Ridge underlain by complexly folded heterogenous sediments of Upper Silurian and Lower Devonian age in middleground. B, Lobate foreset slope south of pit seen in A, grading down into laminated lakebeds at bottom of photograph. C, Rhythmically laminated and graded, fine sand, silt, and clay seen at bottom of B. These are fine lake-bottom sediments (bottomset beds) deposited in Lake Sciota beyond the foreset slope of the delta. Occasional pebbles that disrupt the laminae were seen. These were probably ice-rafted into the lake. Scale is about 2½ feet long. D, Topset and foreset beds exposed in pit.

Figure 42.   Panoramic view looking west from Buckwha Creek valley (at mile 20.6, first day of road log).   From left to right: Blue Mountain, high ridge in far background (underlain by Shawangunk Conglomerate and Bloomsburg Red Beds of Silurian age); Chestnut Ridge in left middle ground (underlain by complexly folded uppermost Silurian to lower Middle Devonian rocks); conical hill in center is capped by a fine-grained sandstone in the Mahantango Formation (W. D. Sevon, oral commun.); ridge on right underlain by the Trimmers Rock Sandstone and lies on the northwest-dipping south limb of the Weir Mountain syncline.

Mileage

24.2   Fork in road, keep LEFT toward Palmerton.

25.5   Note widening of Buckwha Creek valley on left.

26.3   On right well-developed slaty cleavage in Mahantango Formation.   These rocks have been quarried for slate at Aquashicola, 6 miles to the west.

27.0   Slate on right.

27.7   Village of Little Gap.   Continue straight ahead.

28.0   Roadcut through Stony Ridge.   Exposed on right is an antiform and synform in Lower and Middle Devonian rocks (fig. 43).   Sand pits in Palmerton Sandstone of Swartz (1939) and Ridgeley Sandstone on left.   Buckwha Creek joins Aquashicola Creek at this point.

28.8   Broad valley on left is underlain by outwash deposits of Wisconsin age.   More deeply weathered sand and gravel, believed to be remnants of a higher level Illinoian(?) outwash terrace, occur on right along road to Palmerton.

30.0   On left, Blue Mountain.   Gray Shawangunk Conglomerate forms crest; Bloomsburg Red Beds below.

30.5   On left, large waste piles from the New Jersey Zinc Company smelting plant in Palmerton, rise more than 200 feet above valley floor.

31.4   On right, overturned beds of Palmerton Sandstone in Stony Ridge dip steeply southeast.

31.6   On left, The New Jersey Zinc Company smelting plant.

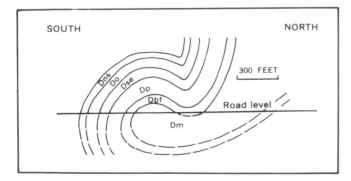

Figure 43.   Diagrammatic geologic section through Stony Ridge at Little Gap.   Dns, New Scotland Formation; Do, Oriskany Group; Dse, Schoharie and Esopus Formations, undivided; Dp, Palmerton Sandstone of Swartz (1939); Dbf, Buttermilk Falls Limestone of Willard (1938); Dm, Marcellus Shale.

The following description of the New Jersey Zinc Company, Palmerton, Pa., operation is supplied by Donald W. Kohls, geologist, the New Jersey Zinc Company:

Proximity of the anthracite fields of Pennsylvania, the zinc deposits of New Jersey, and the large surrounding marketing area prompted the New Jersey Zinc Company, in the late 1890's, to select the area immediately north of the Lehigh Gap as a location admirably suited for the manufacture of zinc.   As a consequence, the Borough of Palmerton was founded, named for Stephen S. Palmer, then president of the company, deriving its life blood from the world-famous Franklin mine.

Until its depletion in 1954, Franklin was the major source of the ore smelted at Palmerton. The majority of ore now smelted in sphalerite, brought from the company's mines at Friedensville, Pa.; Austinville and Ivanhoe, Va.; and Jefferson City and Flat Gap, Tenn. This supply is supplemented by purchased sulfide ore as well as oxide ore from the company's mine at Sterling Hill, N. J.

The zinc is smelted by a pyrometallurgical process. Sphalerite is first roasted to free the zinc from the sulfur, producing impure zinc oxide and sulfuric acid. The impure zinc oxide is mixed with bituminous and anthracite coal, coked, and transferred to a vertical retort where metallic zinc is recovered. The zinc is further refined by fractional distillation to produce 99.99 percent pure zinc.

This zinc is used in die castings rather than in galvanizing, where high-purity zinc is not required.

In addition to metallic zinc, large quantities of zinc oxide are produced here. This pigment is used in paint, rubber, plastic, photoconductive paper, medicine, and a variety of other products.

Spiegeleisen, an iron alloy containing 20 percent manganese, is an important byproduct of smelting the Sterling Hill ore. This material is used in steel manufacturing.

Cadmium, always found as a trace element in sphalerite where it substitutes for zinc, is produced as a byproduct in the treatment of sulfide ores.

Mileage

32.1   Stop sign, Palmerton, proceed straight ahead. We will be traveling on a Wisconsin outwash terrace.

33.3   Railroad underpass. Road rises above Wisconsin outwash terrace at an altitude of 400 feet and onto remnants of an Illinoian(?) outwash terrace at an altitude of 445 feet.

33.8   Junction with Pa. Rte. 248E, bear LEFT.

34.2   Lehigh Gap straight ahead.

34.3   On left, Bloomsburg Red Beds (stop 4).

34.6   On left, quartzite-argillite unit of Shawangunk Conglomerate.

35.4   RIGHT turn onto Pa. Rte. 873S and cross Lehigh River.

35.7   RIGHT turn at fork in road onto Mountain Road. Illinoian(?) drift on right.

36.0   On right, shale pit in Martinsburg Formation.

36.5   Blue Mountain on right. Road crosses sloping plain of Illinoian(?) drift (probably till mixed with colluvium from the mountain).

38.0   Pass over Pennsylvania Turnpike. On right, Shawangunk Conglomerate in Blue Mountain is steeply overturned to southeast.

39.0   LEFT turn.

39.8   On right, Penn Big Bed Slate Company quarry. STOP 2.

STOP 2. **PENN BIG BED SLATE COMPANY QUARRY**
    I, ILLINOIAN(?) DRIFT AND CONGELITURBATE
    II, STRUCTURE, STRATIGRAPHY, AND ECONOMIC GEOLOGY OF THE MARTINSBURG FORMATION

CAUTION: The quarry walls are steep-sided—do not get too close to the edge!

In the north end of the quarry, the overburden consists of 14 feet of generally flat lying tabular chips and fragments of slate as much as 6 inches long and ¼-inch thick. Pebbles and boulders from rocks north of the Martinsburg outcrop belt are scattered throughout this shale-chip gravel (fig. 44). The gravel is overlain by about 10 feet of horizontally stratified clay and pebbly clay which grades up into nearly 20 feet of unsorted and unstratified yellowish- and grayish-orange bouldery clay till. The deep oxidation and leaching of the till shows that it is older than the fresher Wisconsin drift seen earlier. It is evidence of an earlier glaciation, tentatively assigned to the Illinoian Glaciation. The shale-chip gravel is interpreted to have been derived from bedrock hills a few hundred feet to the north and was probably emplaced by solifluction in a periglacial environment. It can be termed a congeliturbate. The scattered erratics in the gravel may have been brought in by streams, or more likely, because the larger boulders are angular and as much as 14 inches long, by the advancing Illinoian(?) glacier. The clay overlying the congeliturbate may be lacustrine sediments deposited in temporarily ponded areas as the ice advanced (note the quarry area apparently occupies the site of a small buried valley). The overlying till was deposited by the

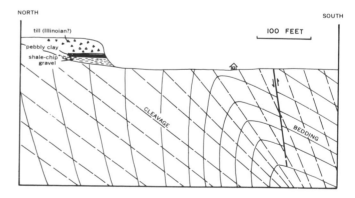

Figure 44. Diagrammatic geologic section of the Penn Big Bed Slate Company quarry, looking east, showing Illinoian(?) drift overlying Pen Argyl Member of Martinsburg Formation. Vertical scale same as horizontal.

south- or southwest-advancing Illinoian(?) glacier as it overran the area.

This slate quarry is the only one active in the Lehigh Valley at present. There are hundreds of abandoned quarries in this area. The quarry is in the Pen Argyl Member (upper member) of the Martinsburg Formation. Behre (1927, 1933) divided the Martinsburg into three members, but Stose (1930) believed that the upper member was the lower member repeated by folding. Stratigraphic and structural evidence confirms

Behre's interpretation. The Martinsburg was divided into three mapable members (Drake and Epstein, 1967) in almost the same way as Behre first defined them (see table 1).

The Martinsburg Formation is thick bedded in the quarry (fig. 45), typical of the Pen Argyl Member. One "big bed" seen near the top of the quarry is 12 feet thick (measured orthogonally) and is 21 feet thick along the "split" (cleavage). The rock is cyclically bedded, with medium-gray slate grading up into grayish-black

Figure 45. Thick-bedded slate in the Pen Argyl Member of the Martinsburg Formation, quarry of Penn Big Bed Slate Company, 2,000 feet north of Slatedale, Pa. View looking to the northeast. Cleavage dips 55°SE. in the axial plane, but fans the fold on the limbs of the overturned anticline (compare with fig. 44). The thickest bed in the quarry, the Penn Big Bed, seen in the photograph, is 12 feet thick measured perpendicular to bedding. Note the steeply dipping fault on the gently dipping southeast limb of the fold. A large piece of slate, about 10 feet long, is being hoisted by the overhead cable towards the mill. The overhead cables have a breaking strength of 490 tons. The smooth-faced slate beds on the right side of the photo have been cut with a wire saw. Note the saw marks nearly perpendicular to bedding. Compare these to the shattered beds on the left, which have been blasted. Blasting is one technique of working the slate which results in tremendous waste that may be more than 50 percent in some quarries. The shed is the signalman's house, who directs the hoisting upon vocal instructions from men in the quarry. The engine houses are located to the south beyond the brink of the quarry. The blocks of slate are hauled to small cars on tracks where they are carried to the mill. Note the glacial deposits overlying the slate.

Figure 46.   Calcite-quartz bedding-plane slickensides (white) offset by cleavage, in Pen Argyl Member of Martinsburg Formation.   Exposed on wall of Albion quarry of the Stephans-Jackson Slate Company, one-half mile south of Pen Argyl, Pa.   Height of area shown in photograph about one foot.   Similar structures are seen in the Penn Big Bed quarry.

carbonaceous slate.   Thinner beds of graywacke may form the base of some of the cycles, but they are not generally present.   The graywacke contains primary convolutions and other penecontemporaneous sedimentary structures.   Van Houten (1954) and McBride (1962) interpreted these structures as indicators of turbidity-current deposition.

The quarry is presently about 200 feet deep. An abandoned quarry about 300 feet along strike to the west is about 300 feet deep but is filled with water to a depth of 200 feet.   The slate taken from the quarry is used for roofing, blackboards, bulletin boards, floor tiles, structural slate, flagging, aquaria bottoms, sills and treads, and billiard table tops.   About 50 percent of the rock removed ends up on the waste pile.   Better quarrying procedures might cut this loss in half. The operation of this quarry, including the use of the wire saw, will be explained at the stop. Behre (1933) presents a complete discussion of slate-quarry operations.

The quarry is an overturned anticline (figs. 27, 44, and 45).   The mechanism of deformation was predominantly passive folding, although the many slickensided surfaces in the quarry indicate some flexural slip movement.   Cleavage wrinkles these surfaces, indicating that bedding slippage preceded passive folding (fig. 46).   The cleavage is generally a flow cleavage, but small-scale displacements on some bedding surfaces indicates that there has been recognizable movement along some cleavage planes.   The cleavage is an axial-plane cleavage and forms a fan with an angle of about 24°.

Mileage
39.8   Retrace route to Lehigh Gap.
40.7   Stop sign, RIGHT turn onto Mountain Road.
43.9   Intersect Pa. Rte. 873 and proceed across Lehigh River.
44.2   Stop sign, RIGHT turn onto Pa. Rte. 248E.
44.3   Fork in road, bear RIGHT on Pa. Rte. 45S.
44.4   RIGHT turn into parking area west of Atlantic gas station.

STOP 3.   I, MARTINSBURG-SHAWAN-
                    GUNK CONTACT ALONG
                    ABANDONED RAILROAD
                    GRADE IN LEHIGH GAP
            II, STRATIGRAPHY AND SEDI-
                    MENTATION OF THE LOWER
                    SHAWANGUNK CON-
                    GLOMERATE

The bedrock structure near the Lehigh River is shown in figure 47.   A syncline and anticline occur in Blue Mountain at Lehigh Gap.   These die out rapidly to the west.   At this stop we will see the unconformable contact between the Martinsburg Formation and Shawangunk Conglomerate.   Detailed structural relations are shown in figure 48.   Southeast-dipping slaty cleavage is well developed in the Martinsburg, 200 feet (stratigraphically) below the contact. The cleavage disappears 120 feet below the contact and bedding-plane slickensides become prominent.   Steps on the slickensides indicate northward movement of the overlying beds. The uppermost 8 inches of the Martinsburg is heavily slickensided and contains fault gouge and breccia with internal structures also indicating northwest movement (fig. 22).   Evidence, listed in the section on the Taconic orogeny in eastern Pennsylvania, indicates that the prominent cleavage in the Martinsburg was not produced during the Taconic orogeny and is an Appalachian feature.   A steep northwest-dipping faint cleavage is seen in thin section in the area where the prominent southeast-dipping cleavage is absent.   It is not certain what the age or origin

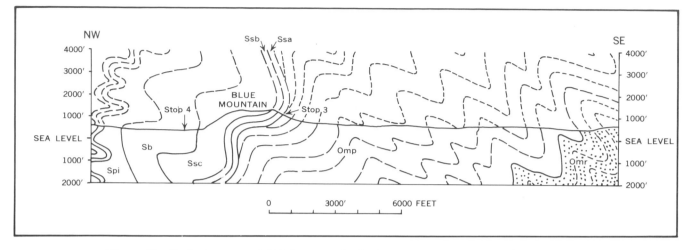

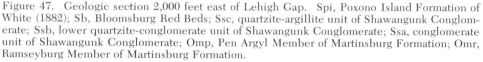

Figure 47. Geologic section 2,000 feet east of Lehigh Gap. Spi, Poxono Island Formation of White (1882); Sb, Bloomsburg Red Beds; Ssc, quartzite-argillite unit of Shawangunk Conglomerate; Ssb, lower quartzite-conglomerate unit of Shawangunk Conglomerate; Ssa, conglomerate unit of Shawangunk Conglomerate; Omp, Pen Argyl Member of Martinsburg Formation; Omr, Ramseyburg Member of Martinsburg Formation.

of the cleavage is, but recent work by I. B. Alterman (written commun., October 1968) has shown that in other places there is an early cleavage, generally obliterated by later deformation that she believes developed during the emplacement of the nappe of Taconic age in the Great Valley discussed earlier (section on the Taconic orogeny in eastern Pennsylvania). Perhaps the northwest-dipping cleavage at this stop is the one described by Alterman. The Martinsburg-Shawangunk contact, which also separates lithotectonic units 1 and 2 (see section on structural geology and fig. 27), is interpreted as a décollement.

Quartz, chert, and quartzite pebbles in the basal conglomerate unit of the Shawangunk indicate that the Martinsburg was breached and underlying rocks were being eroded and supplying pebbles during early Shawangunk time. The sharp contact and angular relationship between the Shawangunk and Martinsburg, and the very different environments of deposition under which these two units accumulated, indicate an unconformable relationship. The contact is a folded and faulted unconformity.

Two cleavages (one a slip cleavage) occur in a 2-inch-thick argillite bed in the Shawangunk Conglomerate about 50 feet north of the contact (fig. 49).

The contact between the conglomerate unit and overlying lower quartzite-conglomerate unit may be seen about 250 feet north of the Shawangunk-Martinsburg contact. The planar-bedded conglomerates and crossbedded sandstones seen in the lower Shawangunk is in-

dicative of fluviatile sediments deposited by streams of great competency, high gradient, and low sinuosity (see section on geographic and tectonic environments of deposition). These streams flowed to the northwest off the highlands uplifted during the Taconic orogeny. A possible antidune will be seen in the conglomerate unit (fig. 50).

A few miles to the south, a partly dissected Illinoian(?) outwash terrace along the east side of Lehigh River can be seen from this stop.

Mileage

44.4   LEFT turn onto Pa. Rte. 248W, heading toward Palmerton.

45.3   RIGHT turn onto service road.

45.4   Cross Aquashicola Creek, RIGHT turn. Quartzite-argillite unit of Shawangunk Conglomerate to left.

45.5   Shawangunk-Bloomsburg contact on left.

45.7   Mud cracks in Bloomsburg Red Beds on left.

46.0   Wisconsin valley train (outwash) to right.

46.1   RIGHT turn onto Lehigh Avenue, road follows Wisconsin valley train.

46.5   Travel two blocks, LEFT turn onto Fifth Street.

46.6   Travel one block, traffic light. LEFT turn onto Delaware Avenue.

46.7   Parallel park on right at First United Church of Christ. Lunch will be served inside.

LUNCH STOP

46.7   Continue west on Delaware Avenue.

47.0   LEFT turn onto Third Street (just past traffic light, and immediately before railroad underpass).

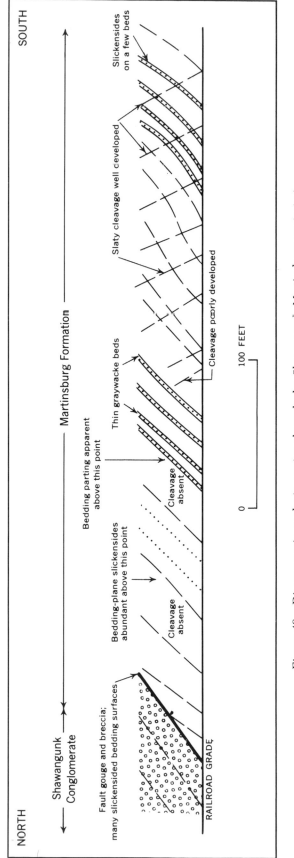

Figure 48.   Diagrammatic geologic section through the Shawangunk–Martinsburg contact at Lehigh Gap, showing the dying out of the prominent slaty cleavage in the Martinsburg as the contact is approached.

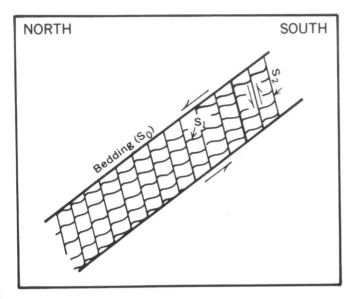

Figure 49. Diagrammatic sketch showing a 2-inch-thick argillite bed in conglomerate unit of Shawangunk Conglomerate. S_0, bedding; S_1, slaty cleavage; S_2, slip cleavage. The first cleavage (S_1) may have formed in response to interbed shear with overriding beds moving to the north as indicated by bedding-plane slickensides. Compare with figure 29B.

A

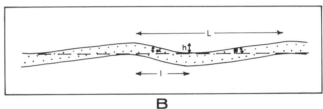

B

Figure 50. Possible antidune in conglomerate unit of Shawangunk Conglomerate, Lehigh Gap, indicative of upper flow regime. A, Bedding surface of bedform. Current movement was to the northwest (right) as shown by crossbedding trends in surrounding rocks. Note the steeper southeast slope. Laminations parallel the bedform, as shown in the drawing, B. The asymmetry of the bedform is generally not considered to be typical of antidunes, but the parallelism of laminations has been figured by Allen (1966, p. 165). Rippled water at top of photograph is the Lehigh River. B, Dimensions of possible antidune in laminated medium to very coarse grained conglomeratic quartzite. It overlies a 1½-foot conglomerate bed. $L = 2'$, $l = 7''$, $h = 1½''$, $\alpha = 13°$, $\beta = 6°$.

47.2 On left, Wisconsin outwash plain; on right Bloomsburg Red Beds.

47.5 Mud cracks in steeply-dipping Bloomsburg Red Beds to right.

47.9 RIGHT turn onto service road.

48.0 Park along soft shoulder of road. STOP 4.

STOP 4. **I, ILLINOIAN(?) OUTWASH**
II, STRATIGRAPHY AND SEDIMENTATION OF THE UPPER SHAWANGUNK AND BLOOMSBURG RED BEDS
III, BEDDING SLIPPAGE, WEDGING, AND TELESCOPING IN THE BLOOMSBURG RED BEDS

In the small valley above the transitional contact between the Bloomsburg Red Beds and the quartzite-argillite unit of the Shawangunk Conglomerate is a deeply weathered deposit consisting of rounded pebbles and boulders in a clay to very fine sand matrix. Two miles to the south, along the Lehigh River, a similar but better exposed deposit is horizontally stratified. The terrace form, stratification, and rounded pebbles show that this is an alluvial deposit (an outwash terrace or valley train), possibly of Illinoian age. A pebble count indicates a north and northeast source area.

Note the fining-upward cycles of crossbedded sandstones with mud clasts at the base, rippled to unevenly bedded shaly siltstones and sandstones in the middle, and indistinctly mudcracked bioturbated shaly siltstones with dolomite concretions at the top. These have been interpreted as meandering stream deposits of a low alluvial coastal plain (see section on geographic and tectonic environments of deposition at the beginning of this report). The transitional continental-marine environment of the argillite-quartzite unit of the Shawangunk Conglomerate will also be discussed. Note, in particular, the rather unique beds of nodular phosphate, siderite, and chlorite.

The near-vertical beds at this stop are in the northwest limb of an overturned anticline (fig. 47). Many bedding planes are slickensided, and the steps indicate northwest movement of the overriding beds. Small-scale wedging corroborates this direction of movement (figs. 25, 51). The net telescoping must have been considerable.

At this locality 900 feet of red beds are exposed. The Bloomsburg is estimated to be 1,500 feet thick. In the exposed section, 128 planes of bedding slippage were counted, so that the total number of such planes in the Bloomsburg may be 1,500/900 x 128, or 213. For purposes of discussion, it is estimated that there has been about 2 feet of downdip displacement along 100 feet of slippage surface. Therefore, for each 100 feet, in a northwest direction parallel to bedding in the Bloomsburg, there could have been 213 x 2, or 426 feet of displacement. Because northwest-dipping beds are at least 6,000 feet deep at this locality (fig. 47), the minimum estimated telescoping is 426 x 6,000/100, or 25,560 feet, or nearly 5 miles. The actual depth of northwest-dipping beds may be more than 10,000 feet, so that total telescoping could be more than 8 miles. If similar wedging and bedding slippage extends up into the Poxono Island Formation of White (1882), where the structural style between lithotectonic units 2 and 3 changes, then the amount of displacement could be even greater.

The estimated displacement parallel to bedding may not be unreasonable. Figure 27 shows that the rocks in lithotectonic unit 3 have been shortened 50 percent. Shortening taken up by folding is at least 3 miles. The entire unit probably moved considerable distances while shortening within the unit took place.

The glide zones between all the lithotectonic units in the field-trip area may extend to the northwest where they may shear upward, producing folds similar to those described by Gwinn (1964).

From the bend in the road, three sags can be seen in Blue Mountain above the Pennsylvania Turnpike Extension 2 miles to the west. The sags occur where a northwest-dipping fault emerges from the tightened core of an overturned anticline (figs. 27, 52), showing that the intensity of deformation in lithotectonic unit 2 increases to the west. The antiforms and synforms shown in figure 52 were first described by Dyson (1956).

Mileage

48.0   Follow service road to stop sign.

48.3   Stop sign. LEFT turn onto Pa. Rte. 248.

48.5   Bear LEFT on Pa. Rte. 248E and pass through Lehigh Gap.

50.2   Bear LEFT and follow Pa. Rte. 248E.

50.4   Illinoian(?) till in roadcuts to right and left.

52.2   Berlinsville, LEFT turn onto Pa. Rte. 946.

52.6   We are riding on the upper member (Pen Argyl Member) of the Martinsburg Formation. Hills to right (south) are supported by sandstones in the middle (Ramseyburg) member of the Martinsburg Formation.

55.3   Intersection in Danielsville. LEFT turn toward Little Gap. Ascend Blue Mountain at Little Gap and cross overturned, nearly recumbent rocks of the Shawangunk Conglomerate and Bloomsburg Red Beds (fig. 35).

56.1   On left, roadcuts in conglomerate unit of Shaw-

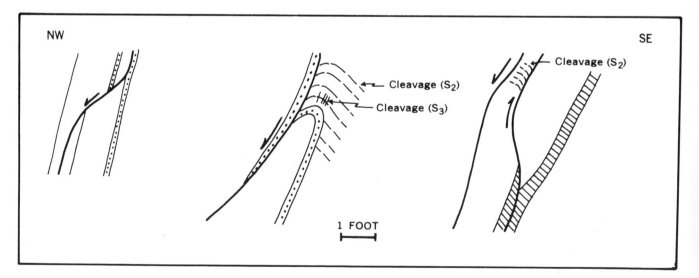

Figure 51. Diagrammatic sketch of wedges and bedding-plane slips in the Bloomsburg Red Beds, Lehigh Gap, stop 4, second day.

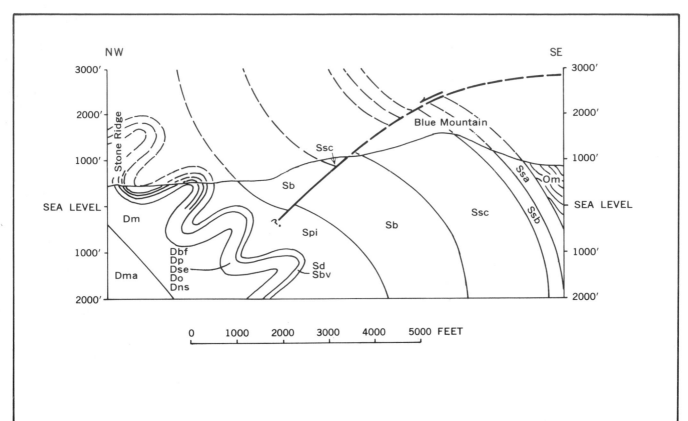

Figure 52. Geologic section through Blue Mountain, 1,500 feet west of the Pennsylvania Turnpike Extension tunnel. Dma, Mahantango Formation; Dm, Marcellus Shale; Dbf, Buttermilk Falls Limestone of Willard (1938); Dp, Palmerton Sandstone of Swartz (1939); Dse, Schoharie and Esopus Formations, undivided; Do, Oriskany Group; Dns, New Scotland Formation; Sd, Decker Formation; Sbv, Bossardville Limestone; Spi, Poxono Island Formation of White (1882); Sb, Bloomsburg Red Beds; Ssc, quartzite-argillite unit of Shawangunk Conglomerate; Ssb, lower quartzite-conglomerate unit of Shawangunk Conglomerate; Ssa, conglomerate unit of Shawangunk Conglomerate; Om, Martinsburg Formation.

Mileage

angunk Conglomerate; quartz pebbles as much as 4 inches long; beds overturned steeply to southeast about 30°.

56.7   Blockfield to right in Little Gap derived from surrounding bedrock and emplaced by solifluction and frost heaving.

57.2   On right, quartzite-argillite unit of Shawangunk Conglomerate overturned and dipping gently to southeast.

57.7   Nearly recumbent Bloomsburg Red Beds to right. Sand pits in Palmerton Sandstone in Chestnut Ridge to north.

58.5   Base of mountain. Valley filled with Wisconsin outwash.

58.8   RIGHT turn toward Kunkletown.

59.6   Valley of Buckwha Creek to right, underlain by Wisconsin outwash. Valley narrows to the east approaching the pre-Wisconsin Lehigh River-Delaware River drainage divide.

62.3   On left, roadcut in Bloomsburg Red Beds.

62.6   Narrowest part of Aquashicola Creek Valley. Former Delaware River-Lehigh River drainage divide.

62.9   Cross Aquashicola Creek. Bloomsburg Red Beds on left. Note widening of valley as we proceed eastward.

63.7   LEFT turn, cross Aquashicola Creek.

64.0   Outcrop on left in Poxono Island Formation of White (1882).

64.4   RIGHT turn into quarries of Universal Atlas Cement Division, U. S. Steel. Route follows ridgetop road.

67.4   Fork in road, bear LEFT and continue ascent of ridge.

68.3   Small road on left, keep RIGHT.

68.5   Clay pit on left, follow blacktop road to right.

68.8   Cross clay pit.

69.0   Buses park at red shed. STOP 5.

STOP 5.   **CLAY AND SAND PITS OF THE UNIVERSAL ATLAS CEMENT COMPANY; STRATIGRAPHY AND STRUCTURE OF CHESTNUT RIDGE**

Residual clay deposits of commercial importance are developed in limy shales and shaly limestones in Cherry and Chestnut Ridges between Kunkletown and Saylorsburg, Monroe County, Pa. The deposits are located southwest of the Wisconsin terminal moraine and are believed to have formed during pre-Wisconsin weathering. No clay deposits are present northeast of the moraine, and it is presumed that any clay that formed there was stripped away during the Wisconsin glacial advance.

The history of clay mining in the area goes back to 1891 (Peck, 1922), and since that time, the clay has had a variety of uses, including white cement, ceramics, porcelain, electrical insulators, and paint. In the pits near Kunkletown, the clay is currently being removed from deeply leached rocks (sedimentary rock saprolites) of the New Scotland Formation, Shriver Chert, and Buttermilk Falls Limestone of Willard (1938). The stratigraphic units exposed in the quarry are shown in figure 53. The clay minerals are chiefly illite (Md muscovite) and kaolinite. The clay is shipped to the company's cement plant at Northampton, Pa., for use in the manufacture of white cement. The mineralogy, structures, and origin of the clays have been discussed in detail by Epstein and Hosterman (1969).

The following description of the white clay operations is supplied by G. F. Smith, geologist, Universal Atlas Cement Division, United States Steel Corporation:

Since 1932, Universal Atlas has used clay from the Kunkletown property as a raw material in the manufacture of White Cement at Northampton, Pa. Cement raw materials must be blended to reach certain fixed ratios of SiO_2, Al_2O_3, Fe_2O_3, and CaO before calcining. The silica, alumina, and ferric oxide components of White Cement are obtained from the Kunkletown pits by proper proportioning of the clays found there. These clays average: 80.8 percent SiO_2, 12.1 percent Al_2O_3, 0.6 percent Fe_2O_3, and 0.3 percent MgO. White Cement raw-material requirements include a very low Fe_2O_3 content; the Kunkletown clays meet this requirement.

Prospecting starts with air-photo-based contour maps and selection of topographically favorable areas, followed by self-potential and resistivity studies to pinpoint probable low-iron clay bodies. Test drilling, using rotary or driven coring methods is then used to outline the clay bodies and to determine their actual composition.

Development of the pits requires test drilling or test pits, but day-to-day operations are guided largely by the experience of the supervisors, who use visual clues for silica and iron content. The supervisors are also aided by daily sample reports from the chemical laboratory at Northampton.

These clay pits provide 200 tons of raw material per day. One small power shovel loads the 15-ton trucks for the 22-mile haul to Northampton, where the clay is crushed, ground, made into a slurry, and blended with limestone into cement kiln feed.

Sand is quarried from the Palmerton Sandstone of Swartz (1939) and Ridgeley Sandstone and is trucked to the crusher at Kunkletown a few miles to the west.

We will stop at the upper two pits in which rock units from the upper part of the Decker Formation to the lower part of the Schoharie and Esopus Formations are exposed (fig. 54). If time permits, we will visit the lower pit and see the Palmerton Sandstone, Buttermilk Falls Limestone, and Marcellus Shale.

Interpretation of the environments of deposition has been discussed in the section on stratigraphy in the beginning of this report. The rocks in the pits are so weathered that many of the sedimentary details are obscured. The most conspicuous sedimentary structures are to be found in the Ridgeley Sandstone of the Oriskany Group. Crossbedding, planar bedding and laminations (foreshore laminations, fig. 20C), and thick-shelled brachiopods (*Acrospirifer murchisoni*) are well exposed and aid in interpreting this unit as an ancient beach deposit.

Four tight, nearly isoclinal folds with axial planes that dip steeply to the southeast can be traced on the floors of the pits (fig. 54). The folds have wavelengths of about 1,000 feet and are believed to be superposed on an anticlinal crest or gently dipping limb of a larger fold in underlying lithotectonic unit 2 (fig. 27).

The lower pit is in an overturned syncline in the Palmerton Sandstone, Buttermilk Falls Limestone, and Marcellus Shale (fig. 54). In places the upright north limb of the fold dips more steeply than the overturned south limb, so the fold is "tighter" than isoclinal. Quarrying during November 1968, exposed gently northwest-dipping doubly overturned Buttermilk Falls Limestone about 160 feet south of the south contact with the Palmerton Sandstone shown in the cross section in figure 54. Thus, the trough of the syncline in the Buttermilk Falls extends to the south where it becomes antiformal and resembles the structure shown in figure 43. In

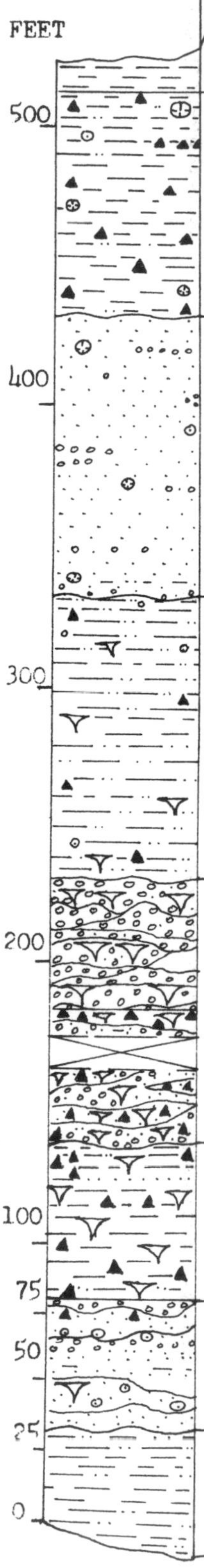

	UNIT	DESCRIPTION
FEET	Union Springs Shale Member of the Marcellus Shale	Shale, grayish-black, fissile, carbonaceous.
500	Buttermilk Falls Limestone of Willard (1938)	Clay, silty clay, and chert. Weathered; white and varicolored. Chert in pods and irregular beds. Hematitic near top. Fossiliferous. Thickness approximate.
400	Palmerton Sandstone of Swartz (1939)	Sandstone and conglomeratic sandstone. Sandstone, very coarse to coarse-grained. Quartz pebbles as much as 3/4 inch long. Medium-dark-gray; weathers yellowish-orange. Scattered coral molds. Thickness approximate.
300	Schoharie and Esopus Formations, undifferentiated	Siltstone and silty shale. Weathered; varicolored; minor chert; fossiliferous. Thickness approximate.
	Ridgeley Sandstone (Oriskany Gp.)	Quartz-pebble conglomerate. Pebbles rounded to well rounded; as much as 1 inch long. Crossbedded. Abundant lenses and beds of brachiopod shell hash.
200	Shriver Chert (Oriskany Gp.)	Chert, conglomerate, sandstone, and minor clay. Weathered; very light gray to light-gray. Chert replaces quartz sand and pebbles. _Acrospirifer murchisoni_ abundant.
100	New Scotland Formation	Chert, clay, and silty and arenaceous clay. Weathered; white to light-gray, pale-yellowish-orange to light-brown. Lenses and thin beds of coarse-grained sandstone near top. Fossils, mainly brachiopods, abundant, especially in chert.
75, 50	Stormville Member of the Coeymans Formation	Sandstone and minor beds and lenses of conglomerate. Sandstone, very fine grained to coarse-grained. Conglomerate containing pebbles as much as 1 inch long. Weathered; very pale orange to dark-yellowish-orange; crossbedded in places; scattered molds of crinoid columnals and brachiopods.
25, 0	Decker Formation	Siltstone, sandstone, and shale. Weathered; pale-yellowish-orange to moderate-brown; limonitic and hematitic.

Figure 53. Stratigraphic units exposed in clay- and sandpits of the Universal Atlas Cement Company, stop 5.

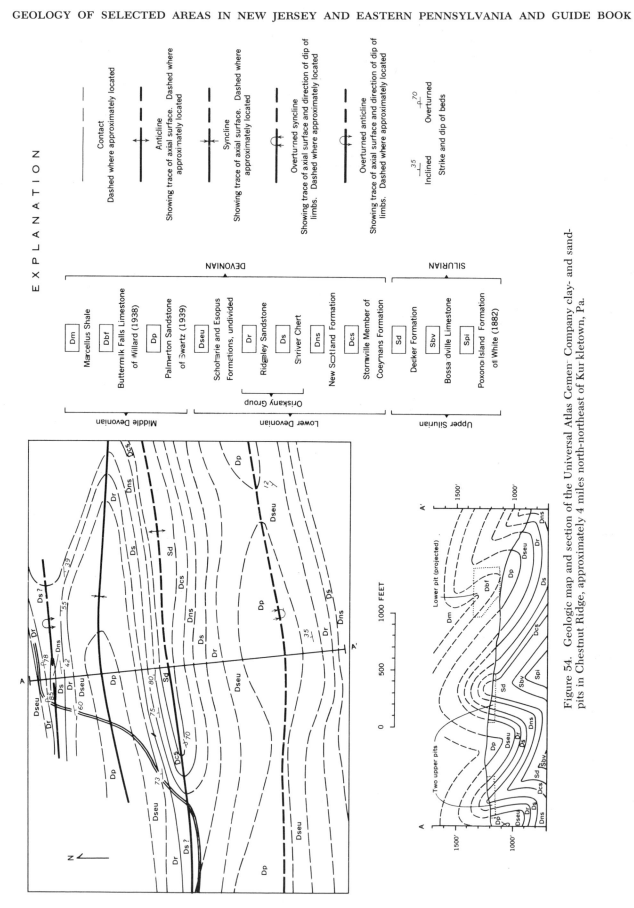

Figure 54. Geologic map and section of the Universal Atlas Cement Company clay- and sand-pits in Chestnut Ridge, approximately 4 miles north-northeast of Kurkletown, Pa.

Figure 55. Spheroidal weathering in massive Palmerton Sandstone seen in lower pit. Note vertical and horizontal joints that control the shape of the weathered masses.

the southeast section of the pit, note the spheroidal weathering developed between joints in the massive Palmerton Sandstone (fig. 55). The Palmerton is similar to the Ridgeley Sandstone except that it is poorly bedded and lacks spiriferid molds but has rare favositid and crinoid columnal molds.

Cascade folds, similar to those developed in lithotectonic unit 3, but on a much smaller scale, can be seen in the normal southeast-dipping limb of the syncline in the Buttermilk Falls Limestone (fig. 56).

Figure 56. Small cascade folds in the Buttermilk Falls Limestone on the normal southeast-dipping limb of a syncline, showing movement of overlying beds away from anticlinal crest. Hammer in center of exposure gives scale.

Mileage

69.0	Retrace route for 1.6 miles.
70.6	LEFT turn toward lower pit.
71.2	Stop in pit.
71.2	Return to intersection with upper pit road and continue on road to intersection with state road.
74.8	RIGHT turn and descend ridge on north side and enter Kunkletown.
75.5	Stop sign, Kunkletown. RIGHT turn and proceed to Saylorsburg.
75.8	Bear RIGHT at fork in road.
78.3	Clay pits of stop 5 to right on Chestnut Ridge.
82.1	Wisconsin terminal moraine which continues into Saylorsburg.
83.5	Stop sign, Saylorsburg. RIGHT turn toward Wind Gap.
83.6	Outlet of glacial Lake Sciota in terminal moraine. Road ascends Cherry Ridge.
84.0	LEFT turn at crest of ridge towards Bossardsville.
84.1	Pass over Pa. Rte. 115 and U. S. Rte. 209. Esopus through Buttermilk Falls interval exposed in roadcuts to right.
84.3	LEFT turn onto Pa. Rte. 115 and U. S. Rte. 209 towards Stroudsburg.
84.5	Wisconsin terminal moraine to left.
86.1	Hills to right underlain by Buttermilk Falls Limestone of Willard (1938).
86.6	Flat valley floor to right and left underlain by rhythmically laminated glacial lake clays, silts, and sands.
86.7	Foreset slope of delta (stop 1) to left.
87.3	Roadcuts in Schoharie Formation capped in places by the lower cherty limestone member of the Buttermilk Falls.
87.5	Eureka Stone Company quarry in the Schoharie Formation to right.
89.0	Intersection of U. S. Interstate 80W and U. S. 209N. Bear RIGHT on U. S. 209N.
93.2	Junction with U. S. Interstate 80E, keep RIGHT.
97.0	Turn RIGHT toward U. S. 209N (exit 52).
97.7	Turn LEFT into Holiday Inn Motel parking lot.

END OF ROAD LOG, FIRST DAY

SECOND DAY

Departure from the Holiday Inn Motel, East Stroudsburg, at 8:00 A.M. Field trip will follow route shown on figure 2.

Mileage

0.0	Leave Holiday Inn parking lot and proceed south on U. S. Rte. 209.
0.3	RIGHT turn onto U. S. Interstate 80.
2.1	Cross Brodhead Creek.
3.4	RIGHT turn to Dreher Avenue exit, exit 49.
3.5	Cemetery to right on Wisconsin outwash terrace; mausoleum on higher kame.

3.6   LEFT turn onto Dreher Avenue.
3.8   Road rises onto Wisconsin outwash terrace.
4.3   Follow Dreher Avenue which forks to right.
4.5   RIGHT turn onto Tanite Road.
4.9   RIGHT turn into sand and gravel pit. STOP 1.

STOP 1. WISCONSIN KAME DEPOSIT IN ABANDONED SAND AND GRAVEL PIT

In the lowest pit the kame deposit consists of more than 50 feet of cross-stratified lenticular beds of sandy pebble gravel with boulders as much as 3 feet long. Gravel composes about 80 percent and sand 20 percent of the deposit. Pebble counts indicate a north to northeast source for the material (Epstein, 1969).

In the higher pit the gravel is overlain predominantly by sand, silt, and clay which were probably deposited, in part, in a locally ponded area (fig. 57). Several interesting sedimentary structures may be seen about midway up the north wall of the pit. Several sand beds contain climbing ripple laminae (fig. 58) indicating rapid deposition of large quantities of sand (McKee, 1965), a characteristic of glacial melt-water deposits. Involutions are exposed in a 4-foot-thick interval under the rippled beds and consist of aimless contortions resulting from intrusion of clay, silt, and sand into overlying coarser beds (fig. 59). These structures are characteristic of load casting. The involutions are

Figure 58. Climbing current ripples (ripple drift) in laminated fine to medium sand in kame deposit. Current from left to right. Note that the bounding surfaces dip to the left, in an upstream direction.

similar to features that have been ascribed to periglacial frost action (see, for example, Schafer, 1949, pl. 1B). Schafer (written commun., 1965), however, agrees that these structures, so common in many glacial deposits, are load casts unrelated to frost action. The structures have also been termed ball-and-pillow by Pettijohn and Potter (1964, pl. 100A).

Figure 57. Kame deposit at stop 1, second day, about 1 mile west-southwest of Stroudsburg, Pa. Horizontally bedded sand, silt, and clay overlie inclined gravel beds. Arrow points to bed containing involutions shown in figure 59.

Figure 59. Involutions (load casts) resulting from intrusion of fine sediments into gravel beds in kame deposit. Height of area shown, 2.5 feet.

ALTERNATE STOP 1

Mileage

0.0 Leave Holiday Inn parking lot and proceed south on U. S. Rte. 209.

0.3 RIGHT turn onto U. S. Interstate 80.

2.1 Cross Brodhead Creek.

3.5 Cemetery to right is on Wisconsin outwash terrace; mausoleum on higher kame.

4.1 RIGHT turn to exit for U. S. 209.

4.3 RIGHT turn onto U. S. 209 heading south.

6.8 LEFT turn onto county road and pass under U. S. Interstate 80.

7.0 LEFT turn into sand and gravel pit of Pocono Industries, Inc. ALTERNATE STOP 1.

ALTERNATE STOP 1. WISCONSIN DELTA IN SAND AND GRAVEL PIT OF POCONO INDUSTRIES, INC.

The pit is in the western slope of a flat-topped 250-foot-high hill, whose top is at an altitude of about 680 feet. The hill is composed of medium to very coarse sand and gravel; pebbles, cobbles, and boulders are well rounded and generally do not exceed about 10 inches in length, although some are more than 1.5 feet long. Horizontal and inclined stratification are prominent (fig. 60). Two topset-foreset couples were exposed in the fall of 1968. Before this time, only the highest topset and scattered foreset beds were exposed, calling for an interpretation of an enlarging Lake Sciota as the ice retreated northeast after deposition of the delta near Sciota (stop 1, first day) (Field Conference of Pa. Geologists, 1967). The new exposures indicate that either there was an earlier lower outlet for Lake Sciota to the northeast at an altitude of about 535 feet or the deltas were deposited around stagnated blocks of ice in locally ponded areas whose water level changed in time and whose tops are coincidentally graded to the 680-foot level. The foreset beds in the pit dip to the southwest (fig. 60), indicating that the main supply of sediment was from the northeast. Note the radiating direction of dip of the foresets. Splendid examples of ripple laminations and frost-heave phenomena may be exposed.

4.9 Return to Tanite Road, turn LEFT.

5.2 RIGHT turn onto Dreher Avenue.

5.3 LEFT onto Vineyard Drive.

5.4 RIGHT onto Glen Brook Road.

5.7 Hills in golf course to right underlain by Schoharie Formation, Buttermilk Falls Limestone, and Union Springs Shale Member and the Stony Hollow Member of Cooper (1941), undivided, of the Marcellus Shale.

5.8 Cross McMichael Creek.

6.1 Ascend north slope of Godfrey Ridge. Buttermilk Falls Limestone, Schoharie Formation, and Esopus Formation are in scattered exposures in roadcuts to left.

6.8 Esopus Formation in roadcut on right.

7.1 Crest of ridge. Esopus exposed in roadcuts on right below.

7.8 Village of Stormville. Decker Formation exposed in creek on right. Intersection with Cherry Valley Road, bear RIGHT (west).

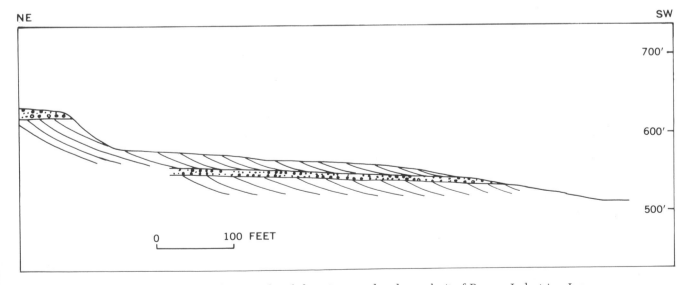

Figure 60. Section showing glacial deposit at sand and gravel pit of Pocono Industries, Inc.

8.7   Flat floor of Cherry Valley on left underlain, in part, by glacial lake clays, silts, and sands. Kittatinny Mountain, underlain by Shawangunk Conglomerate, forms skyline to left.

8.8   Exposure of Bossardville Limestone on right.

9.2   Cemetery behind church in valley to left is on a small kame.

9.7   Poxono Island Formation of White (1882) exposed in creek to left.

10.2   Abandoned quarries in Bossardville Limestone to right and left.

10.4   Village of Bossardsville, bear LEFT.

10.5   LEFT turn into Hamilton Stone Company quarry. Drive to upper working level of quarry.

11.0   STOP 2.

STOP 2. HAMILTON STONE COMPANY QUARRY, STRUCTURE AND STRATIGRAPHY OF THE DECKER FORMATION, BOSSARDVILLE LIMESTONE, AND POXONO ISLAND FORMATION

The Upper Silurian-Lower Devonian strata in easternmost Pennsylvania accumulated in in a mixed carbonate-clastic depositional province. The Upper Silurian rocks, which are well exposed in this quarry (fig. 61), are interpreted to have been deposited in a tidal flat-barrier bar-shallow subtidal environment. The recognition of paleoenvironments within this lithologic complex is based on mineralogy, texture, bedding characteristics, facies fossils, and most important, the vertical and lateral relations of these features.

The oldest rocks exposed in the quarry are the upper beds of the Poxono Island Formation of White (1882), and include shale, siltstone, sandstone, dolomite, and limestone. The uppermost beds are mud-cracked interlaminated calcite, dolomite, and ferran dolomite mudstone (fig. 62A, B). Some mud-cracked columns contain healed mud cracks as well as laminae that are thicker in the center of the curled "microbasin" (figs. 14A, 62B), indicating alternate periods of desiccation and deposition. In addition many bedding surfaces have raindrop imprints or gas pits. The laminae were probably the result of trapping of calcium carbonate muds by algal mats in supratidal flats after transportation from the intertidal zone during storms and spring tides. Between floodings, desiccation of the algal mat and bound sediment produced mud cracks. The mud-crack polygons were protected and perpetrated by the algae, probably in the same manner described by Black (1933) in modern sediments in the Bahamas. Successive algal mats mimicked the underlying mudcrack form and subsequent mud cracking fol-

lowed preexisting shrinkage cracks. (This mechanism may explain the very long mud-crack columns that will be seen in the Bossardville at the next stop on Godfrey Ridge.) During drying, evaporating marine waters in pores in the sediment became enriched in magnesium, and the calcium carbonate mud at and near the surface was dolomitized. (This process has been described in supratidal areas of modern carbonate depositional provinces by Illing and others (1965) in the Persian Gulf; Deffeyes and others (1965), Bonaire, Netherlands Antilles; Shinn and others (1965), the Bahamas.) Dolomite and ferran dolomite in the uppermost beds of the Poxono Island are somewhat irregular and some transect bedding (fig. 14A). Dolomicrite interclasts occur in micrite laminae. These features indicate replacement of calcium carbonate mud by dolomite penecontemporaneous with sedimentation and desiccation.

Overlying the laminated beds is a 1-foot bed of mud-cracked ferran dolomite containing peculiar color contortions (fig. 62C). These contortions may have been produced during dolomitization. (Other suggestions for the origin of these contortions are welcomed!) Electronprobe microanalysis of the color contortions showed no variation in calcium, magnesium, and iron between dark and light bands. Sulfur, however, was much greater in the dark bands than in the light and varied directly with the distribution of pyrite. The dark bands contain much more pyrite than the light bands. The upper part of this bed was eroded during transgression, and clasts derived from this bed were incorporated in the overlying intertidal finegrained limestone (fig. 62D). The overlying Bossardville Limestone, the quarry rock, contains an average of 49 percent CaO and 1.3 percent MgO (R. E. White, written commun., 1965). Converting these oxides to calcite and dolomite yields 82 percent calcite and 6 percent dolomite. Some of the magnesium, however, is in chlorite (as indicated by X-ray diffraction analyses), so that the percent of dolomite is actually less. The remaining minerals are predominantly quartz, muscovite, and pyrite. The Bossardville is a very thin bedded to laminated very fine to fine-grained limestone containing graded laminae and beds of ostracode-rich coarser limestone (fig. 62E, dark laminae) containing approximately 10 percent noncarbonate minerals at base, grading up into finer grained limestone with approximately 30 percent noncarbonate minerals (semiquantitative X-ray diffraction analysis based on peak-height ratios). The quartz content is nearly constant in a graded lamina, but chlorite and muscovite increase from approximately 7 percent at the base to 27 percent near the top and represents concentration of these platy minerals by slower settling. The base of a graded unit is sharp; the top is generally scoured. Rocks of similar description in the Manlius Formation (Devonian) of New York (Laporte, 1967) and in Ordovician and Silurian strata of the Williston basin (Roehl, 1967) have been interpreted as intertidal sediments.

The uppermost beds of the Bossardville are very finely laminated supratidal limestone and dolomite (fig. 62F). Dolomite makes up about 12 percent of the carbonate mineral content.

Within the lower beds of the overlying Decker Formation, supratidal mud-cracked interbedded and interlaminated dolomite and lesser coarser grained limestone are truncated by barrier bar (intertidal) planar-bedded quartzose biosparudite and fossiliferous calcareous sandstone. Overlying these sediments are flaser-bedded calcareous siltstone and sandstone, believed to be the shallow subtidal apron in front of the bar (fig. 63). Younger strata in the Decker are dominantly intertidal-shallow subtidal limestones and sandstones, similar to those described above. Vertical and lateral relations in northeastern Pennsylvania and New Jersey indicate

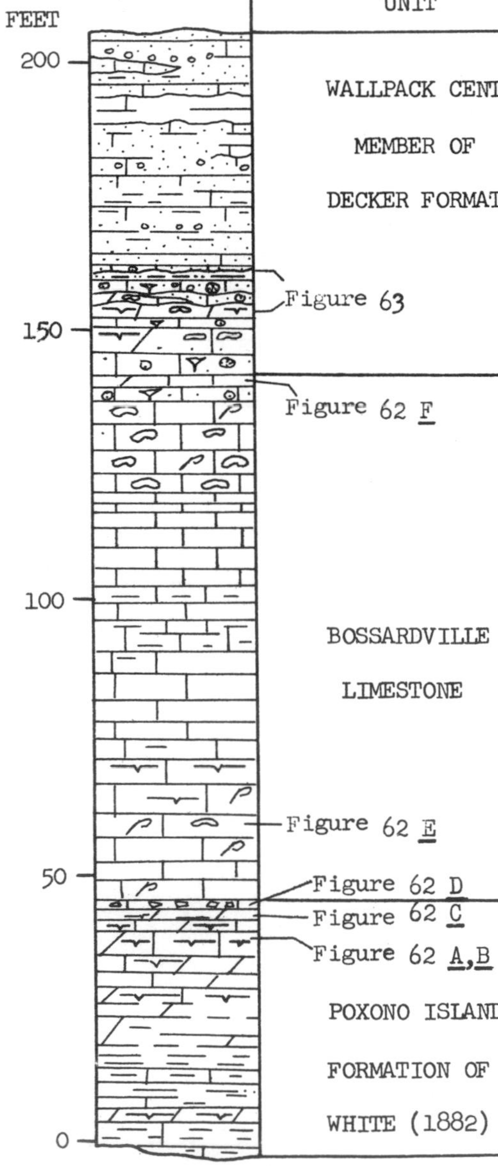

	UNIT	DESCRIPTION
WALLPACK CENTER MEMBER OF DECKER FORMATION — Figure 63		Beds and lenses of medium-gray to medium-light-gray fine- to coarse-grained quartzose limestone, medium-gray fine- to coarse-grained calcareous sandstone, siltstone, and conglomeratic sandstone; quartz pebbles are white and pink. Most abundant fossils include brachiopods, rugose and colonial corals, bryozoans, and crinoid columnals; leperditiid ostracodes abundant in basal beds in close proximity to mud-cracked dolomite beds. Upper contact concealed.
Figure 62 F — BOSSARDVILLE LIMESTONE — Figure 62 E — Figure 62 D — Figure 62 C — Figure 62 A,B		Graded laminae and very thin beds of dark-gray fine- to very fine grained pyritic limestone containing abundant to rare leperditiid ostracodes and scour-and-fill structure at base of graded units. Uppermost beds very finely laminated limestone and lesser dolomite.
POXONO ISLAND FORMATION OF WHITE (1882)		Interbedded and interlaminated medium-gray calcareous dolomite, medium-gray and mottled pinkish-gray-green to medium-greenish-gray to gray very fine grained dolomite, grayish-red limestone, and light-grayish-yellow-green shale and calcareous shale; contains several mud-crack intervals. Upper contact at top of dolomite bed which contains peculiar color contortions.

FEET
200
150
100
50
0

Figure 61. Stratigraphic column showing rocks exposed in Hamilton Stone Company quarry, stop 2, second day.

that most of the Wallpack Center Member of the Decker Formation represents a barrier bar facies (Epstein and others, 1967).

The structure in the quarry area consists of a series of folds in the Poxono Island Formation, Bossardville Limestone, and Decker Formation (fig. 64) whose axial planes strike N.48°–55°E., and axes plunge about 5°–10°SW. These folds are believed to be cascade folds superimposed on the Kemmererville anticline defined by the underlying Bloomsburg Red Beds

in lithotectonic unit 2 (fig. 26). Compare the more upright axial planes of the folds seen here with those seen in lithotectonic unit 3 farther west. The southwest-plunging Kemmererville anticline increases in amplitude to the northeast and nearly dies out in the quarry area. Fifteen folds have been mapped between the ridge crest in the quarry and several abandoned quarries less than 3,000 feet to the northwest. Wavelengths of these folds range from 150–750 feet, somewhat less than the average

A

B

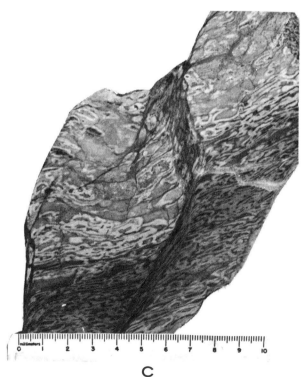

C

D

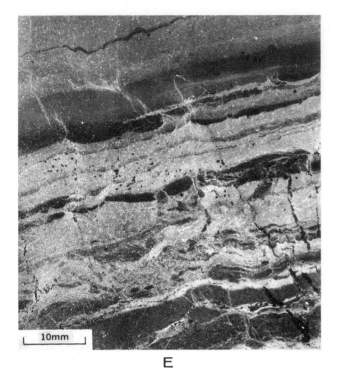

E   F

Figure 62. Sedimentary features in dolomites and limestones, Hamilton Stone Company quarry, Bossardsville, Pa. See figure 61 for stratigraphic location of specimens. Specimens arranged in stratigraphic order, A and B are oldest. A, B, Mud-cracked laminated mudstone containing laminae of ferroan dolomite, dolomite, and calcite. B is a negative print of an acetate peel showing gross mineralogy of one of the curled mud-crack polygons seen in A. Specimen stained with Alizarin Red S and potassium ferricyanide. Dolomite, very dark gray; ferroan dolomite, light to medium gray (upper 7 mm and lower 4 mm of specimen); calcite white to medium gray. Note secondary shallow mud crack (top center). Cleavage, dipping steeply to the left, crinkles and offsets laminae. C, Mud-cracked ferroan dolomite mudstone containing peculiar color contortions. Polished slab. Calcite makes up 5–10 percent of carbonate mineral content. Specimen contains minor (less than 5 percent) quartz and muscovite, rare chlorite, and very rare albite. Consists of light-bluish-gray (5B8/1) contortions in medium-light-bluish-gray (5B6/1) "matrix." Rock weathers a characteristic grayish-orange (10YR7/4). Contortions are well developed and are parallel to bedding at base of sample and are less distinct at top. On a sawed surface, contortions resemble burrows, but (1) serial cuts show that they are continuous surfaces and not tubes, and (2) the mud-crack fillings remain intact (are not bioturbated) even though contortions pass through them. The vertical structure is a desiccation crack along which the development of cleavage has been concentrated. Note that the contortions are found in the sediment filling in the crack and that they are deformed by cleavage. Beds immediately above, D, contain intraclasts derived from the color-contorted beds. D, Interbedded and interlaminated very fine grained limestone and desiccation breccia. Intraclasts are dolomudstone derived from bed shown in C. Note cleavage in lime mud matrix. Specimen stained with Alizarin Red S. Negative print of acetate peel. E, Limestone containing graded beds and laminae of ostracode-bearing coarser grained limestone (dark color) with a carbonate:clastic ratio of 9:1 at base grading up into finer grained limestone with a carbonate:clastic ratio of 7:3. Many beds have small-scale scour features. Specimen stained with Alizarin Red S. Negative print of acetate peel. Note slip cleavage. F, Very finely laminated very fine grained limestone and lesser dolomitic limestone of probable organo(algal)-sedimentary origin. Darker laminae dolomitic. Specimen stained with Alizarin Red S. Negative print of acetate peel.

wavelength of folds in lithotectonic unit 3 (see table 3). Early movement along bedding surfaces (S_0) is indicated by bedding-plane slickensides. These are intersected and wrinkled by a later cleavage (fig. 65A), indicating that flexural slip folding preceded passive folding. The cleavage (S_1) wrinkles and displaces bedding laminae and consists of shear surfaces

that bound microlithons which average less than 1 mm in thickness (fig. 65B). Lineations in a perpendicular to the fold axes, and consisting of fine striae, and small wrinkles in b are locally found on the slip cleavage (S_1) surfaces. In rare exposures, a second slip cleavage (S_2), that wrinkles the earlier slip cleavage, is found. Rocks of different lithic types behaved dif-

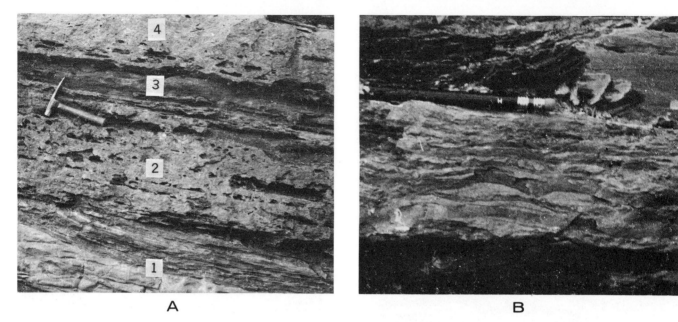

A   B

Figure 63. Transgressive sequence of intertidal (beach) and subtidal facies transecting supratidal facies, base of Decker Formation, Hamilton Stone Company quarry, Bossardsville, Pa.
See figure 61 for stratigraphic location of photographs. A, Units from bottom to top are: (1)
massive (extensively burrowed) mud-cracked dolomite containing remnants of finely laminated
dolomite and dolomite in lower half and interlaminated and interbedded partly burrowed ostracode-rich limestone and massive to finely laminated dolomite in upper half (supratidal to intertidal); (2) planar-bedded quartzose biosparudite and fossiliferous calcareous sandstone (intertidal [beach]) which truncate unit 1; (3) flaser-bedded partly-burrowed calcareous sandstone
and siltstone containing scattered lenses of quartzose biosparudite (subtidal); (4) repeat of
unit 2; note sharp base of unit and scoured upper surface of the underlying unit. B, Close-up
of part of flaser-bedded unit 3. Crossbedded ("starved ripples") calcareous sandstone interbedded with laminated calcareous siltstone, sandstone, and minor shale. Unit contains scattered
burrows.

ferently during deformation, so that more competent rocks, such as sandstones in the Decker
Formation and massive limestones in the Bossardville Limestone, were mainly concentrically
folded, whereas more shaly units deformed by
movement along cleavage (fig. 65C, D), resulting
in disharmonic folding that, on a smaller scale,
probably duplicates the regional structure.

From point A (fig. 64), the Trimmers Rock
Sandstone and Catskill Formation in the southwest-plunging Weir Mountain syncline can be
seen to the west. To the north, Godfrey Ridge,
the Pocono Plateau, and the delta and lakebeds
of stop 1 yesterday can be seen. Eastward, an
anticlinal ridge (the Kemmererville anticline),
glacial deposits in Cherry Valley, and Blue
Mountain are visible (fig. 1). The Decker Formation is well exposed at A. The Bossardville
Limestone and Poxono Island Formation have
been recently uncovered by quarry operations
in the anticline 400 feet southeast of A. In the
northwest limb of this anticline are exposed the
folds shown in figure 65C.

At B, mud cracks in the Poxono Island Formation, and desiccation breccias, other sedimentary structures, slip cleavage, bedding slickensides deformed by slip cleavage, and small-scale
disharmonic folding (shown in fig. 65D) in the
Bossardville Limestone will be examined. Note
the shortening of the mud-crack polygons perpendicular to the fold axes.

At C, the relations of parasitic folds, cleavage,
and mud cracks in the Poxono Island are seen
(fig. 66).

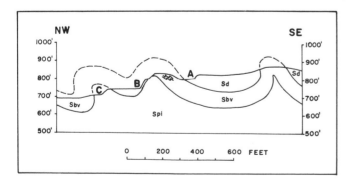

Figure 64. Geologic section of Hamilton Stone Company quarry.
A, B, and C are localities described in text. Sd, Decker Formation; Sbv, Bossardville Limestone; Spi, Poxono Island Formation of White (1882).

Figure 65. Structural features, Hamilton Stone Company quarry, Bossardsville, Pa. A, Slip cleavage intersecting and deforming calcite-smeared slickensided bedding surface indicating flexural slip folding preceded passive slip folding. Basal beds on the Bossardville Limestone. B, Negative print of acetate peel showing slip cleavage developed in laminated dolomite and limestone in uppermost Poxono Island Formation. Shear surfaces separate microlithons averaging less than 1 mm in width. C, Disharmonic folding between concentrically folded (buckling?) calcareous sandstone (more massive beds) and less competent shaly limestone deformed by passive slip folding (buckling has produced fold mullions). Uppermost Bossardville Limestone and lowermost Decker Formation. D, Disharmonic folding in the Bossardville Limestone on the southeast limb of anticline (to right of B, fig. 64). Note convergent fanning of cleavage. Spi, Poxono Island Formation; Sbv, Bossardville Limestone.

Mileage

11.2   Buses meet group at lower level of quarry. Retrace route out of quarry.

11.5   Stop sign, LEFT turn onto Cherry Valley Road towards Saylorsburg.

13.8   We are riding on Cherry Ridge which is held up on the south by the Palmerton Sandstone.

14.0   Scenic view to right (see fig. 67).

14.9   Village of Saylorsburg, RIGHT turn into parking lot of Lake House.

LUNCH STOP

14.9   Retrace route to Bossardsville.

15.7   Scenic view on left (see fig. 67).

18.3   Village of Bossardsville, bear RIGHT towards Cherry Valley.

18.7   Scenic view of Cherry Valley to right (fig. 1).

20.9   Intersection, Village of Stormville. Keep RIGHT towards Delaware Water Gap on Cherry Valley Road.

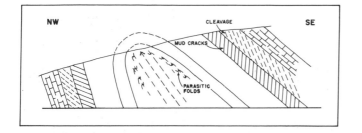

Figure 66. Diagrammatic sketch of overturned anticline in Poxono Island Formation of White (1882), showing structural relationships of cleavage, drag folds, and mud cracks at locality C (fig. 25). Length of exposure about 80 feet.

Mileage

21.2   On left, mud cracks in the Whiteport Dolomite Member of the Rondout Formation.

21.4   Ridge across Cherry Valley to right is Kemmererville anticline which plunges to the southwest. The ridge rises in altitude to the northeast.

21.7   Kame on right.

22.7   Several kames in Cherry Valley to right.

23.5   Stop sign. LEFT turn onto Pa. Rte. 191. To right is Wildcat Hollow (a local name) cut through Kemmererville anticline and exposing Bloomsburg Red Beds. Flat floor of Cherry Valley underlain by glacial lake-bottom clay, silt, and sand.

23.7   Fork in road, bear RIGHT to Cherry Valley. Exposures of the Port Ewen Shale and Oriskany Group in roadcut on left.

24.2   Exposure of sand and gravel in hillside kame on left.

24.4   Kame on right.

25.2   Exposures of sand and gravel in kame on left.

25.7   Cross Cherry Creek.

26.8   Town of Delaware Water Gap, turn LEFT onto Main Street.

26.9   Traffic light. Turn RIGHT to U. S. 209 and Interstate 80. We are riding on terrace deposits of the Delaware River.

27.1   Continue straight on U. S. Interstate 80W toward Stroudsburg.

27.8   Park on right shoulder of highway where feed-on merges with main highway. STOP 3.

STOP 3.   I, STRATIGRAPHY AND STRUCTURE OF GODFREY RIDGE
II, ORIGIN OF WATER GAPS

The route to be followed at this stop is shown in the geologic section, figure 68. This figure is similar to, but more detailed than, the cross section presented by Epstein (1966, fig. 5, C-C'; reprint provided at end of this paper). Figure 5 of the reprint shows a geologic map of the area of this stop.

The rocks that will be seen represent a general transgressive (Bossardville through Port Ewen) and regressive (Port Ewen through Oriskany) sequence. The Esopus unconformably overlies the Oriskany. (See table 1 for the description of these stratigraphic units.)

The folds at this stop are similar in style to the cascade folds seen at Bossardsville (stop 2, today). We will walk up the construction benches on the west side of the highway and over the south side of the ridge to the Croasdale quarry where deep mud cracks in the Bossardville Limestone will be seen.

The flow cleavage in the Port Ewen Shale at this locality is unique in that it fans the fold and opens away from the synclinal trough (fig. 69).

Figure 67. Panoramic view looking north from Cherry Ridge near Saylorsburg, Pa. 1, Wisconsin terminal moraine. 2, Weir Mountain, held up by rocks in the Mahantango Formation and Trimmers Rock Sandstone in the northwest limb of the Weir Mountain syncline. The southeast limb is in the ridges to the left of 1. 3, Camelback Mountain in the Pocono Plateau. 4, Flat-topped glacial delta of stop 1, first day. 5, Lake Creek valley, site of glacial Lake Sciota. 6, Southwest termination of Godfrey Ridge.

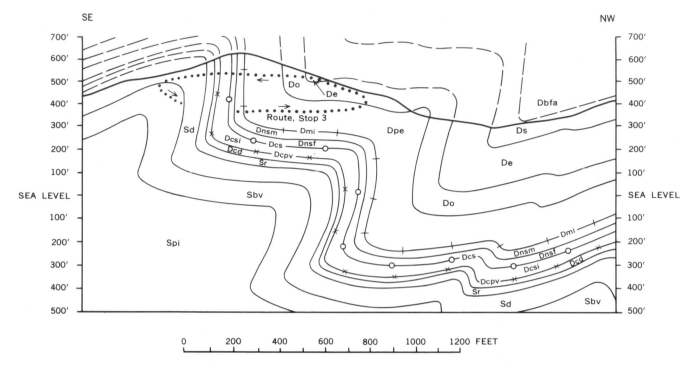

Figure 68. Geologic section through Godfrey Ridge at stop 3. Dbfa, lower member of Buttermilk Falls Limestone of Willard (1938); De, Esopus Formation; Do, Oriskany Group; Dpe, Port Ewen Shale; Dmi, Minisink Limestone; Dnsm, Maskenozha Member of New Scotland Formation; Dnsf, Flatbrookville Member of New Scotland Formation; Dcs, Stormville Member of Coeymans Formation; Dcsi, Shawnee Island Member of Coeymans Formation; Dcpv, Peters Valley Member of Coeymans Formation; Dcd, Depue Limestone Member of Coeymans Formation; Sr, Rondout Formation; Sd, Decker Formation; Sbv, Bossardville Limestone; Spi, Poxono Island Formation of White (1882).

Mileage

27.8   After group leaves buses, buses continue along highway.

28.2   Buses turn RIGHT onto U. S. 209N.

28.6   LEFT turn into Holiday Inn parking lot. Make U-turn and head back toward U. S. Interstate 80E towards Delaware Water Gap.

28.8   Bear LEFT onto Interstate 80E towards Delaware Water Gap.

30.0   Buses pull off on shoulder of road and wait for group to return.

30.0   Continue on exit ramp to U. S. 611S back to village of Delaware Water Gap.

30.5   Mount Tammany on Kittatinny Mountain on New Jersey side of Delaware Water Gap in skyline to left.

30.8   LEFT turn at traffic light.

30.9   RIGHT turn onto Cherry Valley Road.

31.8   LEFT turn at golf course. Note undulating topography with kettle holes in kame. Golf courses are partial to rolling glacial topography.

32.8   Road ascends Bloomsburg Red Beds on the northeast-dipping limb of the Kemmererville anticline.

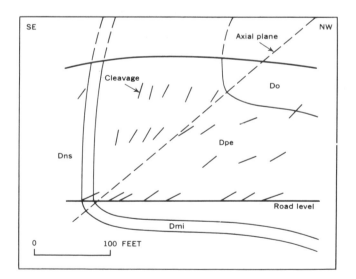

Figure 69. Geologic section showing divergent fanning of cleavage in the Port Ewen Shale in Godfrey Ridge along U. S. Interstate 80, stop 3. Section is perpendicular to strike of beds. Vertical scale same as horizontal. Do, Oriskany Group; Dpe, Port Ewen Shale; Dns, New Scotland Formation; Dmi, Minisink Limestone.

33.3   Crest of Kemmererville anticline.

33.4   Bear LEFT at Y. Road descends along the gently-dipping southeast limb of the Kemmererville anticline in the Bloomsburg Red Beds.

33.7   Trough of Poplar Valley syncline. Note 10-foot-long erratics of Buttermilk Falls Limestone in creek to right.

34.2   Ascend, we hope, steep dirt road to crest of Kittatinny Mountain. Cross Bloomsburg-Shawangunk contact.

34.7   Good view of countryside to north.

36.5   Park at fire tower. Disembark and follow Appalachian Trail out to the edge of the mountain overlooking Delaware Water Gap and surrounding countryside. STOP 4.

STOP 4. TOP OF KITTATINNY MOUNTAIN OVERLOOKING DELAWARE WATER GAP

Walk along the Appalachian Trail to the scenic overlook. At this stop we are in the upper quartzite-conglomerate unit of the Shawangunk Conglomerate. Compare the stratigraphy of the Shawangunk as seen on the northeast slope of the gap in New Jersey with the section seen at Lehigh Gap yesterday (fig. 5, and tables 1 and 2). The coarse conglomerate unit at the base of the Shawangunk in Lehigh Gap is absent here. Apparently, the Lehigh Gap area was closer to the ancient Fall Line. The quartzite-argillite unit, representing a complex marine-nonmarine transitional environment, has thinned to about one-fourth its thickness at Lehigh Gap. At Delaware Water Gap, a fluviatile sequence (upper quartzite-conglomerate unit) overlies the transitional beds. This upper unit probably represents the progradation of a deltaic topset (fluviatile) sequence that is thickest in the Delaware Water Gap area.

The very irregular contact between the Bloomsburg Red Beds and Shawangunk Conglomerate, shown in figure 5, will be discussed.

The structure in Silurian rocks at Delaware Water Gap is described in the reprint provided at the end of this report and shown in Drake and others (1969). Structural control for the location of Delaware Water Gap and the other gaps in this area, as described in the reprint, places the hypothesis of regional superposition in doubt.

Note the satellitic folds in the Shawangunk Conglomerate across the Delaware River in New Jersey.

Mapping in the Delaware Water Gap area shows that the contact between the Pen Argyl and Ramseyburg Members of the Martinsburg Formation and the overlying Shawangunk is covered along the unconformity at the base of the Shawangunk ¾ mile southwest of the gap (fig. 70). The regional strike between the Martinsburg Formation and the Shawangunk Conglomerate differs by about 15° in this area. The southeast-dipping slaty cleavage in the Martinsburg arches and dips to the northwest as it approaches the contact with the Shawangunk. Note that the bedding in the Martinsburg is not arched along with the cleavage, indicating that the arching was not caused by external rotation during a post-Taconic orogeny. It is believed that the arching of the cleavage has resulted from the pressure-shadow mechanism described earlier in this report in the section on structural geology. Thus, a post-Taconic age for the cleavage is indicated.

Three erosion surfaces are seen to the east and south. These are believed to be peneplains by some geologists (the Schooley, Harrisburg, and Somerville peneplains) but are believed by others to represent a state of dynamic equilibrium in the erosion of rocks of varying resistances.

Note the glacial striae and roches moutonnées on joint surfaces of the Shawangunk, indicating that the Wisconsin glacier was thick enough to override Kittatinny Mountain at this point. The mountain is 1,300 feet high here. The striae trend about due south on top of the mountain, but in the valley to the north, the mountain partly deflected the glacier, and striae trend southwest.

The top of the mountain just across the gap in New Jersey is named Mount Tammany (fig. 71), after the great Indian chief Tamanend. During the Revolutionary War, a charitable and patriotic society was established in Philadelphia and New York in his honor. It soon passed out of existence, except in New York City where it became a powerful political machine (Tammany Hall).

Time permitting, we will end the field trip with a recapitulation of the geology between Delaware and Lehigh Water Gaps in eastern Pennsylvania.

Mileage

36.5   Retrace route back to Holiday Inn Motel.

ALTERNATE STOP 4

30.8   LEFT turn at traffic light. Continue straight on U. S. 611S.

31.0   Gray sandstone and siltstone of Shawangunk Conglomerate exposed on right and left and rising to southwest.

31.3   Crest of Cherry Valley anticline. We are going down section as we travel to the southeast. Note

EXPLANATION

—————— · · · · · · · ·

Contact
Dotted where concealed
by surficial deposits

⌐52

Strike and dip of beds

⌐32

Strike and dip of cleavage

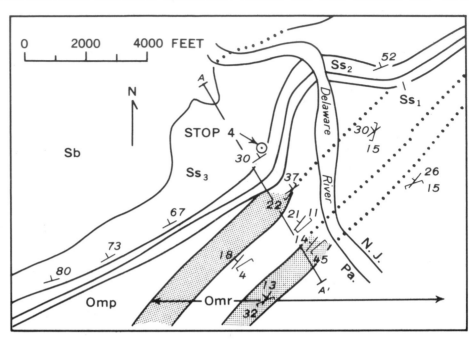

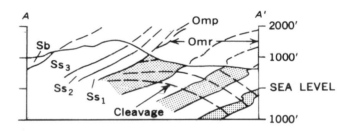

Figure 70. Geologic map and section of the Delaware Water Gap area showing the angular unconformity between the Martinsburg Formation of Ordovician age and the Shawangunk Conglomerate of Silurian age. The contact is also believed to be a zone of movement (décollement). Note the arching of cleavage as the contact is approached. The cleavage is generally steeper than shown in the graywacke beds. Sb, Bloomsburg Red Beds; Ss₃, upper quartzite-conglomerate unit of the Shawangunk Conglomerate; Ss₂, quartzite-argillite unit of the Shawangunk Conglomerate; Ss₁, lower quartzite-conglomerate unit of the Shawangunk Conglomerate; Omp, Pen Argyl Member of the Martinsburg Formation; Omr, Ramseyburg Member of the Martinsburg Formation. Stippled areas are graywacke-bearing intervals in Omr. Surficial deposits not shown.

that gray and green beds change to redbeds parallel to bedding within a few hundred feet. As we continue south on U. S. 611, note complex undulations within the Bloomsburg. These undulations are superimposed on a larger syncline, the Dunnfield Creek syncline which has a half wavelength of about ½ mile. There are at least 32 small folds in the Bloomsburg (Epstein, 1966, fig. 2A). Bedding-plane slips and wedging are common, as they are at Lehigh Gap. Several bedding-plane faults shear off the tops of anticlines. Cleavage is conspicuous and dips to the southeast in all parts of the small folds, showing

that the small folds and cleavage are partly, at least, of different ages. The relative ages of the structures seen in the Bloomsburg seem to be: (1) folding of the Dunnfield Creek syncline; (2) development of the small folds in the syncline, perhaps by sliding; (3) development of cleavage, probably during continued folding of the Dunnfield Creek syncline; (4) bedding-plane slip and wedging. Geometric relations of these features suggest that these are overlapping features and probably developed during one period of deformation.

32.2 Bloomsburg-Shawangunk contact on right.

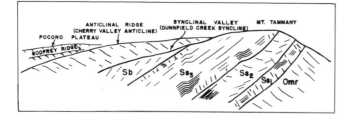

Figure 71. Sketch from photograph looking east into New Jersey showing structure and stratigraphy of Delaware Water Gap in Kittatinny Mountain. A few satellitic folds in the Shawangunk are indicated. Delaware River passes through the gap after making a right-angle bend from between Godfrey Ridge and the anticlinal ridge in Kittatinny Mountain. Sb, Bloomsburg Red Beds; Ss_3, upper quartzite-conglomerate unit of the Shawangunk Conglomerate; Ss_2, quartzite-argillite unit of the Shawangunk Conglomerate; Ss_1, lower quartzite-conglomerate unit of the Shawangunk Conglomerate; Omr, Ramseyburg Member of the Martinsburg Formation, which is talus covered in the area of the sketch.

32.5   Valley of Dunnfield Creek across Delaware River to left. Sand and gravel pit in kame terrace at mouth of Dunnfield Creek.

33.0   RIGHT turn into parking lot. ALTERNATE STOP 4.

REFERENCES CITED

Allen, J. R. L., 1965a, A review of the origin and characteristics of Recent alluvial sediments: Sedimentology, v. 5, p. 89–191.

—— 1965b, Fining-upward cycles in alluvial successions: Liverpool and Manchester Geol. Jour., v. 4, p. 229–246.

—— 1966, On bed forms and paleocurrents: Sedimentology, v. 6, p. 153–190.

Alling, H. L., 1947, Diagenesis of the Clinton hematite ores of New York: Geol. Soc. America Bull., v. 58, p. 991–1018.

Arndt, H. H., and Wood, G. H., 1960, Late Paleozoic orogeny in eastern Pennsylvania consists of five progressive stages: U. S. Geol. Survey Prof. Paper 400-B, p. B182–B184.

Beerbower, J. R., 1956, The Ordovician-Silurian contact, Delaware Water Gap, New Jersey: Pennsylvania Acad. Sci. Proc., v. 30, p. 146–149.

Beerbower, J. R., and Hait, M. H., Jr., 1959, Silurian fish in northeastern Pennsylvania and northern New Jersey: Pennsylvania Acad. Sci. Proc., v. 33, p. 198–203.

Behre, C. H., Jr., 1924, Structures in the slates of northeastern Pennsylvania [abs.]: Geol. Soc. America Bull., v. 35, p. 100–101.

—— 1927, Slate in Northampton County, Pennsylvania: Pennsylvania Geol. Survey, 4th ser., Bull. M9, 308 p.

—— 1933, Slate in Pennsylvania: Pennsylvania Geol. Survey, 4th ser., Bull. M 16, 400 p.

Bernard, H. A., and LeBlanc, R. J., 1965, Résumé of the Quaternary geology of the northwestern Gulf of Mexico province, in Wright, H. W., and Frey, D. G., eds., The Quaternary of the United States: Princeton, N. J., Princeton Univ. Press, p. 137–185.

Bernard, H. A., and Major, C. F., 1963, Recent meander belt deposits of the Brazos River: an alluvial "sand" model [abs.]: Am. Assoc. Petroleum Geologists Bull., v. 47, p. 354.

Black, Maurice, 1933, The algal sediments of Andros Island, Bahamas: Royal Soc., London, Philos, Trans., ser. B, v. 222, p. 165–192.

Brodhead, L. W., 1870, The Delaware Water Gap; its scenery, its legends and early history: 2d ed., Philadelphia, 276 p.

Broughton, J. G., 1946, An example of the development of cleavages: Jour. Geology, v. 54, p. 1–18.

Bushinski, G. I., 1964, On shallow water origin of phosphorite deposits, in Straaten, L. M. J. U. van, ed., Deltaic and shallow marine deposits: Amsterdam, Elsevier, p. 62–70.

Carson, W. P., 1968, Development of flow cleavage in the Martinsburg Shale, Port Jervis south area (northern New Jersey): Tectonophysics, v. 5, p. 531–541.

Clark, T. H., 1921, A review of the evidence for the Taconic revolution: Boston Soc. Nat. History Proc., v. 36, n. 3, p. 135–163.

Clarke, F. W., 1924, The data of geochemistry: U. S. Geol. Survey Bull. 770, 841 p.

Clarke, J. M., and Ruedemann, Rudolf, 1912, The eurypterida of New York: New York State Mus. Mem. 14, 439 p.

Coleman, J. M., and Gagliano, S. M., 1965, Sedimentary structures: Mississippi River deltaic plain, in Middleton, G. V., ed., Primary sedimentary structures and their hydrodynamic interpretation: Soc. Econ. Paleontologists and Mineralogists Spec. Pub. 12, p. 133–148.

Cooper, G. A., 1941, New Devonian stratigraphic units: Washington Acad. Sci. Jour., v. 31, no. 5, p. 179–181.

Craig, G. Y., 1952, A comparative study of the ecology and palaeoecology of Lingula: Edinburgh Geol. Soc. Trans., v. 15, p. 110–120.

Dale, T. N., and others, 1914, Slate in the United States: U. S. Geol. Survey Bull. 586, 220 p.

Davis, R. E., Drake, A. A., Jr., and Epstein, J. B., 1967, Geology of the Bangor quadrangle, Pennsylvania–New Jersey: U. S. Geol. Survey Geol. Quad. Map GQ-665.

Deffeyes, K. S., Lucia, F. J., and Weyl, P. K., 1965, Dolomitization of Recent and Plio-Pleistocene sediments by marine evaporite waters on Bonaire, Netherlands Antilles, in Pray, L. C., and Murray, R. C., eds., Dolomitization and Limestone diagenesis: Soc. Econ. Paleontologists and Mineralogists Spec. Pub. 13, p. 71–88.

Doeglas, D. J., 1962, The structure of sedimentary deposits of braided rivers: Sedimentology, v. 1, p. 167–190.

Donaldson, A. C., 1966, Deltaic sands and sandstones, in Wyoming Geol. Assoc., Symposium on recently developed geologic principles and sedimentation of the Permo-Pennsylvanian of the Rocky Mountains—20th

Ann. Conf., Casper, 1966: Casper, Wyo., Petroleum Inf., p. 31–62.

Donath, F. A., and Parker, R. B., 1964, Folds and folding: Geol. Soc. America Bull., v. 75, p. 45–62.

Drake, A. A., Jr., 1967a, Geologic map of the Easton quadrangle, New Jersey–Pennsylvania: U. S. Geol. Survey Geol. Quad. Map GQ-594.

——— 1967b, Geologic map of the Bloomsbury quadrangle, New Jersey: U. S. Geol. Survey Geol. Quad. Map GQ-595.

Drake, A. A., Jr., Davis, R. E., and Alvord, D. C., 1960, Taconic and post-Taconic folds in eastern Pennsylvania and western New Jersey: U. S. Geol. Survey Prof. Paper 400-B, p. 180–181.

Drake, A. A., Jr., and Epstein, J. B., 1967, The Martinsburg Formation (Middle and Upper Ordovician) in the Delaware Valley, Pennsylvania–New Jersey: U. S. Geol. Survey Bull. 1244-H, p. H1–H16.

Drake, A. A., Jr., Epstein, J. B., and Aaron, J. M., 1969, Geologic map and sections of parts of the Portland and Belvidere quadrangles, New Jersey–Pennsylvania: U. S. Geol. Survey Misc. Geol. Inv. Map I-552.

Drake, A. A., Jr., McLaughlin, D. B., and Davis, R. E., 1961, Geology of the Frenchtown quadrangle, New Jersey–Pennsylvania: U. S. Geol. Survey Geol. Quad. Map GQ-133.

——— 1967, Geologic map of the Riegelsville quadrangle, Pennsylvania–New Jersey: U. S. Geol. Survey Geol. Quad. Map. GQ-593.

Dyson, J. L., 1956, Recumbent folding in the vicinity of Palmerton, Pennsylvania: Pennsylvania Acad. Sci. Proc., v. 30, p. 137–141.

Epstein, A. G., Epstein, J. B., Spink, W. J., and Jennings, D. S., 1967, Upper Silurian and Lower Devonian stratigraphy of northeastern Pennsylvania, New Jersey, and southeasternmost New York: U. S. Geol. Survey Bull. 1243, 74 p.

Epstein, J. B., 1966, Structural control of wind gaps and water gaps and of stream capture in the Stroudsburg area, Pennsylvania and New Jersey: U. S. Geol. Survey Prof. Paper 550-B, p. B80–B86.

——— 1969, Surficial geology of the Stroudsburg quadrangle, Pennsylvania–New Jersey: Pennsylvania Geol. Survey, Bull. G57, 67 p.

Epstein, J. B., and Hosterman, J. W., 1969, Residual clay deposits in rocks of Early and Middle age near Kunkletown, Pennsylvania: U. S. Geol. Survey Prof. Paper 650-D.

Evans, Graham, 1965, Intertidal flat sediments and their environments of deposition in the Wash: Geol. Soc. London Quart. Jour., v. 121, p. 209–245.

Fahnestock, R. K., 1963, Morphology and hydrology of a glacial stream—White River, Mount Rainier, Washington: U. S. Geol. Survey Prof. Paper 422-A, 70 p.

Fahnestock, R. K., and Haushild, W. L., 1962, Flume studies of the transport of pebbles and cobbles on a sand bed: Geol. Soc. America Bull., v. 73, p. 1431–1436.

Field Conference of Pennsylvania Geologists, 31st, Har-

risburg, 1966, Comparative tectonics and stratigraphy of the Cumberland and Lebanon Valleys, by D. B. MacLachlan and S. I. Root: Harrisburg, Pa., Pennsylvania Geol. Survey, 90 p.

Field Conference of Pennsylvania Geologists, 32nd, East Stroudsburg, 1967, Geology in the region of the Delaware to Lehigh Water Gaps, by J. B. Epstein and A. G. Epstein: Harrisburg, Pa., Pennsylvania Geol. Survey, 89 p.

Glaeser, J. D., 1967, Bedding-plane slips, wedge faulting and facies changes in the Parryville syncline, Carbon County, Pennsylvania: Pennsylvania Acad. Sci. Proc., v. 15, p. 95–98.

Grabau, A. W., 1909, Physical and faunal evolution of North America during Ordovicic, Siluric, and early Devonic time: Jour. Geology, v. 17, p. 209–252.

Gray, Carlyle, 1954, Recumbent folding in the Great Valley: Pennsylvania Acad. Sci. Proc., v. 28, p. 96–101.

Gray, Carlyle, and others, 1960, Geologic map of Pennsylvania: Pennsylvania Geol. Survey, 4th ser., scale 1:250,000.

Gwinn, V. E., 1964, Thin-skinned tectonics in the Plateau and northwestern Valley and Ridge provinces of the central Appalachians: Geol. Soc. America Bull., v. 75, p. 863–900.

Hamblin, W. K., 1962, X-ray radiography in the study of structures in homogeneous sediments: Jour. Sed. Petrology, v. 32, p. 201–210.

Harrison, J. V., and Falcon, N. L., 1936, Gravity collapse structures and mountain ranges, as exemplified in southwestern Iran: Geol. Soc. London Quart. Jour., v. 92, p. 91–102.

Häntzschel, Walter, 1939, Tidal flat deposits (Wattenschlick), in Trask, P. D., ed., Recent marine sediments, a symposium: Tulsa, Okla., Am. Assoc. Petroleum Geologists, p. 195–206.

Hess, H. H., 1955, Serpentine, orogeny, and epeirogeny: Geol. Soc. America Spec. Paper 62, p. 391–408.

Hunter, R. E., 1960, Iron sedimentation of the Clinton Group of the central Appalachian Basin: unpub. Ph.D. thesis, Johns Hopkins Univ., Baltimore, Md., 416 p.

Illing, L. V., Wells, A. J., and Taylor, J. C. M., 1965, Penecontemporary dolomite in the Persian Gulf, in Pray, L. C., and Murray, R. C., eds., Dolomitization and limestone diagenesis: Soc. Econ. Paleontologists and Mineralogists Spec. Pub. 13, p. 89–111.

Johnson, Helgi, and Fox, Steven K., Jr., 1968, Dipleurozoa from Lower Silurian of North America: Science, v. 162, no. 3849, p. 119–120.

Johnson, M. E., 1950, Geologic map of New Jersey: New Jersey Dept. Conserv. and Econ. Devel., scale 1:250,000 (revised).

Keith, Arthur, 1923, Outlines of Appalachian structure: Geol. Soc. America Bull., v. 34, p. 309–380.

Klein, G. deVries, 1965, Dynamic significance of primary structures in the Middle Jurassic Great Oolite Series, southern England, in Middleton, G. V., ed., Primary sedimentary structures and their hydrodynamic interpretation: Soc. Econ. Paleontologists and Mineralogists Spec. Pub. 12, p. 173–191.

Land, L. S., and Hoyt, J. H., 1966, Sedimentation in a meandering estuary: Sedimentology, v. 6, p. 191–207.

Laporte, L. F., 1967, Carbonate deposition near mean sea-level and resultant facies mosaic: Manlius Formation (Lower Devonian) of New York State: Am. Assoc. Petroleum Geologists Bull., v. 51, p. 73–101.

Lesley, J. P., 1883, The geology of Lehigh and Northampton Counties: Pennsylvania Geol. Survey, 2d, Rept. D-3, 283 p.

McBride, E. F., 1962, Flysch and associated beds of the Martinsburg Formation (Ordovician), central Appalachians: Jour. Sed. Petrology, v. 32, p. 39–91.

McKee, E. D., 1957, Primary structures in some Recent sediments: Am. Assoc. Petroleum Geologists Bull., v. 41, p. 1704–1747.

——— 1965, Experiments on ripple lamination, in Middleton, G. V., ed., Primary sedimentary structures and their hydrodynamic interpretation: Soc. Econ. Paleontologists and Mineralogists Spec. Pub. 12, p. 66–83.

Mackin, J. H., 1941, Drainage changes near Wind Gap, Pennsylvania—A study in map interpretation: Jour. Geomorphology, v. 4, p. 24–53.

Maxwell, J. C., 1962, Origin of slaty and fracture cleavage in the Delaware Water Gap area, New Jersey and Pennsylvania, in Geol. Soc. America, Petrologic studies: a volume in honor of A. F. Buddington: New York, p. 281–311.

Miller, B. L., 1926, Taconic folding in Pennsylvania: Geol. Soc. America Bull., v. 37, no. 3, p. 497–511.

Miller, B. L., Fraser, C. M., and Miller, R. L., 1939, Northampton County, Pennsylvania, geology and geography: Pennsylvania Geol. Survey, 4th ser., Bull. C48, 496 p.

Moody-Stuart, M., 1966, High- and low-sinuosity stream deposits, with examples from the Devonian of Spitsbergen: Jour. Sed. Petrology, v. 36, p. 1102–1117.

Nichols, M. M., 1962, Hydrography and sedimentology of Sonoran lagoons, Mexico [abs.]: Geol. Soc. America, Ann. mtg., 1962, Houston, Texas, Program, p. 110A–111A.

Oliver, W. A., Jr., 1960, Rugose corals from reef limestones in the Lower Devonian of New York: Jour. Paleontology, v. 34, no. 1, p. 59–100.

Peck, F. B., 1922, White clay deposits at Saylorsburg, Monroe County, Pennsylvania: Pennsylvania Geol. Survey, 4th ser., Bull. 40, 8 p.

Pettijohn, F. J., and Potter, P. E., 1964, Atlas and glossary of primary sedimentary structures: New York, Springer-Verlag, 370 p.

Porrenga, D. H., 1967a, Glauconite and chamosite as depth indicators in the marine environment: Marine Geology, v. 5, p. 495–501.

——— 1967b, Clay mineralogy and geochemistry of Recent marine sediments in tropical areas: Amsterdam, Stolk-Dordt, 145 p.

Reineck, H. E., and Singh, I. B., 1967, Primary sedi-

mentary structures in the recent sediments of the Jade, North Sea: Marine Geology, v. 5, p. 227–235.

Rickard, L. V., 1962, Late Cayugan (Upper Silurian) and Helderbergian (Lower Devonian) stratigraphy in New York: New York State Mus. Bull. 386, 157 p.

Roehl, P. O., 1967, Stony Mountain (Ordovician) and Interlake (Silurian) facies analogs of Recent low-energy marine and subaerial carbonates, Bahamas: Am. Assoc. Petroleum Geologists Bull., v. 51, p. 1979–2032.

Rogers, H. D., 1838, Second annual report on the (1st) geological exploration of the State of Pennsylvania: Harrisburg, 93 p.

——— 1858, The geology of Pennsylvania; a government survey: Philadelphia, 2 v.: 586 and 1046 p.

Sames, C. W., 1966, Morphometric data of some Recent pebble associations and their application to ancient deposits: Jour. Sed. Petrology, v. 36, p. 126–142.

Schafer, J. P., 1949, Some periglacial features in central Montana: Jour. Geology, v. 57, p. 154–174.

Schoen, Robert, 1964, Clay minerals of the Silurian Clinton Ironstones, New York State: Jour. Sed. Petrology, v. 34, p. 855–863.

Seilacher, Adolf, 1967, Bathymetry of trace fossils: Marine Geology, v. 5, p. 413–428.

Sheldon, R. P., 1964, Barrier islands and lagoonal iron sedimentation in the Silurian of Alabama [abs.]: Geol. Soc. America, Ann. mtg., 1964, Miami Beach, Fla., Program, p. 182.

Shepard, F. P., and Lankford, R. R., 1959, Sedimentary facies from shallow borings in lower Mississippi Delta: Am. Assoc. Petroleum Geologists Bull., v. 43, p. 2051–2067.

Shinn, E. A., Ginsburg, R. N., and Lloyd, R. M., 1965, Recent supratidal dolomite from Andros Island, Bahamas, in Pray, L. C., and Murray, R. C., eds., Dolomitization and limestone diagenesis: Soc. Econ. Paleontologists and Mineralogists Spec. Pub. 13, p. 112–123.

Walker, T. R., 1967a, Color of Recent sediments in tropical Mexico—a contribution to the origin of red beds: Geol. Soc. America Bull., v. 78, p. 917–920.

——— 1967b, Formation of red beds in modern and ancient deserts: Geol. Soc. America Bull., v. 78, p. 353–368.

White, I. C., 1882, The geology of Pike and Monroe Counties: Pennsylvania Geol. Survey 2d, Rept. G6, 407 p.

Wietrzychowski, J. R., 1963, The Sweet Arrow fault in northeastern Pennsylvania, in Field Conference of Pennsylvania Geologists, 28th, Stroudsburg, 1963, Stratigraphy and structure of Upper and Middle Devonian rocks in northeastern Pennsylvania: Harrisburg, Pa., Pennsylvania Geol. Survey, p. 43–44.

Willard, Bradford, 1938, A Paleozoic section at Delaware Water Gap: Pennsylvania Geol. Survey, 4th ser., Bull. G11, 35 p.

Willard, Bradford, and Cleaves, A. B., 1939, Ordovician

Silurian relations in Pennsylvania: Geol. Soc. America Bull., v. 50, p. 1165–1198.

Wood, G. H., Jr., Arndt, H. H., and Hoskins, D. M., 1963, Structure and stratigraphy of the southern part of the Pennsylvania Anthracite region: Geol. Soc. America, Ann. mtg., New York City, 1963, Guidebook, Field Trip no. 4, 84 p.

Woodward, H. P., 1944, Copper mines and mining in New Jersey: New Jersey Dept. Conserv. and Devel., Geol. Ser., Bull. 57, 156 p.

—— 1957, Structural elements of northeastern Appalachians: Am. Assoc. Petroleum Geologists Bull., v. 41, p. 1429–1440.

Yeakel, L. S., Jr., 1962, Tuscarora, Juniata, and Bald Eagle paleocurrents and paleogeography in the central Appalachians: Geol. Soc. America Bull., v. 73, p. 1515–1540.

REPRINTED FROM U. S. GEOL. SURVEY PROF. PAPER 550-B, PAGES B80–B86 (1966)

STRUCTURAL CONTROL OF WIND GAPS AND WATER GAPS AND OF STREAM CAPTURE IN THE STROUDSBURG AREA, PENNSYLVANIA AND NEW JERSEY

JACK B. EPSTEIN
Beltsville, Md.

ABSTRACT

Wind gaps and water gaps in the Stroudsburg area, in eastern Pennsylvania and northern New Jersey, are located where resistant rocks dip steeply and have a narrow width of outcrop, where folds die out over short distances, or where folding was more intense locally than nearby. All gaps trend about perpendicular to the strike of the ridges, and parallel to major cross-joint sets. These observations favor the hypothesis of structural control of the location of stream gaps, rather than that of regional superposition of the streams upon the resistant rocks. A second type of structural control, in which more resistant beds were exposed in Wind Gap than in Delaware Water Gap, may explain the capture of Wind Gap River by tributaries of the Delaware River.

WIND GAPS AND WATER GAPS

Numerous and extensive investigations of wind and water gaps in the Appalachians have contributed to the controversies regarding Appalachian geomorphic evolution. Many geologists believe that after the Appalachian orogeny the drainage divide between streams that flowed southeastward into the Atlantic and those that flowed toward the continental interior was located either in the crystalline highlands southeast of the present Great Valley or in the Valley and Ridge province. The divide has since shifted westward to its present position in the Appalachian Plateau. The location of the original divide, the means by which the divide migrated, and the process or processes by which the numerous wind and water gaps were formed are problems that need to be considered in any hypothesis which attempts to explain the drainage development of the Appalachians.

Johnson (1931) believed that the original drainage lines were obliterated during a Cretaceous marine transgression and that the present drainage pattern is mainly the result of superposition from a coastal-plain cover. The location of a gap was purely by chance and is not systematically related to any weakness in the ridge, although there may have been local adjustment to structure.

Meyerhoff and Olmstead (1936) believed that the present drainage descended from the pattern which had been established in Permian time and which had been controlled by structure and topography produced during the Appalachian orogeny. Hence gaps are found along transverse structures or in the northwest limbs of overturned folds.

Thompson (1949) argued that the original divide, which lay on crystalline rocks along the Blue Ridge-Reading Prong axis, was unstable because the southeastward-flowing streams had shorter courses than those that flowed northwestward. As a result, the divide shifted northwestward by normal stream erosion (headward piracy), and the gaps in Kittatinny and Blue Mountains are located at points of rock weakness.

Strahler (1945), who favored Johnson's hypothesis, stressed that the main test substantiating superposition was to show lack of coincidence of gaps and sites of structural weakness (specifically, transverse faults).

Structural characteristics and features other

than transverse faults, however, may influence the resistance to erosion of hard-rock ridges. These include, among others: (1) changes in outcrop width, owing to changes of dip; (2) abrupt changes in strike, owing to dying out of folds; (3) local weakness of otherwise resistant rocks as a result of the overturning of beds and accompanying shearing; (4) closely spaced joints and strong folding resulting from intense local stress; (5) cross folds and attendant fracturing; (6) thinning of resistant units, which reflects the original processes of sedimentation; (7) thinning or elimination of resistant strata by strike faulting; and (8) change in facies.

Detailed structural data from gaps in the Stroudsburg area, presented by previous investigators, generally have been scanty. Detailed mapping of Blue and Kittatinny Mountains and of ridges to the north has shown that there is a correlation of gaps with one or more of the following conditions: (1) steep dips of beds and narrow outcrop widths of resistant units, (2) dying out of folds within short distances, and (3) more intense folding locally than nearby. The parallelism of the gaps is controlled by prominent southeast-trending cross-joint sets present throughout the area. A plot of nearly four hundred joints shows a strong maximum with a strike of N.14°W. and a dip about vertical.

GAPS IN BLUE AND KITTATINNY MOUNTAINS

Blue and Kittatinny Mountains, parts of a single ridge supported by the Shawangunk Conglomerate, are cut by several gaps: Delaware Water Gap, Totts Gap, Fox Gap, and Wind Gap. Figure 1 shows the locations of the gaps and other major physiographic features. Figure 2 shows the distribution of geologic formations and the structural geology of the area, and demonstrates the correlation of gap location with the three structural conditions mentioned above.

Delaware Water Gap. — Many early observers of Delaware Water Gap believed that it was the result of a violent cataclysm. Interesting excerpts of these early discussions are reported by Miller and others (1939, p. 139–142). Rogers (1858, v. 1, p. 283, v. 2, p. 896) noted that the ridge crest is offset 700 feet at the gap. He attributed the displacement to a transverse fault, as did Ashley (1935, p. 1406) and Willard (1938, p. 23). Chance (1882, p. 338), Johnson and others (1933, p. 26), Miller and others (1939, p. 144), and Strahler (1945, p. 58–59) believed that the ridge is offset by a slight flexure. Thompson (1949, p. 56, 59) found many small faults which he suggested might be offshoots of a major transverse fault, and attempted to show that the gap is structurally controlled.

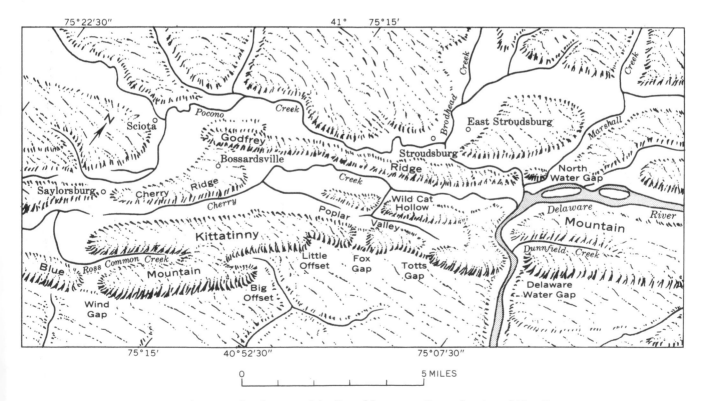

Figure 1. Physiographic diagram of the Stroudsburg area, Pennsylvania and New Jersey.

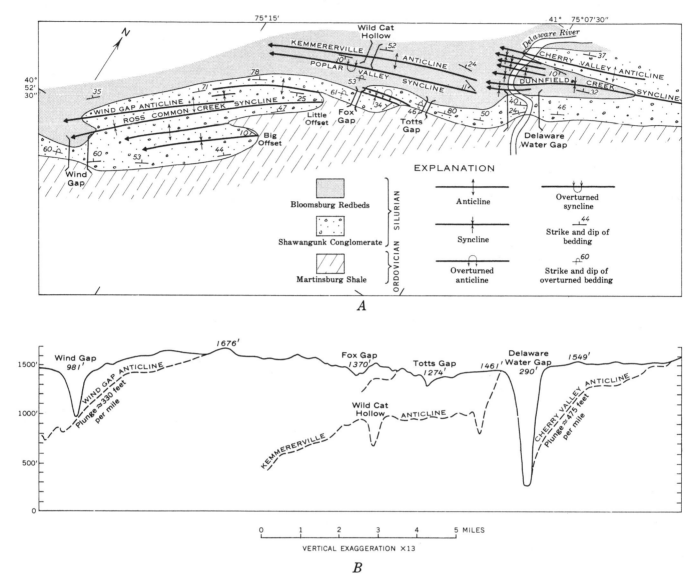

Figure 2.   A, Geologic map of the Stroudsburg area, Pennsylvania and New Jersey.   B, projected longitudinal topographic profile, showing relation of geologic structure to location of gaps in Blue and Kittatinny Mountains (fig. 1).   Profile viewed from the southeast.   Dashed lines indicate topography behind main ridge and correspond approximately to crests of major anticlines. Several small folds north of Wind Gap, similar to those in Delaware Water Gap, are not shown; their extent is not well known because outcrops are poor.   Topography from Wind Gap quadrangle, Pennsylvania–New Jersey (15 min.), and Portland quadrangle, New Jersey (7½ min.).

During the present study no cross fault could be found at Delaware Water Gap.   The Shawangunk Conglomerate consists of three units that match at river level and have contacts that are not displaced (fig. 3).   The bedding dips 35° to 45° to the northwest on both sides of the stream at the bottom of the gap.   At the top of the ridge on the New Jersey side, at Mt. Tammany, the dip is about 50°, whereas on the Pennsylvania side the dip decreases upward toward Mt. Minsi, being less than 25° at a place halfway up the mountain.   Clearly, there is a small

flexure at the gap; the beds on the New Jersey side dip more steeply than those in Pennsylvania, and the ridge crest in New Jersey lies about 700 feet northwest of the axis of the crest on the Pennsylvania side.   The flexure can be seen by looking west from the New Jersey bank.   Consideration of the structural geometry (fig. 4) reveals that there was an abrupt change in strike of the beds that formerly occupied the site of the gap.   As a consequence the brittle Shawangunk must have been weakened by extensive fracturing in the flexure

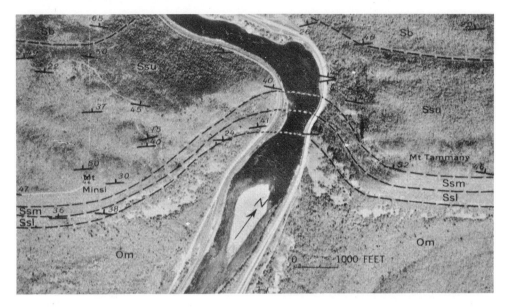

Figure 3. Aerial photograph and geologic map of Delaware Water Gap. The Bloomsburg Red-beds, Sb, of Silurian age are underlain by three members of the Shawangunk Conglomerate: Ssu, upper conglomerate and quartzite member; Ssm, middle black argillite, quartzite, and conglomerate member; and Ssl, lower quartzite and conglomerate member. The Martinsburg Shale, Om, of Ordovician age underlies the Shawangunk.

zone. Structural control is therefore thought to have determined location of the gap.

A series of folds in the Bloomsburg Redbeds along the course of Delaware River north of the gap dies out to the southwest, within a short distance (fig. 2). Probably the rocks are more highly sheared here, and resistance to erosion is less, than in the areas between gaps where similar folds were not observed.

Perhaps equally important in controlling the location of Delaware Water Gap is the fact that the outcrop width of the Shawangunk Conglomerate is narrower at the gap side than to the northeast where the formation is repeated in the southwest-plunging Cherry Valley anticline. The river now flows on the Shawangunk where it crosses the anticline, but undoubtedly it flowed on the weaker Bloomsburg Redbeds earlier in its history before cutting down into the Shawangunk. This is an example of local superposition.

Totts Gap. — The beds at Totts Gap are more strongly overturned than elsewhere along the ridge crest in the area of study, and it seems likely that the rocks here were weakened more than in adjacent areas. Moreover, the Shawangunk Conglomerate has a narrow outcrop width at Totts Gap. Thompson (1949, p. 58) observed that between Totts Gap and Delaware Water Gap the ridge crest is lower where joints are more closely spaced than

elsewhere, and that at Totts Gap, the lowest point along this stretch, the joints are most closely spaced. Whether this reflects greater stress, or whether it is due to chance exposure of different beds in the Shawangunk that possess different structural characteristics, is difficult to determine.

Fox Gap. — Fox Gap is located where two southwest-plunging folds die out over a short distance,

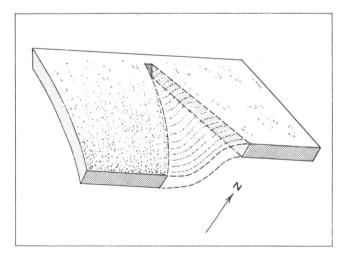

Figure 4. Diagram showing reconstructed flexure at Delaware Water Gap. Dashed lines show form of bed before gap was cut. Presence of the flexure accounts for the topographic offset of the ridge and suggests extensive fracturing of the Shawangunk Conglomerate at the gap site in the flexure zone.

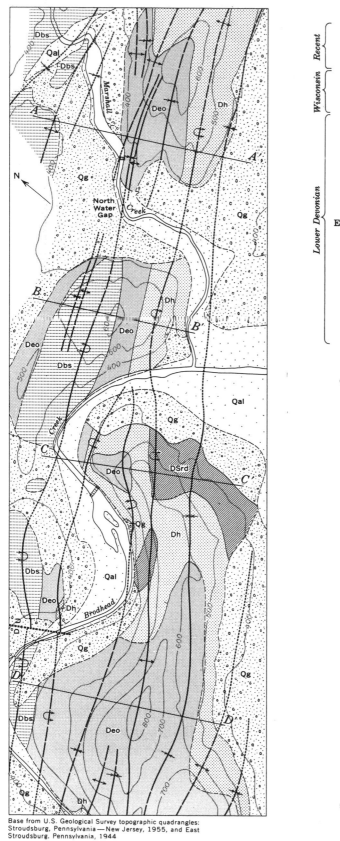

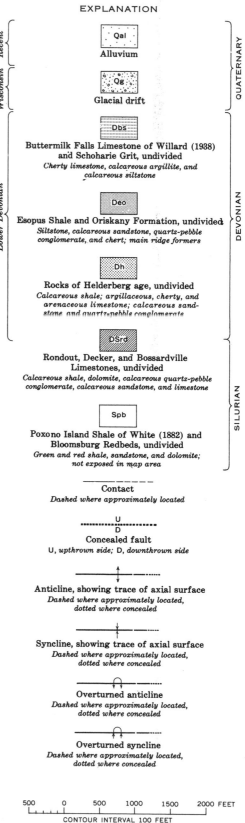

EXPLANATION

QUATERNARY

Wisconsin Recent

Qal
Alluvium

Qg
Glacial drift

Lower Devonian

Dbs
Buttermilk Falls Limestone of Willard (1938) and Schoharie Grit, undivided
Cherty limestone, calcareous argillite, and calcareous siltstone

Deo
Esopus Shale and Oriskany Formation, undivided
Siltstone, calcareous sandstone, quartz-pebble conglomerate, and chert; main ridge formers

Dh
Rocks of Helderberg age, undivided
Calcareous shale; argillaceous, cherty, and arenaceous limestone; calcareous sandstone, and quartz-pebble conglomerate

DSrd
Rondout, Decker, and Bossardville Limestones, undivided
Calcareous shale, dolomite, calcareous quartz-pebble conglomerate, calcareous sandstone, and limestone

Spb
Poxono Island Shale of White (1882) and Bloomsburg Redbeds, undivided
Green and red shale, sandstone, and dolomite; not exposed in map area

DEVONIAN

SILURIAN

– – – – – –
Contact
Dashed where approximately located

········ U/D ········
Concealed fault
U, upthrown side; D, downthrown side

Anticline, showing trace of axial surface
Dashed where approximately located, dotted where concealed

Syncline, showing trace of axial surface
Dashed where approximately located, dotted where concealed

Overturned anticline
Dashed where approximately located, dotted where concealed

Overturned syncline
Dashed where approximately located, dotted where concealed

500 0 500 1000 1500 2000 FEET
CONTOUR INTERVAL 100 FEET

Base from U.S. Geological Survey topographic quadrangles: Stroudsburg, Pennsylvania—New Jersey, 1955, and East Stroudsburg, Pennsylvania, 1944

Figure 5. Geologic map and sections of the area near North Water Gap and the gap of Brodhead Creek. Geology by J. B. Epstein, assisted by A. G. Epstein, 1962–63.

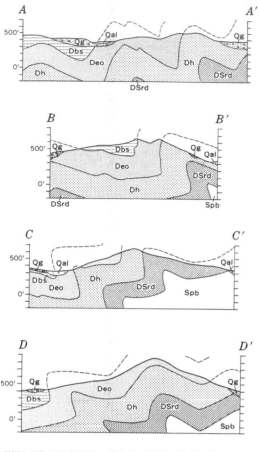

LINES OF SECTIONS ARE SHOWN ON FACING PAGE

much as they do at Delaware Water Gap. Also, the beds are strongly overturned, and the outcrop width is narrow.

Wind Gap.—Geologic structures in Wind Gap duplicate those in the Delaware Water Gap area. Two major folds plunge out north of the gap. The Shawangunk Conglomerate is not repeated to the northwest because the Wind Gap anticline plunges to the southwest. There is a 15° difference in strike in beds in the ridge crest on either side of the gap, indicative of a flexure similar to that at Delaware Water Gap. In addition, several small folds in the Bloomsburg Redbeds are similar to those in the Bloomsburg at Delaware Water Gap. These folds were not included in figure 2A because outcrops are two few to permit tracing of their trends. Thinning of the outcrop width of the Shawangunk at the gap is evident on figure 2A. This is indicative of near-vertical dips which can be seen in cuts along the highway where it passes through the gap.

GAPS IN GODFREY RIDGE

Godfrey Ridge lies about 2½ miles north of Kittatinny Mountain and is supported by complexly folded Upper Silurian and Lower Devonian lime-

stone, shale, sandstone, and conglomerate. Silty shale and sandstone of the Esopus Shale and Oriskany Formation support the higher parts of the ridge (fig. 5). Small folds are numerous and die out rapidly. Sags in the ridge crest are numerous, but it is difficult to relate structural features to them. The two largest gaps in Godfrey Ridge, the gap of Brodhead Creek and North Water Gap, are located about 2½ miles north-northwest of Delaware Water Gap.

Gap of Brodhead Creek.—Brodhead Creek cuts through Godfrey Ridge at an altitude of about 300 feet. No rock crops out in the creek bottom, but bedrock is exposed in the creek bed about 1 mile upstream. Therefore, bedrock cannot be far below creek level in the gap. Folding at the gap is so complex that Willard (1938, p. 23) believed that the gap is the site of a north-south tear fault. Evidence of faulting could not be found, however. Rather, there are four overturned folds in the southwestern part of the ridge near Brodhead Creek, two of which die out and are absent in the northeastern part. The abrupt dying out of folds is well illustrated by sections *B–B'* and *C–C'* in figure 5. Change in strike of the beds that formerly occupied the site of the gap is inferred, as it is at Delaware Water Gap.

North Water Gap.—North Water Gap is located where Marshall Creek cuts through Godfrey Ridge, about half a mile northeast of the gap of Brodhead Creek. It is almost as deep as the latter, but no bedrock is exposed in the creek floor. Glacial debris which may be of great thickness [1] is found along the course of the creek in the gap. Clearly, North Water Gap was much deeper in preglacial times.

The structure at North Water Gap is similar to the structure at the gap of Brodhead Creek. Two folds trending toward the gap from the northeast die out within the valley and do not reappear in the ridge to the southwest (compare sections *A–A'* and *B–B'* in fig. 5). The two gaps are therefore located at sites where folds plunge out abruptly, and implication of structural control is clear. Moreover, because of near-vertical dips, the Oriskany and Esopus, which together make up the main ridge support, have a narrow outcrop width.

STREAM CAPTURE

Structural relations in the Stroudsburg area are believed to have controlled not only the locations but also the history of the gaps. For example, Wind

---

[1] Bedrock was penetrated at a depth of 296 feet in a water well near the west bank of the Delaware River at an altitude of about 300 feet and approximately 2,000 feet southeast of the mouth of the gap, as reported by a local resident.

Gap was cut by a stream that was later captured north of Blue Mountain by tributaries of either the Lehigh River to the west or of the Delaware River to the east (Wright, 1896; Ver Steeg, 1930; Willard, 1939; Mackin, 1941). Mackin showed that the drainage changes in the Wind Gap area were exceedingly complex and implied that evidence of the original captor of the Wind Gap River may be gone. Nevertheless it seems likely on the basis of structural considerations that the tributaries of the Delaware were more agrcssive than those of the Lehigh, and that one captured the headwaters of Wind Gap River. Why was Wind Gap River, rather than the Delaware River tributary, beheaded? Mackin speculated that the present Delaware may have captured a stream which flowed through Culvers Gap in Kittatinny Mountain, 23 miles to the northeast, and greatly increased its own advantage at that time by this addition to its volume. Study of figure 2B suggests an alternative explanation.

The Wind Gap River was captured just after it had cut down to an altitude of about 980 feet, the present altitude of the floor of the gap. At that level the stream had reached the resistant Shawangunk Conglomerate in the Wind Gap anticline at about 980 feet, just northwest of the present gap, and its downcutting was retarded. The Delaware River, if it was at a similar altitude at that time, as seems likely, was still cutting down through weaker Bloomsburg Redbeds in the Cherry Valley anticline where the top of the Shawangunk is several hundred feet lower. The lower altitude of the top of the Shawangunk at Delaware Water Gap is explained by the difference in plunge of the folds: the Cherry Valley anticline plunges about 475 feet per mile while the Wind Gap anticline plunges about 330 feet per mile. Later, after the Delaware had captured the headwaters of Wind Gap River and cut through the Bloomsburg, it was superimposed on the Shawangunk in the Cherry Valley anticline and slipped down the plunge of the fold in a curving course. The curve was accentuated when it migrated downstream and reached the steeply dipping Shawangunk quartzites and conglomerates just west of the gap.

SUMMARY

Each of six major gaps in the Stroudsburg area is located in an area possessing some combination of the following structural features: (1) dying out of folds over a short distance, and associated abrupt changes in strike of bedding; (2) steep dip and narrow width of outcrop of resistant strata; and (3)

folding that was more intense locally than in the surrounding rocks. None of these structural features is present along the ridges where there are no gaps. The presence of the gaps in the Stroudsburg area, at places where structural control seemingly was effective, does not favor the concept of regional superposition. Rather, it favors those hypotheses which maintain that gaps are located in zones of structural weakness where erosion was most effective during the course of stream competition along the ancestral drainage divide. Comparison of the structural settings of Wind Gap and Delaware Water Gap suggests that differences in the underlying rocks are an important factor in the development of gaps. The presence of resistant beds at a markedly higher altitude at Wind Gap favored the abandoning of Wind Gap as a result of stream piracy. At Delaware Water Gap, on the other hand, the resistant beds are several hundred feet lower. The Delaware River continued cutting down through the overlying softer rocks after Wind Gap had been abandoned by its stream.

REFERENCES CITED

Ashley, G. H., 1935, Studies in Appalachian Mountain sculpture: Geol. Soc. America Bull., v. 46, p. 1395–1436.

Chance, H. M., 1882, Special survey of the Delaware Water Gap, in White, I. C., The Geology of Pike and Monroe Counties: Pennsylvania Geol. Survey, 2d ser., Rept. G6, 407 p.

Johnson, D. W., 1931, Stream sculpture on the Atlantic slope, a study in the evolution of Appalachian rivers: New York, Columbia Univ. Press, 142 p.

Johnson, D. W., Bascom, Florence, and Sharp, H. S., 1933, Geomorphology of the central Appalachians: Internat. Geol. Cong., 16th, United States, 1932, Guidebook 7, Excursion A–7, 50 p.

Mackin, J. H., 1941, Drainage changes near Wind Gap, Pennsylvania—a study in map interpretation: Jour. Geomorphology, v. 4, p. 24–52.

Meyerhoff, H. A., and Olmstead, E. W., 1936, The origins of Appalachian drainage: Am. Jour. Sci., 5th ser., v. 32, p. 21–42.

Miller, B. L., Fraser, D. M., and Miller, R. L., 1939, Northampton County, Pennsylvania, geology and geography: Pennsylvania Geol. Survey, 4th ser., Bull. C48, 496 p.

Rogers, H. D., 1858, The geology of Pennsylvania, a government survey: Philadelphia, v. 1, 586 p.; v. 2, 1045 p.

Strahler, A. N., 1945, Hypothesis of stream development in the folded Appalachians of Pennsylvania: Geol. Soc. America Bull., v. 56, p. 45–88.

Thompson, H. D., 1949, Drainage evolution in the Appalachians of Pennsylvania: New York Acad. Sci. Annals, v. 52, art. 2, p. 31–62.

Ver Steeg, Karl, 1930, Wind gaps and water gaps of the northern Appalachians: New York Acad. Sci. Annals, v. 32, p. 87–220.

White, I. C., 1882, The geology of Pike and Monroe Counties: Pennsylvania Geol. Survey (2d), Rept. G6, 407 p.

Willard, Bradford, 1938, A Paleozoic section at Delaware Water Gap: Pennsylvania Geol. Survey, 4th ser., Bull. G11, 35 p.

——— 1939, Stratigraphy and structure of the Kittatinny (Blue) Mountain gaps, in Miller, B. L., Fraser, D. M., and Miller, R. L., Northampton County, Pennsylvania, geology and geography: Pennsylvania Geol. Survey, 4th ser., Bull. C48, p. 145–158.

Wright, F. B., 1896, The origin of the Wind Gap: Am. Geologist, v. 18, p. 120–123.

SEDIMENTOLOGY OF SOME MISSISSIPPIAN AND PLEISTOCENE DEPOSITS OF NORTHEASTERN PENNSYLVANIA

W. D. SEVON

Pennsylvania Geological Survey, Harrisburg, Pennsylvania 17120

INTRODUCTION

This report discusses the sedimentology of two unusual, temporally unrelated deposits in northeastern Pennsylvania. The first is the sedimentologic succession from the Upper Devonian Catskill Formation into the Lower Mississippian Pocono Formation with emphasis placed upon the environmental interpretation of tilloid, pebbly mudstone and laminite lithologies. The second section treats the sedimentology of 11 Pleistocene boulder fields in the region with emphasis on the criteria of their formation.

ACKNOWLEDGMENTS

The writer is indebted to Mr. Alan Adler, College of the Pacific, for generously allowing the inclusion in this report of materials from his thesis research at Pennsylvania State University. Thanks are extended to Drs. G. H. Crowl, E. C. Dapples, R. W. Fairbridge, J. D. Glaeser, R. Parizek, S. Root and J. Veevers for stimulating and helpful discussions at the outcrops. Special thanks are given to T. Berg and J. D. Glaeser for criticism of the manuscript and to R. Sponseller for assistance with the photographs.

LATE DEVONIAN-EARLY MISSISSIPPIAN TRANSGRESSION IN NORTHEASTERN PENNSYLVANIA

A tripartite sequence of rocks composed of tilloid grading upward through pebbly mudstone into laminite has been studied at 30 localities in northeastern Pennsylvania (fig. 1). The distinctiveness of this sequence was recognized by Rogers as early as 1858 and by White in 1883. Both Rogers and White called the sequence a "transition zone" be-

tween the Upper Devonian Catskill Formation and the Lower Mississippian Pocono Formation.

In contrast, Willard (1939, p. 20) suggested a disconformity between the Catskill and Pocono Formations in northeastern Pennsylvania and Trexler and others (1961) indicate an angular unconformity between the two formations in the western part of the Anthracite region in eastern Pennsylvania.

It is the purpose of this section to describe the rocks adjacent to the Catskill-Pocono boundary in northeastern Pennsylvania and to consider a sedimentary model of their origin. The data come from detailed mapping in the Lehighton, Christmans and Hickory Run 7½′ quadrangles (Carbon County) and detailed examination of other good exposures in the region (fig. 1). These data are the result of work done mainly in 1967 and 1968, part of which were previously reported by the writer (Sevon, 1969).

Lithologic Features Above and Below the Catskill-Pocono Boundary

The uppermost member of the Catskill Formation consists mainly of several fining upward cycles. These cycles start with a basal conglomerate or a coarse-grained sandstone and fine upward through cross-bedded sandstones into red siltstones and shales. The uppermost few tens of feet of the Catskill contain no fining upward cycles, but instead alternating red shales or siltstones and poorly sorted gray sandstones. The upper contact with the overlying Pocono Formation is always sharp and conformable (fig. 2, A).

The lowermost lithology of the Pocono Formation is tilloid (a nonglacial conglomeratic mudstone or tillite-like rock). The tilloid is massive, non-

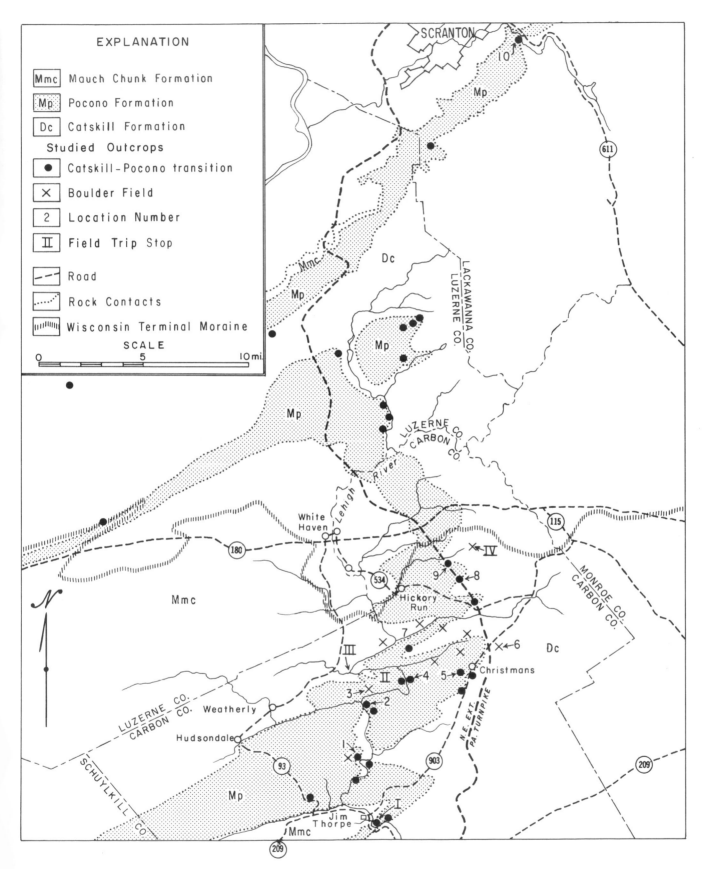

Figure 1. Map of the Pocono and Mauch Chunk Formations in northeastern Pennsylvania with locations of examined outcrops, boulder fields, field trip stops and the Wisconsin terminal moraine. Map modified from Geological Map of Pennsylvania (Gray, C., Shepps, V. C., and others, 1960).

215

Figure 2.   A, Contact of red shale of the Catskill Formation and overlying tilloid of the Pocono Formation occurs just above the pen (location 2, fig. 1).   Pen is 5-inches long.   B, Outcrop of typical tilloid at base of the Pocono Formation (location 2, fig. 1).   C, Thin section of tilloid from the lowermost Pocono Formation (location 2, fig. 1).   D, Thin section of pebbly mudstone from the lower Pocono Formation (location 2, fig. 1).

bedded, unsorted rock and in general character resembles a tillite (fig. 2, B).   Figure 2, C illustrates the lack of sorting, the range of grain sizes and the suspended nature of the grains in the matrix of this tilloid.   Well-rounded quartzite pebbles up to 4-inches in diameter are common in the tilloid and a rounded boulder 2-feet in diameter occurs at stop 1 (fig. 1).

The tilloid grades upward into pebbly mudstone.

The pebbly mudstone is massive in appearance and ranges in dominant grain size from clay to coarse-grained silt (fig. 2, D).   Although this lithology has no apparent bedding in outcrop, laminae seen in thin section indicate the presence of some bedding. Sand grains and rounded pebbles up to 3-inches in diameter are scattered throughout the mudstone (fig. 3, A), but the pebbles are not as abundant as in the underlying tilloid.

The pebbly mudstone grades upward into a laminite (fig. 3, B) which is composed of silt or sand laminae alternating with clay laminae (fig. 3, C). The laminae range from a millimeter to a centimeter in thickness and the coarse-grained laminae show both normal and reverse grading. The laminite contains rare pebbles up to 2-inches in diameter. In some outcrops the coarse-grained laminae show a progressive upward thickening which is culminated by an abrupt and conformable change to well-sorted, planar-bedded sandstones (fig. 4, A). These sandstones commonly have rippled surfaces (fig. 4, B) and some less well-defined structures which are interpreted as burrows. The sorting and structure of one of these sandstones is illustrated in figure 4, C.

The areal distribution and thickness of the tilloid, pebbly mudstone and laminite lithologies are shown in figure 5. To the east of the sections indicated, erosion has removed all Pocono rocks. To the west, the basal tripartite sequence does not occur. The diagram indicates little demonstrable variation except for a suggestion of northward thinning of the tilloid. The planar-bedded sandstones at the top of the basal Pocono sequence are persistent throughout the region and range in thickness from 30 feet at stop I to over 150 feet at Roaring Brook (location 10, fig. 1).

Of the several sections indicated in figure 5, the section at Penn Haven Junction (location 2, fig. 1) is the most complete and most easily interpreted. The basal Pocono sequence in this section (fig. 6) overlies the alternating sandstones and red siltstones of the uppermost Catskill and has a sharp but con-

A   B

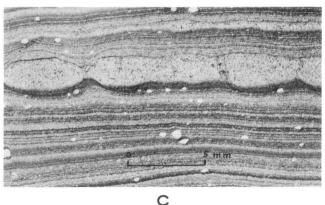

C

Figure 3. A, Pebbles in pebbly mudstone of lower Pocono Formation (location 2, fig. 1). Note variation in orientation of pebbles. Pen is 5-inches long. B, Laminite in the lower Pocono Formation (location 8, fig. 1). Pen is 5-inches long. C, Thin section of laminite from the lower Pocono Formation (location 2, fig. 1).

Figure 4.   A, Planar-bedded sandstones of the lower Pocono Formation (location 10, fig. 1). B, Rippled surface in planar-bedded sandstones of the lower Pocono Formation (location 2, fig. 1). C, Thin section of micro cross-bedded and planar-bedded sandstone from lower Pocono Formation (location 2, fig. 1).

formable basal contact with the topmost underlying red siltstone (fig. 2, A). The Pocono rocks have an upward progression from tilloid through pebbly mudstone to laminite and a sharp conformable contact with the overlying well-sorted, planar-bedded and rippled sandstones and siltstones.

SEDIMENTARY HISTORY

Any attempt to define the environments of deposition of the various rocks described above and shown in figure 6 must consider not only the specific tripartite lithologies, but also the relationship of each lithology to the rocks above and below. In addition, there must be a best fit of each interpretation to a larger sedimentological model.

The following list indicates several depositional environments or mechanisms by which the succession of basal Pocono lithologies could have formed: glacial, subaqueous slump, turbidity current, mudflow and tidal flat. Although the resemblance in the field and in thin section of the tilloid and laminite to tillite and glacial varves is striking, known

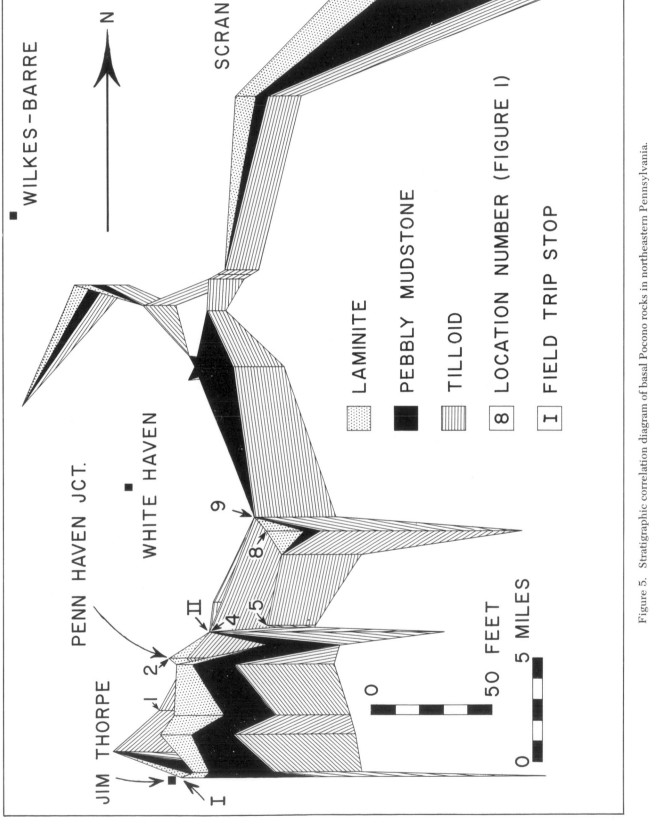

Figure 5. Stratigraphic correlation diagram of basal Pocono rocks in northeastern Pennsylvania.

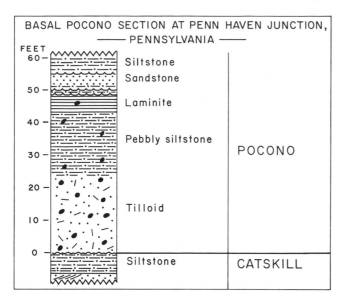

Figure 6. Succession of uppermost Catskill and lowermost Pocono rocks at Penn Haven Junction, Carbon County, Pennsylvania.

climatic conditions in Early Mississippian time (Schwarzbach, 1961, p. 261–265) do not support serious consideration of a glacial origin for these rocks. The gradational upper contact of the tilloid suggests an environmental association of the tilloid and the overlying pebbly mudstone. Both the pebbly mudstone and the laminite appear to require a subaqueous environment for their origin, thus it seems reasonable that the tilloid was also formed in a subaqueous environment. The combination of lithology, subaqueous environment, areal distribution and size of pebbles in the tilloid suggest that a subaqueous mudflow might be the most probable origin of the tilloid. The pebbly mudstone may represent material deposited further offshore in a low energy environment as a lateral equivalent of the tilloid. The laminite may represent (1) an increase in sorting energy possibly associated with shallowing water depth or a stabilized strand line, (2) increasing proximity to a shoreline or (3) a change in detrital influx. The gradational basal contact of the laminite with the underlying pebbly mudstone implies that the laminite was not formed on a tidal flat. Its sharp contact with the overlying rippled and planar-bedded sandstones further supports a more offshore origin for the laminites than these typical nearshore sandstones.

A sedimentological model which best accounts for the sequence of rocks at the Catskill-Pocono boundary in northeastern Pennsylvania is illustrated in figure 7.

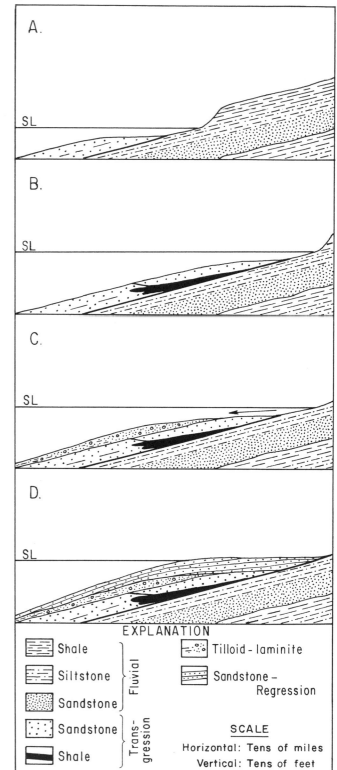

Figure 7. Sedimentological model depicting the origin of the uppermost Catskill-lowermost Pocono rock succession in northeastern Pennsylvania.

A. Transgression by the ravinement process (Stamp, 1922; Swift, 1968) proceeded across a relatively flat alluvial plain composed of fluvial fining upward sediments. Sands and muds eroded at the coastline were deposited offshore and are represented by the poorly sorted sandstones of the uppermost Catskill.

B. A temporary stabilization of sea level transgression resulted in the development of an offshore bar and a lagoon. Renewed transgression resulted in partial preservation of the lagoonal sediments as the red shales and siltstones of the uppermost Catskill.

C. Either periodic excessive influxes of detrital material from rivers or slumping of coastal areas then resulted in offshore deposition of the tilloid material. This process was probably repeated numerous times during the whole period of transgression because of evidence (locations 2 and 4, fig. 1) of multiple tilloid deposits preserved at some localities. Laterally, tilloid probably graded into the pebbly mudstone further offshore and as transgression progressed, this more offshore lithology proceeded to cover first the distal parts of the tilloid and finally the more shoreward parts. Evidently transgression then reached a maximum advance and a stable nearshore environment developed producing nearshore sedimentation.

D. The sedimentation of the planar-bedded and rippled sandstones apparently was widespread over a broad, shallow marine area and its development is reflected in deeper waters by the simultaneous formation of laminite. This nearshore sandstone environment spread across the region producing a sharp upper contact between the laminite and sandstone. This regional sand sedimentation initiated a new regressive phase of deposition which apparently continued throughout Pocono sedimentation.

BOULDER FIELDS IN CARBON COUNTY, PENNSYLVANIA

Although there have been various brief descriptions of boulder fields in northeastern Pennsylvania (Ver Steeg, 1930; Ashley, 1933; and Leverett, 1934), the paper of H. T. U. Smith (1953) on the Hickory Run boulder field was the first to discuss one of these fields in detail. Smith (1953, p. 65) states that the Hickory Run boulder field is "the largest of its kind on record in the Appalachian region," and that "Talus and other types of bouldery accumulations are well known in mountainous areas of strong relief and steep slopes. Accumulations of similar material on low gradients

in areas of moderate relief, however, are much less common."

The writer has found that the record of large boulder fields in northeastern Pennsylvania is not complete and that bouldery accumulations of many types are common. This section presents a general description of some of these boulder fields and discusses their mode of origin.

BOULDER FIELD DESCRIPTION

Boulder fields in northeastern Pennsylvania are mainly flat surfaced, moderate to low gradient boulder accumulations near the heads of small drainage basins or on mountain slopes. No completely barren boulder fields are known and some completely forested boulder fields occur (e.g., location 3, fig. 1). Gradient, surface morphology and lithology are variables common to all boulder fields. These variables can be used to describe boulder fields with almost any degree of precision and practicality desired.

GRADIENT

By definition a boulder field has a gentle slope. The Hickory Run field (stop IV, fig. 1) has a gradient of 1°; the Bowmanstown boulder field (Sevon, 1967) has a gradient of 4½–5°; and other fields in the area (fig. 1) have similar gradients. In all cases the boulder field surface gradient appears to be a reflection of the underlying bedrock surface gradient, although direct proof of this is not available.

SURFACE MORPHOLOGY

Although most boulder fields appear to be flat surfaced (fig. 8, A), they are really areas of complex microrelief. The surface irregularity variations in different parts of a boulder field are often subtle.

The following geomorphic forms can be differentiated on several of the barren boulder fields in Carbon County: stone rings (fig. 8, B) with a central mound of small boulders surrounded by larger boulders; low mounds lacking well developed boulder segregation; elongate mounds (fig. 8, C) which lack boulder segregation but have definite linearity; channels generally oriented parallel to the length of the boulder field; pits of somewhat circular form (fig. 9, A); lobe fronts which show boulder orientation into lobate form; stone rings with small boulder concentrations in a depression surrounded by larger boulders (fig. 9, B); and boulder imbrication (fig. 9, C). Figure 10 shows the surface morphology of part of the Hickory Run boulder field as mapped by Mr. Alan Adler during

A

B   C

Figure 8. A, View to the east across the Hickory Run boulder field (stop IV, fig. 1). B, Stone ring on the Hickory Run boulder field (stop IV, fig. 1). Note central concentration of small boulders. Hammer is indicated by arrow. C, Elongate mound on boulder field (location 6, fig. 1). View is up field. Arrow points to center of mound.

A   B

C

Figure 9.   A, Circular pit on a boulder field (location 6, fig. 1).   B, Stone ring of small boulders concentrated in a depression surrounded by larger boulders (location 6, fig. 1).   C, Imbrication of boulders on boulder field (location 6, fig. 1).   Upfield is to the right.

the course of his thesis research. These patterned features argue for periglacial conditions during and after the formation of the boulder field.

In the upper end of most boulder fields in Carbon County, rock streams can be traced from the boulder field proper to a specific outcrop area. Only rarely is it possible to detect stone strips joining the larger rock streams.

COMPOSITION

The composition of any boulder field comprises two elements: lithology and texture, both a func-

tion of available bedrock even though initial texture is modified during transport.

Most of the boulder fields in Carbon County are composed of rocks derived from Catskill or Pocono rocks. Boulders in the Hickory Run boulder field are derived from the uppermost member of the Catskill Formation (described in the previous section) and comprise hard, reddish sandstones and conglomerates.

Texture varies from one boulder field to another, but is similar to the general texture of the Hickory Run field. At Hickory Run the boulder size ranges

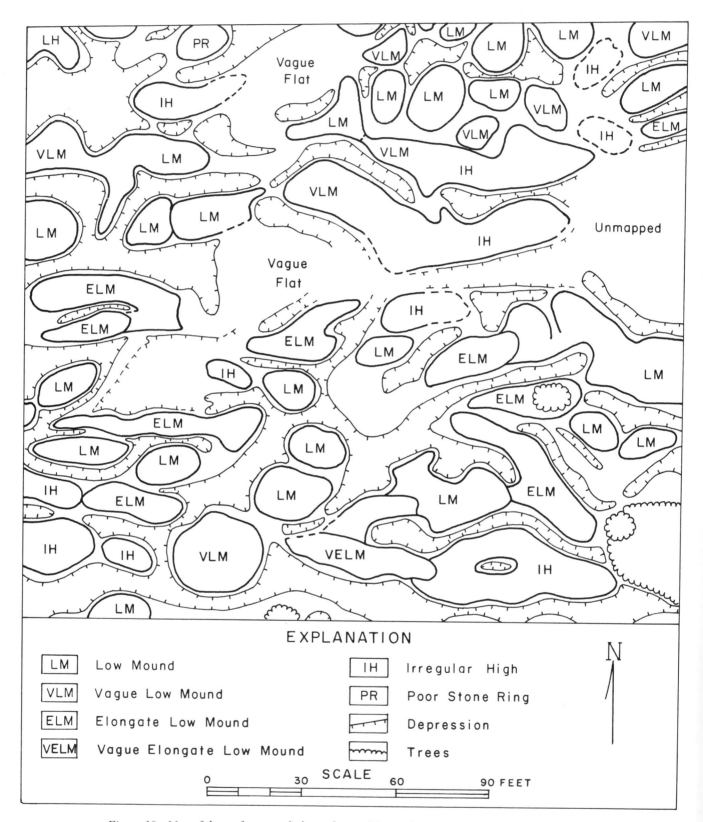

EXPLANATION

LM	Low Mound	IH	Irregular High
VLM	Vague Low Mound	PR	Poor Stone Ring
ELM	Elongate Low Mound		Depression
VELM	Vague Elongate Low Mound		Trees

SCALE

0   30   60   90 FEET

N

Figure 10.   Map of the surface morphology of part of the Hickory Run boulder field (stop IV, fig. 1).   Mapping was done by Mr. Alan Adler, College of the Pacific, and is part of his thesis research at Pennsylvania State University.   Isolated tree near center of right edge in feature ELM is same tree located at right edge of figure 11 in Flow V.

from 4-inch diameter cobbles to 30-foot long angular blocks (found mainly in the upper part of the field). Typical boulder sizes anywhere on the field are 1 to 5-feet in diameter. Generally, cobbles and smaller boulders are confined to high centers of microrelief features and larger boulders occupy lower parts of microrelief features. Although all shapes of boulders are present, equant, tabular and elongated shapes are most common. Boulders decrease in size and increase in roundness both with depth and down field.

Interstitial matrix material such as sand, silt and clay is conspicuously absent in the upper part of all barren boulder fields examined, but present at depths greater than 6 or 8 feet. Although interstitial matrix is generally thought to occur in the forested areas beyond the margins of barren boulder fields (Smith, 1953, p. 640; Sevon, 1967, p. 92), the writer finds that many Carbon County boulder fields (including Hickory Run) do not possess interstitial matrix in the forested marginal areas. Rather, the upper surface comprises a 1 to 2-foot thick humic layer which is underlain by empty interstices.

By using composition and surface morphology, Mr. Alan Adler mapped several distinct flows on the Hickory Run boulder field and a generalized part of his map is shown in figure 11. Some of these flow units can be traced up gradient to source outcrops.

ORIGIN OF BOULDER FIELDS

There seems to be little question today that boulder fields were formed by mass movement of rock material. The flow map of Adler (fig. 11), the structural necessity for movement in the Bowmanstown boulder field (Sevon, 1967, p. 91–92) and the lobate forms of the Blue Rocks field (Potter and Moss, 1968, p. 260) all indicate flow. There remains then, the question of the mechanism and time of flow.

There are three primary criteria necessary for the formation of boulder fields: (1) source rock, (2) mechanism of boulder production, and (3) mechanism of movement.

SOURCE ROCK

A suitable boulder field source rock must possess three qualities: resistant lithology, planes of separation and sufficient outcrop. Any well-indurated rock capable of withstanding rapid disintegration into small particles (sand size or smaller) is suitably resitant as a boulder source. Catskill and Pocono sandstones and conglomerates in

northeastern Pennsylvania meet these qualifications.

Bedding planes, joints and cleavage are the planes of separation essential for rock disintegration. Bedding planes are well developed in most of the Catskill and Pocono rocks of Carbon County and there is a widely spaced joint system oriented approximately normal to bedding and striking N.20°W. A less, but closely spaced fracture cleavage system is oriented approximately normal to bedding and strikes N.65°E. These planes readily allow disintegration of Catskill and Pocono rocks into boulders.

Without adequate outcrop a boulder field cannot develop. Scarps with horizontal rock or rock dipping away from the boulder field are the most suitable outcrops and most boulder fields in Carbon County are derived from such outcrops. Exposures of dip slope rock may also be boulder sources, but the quantity of boulders derived from them tends to be small.

MECHANISM OF DISINTEGRATION

Mechanical disintegration of rocks by frost action (freeze and thaw) is probably the only method capable of mass production of large quantities of boulders required to form a boulder field. Since there is no evidence that any rapid (in terms of volume produced) mechanical disintegration is occurring in Carbon County today, the more extreme climatic conditions associated with Wisconsin glaciation in Carbon County (fig. 1) are thought to be responsible for rock disintegration.

MECHANISM OF MOVEMENT

Boulder fields, scree slopes and mounded, lobe-fronted, bouldery rubble in steep gullies show no evidence of movement at the present time in Carbon County. Undisturbed scree slopes up to 42° generally appear even more stable than might be expected because of creep. Mounded bouldery rubble in the centers of tributary valleys of the Lehigh River (e.g., location 7, fig. 1) show no evidence of recent movement. In general, the present appearance of all debris masses (except floodplain gravel) in Carbon County is that of stability modified only by man or erosion. The lack of mass movement under the present climatic conditions indicates that different climatic conditions are needed for such movement and a periglacial climate seems most suitable.

The exact mechanism of movement for a low gradient boulder field such as Hickory Run (1°) is problematical. Smith (1953) indicated a periglac-

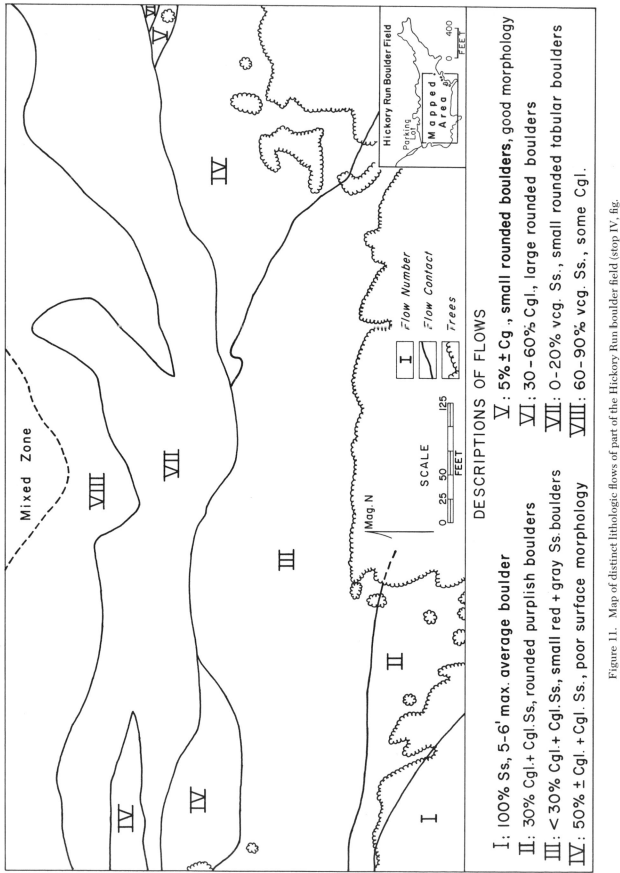

Figure 11.   Map of distinct lithologic flows of part of the Hickory Run boulder field (stop IV, fig. 1).   Mapping done by Mr. Alan Adler, College of the Pacific, and is part of his thesis research at Pennsylvania State University.   A tentative sequence of flows is: V (oldest), VI, IV, II, III, VII and I or VIII (youngest).

ial mass movement. Mr. Alan Adler has interpreted some lobes in the eastern end of the Hickory Run field as possible flowage lobes in which freeze and thaw plus some basal sliding would move masses of boulders, matrix and ice. These lobes occur on steeper gradients than the main field and have a raised central part with imbrication across the front of the lobe. At the western end of the field Adler has detected some possible solifluction lobes. These have a steep frontal end with a concentration of large, partially imbricated boulders and a depressed area with smaller boulders behind the front.

At present, solifluction seems to be the most probable mechanism of origin for most boulder fields in Carbon County. The possible importance of creep as suggested by Potter and Moss (1968, p. 261) cannot be discounted for some of the steeper gradient boulder fields in Carbon County, but the low gradient of the Hickory Run boulder field seems to require solifluction.

MATRIX AND FORESTATION

Characteristic of all barren boulder fields is the absence of interstitial matrix in the upper 6–8 feet of the deposit. Some writers (Smith, 1953; Andersson, 1906) feel that interstitial matrix must be present to aid in movement of the boulders and that this matrix is flushed out either during or after deglaciation of the area. Potter and Moss (1968, p. 261) have suggested the possibility of very little original matrix in the Blue Rocks boulder field. In an earlier section (Source Rock) it was suggested that boulder production by frost action would produce very little fine-grained material. The occasional occurrence of fitted adjacent boulder surfaces and the downfield increase in boulder roundness indicate abrasion during boulder movement which would produce some fine-grained material. However, the writer suggests that the amount of fine-grained material was minimal during the formation of the boulder field and that during deglaciation the fines moved down and formed the matrix now occurring at depths of 6–8 feet.

Earlier writers (e.g., Smith, 1953, p. 630) suggest that marginal sapping of soil at the edges of barren boulder fields such as Hickory Run is removing matrix and causing forest retreat. However, the writer finds that the forested boulder field areas generally have no fine-grained matrix. Instead, a 1–2 foot thick mat of humic debris fills the interstices of the boulder field and serves as a foundation for vegetation. Beneath this humic mat the interstices are empty. In many places this humic mat is actively encroaching onto barren boulder field areas by a process of debris-accumulation in the upper surface hollows followed by establishment of mosses, then small shrubs and eventually trees.

ROAD LOG

(Stratigraphic summaries of most of the rock units mentioned in the road log are given in Section IV).

Mileage

0.0   Start from Holiday Inn, East Stroudsburg, Pennsylvania. Leave Holiday Inn parking lot, turn RIGHT and proceed south on U. S. Rte. 209.

0.3   Turn RIGHT onto U. S. Interstate 80. On left, Godfrey Ridge, underlain by Upper Silurian and Lower Devonian rocks.

2.0   Cross Brodhead Creek.

2.3   Roadcut on left (south) exposes upper cherty limestone member of Buttermilk Falls Limestone on the northwest limb of an anticline.

4.2   Bear RIGHT to U. S. Rte. 209S.

5.0   Road passes through rock-cored kame deposit of Wisconsin age.

6.3   On left, green house in birch grove sits on 1,000-foot-long ridge of stratified sand and gravel, an esker, that leads into a kame deposit which can be viewed back to the left at 6.8 miles.

8.2   Roadcut through Marcellus Shale.

10.0   Roadcuts and Eureka Stone Quarry (on left) expose Buttermilk Falls Limestone.

10.8   EXIT onto U. S. Rte. 209S.

11.4   Roadcut on left exposes shales of Mahantango Formation.

13.5   Exposures of Mahantango Formation on left.

15.0   Village of Brodheadsville.

15.5   Crossing Wisconsin terminal moraine which underlies the village of Brodheadsville.

16.3   LEFT fork following U. S. Rte. 209S.

17.8   Village of Gilbert.

18.0   Road travels upon Wisconsin outwash plain which ends at 20.3 miles. Ridge to left is north limb of Parryville syncline. Rocks in ridge slopes are siltstones and shales of the Trimmers Rock Formation and have shallow (20°–30°) south dips.

Mileage

20.8   Shale-chip sharp stones exposed on right were derived from Mahantango shales by intense frost action during Wisconsin periglacial conditions.

21.3   Village of Kresgeville.

21.9   Cross Middle Creek.

23.55 Small excavation on left is in Mahantango Formation. Road climbs up onto ridge crest of north limb of Parryville syncline. Crest of ridge upheld by sandstones of the Towamensing Member of the Catskill Formation.

23.7   Carbon County line.

23.9   Village of Trachsville.

24.3   Valley to left cut into red siltstones of Walcksville Member of the Catskill Formation.

25.0   Glaeser Sag. Note alignment of gaps to the left. Gap on horizon is Little Gap in Blue Mountain.

27.7   Village of Forest Inn.

28.6   Surface underlain by Illinoian(?) till.

29.2   Exposure on left at road junction is in red sandstones and siltstones of Walcksville Member of the Catskill Formation.

31.55 Exposed along right side of road are sandstones of Towamensing Member of the Catskill Formation.

31.95 Large exposure on right of sandstones at base of Towamensing Member of the Catskill Formation. Exposure on right between here and 32.3 is in siltstones and shales of Trimmers Rock Formation.

32.4   Pass under Pennsylvania Turnpike Northeast Extension.

32.7   Cross Pohopoco Creek. Exposures ahead are in Mahantango shales.

33.8   Exposures on left in Mahantango shales.

34.3   Stop light. Straight ahead across bridge over Weissport and Lehigh River.

34.45 Borough of Lehighton.

34.5   Turn RIGHT on U. S. Rte. 209S. at west end of bridge. Exposure on left is in shales of Mahantango Formation. Fossiliferous horizon occurring at 34.6 miles (opposite bridge over Lehigh River to Weissport) has been correlated with Centerfield of New York State.

35.2   Stop light at center of Lehighton.

35.9   Exposures on left in siltstones of Trimmers Rock Formation.

36.25 Exposures on left in crinoid columnal bearing, very fine-grained sandstones of the Beaverdam Run Member of the Catskill Formation.

36.5   Exposures on left in red, fining upward sandstones and siltstones of Long Run Member of the Catskill Formation.

36.7   Village of Packerton.

37.0   Exposures on left in sandstones and conglomerates of Packerton Member of the Catskill Formation. Ahead, on left, the Big Chief Drive-In sits in the center of a recessed valley of red siltstones and shales of the Sawmill Run Member of the Catskill Formation. At the north end of the valley recess start the brown sandstones and conglomerates of the Berry Run Member of the Catskill Formation. The upper contact of this member is marked by red shales.

37.35 Between 37.35 and 38.2 miles is a long section (on left), mainly along strike, through the Flagstaff Mountain Member of the Catskill Formation.

38.2   Base of the Bear Mountain Member of the Catskill Formation indicated by the appearance of red fining upward cycles. Twelve cycles occur in this section. The town of Jim Thorpe (formerly Mauch Chunk) is viewed to the left.

38.4   Borough of Jim Thorpe.

38.5   Outcrop across the Lehigh River (to right) is location of Stop I.

38.6   Stop light at center of Jim Thorpe. Turn RIGHT, then bear LEFT following U. S. Rte. 209S.

38.75 Base of type locality of Mauch Chunk Formation on left (at north end of service station parking lot).

39.0   Stop light. Turn RIGHT onto Pa. 903N. Cross bridge over Lehigh River.

39.1   Turn RIGHT at east end of bridge. Bear right through supermarket parking lot to access road paralleling railroad tracks. Follow road south. Outcrop on left is Pocono Formation.

39.7   Cross railroad tracks to parking area along the tracks.

STOP I.

An abbreviated description of the rocks exposed at STOP I is presented as Section I. This is the best and most complete exposure of the Pocono Formation in northeastern Pennsylvania and is undoubtedly the proper type locality for the Pocono, although no such designation was ever made by the Pennsylvania Second Geological Survey. Of particular interest at this stop are the sandstones and shales of the uppermost Catskill Formation and the tilloid, laminite and planar-bedded sandstone of the Pocono Formation. The contrasting conglomeratic (upper) and non-conglomeratic (lower) parts of the Pocono Formation are well shown here and the basal conglomerate of the upper member is well exposed. The interbedded upper contact with the Mauch Chunk Formation is exposed along a railroad spur near the supermarket parking lot at the north end of the section.

Leave parking area and return to U. S. Rte. 209S. via same route came. Mileage 40.2 when passing through supermarket parking lot. Turn LEFT onto Pa. 903S. Cross bridge over Lehigh River. Note good view of type locality of Mauch Chunk Formation directly ahead.

Mileage

40.45 Stop light at west end of bridge. Turn RIGHT onto U. S. Rte. 209S.

40.85 Exposure on left of sandstones and conglomerates of interbedded zone of Pottsville and Mauch Chunk Formations. Red siltstones of the Mauch Chunk occur at both ends of the long exposure which ends at 41.1 miles.

43.4 Turn RIGHT onto Pa. 93N. Rock quarry on left is in Mauch Chunk Formation.

44.8 Outcrops are in Pocono Formation.

48.65 Village of Hudsondale.

49.1 Cross Quakake Creek. Exposures of Mauch Chunk Formation in roadcut immediately north of creek.

49.3 Turn RIGHT to Weatherly.

50.95 Borough of Weatherly.

51.5 Outcrop of Mauch Chunk Formation on left.

51.7 Turn RIGHT across bridge. Turn RIGHT immediately after crossing bridge and railroad.

51.8 Turn LEFT on First St.

52.2 Bear LEFT onto Eurana Ave. then turn RIGHT onto E. Main St.

55.4 Turn RIGHT to Rockport. Village of Rockport.

56.0 Road joins from left. Continue straight ahead. Proceed with caution along narrow road with blind curves.

56.5 Turn RIGHT onto old roadbed of New Jersey Central Railroad. Lehigh River on left. Outcrop on right between here and STOP II is all of Pocono Formation.

57.2 To left across Lehigh River is south portal of Rockport Tunnel.

57.3 To left across Lehigh River is large outcrop of uppermost rocks of Upper Member of the Pocono Formation. Top of slope capped by conglomerate and overlain by red siltstones of the Mauch Chunk Formation.

58.95 To left across the Lehigh River is mouth of Drakes Creek. Rocks with north dip on south side of creek are in the lower part of the Upper Member of the Pocono Formation.

59.4 **STOP II.**

An abbreviated description of the rocks exposed at STOP II is presented as Section II. The sequence of rocks includes the uppermost sandstones and shales of the Catskill Formation and the tilloid and planar-bedded sandstone of the Pocono Formation. The tilloid contains some contorted sandstone beds and is exceptionally thick (97′). Of interest is the fact that at the counterpart to this outcrop on the opposite side of the Lehigh River the tilloid is only 13′ thick. The sandstone overlying the tilloid is very well developed here and shows excellent bedding, ripple marks and load casts in the recessed zone.

TURN AROUND and leave STOP II via same route came.

59.8 Mouth of Drakes Creek across the Lehigh River on the right.

61.5 South portal of Rockport Tunnel across the Lehigh River on the right.

62.0 On left is the foundation of a water tank used in the days of steam locomotives.

62.3 Rockport access road enters from left.

STOP III.

Over 450′ of the lowermost Mauch Chunk Formation is exposed here in one of the longest outcrops of Mauch Chunk in northeastern Pennsylvania. At the base of the section is the uppermost part of the interbedded Pocono-Mauch Chunk contact. Channeling, lateral color changes and interbedding are features of interest in this zone. Nodular and discontinuously bedded limestone typical of the Mauch Chunk, particularly the lower half, are well exposed and appear to occur in a filled channel. Extensive planar contacts between beds are characteristic of the Mauch Chunk and show well in this long exposure. The sandstone unit forming the waterfalls near the top of the section has cross-bedding which may be eolian in origin. This cross-bedding is well exposed just north of the waterfalls. Barrell (1907, p. 466) assigned a fluvial floodplain origin to the Mauch Chunk rocks in northeastern Pennsylvania.

Leave STOP III and CONTINUE along railroad roadbed.

62.45 Limestone zone in Mauch Chunk Formation on left.

62.65 Waterfalls on left, bridge and end of section for STOP III.

62.8 Outcrop on left shows deep etching of calcareous sandstone in Mauch Chunk Formation. Horizon is probably the same as the limestone zone viewed in STOP III.

63.2 On left is interbedded red siltstone of Mauch Chunk Formation and gray sandstone of the Upper Member of the Pocono Formation.

63.3 To right across Lehigh River is north portal of Rockport Tunnel.

63.8 Outcrop on left is cross-bedded Pocono Formation (Upper Member).

65.0 On left, about 15 ft. above the road, are a few cobbles of Illinoian(?) outwash gravels.

65.1 On left are some discontinuous siltstones and large foreset beds in sandstones of the Pocono Formation (Upper Member).

65.3 On right across Lehigh River is the mouth of Mud Run.

Mileage

65.4   On left is cleavage dominated red shale of the Mauch Chunk Formation. Bedding is nearly horizontal.

65.95 To right is good view up the Lehigh River valley.

66.2   On left is mouth of Leslie Run.

66.25 On left is sandstone of uppermost Upper Member of the Pocono Formation.

66.4   Ahead is a view of one of the remnant locks of the Lehigh Canal.

66.65 On right is lock of the Lehigh Canal.

67.8   Bridge over Sand Run.

67.9   On left is complex of discontinuous beds and cross-beds in sandstones of the Upper Member of the Pocono Formation.

68.15 On right is lock of Lehigh Canal.

68.6   On left is cross-bedded sandstone of Mauch Chunk Formation.

68.9   Pass between piers of old railroad bridge.

69.45 On left is Mauch Chunk Formation. On right across Lehigh River is Wisconsin kame deposit. To right is good view up the Lehigh River valley.

69.9   On left is Pocono Formation overlain by Wisconsin till.

70.0   On left are cross-bedded and channeled sandstones of the Upper Member of the Pocono Formation.

70.15 STOP at road intersection. PROCEED STRAIGHT AHEAD. (Automobiles may turn RIGHT, cross bridge and turn RIGHT onto Pa. 534, picking up road log at mileage 74.0).

70.4   On left is cross-bedded sandstone of the Mauch Chunk Formation.

71.3   Pass under railroad bridge.

71.5   Pass under U. S. Interstate 80.

71.9   STOP at road intersection in White Haven. Turn RIGHT. Cross bridge over Lehigh River.

72.1   Turn RIGHT to Hickory Run State Park at east end of bridge.

72.5   Pass under U. S. Interstate 80.

73.9   Junction with Pa. 534, CONTINUE STRAIGHT AHEAD.

74.0   Road enters from right (alternate route for automobiles).

74.8   On left is exposure of Wisconsin kame deposit.

75.2   On right is siltstone of the Mauch Chunk Formation.

75.55 Hickory Run State Park.

76.9   On left between here and mileage 77.2 is interbedded contact of the Mauch Chunk and Pocono Formations.

77.3   On right is headquarters of Hickory Run State Park.

77.9   Turn LEFT to boulder field.

78.35 Take LEFT fork to boulder field.

79.7   Gravel pit on left is in sand, silts, clays and gravels of Wisconsin kame deposit.

80.5   Pass under Pennsylvania Turnpike Northeast Extension.

81.1   Exit road to right. Continue straight ahead.

82.0   **STOP IV.**

The unforested part of the Hickory Run boulder field measures roughly 400 by 1,800 feet and has a boulder thickness of at least 12 ft. The boulders are sandstones and conglomerates derived from the uppermost Catskill Formation (Bear Mountain Member). Lithologic differences are best seen by traversing the width of the field, while differences in rounding and size are best seen by traversing the length of the field. Microrelief features are sometimes subtle, but can be viewed anywhere on the field. Figures 10 and 11 in the accompanying text can be oriented by locating the isolated pine tree which is about 1,000 ft. due east from the parking lot entrance to the boulder field. A detached part of the field occurs a few hundred feet southeast from this same isolated tree and contains the largest boulders in the field.

Leave parking lot and return via same route came.

82.9   Turn LEFT on exit road.

83.5   Pass under Pennsylvania Turnpike Northeast Extension.

85.9   Bear RIGHT at picnic area.

87.1   Turn LEFT on Pa. 534E.

90.8   Pass under Pennsylvania Turnpike Northeast Extension.

91.05 On right is exposure of Illinoian(?) till.

92.15 Old Stage Road to right. CONTINUE STRAIGHT AHEAD.

93.5   Turn LEFT onto Pa. 903N.

95.4   On right is conglomerate of the Bear Mountain Member of the Catskill Formation.

96.7   Monroe County line.

96.8   Village of Fern Ridge.

97.35 Turn LEFT on Pa. 115N.

97.65 On left is quarry in sand and till of Wisconsin terminal moraine complex through which the road passes.

99.3   Turn RIGHT onto U. S. Interstate 80E. Proceed G. S. A. meeting in Atlantic City.

SECTION I. STRATIGRAPHIC DESCRIPTION OF THE UPPERMOST CATSKILL, ENTIRE POCONO AND LOWERMOST MAUCH CHUNK FORMATIONS AT JIM THORPE, CARBON COUNTY, PENNSYLVANIA (STOP I). OUTCROP STARTS AT NORTH EDGE OF A STONE RETAINING WALL LOCATED SOUTHEAST OF THE JIM THORPE SEWAGE TREATMENT PLANT AND IS CONTINUOUS NORTH ALONG THE RAILROAD TRACK ON THE EAST BANK OF THE LEHIGH RIVER FOR 0.5 MILE TO A SUPERMARKET PARKING LOT.

Unit	Lithologic Description	Thickness	
	Mauch Chunk Formation		
255–260	Slst, fg-cg, (5R4/2-5Y4/2), gradational contacts.	20'	6"
242–254	Ss, vfg-mg, (5Y6/1), 3 slst interbeds, 2" thick pebble band, sharp contacts.	43'	6"
238–241	Slst, fg-cg, (5R4/2), basal 2' is vfg, ss, gradational base.	29'	
	Pocono Formation, Upper Member		
237	Ss, fg, (5YR5/2), massive.	19'	6"
236	Covered interval, probably ss.	266'	
220–235	Ss, mg, (N7-5Y5/2), ¼-½" quartz pebbles, sharp contacts.	85'	7"
219	Cgl, ½" quartz pebbles, mg-vcg sand matrix, black sh chips.	3'	
218	Ss, cg, (N7), cross-bedded, sharp upper & gradational lower contact.	16'	
217	Cgl, 1/8–1" quartz pebbles, fg-mg sand matrix.	4'	6"
189–216	Ss, fg-mg, plant fossils, scattered quartz pebbles; 10 interbedded cgls which are 5" to 4' 6" thick and have ¼-2½" quartz pebbles, sharp contacts.	267'	1"
188	Cgl, massive, ½-2½" quartz pebbles, mg-cg sand matrix, sharp contacts.	50'	
183–187	Ss, fg, (N5), 1' thick cgl with ½" quartz pebbles, sharp contacts.	50'	6"
182	Sh, (N1), sharp contacts.	5'	
181	Ss, mg, (N7).	8'	
180	Cgl, massive, ½-1" quartz pebbles, mg sand matrix, sharp contacts.	21'	6"
171–179	Ss, mg-veg, (N5-N7), 2 cgls, 1–7' thick, ¼-1" quartz pebbles, sharp contacts.	29'	1"
170	Sh, (N1), interbedded mg ss stringers, plant fossils, sharp contacts.	6'	
165–169	Ss, mg, (N4-N5), 9" thick mg-cg slst, sharp contacts.	36'	3"
164	Slst, mg, (N3).	15'	
163	Ss, mg, (N4-N3), scattered quartz pebbles, sharp contacts.	32'	
162	Sh, (N1).	1'	8"
161	Ss, mg, (N5), sharp contacts.	6'	
160	Cgl, ¼-2" quartz pebbles, mg-cg sand matrix, sharp eroded base.	17'	
	Lower Member		
139–159	Ss (55%) & slst (45%); vfg-mg, (N5) ss beds 7"–21' 10" thick; cg, (5Y5/2-10YR5/4) slst beds 5"–31' thick, sharp contacts.	124'	
138	Sh, (N3).	1'	2"
131–137	Ss, vfg-mg, (N4), 3 slsts, fg-vcg, 1"–2' thick, sharp contacts.	26'	1"
130	Sh, (N1).		2"
50–129	Ss (83%) & slst (17%); vfg-mg, (N3) ss beds 1"–24' thick; 24 fg-vcg, (N4) slst beds, ¼"–8' thick, sharp and gradational contacts.	187'	10"
49	Covered interval, behind stone retaining wall.	10'	
48	Ss, mg, white, quartzite, sharp contact.	33'	
46–47	Ss, clay-mg, (N8-5Y5/1), lower fault (?) contact.	4'	10"
42–45	Slst, cg, (N8-5Y5/2), gradational contacts.	2'	6"
40–41	Laminite, (5Y6/1), varve-like laminations of clay and silt, scattered quartz grains and pebbles, gradational contacts.	9'	2"
39	Sh, (5GY6/1), scattered pebbles up to 2½" diameter.	10'	
38	Fault crush zone.		6"
34–37	Sh, (5GY6/1-5R4/2), scattered pebbles up to 3" diameter, gradational contact.	13'	2"
31–33	Slst, mg-cg, (5YR4/1-5R4/2), scattered pebbles up to 4" diameter, gradational contacts.	44'	
30	Tilloid, clay-vcg, (5R4/2-5Y5/2), scattered pebbles up to 4" diameter, 1 boulder 2' diameter, unsorted, sharp lower contact.	81'	
	Catskill Formation, Bear Mountain Member		
25–29	Slst, mg-cg, (10R4/2); 1 vfg, (5Y5/2) ss 4' 6" thick; 2 (10R4/2) sh, 2' & 2' 6" thick, gradational contacts.	16'	
24	Ss, clay-vcg, (5Y5/2), pebbles up to 3" diameter, sharp lower contact.	27'	
18–23	Ss, fg-cg, (5R4/2), sharp contacts.	38'	2"
17	Slst, cg, (5Y6/1).		10"
16	Ss, fg-mg, (N7), sharp contacts.	17'	6"
15	Sh, (5Y5/2), gradational lower contact.		6"
13–14	Slst, fg-vcg, (5R4/2), gradational internal contacts, sharp lower contact.	18'	6"
11–12	Ss, vfg, (5Y6/1), sharp contacts.	5'	9"
2–10	Ss (67%) & slst (33%); vfg-fg, (5R4/2) ss in 4 beds 2½'–18' thick; fg, (5R4/2) slst in 5 beds 2"–11' thick, gradational contacts.	43'	2"
1	Ss, cg, (5R4/2), disappears behind stone retaining wall.	2' +	
	Total	1749'	6"

SECTION II. STRATIGRAPHIC DESCRIPTION OF THE UPPERMOST CATSKILL FORMATION AND LOWER POCONO FORMATIONS AT BALD MOUNTAIN, CARBON COUNTY, PENNSYLVANIA (STOP II). BASE OF OUTCROP IS 0.45 MILES SOUTH OF THE MOUTH OF DRAKES CREEK ON THE WEST BANK OF THE LEHIGH RIVER.

Unit	Lithologic Description	Thickness
	Pocono Formation	
6	Ss, mg, (N6), planar-bedding, ripple marks, conglomeratic at base, well-sorted sand, 21' above base is 6' thick recess zone with loadcasts and ripple marks in ss.	50' +
5	Tilloid, unsorted sand, silt, clay and pebbles, (5Y5/2), upper half contains contorted masses of vfg-fg, (N5), well-sorted ss, sharp contacts.	97'
	Catskill Formation	
4	Ss, mg, (N5), red sh chips at base, massive.	20'
3	Sh, (5R4/2), massive, silty, sharp eroded upper contact.	55'
2	Ss, mg-cg at base becoming fg-mg at top, (5Y6/1), red sh chips at base and 6' above base, 62' above base is 1' thick vfg-fg, (5R4/2) ss, 66' above base is 3' thick (5R4/2) slst, gradational upper contact.	72'
1	Sh, (5R4/2), silty, massive, sharp eroded upper contact, lower contact in cover.	6' +

SECTION III. STRATIGRAPHIC DESCRIPTION OF THE MAUCH CHUNK FORMATION AT ROCKPORT, CARBON COUNTY, PENNSYLVANIA (STOP III). OUTCROP STARTS AT NORTH SIDE OF MOUTH OF INDIAN RUN AND IS CONTINUOUS FOR 0.3 MILE NORTH ON THE WEST BANK OF THE LEHIGH RIVER.

Unit	Lithologic Description	Thickness
46	Slst, cg, (5RP4/2), massive, some calcite.	20'
45	Covered interval.	15'
44	Ss, vfg, (5RP4/2), massive, some calcite.	22'
43	Ss, fg, (5RP4/2), massive, cross-bedding, sharp contacts.	10'
42	Slst, cg, (5RP4/2), calcareous, calcareous lenses in upper half.	14'
41	Ss, vfg-fg, (5RP4/2), massive, some red sh beds, pits from weathered calcite, gradational contacts.	8'
40	Slst, mg-cg, (5RP4/2), thin red sh beds.	7'
39	Ss, vfg, (5RP4/2), massive, sharp lower & gradational upper contact.	10'
38	Slst, mg-cg, (5RP4/2), massive, 1–2 mm thick red sh beds.	7'
37	Sh, (5R4/2), calcareous nodules and tubes weather to pits, sharp contacts.	6'
36	Slst, cg, (5RP4/2), red sh chips.	1'
35	Sh, (5R4/2), thin mg slst beds, sharp lower & gradational upper contact.	3'
34	Ss, vfg, (5RP4/2), massive, calcareous in lower 1'.	2' 6"
33	Ss, vfg, (5R4/2), calcareous, nodular pitting, sharp-contacts.	16'
32	Ss, fg, (5RP4/2), massive, long steeply inclined cross beds, forms waterfalls.	42'
31	Ss, vfg-fg, (5RP4/2), massive, sh beds, calcareous lenses, laterally variable, sharp contacts.	28'
30	Slst, cg, (5RP4/2), planar-bedding in lower half.	6'
29	Slst, fg-mg, (5R4/2), calcareous nodular zone near base, sharp upper contact.	15' 6"
28	Ss, vfg, (5RP4/2), massive, partly calcareous, gradational upper contact.	5' 6"
27	Slst, fg-mg, (5R4/2), massive, scattered calcareous nodules, lateral variation to sh, sharp contacts.	19'
26	Ss, fg, (5RP4/2), massive, some sh beds and calcareous nodules.	44'
25	Ss, vfg, (5RP4/2), interbedded with red sh and slst, sharp contacts.	3'
24	Ss, vfg, (5RP4/2), hard, prominent bed.	9"
23	Ss, mg, (5RP4/2), calcareous, channel (?) with abundant limestone nodules and discontinuous beds, sharp contacts.	8'
22	Ss, mg, (5RP4/2), red sh chips at base, calcareous nodules, cut out in part by overlying unit.	6'
21	Ss, fg, (5RP4/2), calcareous, massive, sharp contacts.	3'
20	Slst, cg, (5R4/2), massive.	9"
19	Ss, vfg, (5R4/2), gradational upper contact.	2'
18	Covered interval.	18'
17	Sh, (5R4/2), fissile.	2'
16	Slst, cg, (5R4/2), poor bedding, gradational upper contact.	10'
15	Ss, mg, (5RP4/2), cross bedding, gradational upper contact.	18'
14	Sh, (5R4/2), calcareous, some limestone nodules, some laterally discontinuous slst, sharp upper contact.	16'
13	Slst, fg, (5R4/2), massive, gradational upper contact.	13'
12	Slst, cg, (N5), sharp upper contact.	1'
11	Slst, mg, (5R4/2), becomes sh in upper 1', deeply recessed, gradational upper contact.	4'

SECTION III — (Continued)

Unit	Lithologic Description	Thickness
10	Ss, fg-vcg, (N5), reddish in upper third, discontinuous cgl in lower 4', sharp contacts.	9' 4"
9	Sh, (5R4/2), laterally discontinuous.	1' 6"
8	Ss, vcg, (N5), conglomeratic, channel cuts, sharp contacts.	5' 4"
7	Cgl, (N4), quartz pebbles up to 1" diameter, vcg sand matrix, massive, sharp contacts, fault contact at base.	7' 6"
6	Slst, mg, (5R4/2), massive, some small calcareous nodules, sharp upper contact.	10'
5	Sh, silty, (5R4/2), some calcareous nodules, gradational upper contact.	4'
4	Slst, mg, (N5), laterally becomes (5R4/2) sh, gradational upper contact.	2' 6"
3	Sh, (N6), massive, laterally is (5R4/2), gradational upper contact.	7'
2	Ss, fg, (N4), massive gradational upper contact.	4'
1	Ss, mg, (5RP4/2), massive, micaceous, sharp upper contact.	19' +
	Total	477' 2"

At base of section (north side of mouth of Indian Run) a channel cuts out Units 2–6 and part of Unit 1. The sequence in the channel is:

(top)

E	Sh, (5Y4/2), variable thickness, sharp upper contact.	5'
D	Ss, fg, (5Y4/2), gradational upper contact.	5'
C	Sh, (5Y4/2), bedded, sharp upper contact.	3' 6"
B	Ss, fg, (5Y6/1 to N6), massive, gradational upper contact.	8'
A	Sh, (5GY6/1), sharp lower & gradational upper contact.	1'
	Total	22' 6"

SECTION IV. LITHOLOGIC DESCRIPTIONS OF THE UPPER DEVONIAN AND MISSISSIPPIAN ROCK UNITS IN CARBON COUNTY

Mississippian

Mauch Chunk Formation. 2,200'. Mainly red coarse-grained siltstones to fine-grained sandstones. Some red shales. Rare conglomerates. Calcite cement and white nodular to discontinuously bedded limestones. Ripple marks, cross-bedding, mud cracks, burrows and plant impressions. Planar contacts. Transitional interbedded lower and upper contacts.

Pocono Formation

Upper Member. 940'. Light gray conglomerates with quartz pebbles up to 2½" diameter. Light gray, fine- to medium-grained sandstones. Rare black shales. Cross-bedding. Plant fossils. Sharp contacts.

Lower Member. 547'. Very fine- to medium-grained, medium gray sandstones. Olive gray siltstones. White, quartzitic, planar-bedded, ripple marked sandstone overlying laminite and tilloid at base. Sharp base.

Upper Devonian

Catskill Formation.

Bear Mountain Member. 865'. Red conglomerates, sandstones and siltstones. Twelve fining-upward sequences. Massive at base becoming flaggy upward. Cross-beds common. Siltstones and shales, often rippled, mudcracked and fissile. Scour and inter-tongued beds common in lower third.

Flagstaff Mountain Member. 904'. Nonred, medium- to coarse-grained sandstones and conglomerates. Massive, tabular cross beds. Individual beds fine upward. Bed tops often strewn with fine-grained rock fragments.

Berry Run Member. 985'. Nonred medium- to fine-grained sandstones dominate. Few red siltstones up to 40' thick in upper part. Minor conglomerates. Wedging, truncation and lenticularity of cross-beds common.

Sawmill Run Member. 424'. Nonred, dominantly fine- and very fine-grained, well-sorted, sandstones, commonly cross-bedded. Some sandstones calcareous. Basal, middle and upper red siltstones. Minor shale and deeply leached, sponge-like zones of crystalline limestone with high clay content. Few beds with fragments of supratidal dolomite (?).

Packerton Member. 416'. Dominantly medium- to fine-grained, well-sorted sandstones, commonly cross-bedded. Minor fine-grained conglomerates, siltstones and shales. Some calcareous sandstones. Lenticularity, wedging and truncation of beds common. Some channelling.

Long Run Member. 2,362'. Predominantly red, planar-bedded, massive, very fine- to fine-grained sandstones alternating with massive to fissile, red siltstones and fissile shales. Sandstones more prominent upward. Some lenticular zones of fossil debris. Several fining-upward sequences mainly in upper part with basal gray-green sandstone grading upward into red sandstone then to siltstone then to shale. Topmost red shale often reduced green color in top few inches under contact with nonred sandstone.

SECTION IV — (Continued)

Beaverdam Run Member. 1,157'. Nonred. Principally siltstone with some very fine-grained sandstones. Planar-bedded, massive and fissile beds alternate. Ball and pillow structures prominent. Abundant trains and lenses of *Tentaculites* and crinoid columnals.

Walcksville Member. 645'. First red bed sequence above Catskill base. Planar-bedded, massive and fissile, nonred and red siltstones and sandstones. Abrupt lateral color and grain size changes common.

Towamensing Member. 190'. Gray, fine- to very fine-grained sandstones and coarse-grained siltstones. Many undulatory and scoured surfaces. Shale-clast conglomerates. Poorly sorted sandstones contain carbonaceous debris. Churned and burrow mottled siltstones. Some burrowed zones graded, shale capped and rippled or mudcracked with infilling by overlying sediment. Many graded beds.

Trimmers Rock Formation. 1,042'. Turbidite sequence. Many fining upward siltstones and shales. Thickness of beds increases upward. Burrowing common. Brachiopod shell concentrations in lenses in upper half. Some load casts. Micro cross-bedding at tops of some turbidites. Medium to dark gray color.

REFERENCES CITED

Andersson, J. G., 1906, Solifluction, a component of subaerial denudation: Jour. Geology, v. 14, p. 91–112.

Ashley, G. H., 1933, The scenery of Pennsylvania, its origin and development: Pa. Geol. Survey 4th ser., G 6, 91 p.

Barrell, J., 1907, Origin and significance of the Mauch Chunk shale: Geol. Soc. America Bull., v. 18, p. 449–476.

Gray, C., Shepps, V. C., and others, 1960, Geological map of Pennsylvania: Pa. Geol. Survey 4th ser.

Leverett, F., 1934, Glacial deposits outside the Wisconsin terminal moraine in Pennsylvania: Pa. Geol. Survey 4th ser., G 7, 123 p.

Potter, N., Jr., and Moss, J. H., 1968, Origin of the Blue Rocks block field and adjacent deposits, Berks County, Pennsylvania: Geol. Soc. America Bull., v. 79, p. 255–262.

Rogers, H. D., 1858, The geology of Pennsylvania: Philadelphia, v. 2, p. 8.

Schwarzbach, M., 1961, The climatic history of Europe and North America *in* A. E. Nairn (ed.), Descriptive Palaeoclimatology: Interscience Publishers Inc., New York, p. 255–291.

Smith, H. T. U., 1953, The Hickory Run boulder field, Carbon County, Pennsylvania: Amer. Jour. Sci., v. 251, 5th Series, p. 625–642.

Sevon, W. D., 1967, The Bowmanstown boulder field, Carbon County, Pennsylvania: Pa. Acad. Sci. Proc., v. 40, p. 90–94.

———— 1969, The Pocono Formation in northeastern Pennsylvania: Field conference of Pennsylvania geologists, 34th, Hazleton, 1969; Pa. Geol. Survey, Harrisburg.

———— (in press), Subaqueous mudflow origin of basal Pocono rocks in northeastern Pennsylvania *in* The Geol. Soc. Amer., Abstracts for 1968: Geol. Soc. Amer. Special Paper.

Stamp, L. D., 1922, An outline of the Tertiary geology of Burma: Geol. Mag., v. 59, p. 481–501.

Swift, D. J. P., 1968, Coastal erosion and transgressive stratigraphy: Jour. Geol., v. 76, p. 444–456.

Trexler, J. P., Wood, G. H., Jr., and Arndt, H. H., 1961, Angular unconformity separates Catskill and Pocono Formations in western part of Anthracite region, Pennsylvania: Art. 38 *in* U. S. Geol. Survey Prof. Paper 424-B, p. B84–B88.

Ver Steeg, K., 1930, Wind gaps and water gaps of the northern Appalachians, their characteristics and significance: New York Acad. Sci., Annals 32, p. 87–220.

White, I. C., 1883, The geology of the Susquehanna River region in the six counties of Wyoming, Lackawanna, Luzerne, Columbia, Montour, and Northcumberland: Pa. Geol. Survey 2nd, G 7, 464 p.

Willard, B., 1939, The Devonian of Pennsylvania: Pa. Geol. Survey 4th ser., G 19, 481 p.

SHELF AND DELTAIC PALEOENVIRONMENTS IN THE CRETACEOUS-TERTIARY FORMATIONS OF THE NEW JERSEY COASTAL PLAIN [1]

JAMES P. OWENS AND NORMAN F. SOHL

U. S. Geological Survey, Beltsville, Md., and Washington, D. C. 20242

ABSTRACT

The northern Atlantic Coastal Plain consists of sediments deposited in a wide variety of continental, transitional, and marine environments.

Pleistocene erosion of the emerged Coastal Plain has exposed nearly a complete stratigraphic sequence of formations ranging in age from Late Cretaceous (Cenomanian) through Pliocene (?). The Coastal Plain or continental terrace can be separated into four natural rock sequences; these are, in ascending order: (1) largely fluviatile-deltaic deposits of early Late Cretaceous age, represented by the Raritan Formation; (2) chiefly interstratified deltaic and shelf deposits of Late Cretaceous and early Tertiary age; (3) sediments of middle Miocene-Pliocene(?) age; and (4) sediments of Quaternary age.

Deposits of the first three sequences are discussed and demonstrate the contrast between the deposits formed in deltaic and shelf environments. These three sequences are compared with modern analogs from the Gulf of Mexico.

INTRODUCTION

The northern Atlantic Coastal Plain is an eastward-thickening wedge of unconsolidated and partly consolidated sediments which occurs along the continental margin (fig. 1). This wedge, commonly referred to as the continental terrace, consists of interleaved marine, marginal marine, and nonmarine beds.

[1] Publication authorized by the Director, U. S. Geological Survey.

The nonmarine beds are best developed at the base (beds of late Early Cretaceous and early Late Cretaceous age) and top (beds of Quaternary age) of the continental terrace. The marine-marginal and marine beds occupy the middle part of the terrace and were deposited in two distinct periods of sedimentation; Late Cretaceous through middle Eocene and from middle Miocene to Pliocene(?). The marine beds are all shallow-water deposits which have been most frequently called shelf deposits (beds deposited in less than 600 feet of water).

The entire sedimentary succession is similar to deposits found in other miogeosynclinal zones. These sediments, however, accumulated in a dominantly clastic province; carbonates are minor and are always detrital. Authigenic minerals, mostly glauconite, siderite, and chlorite(?) occur in large quantities, particularly in marine and marginal marine units.

REGIONAL SETTING

The Coastal Plain sediments of Maryland, Delaware, and New Jersey lie, with marked unconformity, on a basement consisting of rocks of Precambrian(?), early Paleozoic, and Triassic age. This major unconformity marks the base of the Zuni sequence, a major rock-stratigraphic unit which can be traced over major areas of the continent (Sloss, 1963, p. 104). As defined by Sloss, this sequence ranges in age from Middle Jurassic through middle Paleocene. The best information to date suggests that this sequence in the northern Atlantic

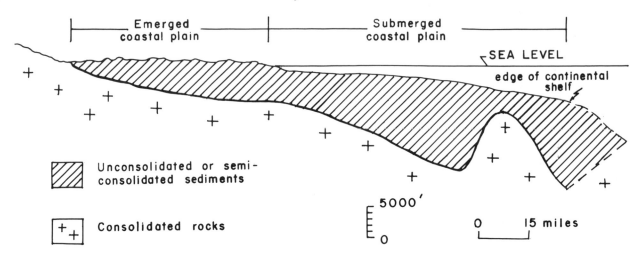

Figure 1. Cross section of the New Jersey Coastal Plain from the basin edge to the continental slope (modified from Drake and others, 1959).

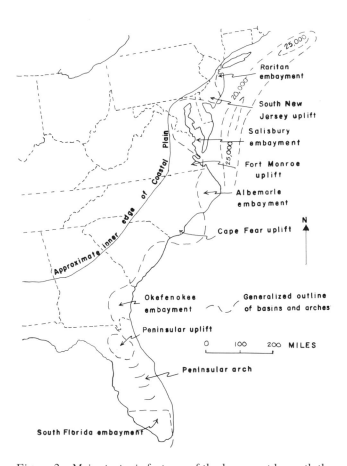

Figure 2. Major tectonic features of the basement beneath the Atlantic Coastal Plain (from Owens and others, 1968). Contours indicate thickness (in feet) of post-Triassic sediments. Of particular interest on this field trip is the Raritan embayment and the south New Jersey uplift.

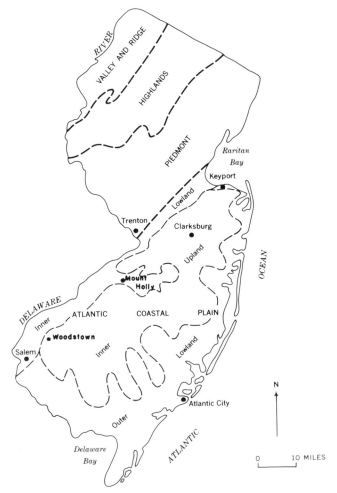

Figure 3. Major physiographic provinces of New Jersey and subprovinces within the Coastal Plain.

Coastal Plain ranges in age from Early Cretaceous (Aptian-Albian, Wolfe and Pakiser, unpub. data) through late Eocene (Richards and others, 1957). A second major rock stratigraphic sequence called the Tejas (Sloss 1963, p. 107) is also present in this area and ranges in age from middle Miocene through Pliocene(?).

Most of the Coastal Plain sediments overlie crystalline rocks. The mineralogy of the Coastal Plain deposits indicates that a large volume of the sediment probably was derived from the crystalline basement.

The surface of the basement underlying the Coastal Plain is not flat, but is highly irregular, apparently due to basement warping. The arrangement of the structural elements in the basement along much of the continental margin is shown diagrammatically in figure 2. These structures apparently controlled the sediment dispersal patterns along the eastern continental margin. In the northern Atlantic Coastal Plain, the dominant structural feature is a large downwarp centered in the vicinity of Chesapeake Bay. This trough is most commonly called the Salisbury embayment (Richards, 1945) and is the site of the thickest Coastal Plain sedimentary section in the emerged part of the northern Atlantic Coastal Plain. Other smaller structural features are present in the northern Coastal Plain. In New Jersey, a major trough or smaller basin is centered in the vicinity of Raritan Bay and a high occurs in southern New Jersey. These features are shown in highly generalized form on figure 2. Both these large structures are

actually a complex of many smaller troughs and arches (U. S. Geol. Survey, 1967). Structures such as these influenced sedimentary patterns in this area, particularly during Late Cretaceous time.

The physiographic features shown in figure 3 developed during Pleistocene time. Fluviatile dissection of the Coastal Plain virtually stripped away large parts of the uppermost marine beds of the continental terrace. The inner upland is largely remnants of these marine beds of chiefly Miocene-Pliocene(?) age. The inner lowland was formed by a great river which once flowed parallel to the inner edge of the Coastal Plain. The deep entrenchment of this stream exposed most of the formations which reach the inner edge of the Coastal Plain and thus affords a three dimensional view of part of the continental terrace.

ROCK UNITS

The pre-Quaternary stratigraphic units in the New Jersey Coastal Plain used by us and most commonly by other workers (Richards, 1967) are listed below.

Sixteen lithostratigraphic units have been mapped in the New Jersey Coastal Plain. All the units are thin and the total thickness of the section along the basin edge is only slightly more than 1,000 feet. Not all units persist throughout the New Jersey Coastal Plain (fig. 4). The Upper Cretaceous beds thin southwestward, and some wedge out, so that, in general, fewer units occur in the southwest part of the New Jersey Coastal Plain (fig. 4). The coarser clastic units are the thickest, but even these

Age	Unit	Maximum known thickness in outcrop (in feet)
Miocene(?)-Pliocene(?)	Cohansey Sand	100
Middle Miocene	Kirkwood Formation	100
Early and Middle Eocene	Manasquan Formation	60
Paleocene	Vincentown Formation	100
Paleocene	Hornerstown Sand	25
Late Cretaceous	Tinton Sand	25
	Red Bank Sand	150
	Navesink Formation	50
	Mount Laurel Sand	100
	Wenonah Formation	70
	Marshalltown Formation	20
	Englishtown Formation	120
	Woodbury Clay	50
	Merchantville Formation	60
	Magothy Formation	150
	Raritan Formation	500(?)
Early and Late Cretaceous	° Potomac-Raritan	250

° Raritan in Delaware Valley is lithologically similar to the older Potomac Group sediments of northeastern Maryland. Because of this, the two units are not differentiated.

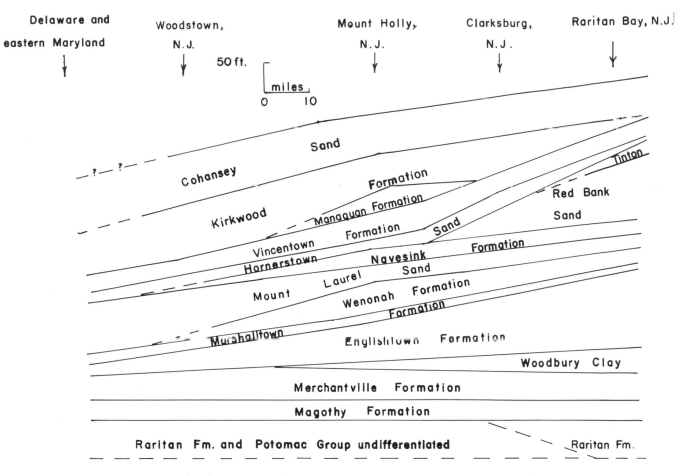

Figure 4. Generalized cross section from Raritan Bay, New Jersey, to eastern Maryland showing the approximate thickness and lateral relations of the Coastal Plain formations.

are less than 200 feet thick. Other units, such as some of the glauconite sands, have an average thickness of only 15–20 feet in outcrop.

Most of the formations are well defined, particularly those of marine origin, and many represent a single lithofacies. The greensands, considered to have accumulated farthest from shore, are of this type. Some of these beds can be traced for more than 120 miles with little change in lithology or thickness. The near-shore marine deposits, the quartz sands, in contrast, vary greatly in lithology and thickness along strike.

The Coastal Plain formations which will be examined, range in age from Late Cretaceous (Cenomanian) through Miocene-Pliocene(?).

RARITAN FORMATION

For convenience, all pre-Magothy sediments will be referred to as the Raritan Formation even though in the Delaware River Valley from Trenton to Salem, N. J., Raritan age sediments are indistin-

guishable from Lower Cretaceous Potomac Group sediments.

At the type locality of the Raritan at Raritan Bay, the formation is exposed in a large number of pits, mainly south of the Raritan River. Ries, Kümmel, and Knapp (1904), later modified by Berry (1906), divided these beds into several units:

The Amboy stoneware clay
No. 3 Sand—chiefly quartz
The South Amboy fire clay
No. 2 Sand—in including beds of so called "feldspar" and "Kaolin"
The Woodbridge clay-fire, stoneware and brick clays
No. 1 Sand—in part fire sand
The Raritan clay-fire and terra-cotta clays

Barksdale and others (1943) named the Number 1, 2, and 3, Sands the Farrington Sand, Sayreville Sand, and Old Bridge Sand Members, respectively, and this is the usage followed by the U. S. Geological Survey. The other names used by Ries, Kümmel, and Knapp (1904) apply to economic beds and are considered informal terms by the U. S. Geological Survey.

The beds listed above are chiefly interstratified light-colored sands and dark or variegated clayey silts or silty clays which vary in texture, composition and thickness within short distances. The maximum thickness, in outcrop, of all these beds is about 500 feet.

Recently, the middle and upper members of the Raritan were reexamined and the Amboy stoneware clay was found to be the same age as the overlying Magothy (Wolfe and Pakiser, unpub. data). In addition, no unconformity or disconformity could be demonstrated between the Amboy stoneware and the lower Magothy (Owens and others, 1968). The top of the Raritan, therefore, as used by us, is the Old Bridge Sand Member.

Some of the members of the redefined Raritan are shown in figures 5A–D. These are interpreted to be deposits of a subaerial deltaic plain (Owens and others, 1968) and as is common in such an environment, beds representing a wide variety of depositional (i.e., energy) conditions are present (Allen, 1965b).

The Woodbridge (fig. 5A) and the South Amboy are both predominantly dark silts and clays; the Woodbridge is more extensive (Barksdale and others, 1943, fig. 3), and is widespread throughout the Raritan Bay area. It is largely a thin- to thick-bedded sequence of micaceous silts and clays, containing an enormous amount of woody fragments and siderite concretions. Locally, at Sayreville, a marine fauna occurs in the middle part of this member. The interbedding of these highly carbonaceous and siderite-rich deposits with deposits containing open-marine fossils indicates the Woodbridge was deposited in close proximity to the sea. In addition, occurrences of the near-shore *Callianassa* borings, are common in these beds. It is likely that the thick accumulations of carbonaceous-rich beds of the Woodbridge were deposited in marginal-marine mangrove-type swamps.

The South Amboy clay is thinner (maximum thickness of 35 feet) and less widespread (lenses rapidly) than the Woodbridge clay and lacks siderite concretions as well as marine fossils.

The Sayreville and Old Bridge Members are largely light-colored medium-grained sands which locally have interbeds of light- to dark-colored clayey silts. Not uncommonly, the silty strata are disrupted forming intraformational breccias. Coarse sandy beds are abundant in both members, but gravel is rare. The sandy beds in both members are extensively cross-stratified with the scale of the planar beds varying from a few inches to several feet. The larger scale cross-strata appear to be more common in the Old Bridge than in the Sayreville. The Old Bridge is thicker and more persistent at the surface and in the subsurface than the Sayreville (Barksdale and others, 1943, fig. 3). The Sayreville-Old Bridge Members were deposited in a higher energy environment than the South Amboy and Woodbridge clays. The widespread development and type of cross-stratification suggests they were deposited in river channels, most likely as point bar sands (Harms and other, 1963; Frazier and Osanik, 1961; and Allen, 1965a).

A different facies of the Raritan occurs in the Delaware River Valley. Here the formation consists largely of thick interbeds of light-colored sands (similar to the sand of the Old Bridge and Sayreville Members) and massive to thick-bedded variegated (shades of red, white, and yellow) silty clays (Greenman and others, 1961). Thick black clays containing abundant siderite concretions (Woodbridge clay) are absent. Figure 4D shows one of the sands beds near the top of the Raritan. The sands have large-scale cross-stratification and abundant gravel.

MAGOTHY FORMATION

The Magothy, like parts of the Raritan Formation (upper Woodbridge clay specifically) at Raritan Bay, consists of intercalated dark carbonaceous-rich silty clays or clayey silts and light-colored sands. It is, however, more laterally persistent than the Raritan and does not have thick brightly colored silts and clays. The Magothy can be traced from Cliffwood Beach, on Raritan Bay, more or less continuously, to the western shore of Chesapeake Bay in Maryland. The type locality for this formation is along the Magothy River near Annapolis, Md.

In New Jersey, the Magothy is best exposed in the Raritan Bay region and it was here that Clark (1904) first recognized this unit (he called it the Cliffwood-Magothy Formation). Prior to that time, these beds were included as the uppermost beds of the Raritan Formation (Kümmel and Knapp, *in* Ries and others, 1904).

Owens, Minard, and Sohl (1968) reexamined the Magothy and underlying Raritan in the Raritan Bay area. Palynological analyses of both units revealed that the Amboy stoneware clay of the Raritan Formation was more closely allied to the Magothy than to the Raritan (Wolfe and Pakiser,

Figure 5. The Raritan Formation. A, Woodbridge clay of the Raritan Formation in the Sayre and Fisher Co. pits at Sayreville, N. J., showing the dark-colored and thin- to thick-bedded character of this deposit Most of the banding is due to variations in carbonaceous matter. B, Sayreville Sand Member overlying the Woodbridge clay in the pits at Sayreville. Note the sandy character and well-developed planar cross-stratification. C, Old Bridge Sand Member in Sayre and Fisher Co. pits at Phoenix, N. J. Scale and type of cross-stratification shown in photograph is similar to that in the Sayreville although larger scale cross-stratification is very common in this member. D, Large scale cross-stratification in the Raritan Formation at Palmyra, N. J., in the Delaware River Valley. Pollen and spores from the clays underlying this deposit indicate this sand is at the top of the formation. Note the gravel, a feature not observed in the pits near Raritan Bay.

unpub. data). For this reason and because no unconformity could be demonstrated between the Amboy and the lower part of the Magothy, the Amboy has been included with the Magothy.

The Magothy as defined, therefore, consists of three units; a basal massive to laminated unit, the Amboy stoneware clay; a middle thin-bedded unit, the Morgan beds; and an upper cross-stratified unit, the Cliffwood beds.

The Amboy clay is only exposed at a few localities. Detailed studies of the area by Kümmel (*in* Ries, Kümmel, and Knapp, 1904) and of the subsurface by Barksdale and others (1943) have shown this unit to vary in thickness from 0 to 35 feet. It is a very dark colored micaceous silt (fig. 6A) which contains abundant, mostly finely comminuted carbonaceous matter. Plant impressions are common especially near the upper part of the bed. The basal contact with the underlying Raritan was not observed but this unit is reported to overlie a white to pale-blue clay (Kümmel and Knapp, *in* Ries and others, 1904, p. 169). This light bed most likely an underclay, is probably part of

the Amboy clay. The contact between Magothy and the Raritan is, therefore, probably between the light clay and the underlying Old Bridge Sand Member.

The Amboy clay is conformably overlain by the Morgan beds. The Morgan beds are typically thin-bedded intercalated dark clays and silts and light-colored sands (fig. 6B) which grade laterally into light-colored cross-stratified sand containing beds

Figure 6. The Magothy Formation. A, Massive to very finely laminated dark-colored Amboy stoneware clay exposed in a pit along Melvins Creek, 1 mile north of Cheesequake, N. J. B, Horizontally stratified thin interbeds of dark silts and light-colored sands of the Morgan beds, in the same pit. Note the variability in thickness of the silty and sandy beds upward through the section. C, Cliffwood beds of the Magothy at the type locality Cliffwood Beach, N. J. These beds are largely sand interstratified with thin dark-colored carbonaceous-rich clay beds. The sands have small-scale low-angle cross-stratification. D, Channel-fill sand in the Magothy Formation near Bordentown, N. J., Delaware River Valley. Cross-stratification in these sands is similar to cross-stratification in the Sayreville and Old Bridge Sand Members of the Raritan (fig. 4B, C).

or lenses of coarse carbonaceous matter. The Morgan beds are about 100 feet thick.

The contact between the Morgan beds and the overlying Cliffwood beds was not observed. A light-gray clayey silt to very fine grained sand containing a large concentration of small (as much as 2 inches in diameter) rounded siderite concretions occurs at the base of the Cliffwood bed. Many of these concretions contain open-marine fossils (Stephenson, 1954). The bed containing these fossils is overlain by 40 feet of largely cross-stratified sand (fig. 6C). Interstratified with the sand are thin dark-colored layers of predominantly fine to coarse carbonaceous hash. Float blocks of hard sandstone, apparently derived from the upper part of this unit, contain an abundant marine fauna. Thus, the Cliffwood beds are largely marine in origin.

Southwest of the Raritan Bay area, in the vicinity of Trenton, the Magothy thins to approximately 50 feet and the three subdivisions of the Magothy are no longer evident. Here the formation consists of rapidly lensing dark clays and light-colored sands. The light-colored cross-stratified sands shown in figure 6D are particularly abundant in this area. Based on paleobotanical evidence, Berry (1906) correlated these beds with the upper Magothy (Cliffwood beds).

MERCHANTVILLE FORMATION

The Merchantville Formation is the oldest glauconitic unit which crops out in the New Jersey Coastal Plain. It overlies the Magothy throughout New Jersey and the Delmarva Peninsula. The contact between the two units is always sharp and disconformable. A thin bed of reworked sediment, usually about one foot thick, occurs along this boundary.

The Merchantville is consistently about 50 feet thick throughout the Coastal Plain, but varies in lithology along strike. In the northern New Jersey Coastal Plain, in the Raritan Bay area, it is largely interstratified massive thick glauconite sands and thin-bedded very micaceous carbonaceous-rich clayey silts (fig. 7A). Siderite concretions are abundant in the thinly bedded silts, and resemble the dense network of siderite concretions found in the Woodbridge clay of the Raritan Formation.

The glauconite deposits occur as massive, approximately 4–5 feet thick beds that typically contain numerous borings. The glauconite grains are mostly fine and many are accordion-shaped. The glauconite is associated with abundant small pieces of carbonized wood. Molds of well-preserved marine fossils, such as the ammonite *Scaphites hippocrepus,* have been found in these beds.

Southwest of the Raritan Bay area, in the vicinity of Trenton, the glauconite beds increase in thickness, and the thin-bedded sequences and abundant siderite concretions are absent. Overall, the Merchantville in this area is much thicker bedded and more glauconitic than near Raritan Bay. The massive character of the Merchantville is shown in figure 7B.

Exposures of the Merchantville in New Jersey southwest of Camden are rare. In Delaware and eastern Maryland, however, the formation is exposed at a number of localities, particularly along the Chesapeake and Delaware Canal and the Sassafras River. It is still predominantly thick bedded, but glauconite, except in the basal few feet, is only a minor sand constitutent. Typically, the Merchantville is a dark-colored micaceous very silty fine quartz sand. Siderite concretions and large carbonized pieces of wood are common, particularly in the base, but these concentrations do not approach those found in this unit at Raritan Bay. Casts and molds of marine fossils, particularly in siderite concretions, occur within the formation in this region. The Merchantville is clearly a marine deposit throughout its entire area of outcrop.

WOODBURY CLAY

Throughout most of the New Jersey Coastal Plain, the Merchantville is overlain by a thick massive clayey silt, the Woodbury Clay. Until recently, the Woodbury, was thought to persist to the Delaware River in southern New Jersey, but, detailed mapping by Minard (1965) in the vicinity of Woodstown has shown that this formation pinches out somewhere to the northeast. The exact point of pinchout, however, is still unknown but must lie between Woodstown and Woodbury, N. J.

The Woodbury grades into the underlying Merchantville. It contains larger concentrations of clay and much less glauconite sand than the Merchantville.

The Woodbury is well exposed near Trenton where it is approximately 50 feet thick. Kümmel and Knapp (*in* Ries, and others, 1904) indicate a similar thickness to the northeast in Monmouth County.

Throughout most of its outcrop area, the Woodbury is a massive, monotonous dark-gray, micaceous clayey silt (fig. 7D). Locally thin lentils of

Figure 7.   Merchantville and Woodbury Formations.   A, Thinly stratified to laminated beds in the Merchantville Formation, Oschwald Brick Co. pit at Cliffwood, N. J. These beds are largely very fine grained sand or silts.   Although not seen here, siderite concretions are very abundant in these beds. B, Massive dark-colored glauconite beds in the Merchantville at Bordentown, N. J. Massive bedding is typical in the formation throughout the Delaware River Valley.   C, Woodbury Clay-Merchantville contact in an abandoned pit on the west edge of Crosswicks, N. J. The darker colored beds at the base are the Merchantville Formation, but the color difference is only due to differential wetting. D, Basal Woodbury Clay in pit at Bordentown, N. J., showing the massive character typical of this unit throughout the Trenton area.

glauconite sand are associated with large discoidal gray siderite concretions in the upper part of this formation.   Siderite concretions, however, are rare in most of the Woodbury except at Raritan Bay where this unit contains abundant concretions similar to those in the Merchantville (Weller and Knapp, 1907).

Fossil impressions are common in the Woodbury and locally a well-preserved calcareous marine fauna is present at Haddonfield, N. J., thus establishing a marine origin for this formation.

ENGLISHTOWN FORMATION

The Englishtown Formation overlies the Woodbury in the northern and west-central New Jersey and the Merchantville in southern New Jersey.   It

Figure 8.   The Englishtown and Marshalltown Formations.   A, Cross-stratified sands in the Englishtown Formation at Columbus, N. J.   The cross-stratified sands (A) resemble the channel-
fill sands in the Magothy Formation at Bordentown (fig. 6C) except that the type of cross-stratification is trough rather than planar.   B, Interbedded dark silts and light-colored sands in the
Englishtown Formation, 1.7 miles southeast of Crosswicks, N. J.   Englishtown in the area resembles the Cliffwood beds of the Magothy Formation.   C, Massive-bedded dark-colored Marshalltown Formation exposed in a small gully, 1.5 miles east of Extonville, N. J.   Note abundant
borings, a common feature in this and other glauconite sands in this area.

resembles parts of the Magothy Formation, and
characteristically has large concentrations of quartz
sand.   The contact between the Englishtown and
underlying units is gradational hence conformable.
The transition from the Woodbury to the Englishtown is characterized by a gradual increase in sand-
sized quartz and a decrease in silt and clay.   Large
concentrations of fossiliferous siderite concretions
are abundant in this interval.

Several lithofacies can be recognized within this
formation along strike, but three lithofacies are

most common.   In the northern Coastal Plain,
where the formation is thickest (about 120 feet
thick), cross-stratified sands of the type shown in
figure 8A are most abundant.   These are interstratified with the dark carbonaceous-rich silts (fig.
8B) similar to the Cliffwood beds of the Magothy.
The lithofacies shown in figure 8B is best developed in the west-central part of the Coastal Plain.
In the south, where the formation is only 40 feet
thick, the Englishtown is a massive dark-colored
silty sand which resembles the nonglauconitic beds

of the Merchantville. Although fossils are rare, especially in the north and west-central parts of New Jersey Coastal Plain, a sufficient number have been found to establish a marine or marginal marine origin for parts of the formation. In addition, burrows of possible marine origin are very common locally in the thin-bedded silts and sands. Marine fossils are abundant in the massive dark-colored beds of the Englishtown, in southern New Jersey.

The bulk of the sediment in the Englishtown is largely fine to medium sand, silt, and clay. Gravel, except locally, is rare in the Englishtown. The cross-stratified sandy lithofacies of the Englishtown probably represents alluvial deposits (most likely channel-fill deposits). The small-scale cross-stratification resembles that found in the Magothy near Trenton. It does not resemble the large-scale cross-stratification of the fluviatile deposits, of the Raritan. The cross-stratified sandy lithofacies of the Englishtown probably represents deposits of streams of somewhat lower velocity than those responsible for deposition of the Raritan sands.

MARSHALLTOWN FORMATION

A thin (15–20 feet thick) bed of micaceous, silty glauconite sand, the Marshalltown Formation, overlies the Englishtown throughout the New Jersey Coastal Plain and in the northern Delmarva Peninsula.

This sand abruptly overlies the Englishtown, suggesting a disconformable relationship. The basal few inches of the Marshalltown contain coarse quartz sand and large woody fragments reworked from the underlying Englishtown.

The Marshalltown is massive, dark-gray, and extensively burrowed (fig. 8C). Many of the borings are filled with glauconite sand and typically the formation is mottled. Most of the formation is a very poorly sorted micaceous quartz-glauconite sand. Casts of fossils are common in most outcrops and in the southern part of New Jersey near Auburn, large concentrations of the thick-shelled pelecypod, *Exogyra ponderosa* are present.

WENONAH FORMATION

The Marshalltown grades into the overlying Wenonah Formation, a dark-gray, poorly sorted very micaceous silty fine quartz sand. The Wenonah crops out throughout the New Jersey Coastal Plain but is not present in the Delmarva Peninsula to the southwest (fig. 4).

The Wenonah is a maximum of 70 feet thick near Trenton and thins both to the northeast and southwest. The thinning to the southwest is accompanied by a general increase in grain size to the point where it becomes indistinguishable from the overlying Mount Laurel. The Wenonah resembles the Woodbury in its massive character and like that unit has large concentrations of fine sediment: mica, and sand-sized woody fragments (fig. 9A). Glauconite sand is most abundant in the lower part of the formation but diminishes rapidly upwards.

Casts of marine fossils are common throughout the entire formation thus establishing its marine origin.

MOUNT LAUREL SAND

The Wenonah is conformably overlain by a coarser clastic unit, the Mount Laurel Sand. The transition from the Mount Laurel to the Wenonah is generally marked by an increase in average grain size, a change from dark-gray to lighter gray and a noticeable decrease in mica (fig. 9B). All these changes may not occur; in the northern New Jersey Coastal Plain some of the Mount Laurel beds are locally very micaceous and contain abundant thin layers of clay.

Like the other coarse clastic units, the Mount Laurel is not a single lithofacies throughout its area of outcrop. Three distinct bedding types are present: 1) a thin-bedded intercalated dark clay-light sand sequence (fig. 10A), 2) massive sand beds (fig. 10B), and 3) an upper thin massive pebble sand (fig. 10A). The thin massive pebble sand occurs in nearly every exposure of the upper beds of the Mount Laurel.

The thin-bedded dark clay and light sand sequence is the major lithofacies in northern New Jersey and resembles part of the Englishtown and the Cliffwood beds of the Magothy. The sands in this sequence are commonly cross-stratified as are the sands in the Magothy (Cliffwood beds) and Englishtown. The dark beds contain large concentrations of mica, sand-sized carbonaceous matter, and, locally, thin beds containing siderite concretions.

The thin-bedded sequence interfingers along strike to the southwest with massive sands, a facies not observed in any of the underlying formations. This lithofacies is chiefly a medium quartz sand containing small pebbles and granules scattered throughout. In the vicinity of Trenton, shell layers containing the thick-shelled pelecypod *Exogyra cancellata* and the ammonite *Belemnitella americana* occur in the upper part of the sand. This thick-bedded sequence thickens to the southwest and reaches a maximum thickness of 80 feet.

The pebbly massive sand found everywhere at

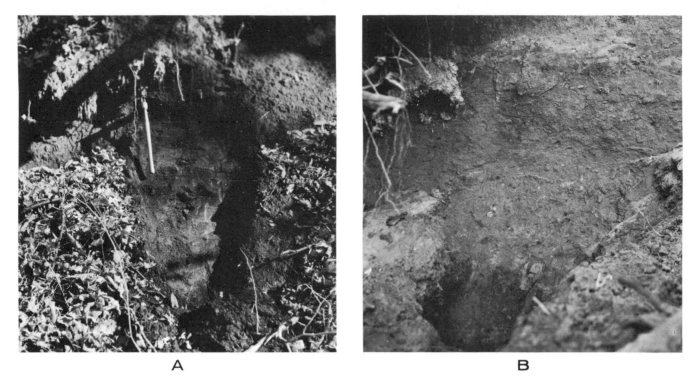

A   B

Figure 9. The Wenonah Formation. A, Dark-colored massive beds of the lower part of the Wenonah Formation at Walnford, N. J. B, Gradational contact between the Mount Laurel Sand (light-colored) and the Wenonah Formation (dark-colored) along Crosswicks Creek, near Homes Mills, N. J.

the top of the formation, is an interesting deposit. It is typically 4–8 feet thick, but may be as much as 12 feet thick locally. It consists largely of coarse sand, fine gravel, and abundant disoriented fossils. Many borings filled with glauconite sand extend down from the overlying Navesink Formation. Although this unit is mainly a clastic sand similar to the underlying beds, it has all the appearances of a interval of reworked sediments, related to the deposition of the overlying Navesink.

NAVESINK FORMATION

The Navesink is another of the dominantly greensand units which crop out in the New Jersey Coastal Plain. It is found only in New Jersey and does not crop out in the Delmarva Peninsula to the southwest (fig. 4).

The Navesink is a maximum of 50 feet thick, but is commonly only about 30 feet thick. Most of the Navesink is a massive dark-greenish-gray very poorly sorted clayey glauconite sand (fig. 10D). Both the Marshalltown Formation and the Navesink are greensands, the Navesink, however, is less micaceous, coarser grained, and has less clastic sand and woody fragments than the Marshalltown. The coarser grained glauconite sand in the Navesink distinguishes it from the older glauconitic sands.

The Navesink abruptly overlies the Mount Laurel Sand and, locally, a fossil-rich layer containing the thick-shelled pelecypod *Exogyra costata* occurs at the base of the Navesink. In the northern New Jersey Coastal Plain, other similar fossil-rich layers occur in the middle of the formation, but this is the only area in New Jersey in which these fossil layers occur above the base of the Navesink.

RED BANK SAND

The Navesink Formation in the northern New Jersey Coastal Plain is overlain by the Red Bank Sand, a complex unit containing several lithofacies. The Red Bank has a very restricted distribution in New Jersey, cropping out only from northeast of Trenton to Raritan Bay. Within this area, however, it is as much as 150 feet thick.

The Red Bank represents a return to coarse sandy deposition similar to that associated with the Englishtown Formation and Mount Laurel Sand. Unlike the other near-shore clastic units, however, the Red Bank also contains a deeper water facies. Olsson (1960, table 1) subdivided the Red Bank into two members. The lower member is dominantly a massive dark-gray very fossiliferous silty sand, the Shrewsbury, and the upper member, the Sandy Hook, is a light-colored dominantly quartz sand. Detailed mapping in west-central New Jersey has shown that the lower unit changes to a very glauconitic sand and a persistent massive very clayey quartz sand (the transitional bed of Minard, 1964) which occurs beneath the Sandy Hook Member, and above the glauconite sand.

The Sandy Hook Member is approximately 120 feet thick where the formation is thickest, at Raritan Bay, but thins to the southwest, and is absent southwest of Arneytown. It strongly resem-

bles the massive quartz sands of the Mount Laurel Sand of southern New Jersey and the northern Delmarva Peninsula (fig. 11A). The Sandy Hook Member grades into the underlying Shrewsbury Member with no perceptible break. The Shrewsbury is predominantly a massive very micaceous silty fine sand which resembles the Wenonah Formation. It differs from the Wenonah, however, in that it contains an abundance of calcareous fossils in the Raritan Bay region. Southwest of the

A

B

C

D

Figure 10. Mount Laurel Sand and Navesink Formation. A, Exposure showing the sharp contact between the Navesink Formation (dark-colored) and the underlying Mount Laurel Sand (light-colored) in a pit on State Route 34, 1.6 miles southeast of Matawan, N. J. Two of the facies of the Mount Laurel, the upper massive pebbly bed and the lower thinly bedded dark clay and light sand sequence are exposed in this cut bank. B, massive bedded quartz sand facies of the Raritan Formation along Crosswicks Creek near Kirbys Mills, N. J. This is the major facies from Trenton to the southwest. C, *Callianassa* borings in the massive sandy facies of the Mount Laurel on U. S. Rte. 295 at Runnemeade, N. J. Burrows of this type are generally found in beach and shallow-water marine environments. D, Dark-colored massive character of the very glauconitic Navesink Formation exposed in a roadcut east of Cream Ridge, N. J.

Figure 11. A, Bluffs at Waterwitch, N. J., on Raritan Bay near the Atlantic Highlands. The entire thickness of the Sandy Hook Member of the Red Bank Sand is exposed. B, Massive-bedded transitional facies of the Red Bank Sand exposed at Emleys Corner, N. J. C, Lower dark-colored glauconitic sand facies of the Red Bank Sand exposed along Long Bog Run, near Hornerstown, N. J. D, Exposure of the Cohansey (iron-oxide cemented), Vincentown (massive), and Tinton (containing concretionary masses) at base of cut. Finely crystalline siderite, now partly oxidized, cements the Tinton.

Raritan Bay area, the glauconite sand in these lower beds increases gradually, and, in the vicinity of New Egypt, the glauconite facies of the Red Bank reaches a maximum thickness of 30 feet. This glauconitic unit grades into the underlying Navesink and strongly resembles it, but the two units can be distinguished on the basis of the greater abundance of coarse mica in the Red Bank. The very fossiliferous beds found in the Shrewsbury Member at Raritan Bay do not occur in the glauconite beds near New Egypt.

No glauconite beds referrable to the Red Bank

were observed southwest of Sykesville, that is, approximately 4 miles beyond the southwestern-most outcrop of the Sandy Hook Member.

TINTON SAND

A massive dark-grayish to reddish-brown glauconitic quartz to quartz-glauconite sand overlies the Red Bank Sand near Raritan Bay. This unit, the Tinton Sand, has a maximum thickness of 25 feet and crops out only in the northern New Jersey Coastal Plain from Roosevelt to Sandy Hook. The Tinton has a variable lithology and resembles the transitional beds of the Red Bank. The contact between the Red Bank and Tinton is gradational and is characterized by a gradual upward increase of glauconite sand in the Tinton. Although the Tinton indicates a return to somewhat deeper water deposition and is classed with the greensands, it has a much higher percentage of coarse-grained detritus than any of the underlying greensands. Pebbles as much as ¼ inch in diameter and coarse sand are abundant. Another unique feature of this formation is its widespread cementation by finely crystalline siderite; locally, the Tinton is a compact hard sandstone (fig. 11D) that is so resistant to erosion that at the type locality a small waterfall is developed on it.

Locally, thick-shelled pelecypods, *Exogyra costata,* are concentrated in this unit, as well as large numbers of well-preserved crab claws. Weller and Knapp (1907) recognized the Tinton as faunally distinct from the underlying Red Bank. Cooke and Stephenson (1928) established an early Tertiary age for the overlying Hornerstown Sand and the Tinton, therefore, is the youngest Cretaceous formation in the New Jersey Coastal Plain.

HORNERSTOWN SAND

The Tinton Sand is overlain unconformably by a glauconite-rich sand, the Hornerstown. The Hornerstown overlies successively older units to the southwest (Red Bank, Navesink, and Mount Laurel), therefore, an unconformity separates Cretaceous and Tertiary deposits in the northern Atlantic Coastal Plain. The Hornerstown is a persistent unit that crops out from Raritan Bay to the Sassafras River in eastern Maryland. There is little change in lithology of thickness (20 feet) along much of its outcrop, except in the updip areas where it is thinned by erosion prior to the deposition of overlying formations.

In outcrop, like all the other greensands, the Hornerstown (fig. 12A) is massive and extensively burrowed. This unit is unique in having a bright-green cement. Its deep green color distinguishes it from the typically dark-gray older glauconite sands.

Invertebrate fossils, even as casts or molds, are rare in the Hornerstown. Vertebrate remains, however, are abundant, but most are worn and of very little use in establishing the age of the formation.

VINCENTOWN FORMATION

The Hornerstown Sand is conformably overlain by the Vincentown, another of the many quartz sand units which crop out throughout New Jersey and northern Delmarva Peninsula.

The Vincentown, like the Red Bank, consists of more than one lithofacies. In the northern New Jersey Coastal Plain, this unit is largely a thick (100 feet) massive quartz sand which resembles the Sandy Hook Member of the Red Bank Sand and the Mount Laurel Sand in southern New Jersey. To the southwest, in the vicinity of Trenton, the Vincentown thins to approximately 50 feet and changes facies; the lower part of the formation is a dark-greenish-gray bed containing large concentrations of calcareous fossils, chiefly the brachiopod *Oleneothyris harlani* and *Gryphaea* sp. (fig. 12B). This bed is predominantly glauconite sand, but the glauconite decreases rapidly upward. This lower dark-colored glauconite bed resembles the transitional beds of the Red Bank in sorting, texture, and composition.

Overlying this dark-colored layer is an intercalated thin- to thick-bedded sequence of light-colored quartz calcarenite and calcareous quartz sand layers. These fossiliferous beds (fig. 12C) have been intensely studied by paleontologists. The two subunits of the Vincentown extend from Trenton to the Delaware River. The calcarenite facies has not been recognized in the northern Delmarva Peninsula where the unit is largely a glauconite quartz sand.

MANASQUAN FORMATION

The Vincentown Formation is overlain in the downdip areas by the Manasquan Formation. The Manasquan is exposed only where overlying middle Miocene sediments have been eroded away, such as in major river valleys which cross the strike of the formations. Because of this, the Manasquan is exposed only at a few localities in the New Jersey Coastal Plain.

The Manasquan, where mapped in detail, consists of a lower dark-colored poorly sorted quartz

Figure 12.   The Hornerstown Sand and Vincentown and Manasquan Formations. A, Massive-bedded Hornerstown Sand in a pit at Hornerstown, N. J. The formation is actually a bright green in the outcrop.   B, Basal dark-gray Vincentown Formation containing abundant *Oleneo-thyris harlani* shells (light colored).   This concentration of shells occurs only locally in the base of the formation.   C, Light-colored, massive calcarenite sand facies of the Vincentown Formation, exposed in the west central New Jersey Coastal Plain at Pemberton, N. J. D, Massive dark-gray glauconite quartz sand of the basal Manasquan Formation.   Exposed along west bank of Crosswicks Creek, 1.8 miles south of New Egypt, N. J.

sand and an upper green very clayey glauconite sand.   It is the youngest of the thick glauconitic sands that crop out in the New Jersey Coastal Plain.

South of New Egypt in west-central New Jersey, the Manasquan is a maximum of 40 feet thick at the surface.   It abruptly overlies the quartz calcarenite beds of the Vincentown, and frequently small reworked pieces of calcarenite and phosphatized fossils occur in the base of the formation.

This lower unit is similar to the dark-colored beds found in the base of the Vincentown, but lacks fossiliferous greensand (fig. 12D).

These basal quartz sands grade up into the upper unit with an increase in pale-gray clay and locally large concentrations of glauconite sand (fig. 13A). These upper greensands contain abundant glauconite and resemble the Navesink in abundance and grain size of the glauconite.   A unique feature of the Manasquan, is large outcrops of pale-gray silty clay containing sparsely distributed glauconite and quartz sand throughout.

KIRKWOOD FORMATION

The Kirkwood Formation is a middle Miocene sand which covers most of the New Jersey Coastal Plain. According to Sloss' definition (1963), this unit would mark the base of his Tejas sequence.

Updip, the Kirkwood progressively overlaps older units (Manasquan, Vincentown, and Hornerstown) throughout the New Jersey Coastal Plain and is the basal unit of a major transgression.

The Kirkwood, as defined in New Jersey, applies only to beds at the surface. At the surface, the formation is approximately 100 feet thick, but downdip, in the subsurface, the middle Miocene beds thicken rapidly to over 1,000 feet at Atlantic

Figure 13. The Manasquan and Kirkwood Formations. A, Massive-bedded glauconite sand facies of the Manasquan Formation exposed in a cut bank on Rancocas Creek at Pemberton, N. J. The Manasquan is overlain by cross-bedded alluvial gravels of Pleistocene(?) age at this locality. B, Updip light-colored, thick-bedded facies of the Kirkwood Formation overlain by cross-stratified sands of the Cohansey Sand, exposed in a now abandoned pit on the Fort Dix Army Reservation near Trenton, N. J. C, *Callianassa* borings in the Kirkwood Formation in the Fort Dix pit indicating a marginal marine depositional environment for these beds. D, Interbedded light-colored gravelly sands and massive fine sands of the Kirkwood Formation along the inner edge of the Kirkwood outcrop belt. Exposed in a small pit along State Rte. 537 south of Jacobstown, N. J.

City (Richards and Harbison, 1942, fig. 7). In a less restricted sense, therefore, the Kirkwood is a group in the subsurface possibly correlative with the Calvert and St. Marys Formations of the Chesapeake Group in Maryland.

In the west-central New Jersey Coastal Plain where this formation was mapped in detail, the Kirkwood, at higher elevations, is mostly a light-colored massive- to thick-bedded micaceous very fine quartz sand (fig. 13B). These beds are locally extensively burrowed (fig. 13C), thus suggesting deposition in a shallow marine or beach environment. Thin beds of sandy gravel (fig. 13D) and crossbedded sands occur along the inner edge of the outcrop belt, and these also indicate marginal-marine, high-energy conditions. Down-dip, these light-colored strata interfinger with darker more silty beds containing significant concentrations of finely comminuted carbonaceous matter and mica. This apparently is the major lithofacies in the shallow subsurface throughout this general area. In general, these beds are similar to the massive dark-colored beds in the Wenonah Formation.

To the southwest at Woodstown and in the down-dip areas, the Kirkwood contains thick beds of clayey silt. The up-dip facies is light-colored sands similar to the Kirkwood near Trenton.

Fossils are rare in the sandy updip facies of the Kirkwood but are common in the dark-colored silts. The interfingering of the light-colored sands with dark fossiliferous silts establish the marine and marginal-marine nature of this formation.

COHANSEY SAND

The Kirkwood Formation is overlain by a coarser clastic unit, the Cohansey Sand. The Cohansey, like the Kirkwood, is extensively dissected, and outliers capped by this unit are common. The Cohansey is reported only in New Jersey.

The Cohansey is approximately 100 feet thick. Like all the other marginal marine clastic units, the Cohansey is composed of a large number of lithofacies. In spite of its very extensive outcrop area, however, the spatial distribution of these lithofacies is only poorly understood.

The Cohansey has been examined along the inner edge of the Coastal Plain from Sandy Hook to Woodstown, N. J., where it is largely an alluvial deposit. The gravelly sands, illustrated in figures 14A and 14B, are common in the extreme up-dip areas, whereas down-dip this unit is mainly sand (fig. 14C, D). The sands contain a wide variety of bedding types. Trough-type stratification is well developed at the contact with the Kirkwood Formation (fig. 14C), and may be related to barrier-bar development. The long sloping thin parallel beds (fig. 14D) commonly formed in the beach environment are well developed in the Cohansey in the same general area. Other cross-stratified sands appear to be fluviatile deposits (point bar).

In southern New Jersey downdip, the Cohansey has many of the characteristics of the Raritan Formation, such as interbedded clean crossbedded sands and thick dark-gray clays. These are interpreted to be deltaic deposits, and the Cohansey, therefore, represents a wide variety of marginal-marine environments.

DEPOSITIONAL ENVIRONMENTS

The foregoing descriptions of formations illustrate the wide variety of lithofacies present in the northern Atlantic Coastal Plain.

Figure 15 is an idealized reconstruction of the sedimentary framework within which the various lithofacies are believed to have formed.

Two large environmental groups (super environments of Allen, 1965a) are distinguishable: deltaic and shelf. These super environments contain many smaller discrete environments.

DELTAIC DEPOSITS

The terminology applied to deltaic deposits varies widely, ranging from the simplistic topset, foreset, bottomset terminology proposed by Gilbert (1914), to the very complex terminology used by many modern investigators such as Fisk and others (1954) for the Mississippi River delta and Allen (1965b) for the Niger River delta. Ancient deltaic deposits may be most conveniently subdivided into subaerial and subaqueous plain deposits, that is, deposits influenced primarily by continental processes, and deposits modified by marine processes.

SUBAERIAL PLAIN DEPOSITS

Subaerial plains are characterized by fluviatile and marsh (or mangrove swamp) deposits. The fluviatile deposits can be separated into channel fill and overbank deposits.

Channel-fill deposits are largely extensively cross-stratified sand. The type of cross-strata in channel-fill deposits is related to the flow regime of the depositing stream (Simons and Richardson, 1961; Harms and others, 1963; Harms and Fahnestock, 1965; and many others). The deposits in the New Jersey Coastal Plain are evaluated within this scheme.

Figure 14 The Cohansey Sand. A, Gravelly sand facies exposed in a pit 1 mile west of Jacobstown, N. J. Trough cross-stratification is well developed in these beds. B, Gravelly sand facies exposed in a pit on State Rte. 36 near Hilton, N. J. Planar, rather than trough, is the major type of cross-stratification. C, Contact between the Cohansey Sand and the Kirkwood Formation in the Fort Dix Army Reservation pit. The contact is sharp and fine gravel has been reworked into the base of the Cohansey. D, Thin-bedded sands at Juliustown, N. J. This is one of the finest exposures of the beds interpreted to be beach deposits.

Crossbedded sands interpreted as point- or channel-bar deposits are well developed in the Farrington Sand, Sayreville Sand, and Old Bridge Sand Members of the Raritan Formation, the Magothy and Englishtown Formations and the Cohansey Sand. The cross-stratified sands are particularly well exposed in the Raritan (figs. 5B, C). Nearly all the cross-stratification in the Raritan is large scale (greater than 2 inches thick per set); small-scale cross-strata are rare. Trough and planar cross-strata are common; trough is the most abundant type. As interpreted by Harms and Fahnestock (1965, pl. 1) this type of cross-stratification suggests deposition in the lower flow regime of a stream. High-angle tabular or planar cross-stratification (fig. 5D) also indicate deposition in the lower flow regime but in this case the lower part. Both these interpretations support a deltaic origin (meandering streams) for the Raritan Formation.

The Cohansey (fig. 14A, B) and Raritan have cross-

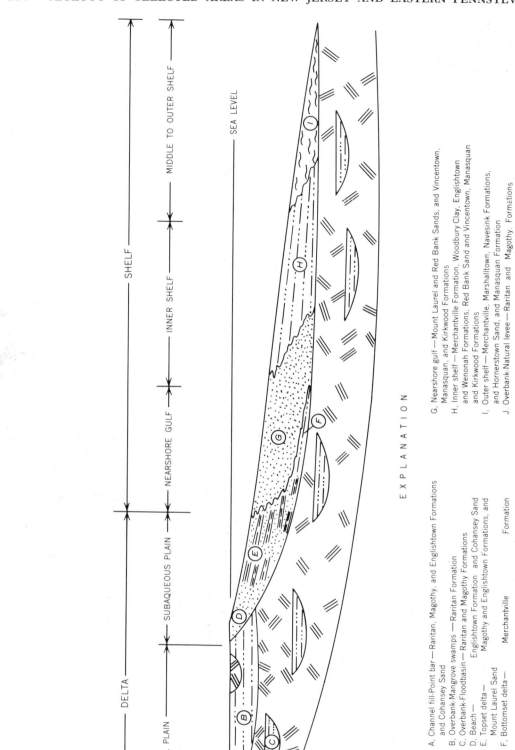

EXPLANATION

A, Channel fill-Point bar — Raritan, Magothy, and Englishtown Formations
    and Cohansey Sand
B, Overbank-Mangrove swamps — Raritan Formation
C, Overbank-Floodbasin — Raritan and Magothy Formations
D, Beach —   Englishtown Formation and Cohansey Sand
E, Topset delta — Magothy and Englishtown Formations, and
    Mount Laurel Sand
F, Bottomset delta — Merchantville   Formation

G, Nearshore gulf — Mount Laurel and Red Bank Sands, and Vincentown,
    Manasquan, and Kirkwood Formations
H, Inner shelf — Merchantville Formation, Woodbury Clay, Englishtown
    and Wenonah Formations, Red Bank Sand and Vincentown, Manasquan
    and Kirkwood Formations
I, Outer shelf — Merchantville, Marshalltown, Navesink Formations,
    and Hornerstown Sand, and Manasquan Formation
J, Overbank-Natural levee — Raritan and Magothy Formations

Figure 15.   Composite diagram of the major depositional environments within the Cretaceous
and Tertiary formations in the New Jersey Coastal Plain.

stratification that is similar to but in sets that are thicker than cross-stratification in the Magothy (fig. 6) and Englishtown (fig. 8A).

In meandering rivers, abandoned channels are typically filled with fining-upwards dark organic-rich clays and silts. Such slack-water deposits are abundant in subaerial deltaic plains and are represented in the New Jersey Coastal Plain by the South Amboy clay of the Raritan and the Amboy clay of the Magothy.

Overbank deposits are characteristically much finer grained and more poorly stratified than the sandy channel-fill deposits. Allen (1965a) has summarized the characteristics of overbank deposits.

One of the most interesting Coastal Plain deposits interpreted to be of overbank or topstratum origin is the horizontally stratified interbedded sequence of thin dark silts and light sands. Beds of this type are common in the Raritan (in the upper part of the Woodbridge clay, Kümmel and Knapp, *in* Ries and others, 1904, pl. 46) and the Old Bridge Sand Member at Old Bridge, N. J. Perhaps the best development of overbank deposits, however, is in the Morgan beds of the Magothy Formation (fig. 6B). Beds within a fluviatile sequence, having rapid alternations of coarse and fine sediment are typical of natural levee deposits (Allen, 1965a, p. 146) or deposits formed on an unimpeded flood plain (p. 153). In modern studies of both these environments, however, the dark-colored finer grained deposits are commonly oxidized to red and brown, or are mottled. This is not the case with beds of this type in the Raritan or Magothy Formations. In addition, the sands in deposits formed on unimpeded flood plains are extensively cross-stratified (Allen, 1965a, fig. 29). Figure 6B shows inconspicuously cross-stratified sands. Field relations have shown that the thin-bedded sequences in the Magothy and Raritan grade into cross-stratified point- or channel-bar sands. This suggests near-channel deposition and for this reason, a natural levee origin is favored for this type of bed.

The Woodbridge clay is also a predominantly fine-grained sediment. It differs, however, from the Amboy clay and the South Amboy clay in its greater lateral extent, greater thickness (as much as 80 feet thick), crudely developed horizontal stratification (fig. 5A), and, perhaps most significantly, its high concentration of sand-sized mica, quartz silt, and many layers containing large siderite concretions. Cobble-sized iron concretions develop in swampy environments (Coleman and Gagliano, 1965, p. 146). It is likely, therefore, that the Woodbridge clay is a mangrove swamp deposit

formed near the delta edge where occasional storms swept marine organisms into this environment. Swamp deposits of this type are widespread in the Mississippi and Niger River deltas, in fact, they make up one of the major overbank environments.

SUBAQUEOUS PLAIN DEPOSITS

Coleman and Gagliano (1965, fig. 9) subdivided the subaqueous deltaic plain into four subenvironments; subaqueous levee, distributary channel, distributary mouth bar, and distal bar. Collectively they refer to these as delta-front deposits. We include beach deposits in this group much as Allen (1965a) did in the Niger delta.

Considering the large number of sandy deposits in the New Jersey Coastal Plain, and the many deposits which might be construed to be marginal-marine in origin, it is surprising that beach deposits are so rare. Beach sands have very diagnostic gently inclined foreshore laminations. Deposits of this type were found only in the Englishtown Formation (central New Jersey and Delaware) and in the Cohansey Sand (fig. 14D). None of the other formations in the Coastal Plain contain recognizable beach deposits.

Subaqueous levee deposits superficially resemble the subaerial levee deposits, i.e., they are interstratified thin-bedded dark silts and light sands. Subaqueous levee deposits have well-developed cross-stratification, wavy laminations, scattered burrows, clay inclusions, and convolute laminations. The Englishtown Formation and Mount Laurel Sand, near Trenton, have many of these features. Convolute laminations, for example, occur locally in the Mount Laurel, and wavy laminations in the Englishtown. Layers containing siderite concretions are characteristic of subaqueous levee deposits in the New Jersey Coastal Plain but have not been described by Coleman and Gagliano (1965, p. 145) in the Mississippi delta.

Distributary channel deposits are important in the subaqueous part of the delta. These deposits differ from ordinary channel-fill deposits in that they contain an abundance of interstratified clays. The Cliffwood beds of the Magothy Formation (fig. 6C) exemplify this type of deposit. They are extensively cross-stratified and contain light-colored sands interbedded with black laminae of finely comminuted plant debris. As described previously, thin beds containing marine fossils interfinger with these sands.

Distributary mouth bar deposits are widespread in the delta front environment (Coleman and Gagliano, 1965, fig. 9; Allen, 1965b, fig. 7). These deposits are regularly interstratified silts and sands, similar to subaqueous delta deposits described by Moore and Scruton (1957, fig. 8). The Mount Laurel Sand of northern New Jersey (fig. 10A) and part of the Englishtown Formation near Trenton (fig. 8B) are excellent examples of this type of deposit.

Distal bar deposits are the farthest seaward part of the delta front. Typically, these are laminated silts and clays, in which small-scale cross laminations are locally well developed, and in which small concentrations of woody fragments are common in many places. Parts of the lower Merchantville Formation (fig. 7A) near Raritan Bay have intercalated clayey silts and silts that were formed in a distal bar environment. Burrows occur locally, and layers of siderite are abundant. Abundant siderite is to be expected here, because this type of environment is especially favorable for its development. Strakhov and others (1959) found that siderite is a late diagenetic mineral diagnostic of river-influenced sediments in deposits in the Russian platform. The basal Merchantville beds in eastern Maryland locally have similar type beds. The Merchantville, however, is the only formation in which distal bar type deposits have been observed.

In summary, deltaic sediments are particularly thick and widespread in the Raritan embayment region in the northern New Jersey Coastal Plain (fig. 2). Apparently this embayment was a locus of clastic sedimentation during a long period of time, from Late Cretaceous (Cenomanian-Raritan Formation) through Campanian (Mount Laurel). Deposits representing many depositional environments can be recognized in these formations.

SHELF DEPOSITS

Excluding the Raritan embayment, the Coastal Plain formations have lithologies similar to shelf deposits of other regions. The classification of Shepard and Moore (1955) for the modern western Gulf of Mexico seems most applicable to the marine, non-deltaic deposits of the northern Atlantic Coastal Plain, at least to shelf deposits of Late Cretaceous through Pliocene(?) age. Shepard and Moore described three major shelf environments; near-shore gulf, inner shelf, and outer shelf. The outer shelf is herein designated a middle-outer shelf.

NEARSHORE GULF DEPOSITS

Most of the inshore areas in the western Gulf of Mexico are covered by sand to depths of as much as 30 feet, and locally, to depths of 90 feet (Shepard and Moore, 1955). Even though these sands are very shallow water deposits, they apparently are massive (classed as homogeneous coarse according to Moore and Scruton, 1957, p. 2731). Similar but coarser deposits (medium-grained in the northern Atlantic Coastal Plain vs. fine- to very fine-grained in the western Gulf) are present in the Mount Laurel Sand (from Trenton to northern Delmarva Peninsula), Red Bank Sand (Sandy Hook Member), Vincentown Formation (particularly in northern New Jersey), and the up-dip facies of the Kirkwood Formation. These deposits are all thick, and where they are best developed, they are the thickest of the marine shelf lithofacies (fig. 15). One characteristic which distinguishes these sandy deposits from delta or back bar sands is the presence of glauconite. All the New Jersey Coastal Plain formations containing this lithofacies, except the Kirkwood, contain small amounts of glauconite. It seems anomalous that thick sands, deposited in shallow water are massive. This massiveness has been attributed to disruption or churning of the sediment by burrowing animals (bioturbation). There is evidence of bioturbation, *Callianassa* borings, in the Mount Laurel (fig. 10C) and in the Kirkwood (fig. 13C), but the number of burrows is not great. This type of boring, however, supports a very shallow water origin for these sands. No borings were found in the Sandy Hook Member of the Red Bank or in the Vincentown.

Marine fossils occur in all the massive sands. They are most abundant in the Mount Laurel (Minard and others, 1961) where they occur in distinct beds and in the calcarenite facies of the Vincentown where the fossils occur in thin to thick beds or in a bioherm (*Oleneothyris* beds at the base of the Vincentown). Fossils are uncommon in the Kirkwood and in the Sandy Hook Member of the Red Bank. Some fossils, however, have been reported from both these sands (Sandy Hook Member of the Red Bank Sand—Weller and Knapp, 1907; Kirkwood Formation—Richards and Harbison, 1942).

INNER SHELF DEPOSITS

Locally, dark-colored very micaceous very silty sands interfinger with the nearshore gulf deposits (as in the Mount Laurel near Trenton). These dark beds are interpreted to be inner shelf deposits. Included in this group are the Merchantville strata in southern New Jersey and northern Delmarva Peninsula, Woodbury, Englishtown (in southern New Jersey), Wenonah, Shrewsbury Member of the Red Bank, basal Vincentown near Trenton, and the down-dip facies of the Kirkwood.

These dark beds are massive and are extensively bioturbated. These burrows are more abundant and are much smaller in average diameter than the *Callianassa* borings of the nearshore gulf sands. Moore and Scruton (1957, p. 2736) classed these deposits as homogeneous fine in their studies of bedding characteristics in sediments of shelf origin.

The following characteristics may be used to differentiate inner shelf deposits from nearshore sands: Inner shelf deposits contain (1) more sand-sized mica and silt, (2) more sand-sized and finely comminuted carbonaceous matter, (3) generally more glauconite, and (4) more fossils (usually in biostromal rather than biohermal accumulation; the fossils occur as casts or in siderite concretions and, only locally, are the actual shells preserved).

MIDDLE TO OUTER SHELF DEPOSITS

Shepard and Moore (1955, p. 1539) discussed the difficulties in separating, on the basis of physical evidence only, deposits of the inner shelf from deposits of the outer shelf. Faunas, however, provide the best criteria.

Deposits of middle to outer shelf origin in the New Jersey Coastal Plain are recognized by their concentration of glauconite sand. It is assumed that glauconite forms in areas of slowest sedimentation and thus is formed farthest from areas of clastic influx. The distribution of the zone of high glauconite accumulation is irregular with respect to distance from shore. The greensands are mainly indicators of slow sedimentation and not of distance from shore or depth of water. They are, however, interpreted as always lying seaward of the massive very micaceous silty sands. The greensands are the thinnest of all the marine lithofacies, but they are the most persistent and easily recognizable beds, suggesting that uniform depositional conditions prevailed over wide areas.

In outcrop, all the glauconite sands are massive except for the northernmost outcrops of the Merchantville which are thick bedded. Even in these beds, however, each bed of glauconite sand is massive (fig. 7B). The glauconite beds and inner shelf deposits are both extensively and similarly burrowed. Where these beds are highly carbonaceous, borings are commonly filled with light brighter green glauconite sand giving rise to a mottled appearance.

Inner and middle to outer shelf beds interfinger over a broad area. Some of the glauconitic sands contain large though varying amounts of terrigenous debris such as quartz, mica, and woody fragments (Tinton, basal Red Bank near New Egypt,

Marshalltown, and the glauconite beds in the Merchantville). In other greensands, most of the terrigenous debris is of clay to silt size and the total sand fraction is nearly all glauconite (Navesink, Hornerstown, and some beds in the Manasquan). Glauconite in these beds is typically coarser than glauconite found in clastic-rich beds.

If the amount of sand-sized terrigenous material is a crude indicator of closeness to source, then glauconite can apparently form over a rather broad range of environments.

Shepard and Moore (1955) thought that glauconite sand was more diagnostic of inner rather than outer shelf deposits. The shelf deposits of the Coastal Plain in New Jersey and Delaware do not support this interpretation.

In general, there is a similarity between the modern shelf deposits in the western Gulf of Mexico and the Upper Cretaceous to lower Tertiary shelf deposits in the northern Atlantic Coastal Plain. As in the case in almost every sedimentary basin, however, each has characteristics unique to that basin. The northern Atlantic Coastal Plain shelf deposits appear to be, on the average, coarser grained than their modern gulf coast analogs. In addition, glauconite is much more abundant than in any of the gulf coast deposits and is even more abundant than in any known Holocene shelf deposits.

CYCLIC SEDIMENTATION

Figure 16 shows a composite stratigraphic section of the pre-Quaternary formations in the New Jersey Coastal Plain. The vertical succession of marine and marginal-marine lithofacies (Merchantville through Manasquan Formations) follows a pattern. The systematic recurrence of similar lithologies through this stratigraphic interval illustrate the cyclic sedimentation in this region.

The cyclic pattern is most evident where the greensands are part of a sedimentary cycle. By choosing any of the glauconite sands as the base of a cycle, it can be seen that typically greensands are overlain by a silty unit which in turn is overlain by a dominantly quartz sand unit. In all, four ideal cycles can be recognized in the exposed northern Atlantic Coastal Plain deposits, and if the Tinton and the Manasquan are considered partial cycles, six cycles of sedimentation are present from Late Cretaceous (Campanian) through early Eocene time. Cycles of this type are common within the alternating marine and nonmarine sequences of the gulf coast (Fisher, 1964, p. 158–163, and especially fig. 4).

Cyclic deposits in the Raritan, Magothy, Kirk-

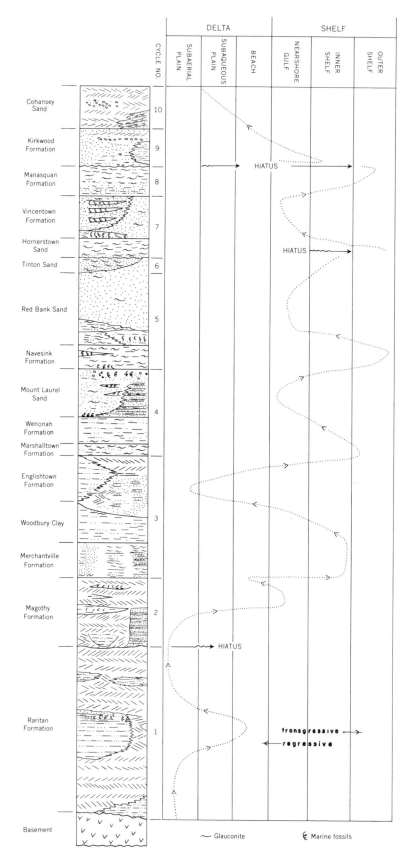

Figure 16. Composite stratigraphic section showing cyclic sedimentation and changes in environment in the Coastal Plain deposits of New Jersey.

wood, and Cohansey are difficult to recognize, and for convenience, they have been assigned individual numbers (fig. 16). In many ways they are similar to the cycles within predominantly regressive sequences outlined by Fisher (1964, p. 163–166).

The transgressions and regressions diagrammed in figure 16 are based upon environmental interpretations. In this analysis, all the glauconite sands are interpreted as transgressive beds, and are regionally disconformable on nearshore marine or nonmarine beds. The silts and overlying sands are largely regressive deposits in which the gradual coarsening of the clastics indicates a shoaling of the sea.

The cycle just discussed is asymmetrical. In such a cycle the deep-water facies overlies very shallow water deposits without intervening inner shelf deposits. This is not, however, always true, especially in the Navesink-Red Bank-Tinton and the Hornerstown-Vincentown-Manasquan cycles. In both these cycles, outer shelf glauconite sands are separated from the nearshore gulf sands by an inner shelf very silty sand.

The probable relations of the sedimentary cycles to tectonic and (or) eustatic movements will be discussed in the following section on petrology.

Recently, Duff, Hallam, and Walton (1967) evaluated the many proposed causes for cyclic deposits—sedimentary, tectonic, eustatic, and climatic. These authors believe sedimentary processes are responsible for most cyclicity, but others, notably Wells (as cited by Duff and others, 1967) favor eustatic control. In the continental margin environment, within which the Coastal Plain lithofacies were deposited, we believe that tectonism is responsible for the cyclic sedimentation.

PETROLOGIC INVESTIGATIONS

The lithofacies outlined in figure 15 were analyzed for their texture, mineralogy, and chemistry. The Coastal Plain sediments are particularly well suited for grain-size and mineral analyses because they are unconsolidated. Bulk chemical analyses were made for each of the environmentally controlled lithofacies to ascertain chemical variations within the depositional basin. Chemical variations are important in this basin where diagenesis is so prevalent.

At the surface, the Coastal Plain deposits can be separated into four rock sequences:

(1) Beds of Albian-Aptian(?) or Cenomanian age. Sediments in this sequence are nearly all nonmarine. These beds which contain thick red clay deposits and are arkosic to the southwest in Virginia are a red bed assemblage (Pettijohn, 1957, p. 628–630).

(2) Marine and continental beds of Santonian (Late Cretaceous) through early to late Eocene age. Largely because of high sustained terrigenous supply, carbonate deposition was inhibited and the association is a typical black shale-subgraywacke assemblage. Glauconite is prevalent in this sequence.

(3) Deposits of middle Miocene to early(?) Pliocene age. A black shale-subgraywacke assemblage but lacks the glauconite of the underlying sequence.

(4) Quaternary deposits. Quaternary fluviatile deposits occur along the basin edge. They are extremely heterogeneous (reworked Coastal Plain and glacial deposits and first-cycle sediments from Piedmont rocks) that they do not fit any specific rock classification.

Collectively these four sequences are similar to the sediment types found in molasse deposits. The rock associations, black shale-subgraywacke are typical of molasse deposits accumulating on an unstable shelf (Krumbein and Sloss, 1951).

Sequences 1 and 2 make up the Zuni sequence of Sloss (1963). Sequence 3 makes up the Tejas sequence.

TEXTURAL STUDIES

The many textural studies by Shepard and Moore (1955) of the deltaic-shelf deposits of the gulf coast provide excellent references with which to compare northern Atlantic Coastal Plain sediments.

Samples from each of the lithofacies shown were sieved. The results were plotted on triangular diagrams (fig. 17) similar to those used by Shepard and Moore (1955, fig. 26).

The beach and nearshore gulf lithofacies from the northern Atlantic Coastal Plain and their modern analogs from the gulf coast have virtually the same size distributions. Actually the northern Atlantic Coastal Plain sediments in both these environments are somewhat coarser on the average than those in the gulf but this fact is not reflected in these diagrams.

The inner and middle to outer shelf lithofacies of the Atlantic Coastal Plain, however, have a distributional pattern decidedly different from that of the comparable gulf coast lithofacies. The inner shelf samples have more silt and less clay than the gulf coast samples. The abundant silt in the northern Atlantic Coastal Plain samples is consistent in all the formations studied. This indicates a sustained

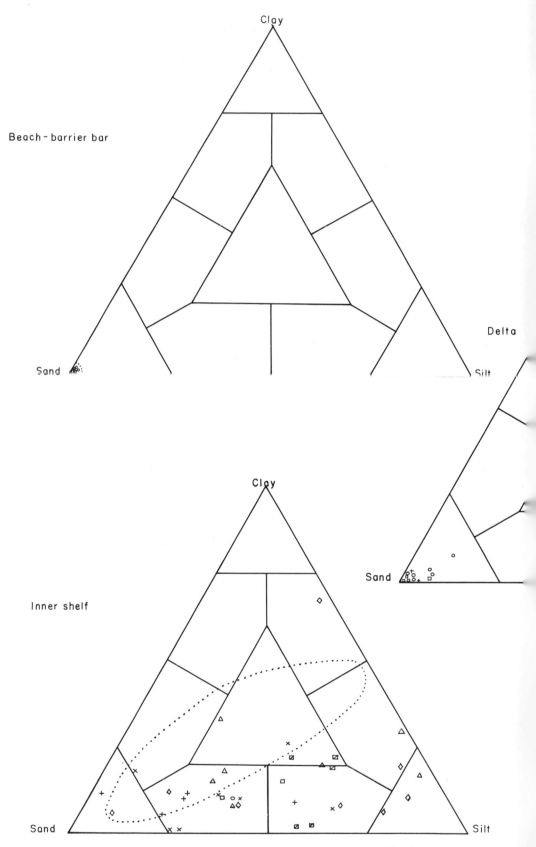

Figure 17. Sand-silt-clay distribution in the major environments of the Cretaceous-Tertiary formations of the northern Atlantic Coastal Plain. Beach-barrier bar: circles—Cohansey and Englishtown.   Delta: squares—Magothy;   circles—Englishtown;   triangles—Merchantville; crosses—Raritan. Nearshore gulf: diamond—Kirkwood; circles—Vincentown; cross—Red Bank; square—Englishtown.   Inner shelf: diamond—Kirkwood;   circle—Vincentown;   cross—Red Bank; x—Wenonah; square—Englishtown; square with slash—Woodbury; triangles—Merchant-ville. Middle to outer shelf: triangles—Hornerstown; squares with slash—Tinton; circle—Red Bank; cross—Navesink; x—Marshalltown.   Area bounded by dotted lines indicates field for samples from modern gulf coast analogs.

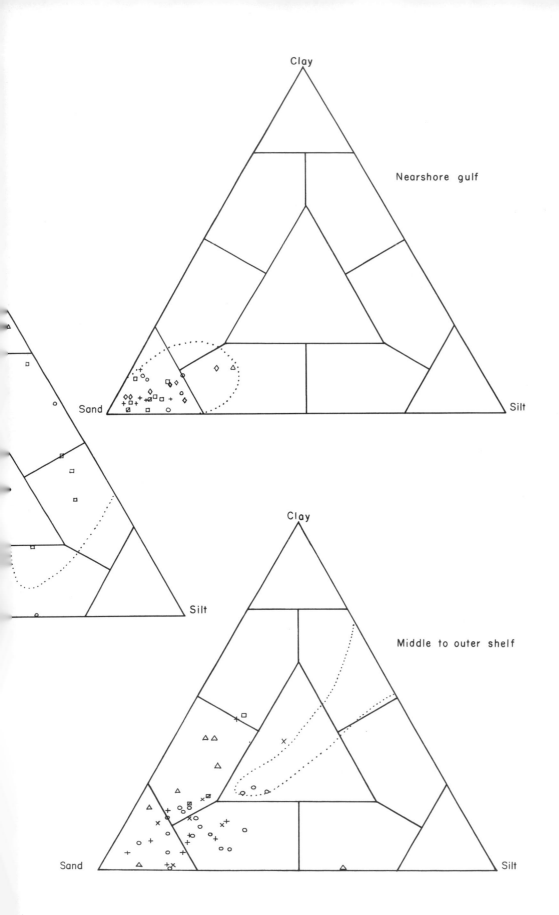

261

higher supply of silt to the depositional basin in this area than to the modern western Gulf of Mexico.

The middle to outer shelf samples in the northern Atlantic Coastal Plain have a distributional pattern comparable to that of the nearshore gulf samples (high sand content). This, however, is caused by high concentrations of glauconite sand in these beds, rather than by abundance of sand-sized terrigenous clastics. The glauconite typically oc-

curs as rounded grains imbedded in a much finer grained matrix, suggesting that the grains have been transported some distance, and may be of detrital origin.

The deltaic beds also have a distributional pattern significantly different from that of the smaller deltas in the Gulf of Mexico (Shepard and Moore, 1955, fig. 25).

Figure 18 shows photomicrographs of thin sec-

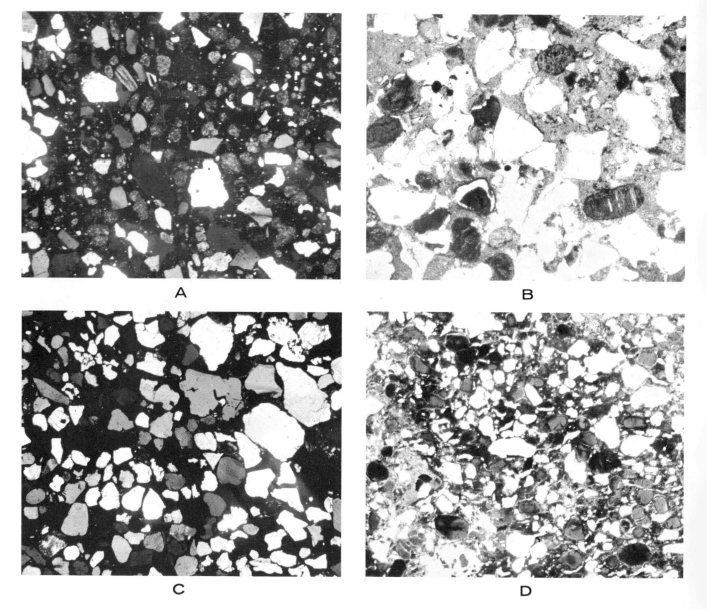

A

B

C

D

Figure 18. Photomicrographs of thin sections of nearshore gulf deposits. A, Massive pebbly quartz sand of the Mount Laurel Sand showing size and angularity of quartz sand. Dark speckled sand grains are glauconite (×15, crossed polarizers). B, Transitional facies of Red Bank Sand at Emleys Hill, N. J., showing admixture of quartz and glauconite sand. Note angularity of quartz and roundness of glauconite grains. Note also the concentric rims on the glauconite grains (×30, plane polarized light). C, Sand from beach facies of the Cohansey Sand showing mixture of angular and subrounded grains. Sample from Juliustown, N. J. (×15, crossed polarizer). D, Basal beds of Manasquan Formation from Pemberton, N. J. Note abundance of glauconite sand associated with quartz. Quartz grains are angular to subrounded (×15, crossed polarizers).

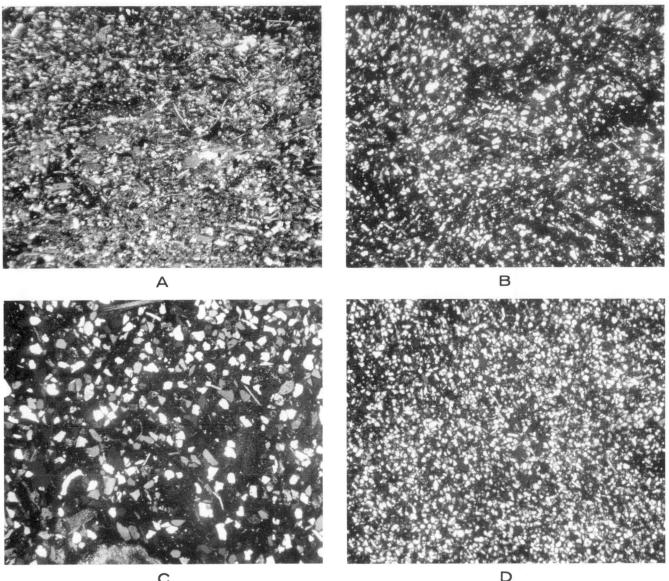

Figure 19. Photomicrographs of inner shelf sediments. A, Fine-grained micaceous clayey silt of the Woodbury Clay from Bordentown, N. J. Note abundant mica (×30, crossed polarizers). B, Fine-grained Wenonah Formation from Kirbys Mills, N. J. (×30, crossed polarizers). C, Shrewsbury Member of the Red Bank Sand from Waterwitch, N. J. (×15, crossed polarizers). D, Downdip facies of Kirkwood Formation from near New Egypt. Note silty character of sediment (×15, crossed polarizers).

tions from sandy marginal marine and marine lithofacies. Because it is difficult to obtain coherent samples, most of the photographs are from the more clayey and silty nearshore gulf sands. The size and angularity of the clasts in this environment are evident in these photographs. Angular grains of this size indicate first-cycle sediments, hence a continued fresh supply from the source lands, rather than a recycling of sediments, suggesting continued uplift of the source area.

There is a pronounced increase in the clay-silt content in the inner shelf lithofacies (fig. 17). This is the environment in which the fines (very fine grained sand through clay) largely accumulate. It is also the environment in which sand-sized mica and woody fragments are concentrated. Figure 19 shows the texture and fine-grained, carbonaceous and micaceous character of the clastics sediments within this lithofacies. Nevertheless, figure 18 shows the considerable range in grain size in this lithofacies. The grain-size distribution in this lithofacies is much broader than that found in the nearshore gulf lithofacies. Shepard and Moore (1955) reported similar relations in the western gulf coast (fig. 17, this report). Curray (1965, fig. 8) plotted the distribution of shelf lithofacies in the Gulf of Mexico. The variability of the deeper water (inner and outer shelf) deposits within the shelf environment is great. Apparently, local factors such as rate and type of sediment supply are more important in determining the local nature of the sediment in this deeper water environments than in the higher energy beach and nearshore gulf environments where sand is concentrated.

Although the inner shelf facies of the northern Atlantic Coastal Plain is clayey and silty, it is less so than the inner shelf facies of the Gulf of Mexico (Shepard and Moore, 1955). Most of the samples analyzed from the northern Atlantic Coastal Plain plot closer to the sand-silt join than to the clay pole. The average grain size of sands of the nearshore gulf lithofacies of the northern Atlantic Coastal Plain is larger than the average grain size of the same lithofacies in the western gulf coast deposits.

The inner shelf sediments of the northern Atlantic Coastal Plain are only a slightly coarser than the gulf coast deposits, but the middle to outer shelf lithofacies is markedly coarser than the same facies in the gulf coast (fig. 17). The deposits from this environment have a textural distribution similar to those of nearshore gulf deposits. The middle to outer shelf deposits in the Gulf of Mexico (fig. 17), are much more clayey than middle to outer shelf deposits of the northern Atlantic Coastal Plain.

Although the middle to outer shelf deposits in the northern Atlantic Coastal Plain are sandy, most of the sand is glauconite rather than quartz. Apparently, glauconite formed readily in areas of slow sedimentation in the distal parts of the basin. Even though it is authigenic, the indications are that the glauconite has been transported. The distance of transport, however, is difficult to evaluate, but the occurrence of sand-sized glauconite in a clay-silt matrix indicates mixing of sediment of different hydraulic equivalents. Part of the mixing is due to animal burrowing, but in other cases the glauconite has been redistributed by current activity (fig. 20A).

Grain-size distribution within the deltaic deposits have the same wide range as grain-size distribution in shelf deposits. Some of the deposits of the subaerial plain, however, are much more clayey than any of the samples analyzed from the shelf environment (fig. 17).

The large number of subenvironments in the subaerial and subaqueous parts of the delta were discussed earlier. The overbank deposits (natural levee, marsh, and floodbasin(?)) and the substratum channel-fill deposits of the subaerial plain are very silty and clayey, as are the distal bar and river-mouth bar deposits of the subaqueous plain. All of these lithofacies lie near the clay-silt join in the textural diagram.

Well-defined microlaminations can be seen in thin sections of the clayey and silty strata (figs. 21A and B). This feature distinguishes these deposits, in thin section, from the very fine grained shelf deposits in which microlaminations were not observed.

The point- and channel-bar deposits of the subaerial plain are very sandy, and these plot near the sand pole in a sand-silt-clay diagram (fig. 17). These substratum deposits appear to be better sorted than those analyzed from the shoal water Guadalupe River delta (Shepard and Moore, 1955, fig. 25). In this respect, these channel sands resemble those from the subaerial plain in the Mississippi and Niger deltas.

Because they are unconsolidated, thin sections of these sands are difficult to obtain. Two thin sections of the Englishtown and Magothy Formations shown in figures 21C, and D, are indurated by pyrite, or goethite, and these could be cut without disrupting the sample. Both these subaerial sands are very similar in grain-size distribution and angularity of the grains. The angularity of the grains suggests that both sands are first cycle sediments.

In summary, textural variations within deltaic and shelf deposits in the northern Atlantic Coastal Plain are closely comparable to their modern gulf coast analogs. In general, the ancient deposits of the northern Atlantic Coastal Plain are coarser, but the same general textural trends found in the Gulf of Mexico marine deposits (fining of the sediment seaward) can be demonstrated in their ancient analogs. The similarity, however, does not hold true in the deeper water facies where in the northern Atlantic Coastal Plain the glauconite sands are much more abundant and anomalously coarser than in the equivalent facies in the gulf coast.

The deltaic deposits also follow the predicted size fractionation. The overbank deposits are much finer grained than the channel-fill deposits (point and channel bar). Again as was evident in the shelf deposits, the deltaic deposits are on the average coarser in the ancient deposits, especially in the channel-fill deposits where sand is much more abundant than in the channel-fill deposits in the Guadalupe delta.

MINERALOGY

Mineralogic studies of the various size fractions in each of the lithofacies have been made but evaluation of the results are still in the preliminary stages.

The heavy mineral (specific gravity >280) and silt-clay fraction have been examined in more detail than the light-mineral sand fraction.

HEAVY MINERALS

The bulk of this discussion will be directed toward an evaluation of the limited versus full suite heavy mineral concept as proposed by Dryden and Dryden (1960). According to Dryden and Dryden,

the nonmarine deposits have a suite of heavy minerals impoverished in all but the more stable heavy minerals. As specifically defined, this limited suite does not contain significant concentrations of any two of the following minerals: epidote, chloritoid, garnet, or hornblende. A full suite would have two or more of these minerals. Groot and Glass (1960, p. 273), although presumably using the scheme of Dryden and Dryden (1960) defined a full suite as one which has an abundance of epidote and consistent occurrence of staurolite, chloritoid, silli-

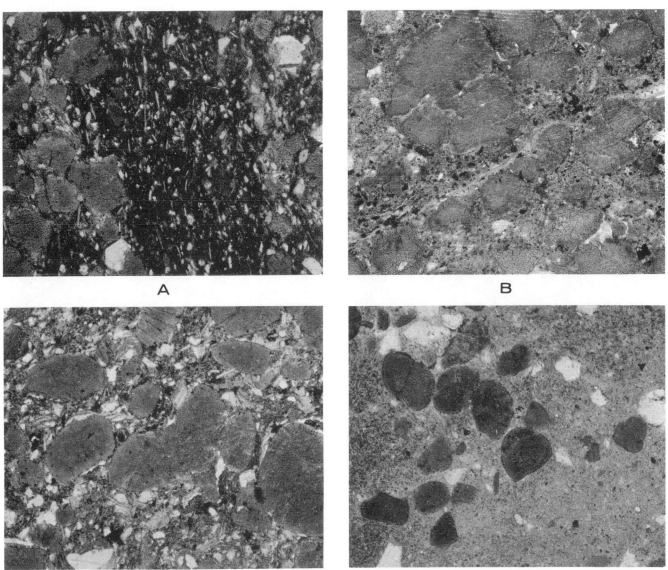

Figure 20. Photomicrographs of thin section from middle to outer shelf sediments. A, Merchantville Formation showing interbedded micaceous silts and glauconite sands (×30, crossed polarizers). B, Coarse glauconite sand in the Navesink Formation (×30, plane polarized light). C, Glauconite sand facies of the Red Bank Sand. Note the higher incidence of mica and silt to fine sand in the matrix as compared with the Navesink (×30, plane polarized light). D, Manasquan Formation showing the sparse distribution of glauconite and quartz sand embedded in a silt-clay matrix (×30, crossed polarizers).

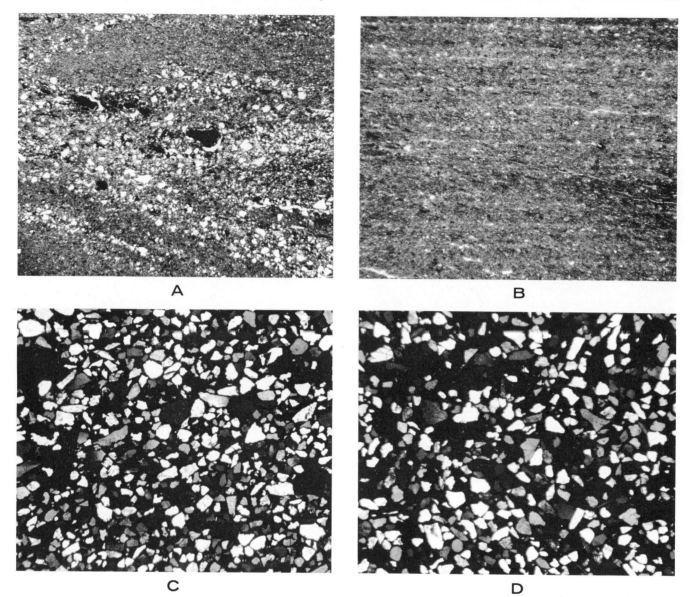

Figure 21. Photomicrographs of thin sections of coarse and fine sediments from deltaic environments. A, Interlayered clays and silts of the Magothy Formation (×17, plane polarized light). B, Finely laminated carbonaceous clayey silts of the Englishtown Formation. Specimen came from thin bedded sequences (×30, plane polarized light). C, Pyrite-cemented sands from the fluviatile facies of the Magothy near Bordentown, N. J. Note angularity of the grains (×17, plane polarized light). D, Goethite-cemented sand from the fluviatile facies of the Englishtown from Rancocas, N. J. Note similarity to sand from the Magothy in figure 21C (×17, plane polarized light).

manite, kyanite, and andalusite in addition to the more stable mineral species. This ambiguity diminishes the use of this concept in heavy minerals studies in the northern Atlantic Coastal Plain.

Heavy minerals are abundant and varied in the Coastal Plain formations. Figure 22 is a series of bar diagrams showing the heavy mineral distribution in all the pre-Quaternary formations at the surface in the New Jersey Coastal Plain. The re-

sults, however, are biased toward distribution in the west-central New Jersey Coastal Plain where most of the samples were collected. In addition, the results are averages for the formation as a whole and do not reflect changes in heavy mineral composition along strike or where a formation consists of more than one lithofacies and differences in heavy mineral distribution may be due to varying hydrodynamic conditions in the depositional en-

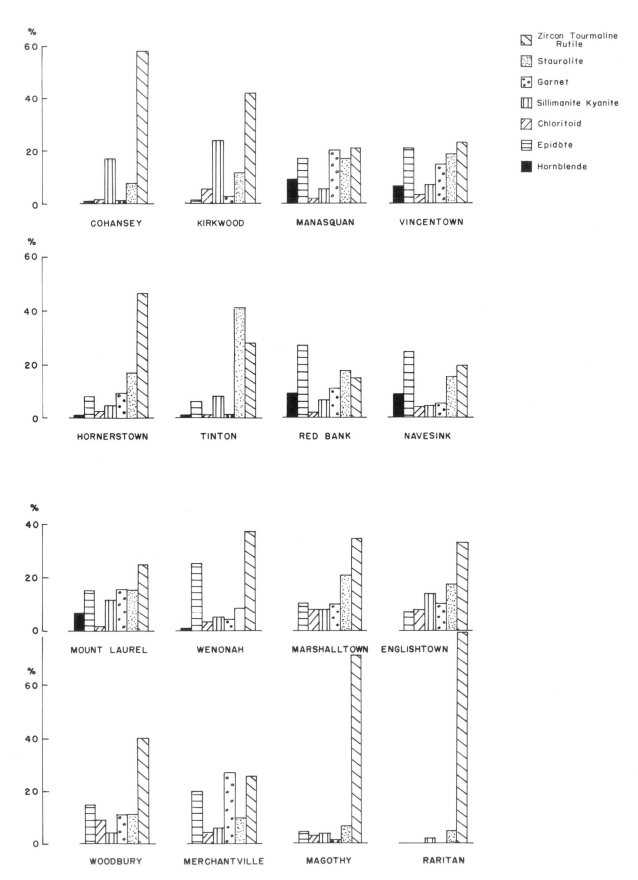

Figure 22. Bar diagram showing sand-sized heavy mineral distribution in the New Jersey Coastal Plain formations.

vironment. Both these factors are significant in the Coastal Plain formations. Without going into details, however, our studies showed that heavy mineral assemblages are of little value for stratigraphic correlations.

Heavy mineral distribution in the New Jersey Coastal Plain formation (fig. 22) can, in most cases, be classed as full or limited suite if Dryden and Dryden's (1960) "appreciable amounts" is considered to be 5 percent or more of an analyzed sample. Using this arbitrary 5 percent division point, the Raritan, Kirkwood, and Cohansey would have limited suites, and all the other formations full suites. If however, a genetic association is applied (limited = nonmarine) then all these units would be nonmarine. As discussed earlier in this manuscript, the Kirkwood is marine and parts of the Cohansey are marginal marine (beach), thus this relationship is not valid. In addition, the Englishtown is interpreted to be a largely nonmarine unit but it contains a full suite of heavy minerals. Reed (1960) considered the Englishtown to have a limited suite but using the arbitrary limits established above it falls within the full suite category.

Rather than use the full vs. limited suite concept with its implications of two source areas for the Coastal Plain sediments, it might be well to consider the sediments as coming from a single source. Pettijohn (1957, p. 518) discusses the vertical zonation of heavy minerals in a number of Cretaceous-Tertiary sections. He observed that the minerals follow a pattern, the number of mineral species increases in the younger rocks. He then considered three possibilities to explain this phenomenon; provenance, tectonism, and intrastratal solution. He concludes that intrastratal solution is the most likely explanation. Hubert (1962) discussed the maturity of sandstones and specifically their relationship to heavy mineral assemblages. He favors tectonism as the major factor in determining the type of assemblage. In view of the pronounced cyclic nature of the Coastal Plain sediments, we favor this interpretation, and the heavy mineral data will be analyzed accordingly.

The Raritan has an impoverished heavy mineral suite containing only the most stable minerals — zircon, tourmaline, and rutile (high ZTR-index assemblage of Hubert, 1962). The Magothy, also containing a high ZTR-index suite of heavy minerals, has, however, more unstable heavy mineral species. Although not shown, the number of heavy mineral species increase from the base of the Magothy (Morgan beds) through the upper Magothy (Cliffwood beds). Groot and Glass (1960) consid-

ered the upper Magothy to be marine (it contains marine fossils). Nevertheless, these beds are rivermouth bar deposits and therefore reflect the local source. They are not sediments brought in by long shore currents, as was suggested by Glass and Groot (1960). Three of the four minerals considered by Dryden and Dryden (1960) to be diagnostic of their full suite assemblages are present in the Magothy.

The Merchantville through Marshalltown deposits have virtually the same number of heavy mineral species as the Magothy but their ZTR index is much lower.

In the Wenonah-Manasquan interval, one of the least stable common heavy minerals, hornblende, appears in abundance for the first time. Its presence suggests the stripping of fresher rock in the source area.

The Kirkwood was deposited following a major unconformity. Hornblende which is characteristic of the Upper Cretaceous-lower Tertiary beds is absent in the Kirkwood and in the overlying Cohansey. Presumably, a period of weathering preceded Kirkwood deposition (roughly equivalent to Oligocene time) during which time hornblende and some of the other iron-bearing minerals were selectively destroyed in the source area. Pleistocene sediments in this area are characterized by the introduction of large concentrations of hornblende, epidote, and garnet.

Thus heavy mineral assemblages appear to be clearly related to tectonic movements in the source lands rather than to intrastratal solution or to provenance. This data is compatible with the cyclic nature of the beds and their mineralogic maturity (black shale-subgraywacke assemblage).

SILT-CLAY FRACTION

The silt-clay fractions of all the pre-Quaternary formations are shown in figure 23. A large number of minerals is present in this size fraction, including, in addition to all the major clay minerals, several other common nonclay minerals, such as quartz.

Groot and Glass (1960) studied the clay minerals from all the pre-Quaternary formations. They noted that there were fewer varieties of clays in the limited heavy mineral suites than in the full suites. Four clay minerals, kaolinite, montmorillonite, chlorite, and illite, were considered the most important minerals within the clay fractions. The limited heavy mineral suite, found in the nonmarine beds, contained only kaolinite and illite but the marine beds had all four of the important clay minerals.

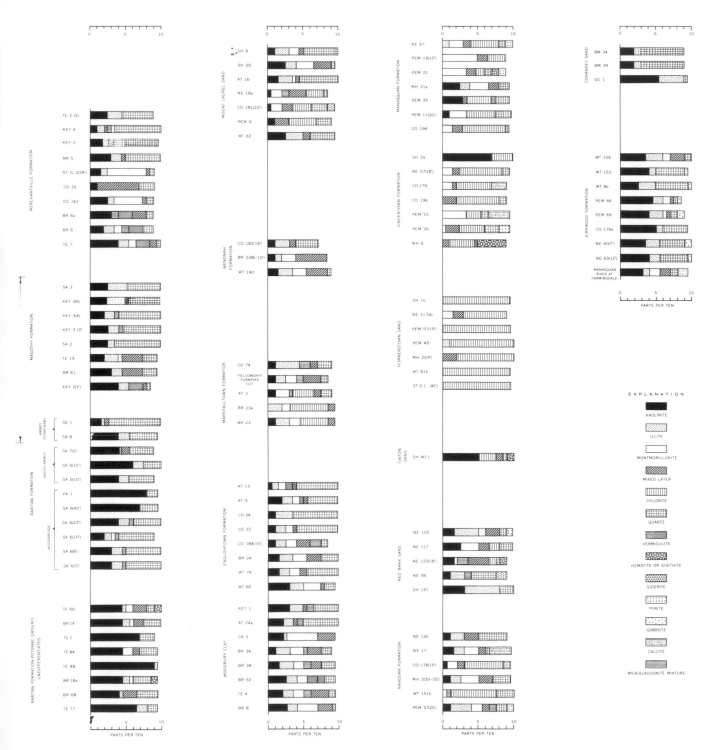

Figure 23. Distribution of minerals in the clay-silt fraction of the Cretaceous-Tertiary formations of the northern Atlantic Coastal Plain. Field numbers are listed with each of the samples, but these numbers are not referred to or discussed in the text.

Differences in the clay assemblages were interpreted as resulting from differences inherited from the source area (provenance), and from modification of these sediments by differential sorting in the various environments through which the clays were transported. Even in those units which had a full suite of heavy minerals, it was deduced that the nonmarine sequences would show a preference toward kaolinite-mica assemblages.

Diagenesis was considered unimportant in modifying the clay minerals within the various depositional environments.

The clay-silt mineral assemblages shown in figure 23, show the distribution of clays in the Coastal Plain formations. The pattern is not as simple as that proposed by Groot and Glass (1960). For example, montmorillonite occurs in the Raritan near Trenton even though these beds have a limited heavy mineral assemblage. The proposed relationship of montmorillonite-marine environment, therefore, is doubtful. The Potomac Group sediments (not shown in fig. 23), west of the Potomac River in Virginia, are arkosic sands with a full suite of heavy minerals (largely epidote bearing). These sediments also contain thick clay beds, in which montmorillonite is the major clay mineral (Owens, 1969). These deposits are clearly nonmarine, but they do contain montmorillonite. The presence of this full-suite heavy-mineral assemblage in a nonmarine sequence in the northern Atlantic Coastal Plain casts further doubts on the validity of the montmorillonite-marine environment hypothesis. The presence of montmorillonite is probably related to the immaturity of the associated sand, rather than to the depositional environment.

Although the clay-silt mineral assemblages shown in figure 23 are complex, a decided trend toward kaolinite depletion, and toward montmorillonite enrichment, is apparent in the sediments from the Raritan through the Manasquan. This trend is compatible with the heavy mineral data shown in figure 22, which shows that the more unstable mineral species became more abundant upward through this same interval.

CHEMICAL COMPOSITION

The lithofacies shown in figure 17 were analyzed to determine the chemical composition within the various deposits and to collate this data with their mineralogy. The chemical data are particularly significant in the marginal marine-marine lithofacies where the effects of mechanical sorting and diagenesis are reflected in the bulk chemical composition. Only the chemical data from these marine influenced lithofacies will be discussed below.

Figure 24 shows the changes in concentration of the major chemical elements with distance from shore. The graphs were constructed using the maximum and minimum values for each element. Additional values lying between these two extremes are also shown in order to indicate the scatter.

As might be expected, the beach-barrier bar lithofacies is very high in silica and is low in the other oxides. This, of course, reflects the high concentrations of quartz sand within this environment. The lower silica values are the result of local concentrations of mica in these beds.

The nearshore gulf deposits contain more silt-clay and glauconite than barrier bar-beach deposits and therefore are lower in silica and have more Al_2O_3 and Fe_2O_3. All the other major oxides have increased relative to the beach environment.

The progressive increase in oxides of all the elements except silica is continued into the inner shelf zone. Again this is in large part due to the increase in the silt-clay fraction but is also a reflection of the concentrations of sand-sized mica. Al_2O_3 shows the most pronounced increase as would be expected where there are higher concentrations of clay minerals and mica.

In the middle and outer shelf zones, Al_2O_3 and SiO_2 both tend to decrease. Fe_2O_3, K_2O, and MgO all show a sharp increase which can be directly related to the very large concentrations of glauconite sand.

Thus, across the marine influenced parts of the depositional basin, the chemical trends can be related to the mechanical sorting of the sediments and the concentrations of clay. This is virtually the same observation made by Shepard and Moore (1955, p. 1524) for modern sediments in the Gulf of Mexico. In the middle to outer shelf deposits, however, there is a marked enrichment in K_2O, Fe_2O_3, and MgO. This is the zone of maximum chemical activity or diagenesis in this basin and the chemical composition of these deposits indicates an iron- and potassium-rich source area.

ROAD LOG

FIRST DAY

First day of trip will be devoted to an examination of the deltaic sediments in the Raritan embayment (fig. 25). Several of the stops will be in active pits so the sections described in road log may have changed by the time of the field trip.

Mileage

0.0 Proceed north from Atlantic City; exit from the Garden State Parkway (GSP) at Interchange 123 on Rte. 9.

0.3 Right on South Amboy Road.

1.0 Right on Ernston Road.

1.6 Left on Washington Road.

3.3 Right on first through road.

4.2 Left on Main Street.

4.5 Turn left into large pit. Walk to back of pit.

STOP 1.

Woodbridge clay of the Raritan Formation. This member is an excellent example of a marginal marine swamp deposit (mangrove). Section as follows:

0–10 feet of sand and gravel of Quaternary age.

30–40 feet of interbedded silts and clays contains many layers of siderite- and iron-oxide-cemented sand. Some sand layers, particularly near the top, contain marine fossils. Note the abundance of mica in these beds.

5–6 feet of sand and clay containing much wood. Some of the wood appears to be in an upright position.

20 feet of dark-gray clay, weathered at the top. This weathered bed is probably an underclay. In the rear of the pit, farther to the east, 12 to 14 feet of yellow to white extensively cross-stratified sand is exposed. This is apparently the lower part of the Sayreville Sand Member of the Raritan Formation.

Return east along Main Street.
Turn left into Phoenix pit. Drive north about
8.0 0.5 mile into main pit area.

STOP 2.

Sayreville Sand Member of the Raritan Formation. Section as follows:

20–40 feet of cross-stratified sand exposed in the pit walls. Interpreted as channel-fill facies (most likely point bar deposits).

Return to Main Street, turn right and proceed west.

10.5 Turn right on Jernee Mill Road.

13.9 Turn left on South Amboy Road.

15.4 Turn right on River Road.

17.4 Turn right into pit.

STOP 3.

Laminated sand and clay of the Morgan beds (Berry, 1906) in the lower part of the Magothy Formation. Morgan beds are an excellent example of natural level deposits. Section as follows:

15–25 feet of alternating beds of yellow-brown to light-gray micaceous quartz sand and clay containing abundant carbonaceous matter throughout.

10–15 feet of light-colored cross-stratified sand.

Return to River Road; turn left (south).

17.6 Turn right onto South Amboy Road.

17.8 Turn left proceed south.

18.5 Junction with Rte. 18.

21.8 Turn left.

23.2 Turn right on Rte. 34.

23.9 Junction with Morristown Road.

24.9 Turn right (east) after rounding 90-degree turn to left.

25.2 Angle slightly left at fork in road. Continue on Cliffwood Avenue crossing GSP and Rte. 35.

27.8 End of road at beach. Walk to left (northwest) along beach about 100 yards to bluff.

STOP 4.

Magothy Formation overlain by several feet of Merchantville Formation and Pleistocene sand and gravel. Section as follows:

5–6 feet of Pleistocene silt, sand, and gravel.

5–6 feet of dark-greenish-black quartz-glauconite sand of the Merchantville Formation. Interpreted as open shelf deposits.

8 feet of dark-gray clay containing a few sand partings.

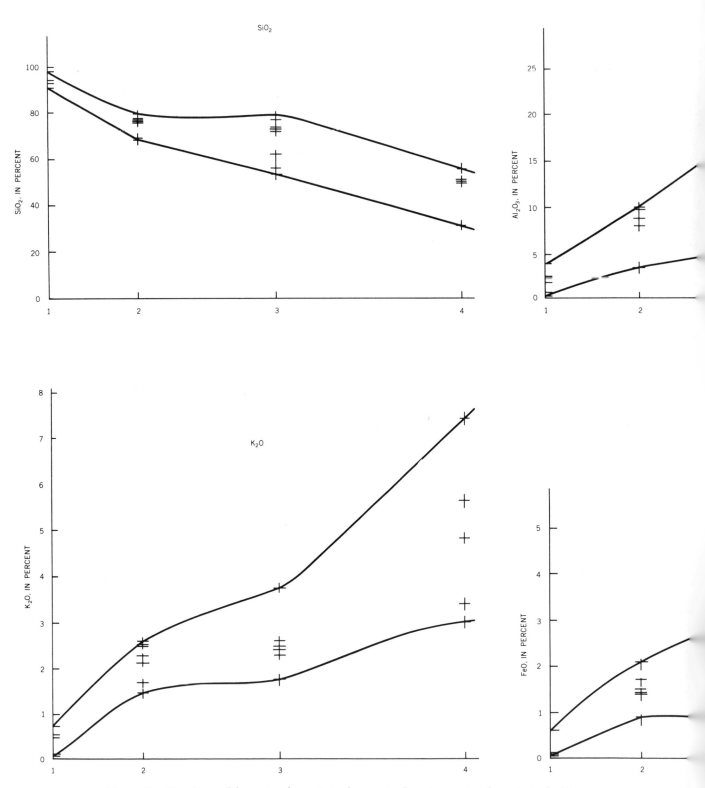

Figure 24. Variations of the major elements in the marginal marine-marine deposits in the New Jersey–Delaware Coastal Plain.

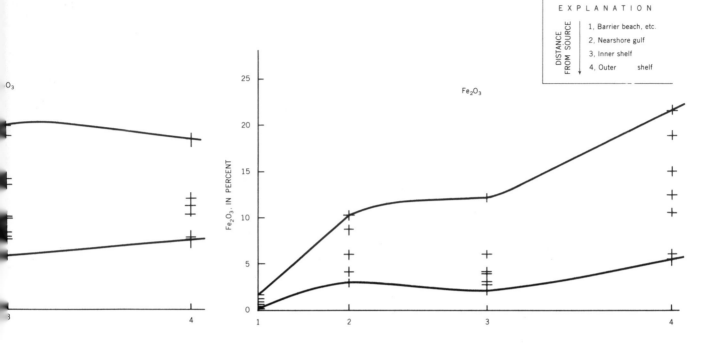

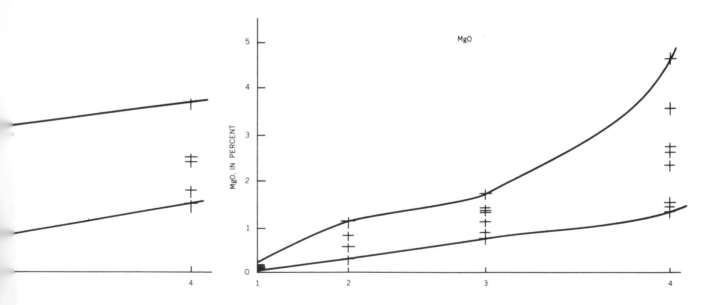

Figure 25. Sketch map showing field-trip route and stops.

Turn around at dead end and return along Cliff-wood Avenue.

Mileage

30.4   Turn right (north) after crossing Rte. 35 and the GSP.

31.0   Angle right after recrossing over the GSP and immediately turn right into the Oschwald pit.

STOP 5.

Merchantville Formation overlain by several feet of Pleistocene sand and gravel and spoil from the Woodbury Clay. Section as follows:

5–10 feet or more of Pleistocene sand gravel and disturbed Woodbury Clay (disturbed during pit operations).

20± feet of dark-greenish and grayish-black sandy silt and clay and sandy clay of the Merchantville Formation. Contains abundant mica, glauconite, and layers of siderite. Interpreted as a distal bar deposit.

Several feet of the upper, weathered sand and clay layers of the Magothy can be seen near the entrance of the pit.

Return to road, turn left and cross over the GSP.

31.6   Turn left on Cliffwood Avenue; recross the GSP.

33.2   Turn right (southeast) on Rte. 35.

37.4   Turn right (south) on Kings Highway.

38.3   Turn right (south) on Middletown Road. Pass railroad station.

40.1   Park in space at right side of road just short of Poricy Brook (northwest corner of the Long Branch quadrangle). Walk downstream along Poricy Brook about 100 yards to a high bank on the southwest side of the brook.

STOP 6.

Upper Navesink Formation overlain by lower part of the Red Bank Sand. Section as follows:

25 feet of the lower part of the Red Bank Sand (Shrewsbury Member). Mostly fine to medium quartz sand containing abundant feldspar, much mica (green and colorless), and some glauconite, particularly at the base. The upper part of the section is weathered to brown; the lower part is dark gray. Microfossils and small megafossils and shell fragments are abundant in the dark basal part. Interpreted to be an inner shelf deposit.

15 feet of greenish-black clayey glauconite sand of the Navesink Formation containing several shell beds. *Exogyra, Pycnodonte, Ostrea, Choristothyris,* and *Belemnitella* are common.

12 feet of alternating layers of dark-gray to black clay and light-gray quartz sand containing pyrite nodules throughout. Layers are about ¼- to ½-inch thick.

12 feet of white to light-gray quartz sand with some lignite and clay layers. The above 32 feet of the Magothy is interpreted to be a distributary channel deposit.

1 foot of gray-black clay with light-gray quartz sand containing siderite concretions (exposed at low tide). Fossils occur in the siderite concretions. Interpreted to be an open-shelf deposit.

(The total thickness of the Navesink at this stop is about 25 feet.)

These sediments are interpreted to be a middle to outer shelf deposit. Except for the shell beds, bedding is not conspicuous. The abundance of glauconite suggests a deep-water origin where conditions of sedimentation were uniform.

END OF FIRST DAY

SECOND DAY

Most of the second day will be spent examining shelf deposits. Leave Freehold about 8:00 A.M. and proceed southwest on N. J. Rte. 537.

Mileage

0.0 Right on Rte. 69 (Fort Dix Highway).
4.9 Right on Rte. 206.
5.8 Left on paved road.
6.0 Left into pit.

STOP 7.

Woodbury Clay overlying Merchantville Formation. The contact between the two units illustrates the gradational change from the middle or outer shelf to the inner shelf. *Section measured in southwall of pit:*

0–6 feet of light-brown sand and gravel of Quaternary age.
6–13 feet of dark-gray massive micaceous clayey silt to silty clay (Woodbury Clay).
13–14 feet of dark-gray to greenish-gray very silty and clayey micaceous quartz-glauconite sand. Glauconite sand is typically concentrated in small burrows (Merchantville Formation).

6.2 Return to Rte. 206, turn right.
7.0 Turn left on Old York Road.
12.5 Park bus in field adjacent to Crosswicks Creek, walk to north bank of creek. You probably will get your feet wet at this locality.

STOP 8.

This is an exposure of the Englishtown Formation which, in this region, overlies the Woodbury Clay. Section as follows:

0–14 feet of thinly bedded, dark silty clays and white sands of the Englishtown Formation. This lithology is typical of the Englishtown in this area. In general these strata are similar to the Morgan beds of the Magothy Formation. Both have been interpreted to be subaerial plain deposits.

Board buses proceed east on Old York Road.
15.3 Turn left on paved road.
15.7 Turn left in farm yard. Leave bus and walk about 1,500 feet to north bank of Crosswicks Creek. The banks are steep here so be careful.

STOP 9.

Three formations are exposed at this locality: the Englishtown, Marshalltown, and Wenonah. The Englishtown is similar to the beds seen at the last stop but is deeply weathered. The Marshalltown and Wenonah represent deposits formed in the transition zone between the middle shelf and inner shelf.

0–11 feet of dark-grayish-black silty fine quartz sand (Wenonah Formation). Note the high concentrations of mica and sand-sized carbon fragments.
11–21 feet of dark-greenish-black silty very fine quartz-glauconite sand (Marshalltown Formation). A thin bed of quartz granules and coarse lignite fragments occurs in the middle of the formation. Casts of fossils are common in this thin bed. Note also the concentration of glauconite sand in the burrows.
21–30 feet of orange-brown quartz sand containing thin discontinuous stringers of brown clay (Englishtown Formation).

LUNCH

18.1 Return to buses and proceed south on Walnford Road.
18.8 Turn left.
19.4 Turn left, proceed south. Park buses along road. Walk to small tributary of Crosswicks Creek.

STOP 10.

At this locality the Hornerstown Sand overlies another greensand, the basal Red Bank Sand. This contact lies at the Cretaceous-Tertiary boundary.

0–14 feet of dark-green clayey glauconite sand. Covered locally by a thin veneer of orange-brown iron oxides (Hornerstown Sand).

Farther downstream and in a small side gully the contact between the Hornerstown and the Red Bank is exposed. Notice the markedly different lithology of the Red Bank in comparison with the Hornerstown. The Red Bank is a dark-grayish-black silty micaceous quartz-glauconite sand. Locally large flattened siderite concretions are present at the contact.

Mileage

19.8   Return to bus proceed south, turn right on Rte. 537.

26.1   Turn right on Rte. 69 (Fort Dix Highway).

27.9   Turn left on Saylor Pond Road.

29.5   Turn left on Juliustown Road.

30.8   Turn left on Arneys Mount Road.

31.2   Turn left into pit.

STOP 11.

This pit is operated intermittently so that good exposures are not always available. Two formations are exposed here, the Cohansey Sand and Kirkwood Formation. The Cohansey Sand is one of the rare examples of a preserved beach deposit in the New Jersey Coastal Plain. The Kirkwood here is a nearshore gulf deposit.

0–45 feet of light-colored, thin- to thick-bedded extensively cross-stratified or horizontally stratified quartz sand. Fine gravelly sand beds are common but most of the deposit is quartz sand. A bed containing abundant ilmenite and other heavy minerals occurs near the top of this cut bank.

Small but well-developed *Callianassa* borings have been found in the horizontally stratified sands (foreshore deposits). Cohansey Sand.

45–50 feet of light-colored massive pebbly quartz sand (Kirkwood Formation). The contact between the two units is sharp.

34.4   Return to bus, proceed west on Arneys Mount Road. Turn left.

36.1   Turn right on Pemberton Road.

38.1   Turn right on Rte. 541.

39.4   Turn left into toll gate 5 of the New Jersey Turnpike. Proceed south.

57.4   Exit Turnpike at toll gate 3. Turn left on Black Horse Highway.

57.6   Turn left on paved road. Continue on dirt road extension into pit. Storms often wash out small culvert along this road, so we may have to hike into pit.

STOP 12.

This is a locality commonly called Irish Hill and is one of the most famous exposures in southern New Jersey. Four formations are exposed: the Marshalltown, Wenonah, Mount Laurel and Navesink. Here a complete cycle of sedimentation discussed earlier in the report can be observed. Section as follows:

Top of cut bank approximately 100 feet.

0–9 feet of dark-green medium-glauconite sand becoming finer grained toward top. Coarse granules and fine pebbles are common in lower three feet (Navesink Formation).

9–14 feet of upper pebbly bed of the Mount Laurel Sand. Glauconite is relatively abundant and concentrated in borings (Mount Laurel).

14–33 feet of light-brown to white massive medium to coarse sand becoming coarser grained toward top. Lower 10 feet somewhat more micaceous and clayey in addition to containing more glauconite than the overlying sands (Mount Laurel).

33–50 feet of light-brown grading downward to dark-grayish-black fine micaceous quartz sand. Numerous large borings have been etched out on water washed surfaces (Wenonah Formation).

50–52 feet of dark-greenish-black silty fine micaceous quartz-glauconite sand. Numerous large casts of fossils have been observed at this locality (Marshalltown Formation).

59.6   Return to bus. Retrace route back to Black Horse Highway.

61.6   Turn left.

103.6   Turn left. Proceed south and merge with Atlantic City Expressway at Turnersville. Return to Atlantic City.

END OF TRIP

REFERENCES CITED

Allen, J. R. L., 1965a, A review of the origin and characteristics of Recent alluvial sediments: Sedimentology, v. 5, p. 89–191.

———— 1965b, Late Quaternary Niger delta and adjacent areas—Sedimentary environments and lithofacies: Am. Assoc. Petroleum Geologists Bull., v. 49, p. 547–600.

Barksdale, H. C., Johnson, M. E., Schaefer, E. J., Baker, R. C., and DeBuchananne, G. D., 1943, The ground-water supplies of Middlesex County, N. J.: New Jersey State Water Supply Policy Comm. Spec. Rept. 8, 160 p.

Berry, E. W., 1906, The flora of the Cliffwood clays: New Jersey Geol. Survey Ann. Rept., 1905, p. 135–172.

Clark, W. B., 1904, The Matawan formation of Maryland, Delaware and New Jersey: Am. Jour. Sci., 4th ser., v. 18, p. 435–440.

Coleman, J. M., and Gagliano, S. M., 1965, Sedimentary structures: Mississippi deltaic plain *in* Middleton, G. V., ed., Primary sedimentary structures and their hydrodynamic interpretation: Soc. Econ. Paleontologists and Mineralogists Spec. Pub. 12, p. 133–148.

Cooke, C. W., and Stephenson, L. W., 1928, The Eocene age of the supposed late Upper Cretaceous greensand marls of New Jersey: Jour. Geology, v. 36, p. 139–148.

Curray, J. R., 1965, Late Quaternary history, continental shelves of the United States *in* Wright, H. W., Jr., and Frey, D. G., eds., The Quaternary of the United States: Princeton, N. J., Princeton Univ. Press, p. 723–735.

Drake, C. L., Ewing, W. M., and Sutton, G. H., 1959, Continental margins and geosynclines—the east coast of North America north of Cape Hatteras [Chap.] 3, *in* Ahrens, L. H., and others, eds., Physics and chemistry of the earth: New York, Pergamon Press, p. 110–198.

Dryden, A. L., Jr., and Dryden, Clarissa, 1960, Atlantic Coastal Plain heavy minerals—a speculative summary: Internat. Geol. Congress, 20th, Mexico City, 1956.

Duff, P. M. D., Hallam, A., and Walton, E. K., 1967, Cyclic Sedimentation: Amsterdam, Elsevier, 280 p.

Fisher, W. L., 1964, Sedimentary patterns in Eocene cyclic deposits, northern Gulf Coast region *in* Merriam, D. F., ed., Symposium on cyclic sedimentation: Kansas Geol. Survey Bull. 169, v. 1, p. 151–170.

Fisk, H. N., McFarlan, E., Kulb, C. R., and Wilbert, L. J., 1954, Sedimentary framework of the modern Mississippi delta: Jour. Sedimentary Petrology, v. 24, p. 76–99.

Frazier, D. E., and Osanik, A., 1961, Point bar deposits, Old River Locksite, Louisiana: Gulf Coast Assoc. Geol. Socs. Trans., v. 11, p. 127–137.

Gilbert, G. K., 1914, The transportation of debris by running water: U. S. Geol. Survey Prof. Paper 86, 263 p.

Greenman, D. W., Rima, D. R., Lockwood, W. N., Meisler, Harold, 1961, Groundwater resources of the Coastal Plain area of southeastern Pennsylvania: Pennsylvania Geol. Survey Bull., 4th ser., W 13, 375 p.

Groot, J. J., and Glass, H. D., 1960, Some aspects of the mineralogy of the northern Atlantic Coastal Plain *in* Swineford, Ada, ed., Clays and clay minerals: Natl. Conf. Clays and Clay Minerals, 7th, Washington, D. C., Oct. 1958, Proc.: New York, Pergamon Press, p. 271–284.

Harms, J. C., and Fahnestock, R. K., 1965, Stratification, bed forms, and flow phenomena (with an example from the Rio Grande) *in* Middleton, G. V., ed., Primary sedimentary structures and their hydrodynamic interpretation: Soc. Economic Paleontologists and Mineralogists Spec. Pub. 12, p. 84–115.

Harms, J. C., MacKenzie, D. B., and McCubbin, D. G., 1963, Stratification in modern sands of the Red River, Louisiana: Jour. Geology, v. 71, p. 566–580.

Hubert, J. F., 1962, A zircon-tourmaline-rutile maturity index and the interdependence of the composition of heavy mineral assemblages with the gross composition and texture of sandstones: Jour. Sedimentary Petrology, v. 32, p. 440–450.

Krumbien, W. C., and Sloss, L. L., 1951, Stratigraphy and sedimentation: San Francisco, W. H. Freeman and Co., 495 p.

Minard, J. P., 1964, Geology of the Roosevelt quadrangle, New Jersey: U. S. Geol. Survey Geol. Quad. Map GQ-340.

——— 1965, Geology of the Woodstown quadrangle, New Jersey: U. S. Geol. Survey Geol. Quad. Map GQ-404.

Minard, J. P., Owens, J. P., and Todd, Ruth, 1961, Redefinition of the Mount Laurel Sand (Upper Cretaceous) in New Jersey: U. S. Geol. Survey Prof. Paper 424-C, p. C64–C67.

Moore, D. G., and Scruton, P. C., 1957, Minor internal structures of some recent unconsolidated sediments: Am. Assoc. Petroleum Geologists Bull., v. 41, p. 2723–2751.

Olsson, R. K., 1960, Foraminifera of latest Cretaceous and earliest Tertiary age in the New Jersey Coastal Plain: Jour. Paleontology, v. 34, p. 1–58.

Owens, J. P., 1969, Coastal Plain rocks *in* Southwick, D. L., and Owens, J. P., Geology of Harford County, Maryland: Maryland Geol. Survey Bull. (in press).

Owens, J. P., Minard, J. P., and Sohl, N. F., 1968, Cretaceous deltas in the northern New Jersey Coastal Plain *in* Finks, R. L., ed., Guidebook to field excursions at the 40th Annual Meeting of the New York State Geological Association: New York, Queens College of the City University, p. 33–48.

Pettijohn, F. J., 1957, Sedimentary rocks: New York, Harper Bros., 718 p.

Reed, J. C., 1960, Heavy minerals of the Englishtown (Cretaceous) formation of New Jersey: Pennsylvania Acad. Sci. Proc., v. 34, p. 147–154.

Richards, H. G., 1945, Subsurface stratigraphy of the Atlantic Coastal Plain between New Jersey and Georgia: Am. Assoc. Petroleum Geologists Bull., v. 29, p. 885–995.

——— 1967, Stratigraphy of the Atlantic Coastal Plain between Long Island and Georgia—Review: Am. Assoc. Petroleum Geologists Bull., v. 51, p. 2400–2429.

Richards, H. G., Groot, J. J., and Germeroth, R. M., 1957, Cretaceous and Tertiary geology of New Jersey, Delaware and Maryland *in* Geol. Soc. America Guidebook for field trips, Field Trip no. 6, Ann. Mtg. Atlantic City, 1957, p. 183–230.

Richards, H. G., and Harbison, Anne, 1942, Miocene invertebrate fauna of New Jersey: Philadelphia Acad. Nat. Sci. Proc., v. 94, p. 167–250.

Ries, Heinrich, Kümmel, H. B., and Knapp, G. N., 1904, The clays and clay industry of New Jersey: New Jersey Geol. Survey, Final Rept. 6, pt. 2, p. 117–208.

Shepard, F. P., and Moore, D. G., 1955, Central Texas coast sedimentation—characteristics of sedimentary environment, recent history, and diagenesis: Am. Assoc. Petroleum Geologists Bull., v. 39, p. 1463–1593.

Simons, D. B., and Richardson, E. V., 1961, Forms of bed roughness in alluvial channels: Am. Soc. Civil Engineers Proc., v. 87, no. HY3, p. 87–105.

Sloss, L. L., 1963, Sequences in the cratonic interior of North America: Geol. Soc. America Bull., v. 74, p. 93–114.

Stephenson, L. B., 1954, Additions to the fauna of the Raritan Formation: U. S. Geol. Survey Prof. Paper 264-B, p. B25–B43.

Strakhov, N. M., Rodinova, K. F., and Zalmanson, E. S., 1959, A contribution to the geochemistry of the reservoir formations—the lower Frasnian series of the Volga-Urals, Part III: Internat. Geology Review, v. 1, no. 7, p. 41–61.

U. S. Geological Survey, 1967, Engineering geology of the northeast corridor, Washington, D. C. to Boston, Massachusetts—bedrock geology: U. S. Geol. Survey Misc. Geol. Inv. Map I-514A, sheets 2 and 3.

Weller, Stuart, and Knapp, G. N., 1907, A report on the Cretaceous paleontology of New Jersey, based upon stratigraphic studies of George N. Knapp: New Jersey Geol. Survey, paleontology ser. 4, 1106 p.

QUATERNARY GEOLOGY OF PART OF NORTHERN NEW JERSEY AND THE TRENTON AREA [1]

JAMES P. MINARD AND EDWARD C. RHODEHAMEL

U. S. Geological Survey, Washington, D. C. 20242, and Trenton, N. J. 08607

ABSTRACT

The Quaternary geology of New Jersey is varied and complex. Deposits of Quaternary age can be divided into two broad classes; glacial deposits of the northern glaciated part of the State, and fluviatile and marine deposits of the unglaciated southern part of the State. Both regions are briefly discussed, however the Trenton area and the northwest part of the State are emphasized. Moraines recently mapped by the senior author in Sussex County are discussed. Deposits in the Trenton area are mapped and described as lithostratigraphic units. Previously these deposits were mapped according to altitude and inferred age. The road log and stop descriptions include examples of typical landforms and internal structures of end moraines, kames, outwash, deltas, natural levees, and channel deposits.

INTRODUCTION

The Quaternary geology of New Jersey is briefly summarized to provide a background for the field trip. Some local deposits are described in detail. A thorough description of the Quaternary geology of New Jersey is beyond the scope of this report; rather it partly updates and supplements earlier work. The monumental report by Salisbury and others (1902) on the glacial geology of New Jersey was the culmination of many years work and is exhaustive in its detail and discussion of the glacial deposits and history of northern New Jersey. The Quaternary formations and history of the southern part of the State are also discussed at length by Salisbury and Knapp (1917).

The field trip, scheduled for two days, is divided between the glaciated northern part of the State and the unglaciated southern part. The glacial deposits and their more striking features are described and interpretations of their origins are emphasized on the first day. Newly discovered deposits of significant extent are described. Field-trip stops are in Somerset, Morris, Sussex, and Warren Counties.

The field trip on the second day concerns the Quaternary deposits in the unglaciated southern part of the State. These are described chiefly from within the area of ten 7½-minute quadrangles recently mapped by the U. S. Geological Survey in the central part of the State (see fig. 6). The deposits in the Trenton area, south of the terminal moraine, are mapped on lithologic differences. Previously, these deposits were mapped according to altitude and inferred age relations. Their separation based on lithology results in more units, some of which were deposited contemporaneously. An attempt is made to reconstruct some of the earlier drainage patterns and directions of flow, and the relation of the deposits to this drainage.

ACKNOWLEDGMENTS

In addition to the references cited, the authors benefited from field trips and conversations with Meredith Johnson, former New Jersey State Geologist, Paul MacClintock and Sheldon Judson of Princeton University, Kemble Widmer, New Jersey State Geologist, Frank Markewicz, Principal Geologist of New Jersey, Horace Richards of the Philadelphia Academy of Natural Sciences, and George Crowl of Ohio Wesleyan.

Data on the glacial geology of northern New Jersey and for the locations of field-trip stops in that area are based on the literature and on the senior

[1] Publication authorized by the Director, U. S. Geological Survey.

author's participation during 1951–55 in an engineering soil mapping program of New Jersey. This program was staffed by personnel of Rutgers University and sponsored by the State highway department; it was largely under the supervision of A. R. Jumikis of Rutgers University and was aided by the critical review of James Martens and Peter Wolfe, also of Rutgers.

Most of the data on the unglaciated southern region are from detailed geologic mapping by the senior author and James P. Owens, U. S. Geological Survey, in a block of ten 7½-minute quadrangles in the Trenton area and in the Woodstown and Sandy Hook quadrangles (fig. 6). Some data are from the engineering soil mapping program cited above and from a detailed study by the junior author of the Quaternary deposits in the Woodmansie area (see fig. 6).

All photographs, except those credited to Aero Service Corporation, were taken by the senior author.

PREVIOUS INVESTIGATIONS

The glacial geology of New Jersey received little attention during the early years of the State geological survey, authorized in 1834. Some discussion of the glacial deposits was presented in the annual reports of 1877, 1878, and 1880 (Cook, 1877, 1878, 1880). In 1891, R. D. Salisbury began work on the surficial deposits, and a comprehensive report was published in the Annual Report of the State Geologist for the year 1893 (Salisbury and others, 1894). Progress reports on the surficial deposits were made in each annual report for the years 1894 through 1898 (Salisbury and others, 1894; Salisbury, 1895, 1896, 1898, 1899; Salisbury and Knapp, 1897). In 1902 the detailed report by Salisbury and others, on the glacial geology of New Jersey was published.

In 1917, "The Quaternary formations of southern New Jersey," by Salisbury and Knapp, was published. This report was based on field work (largely by Knapp) completed in 1903; much of it was done concurrently with studies of the glacial geology.

A report on an important phase of the glacial geology, a detailed description of glacial Lake Passaic, was given by Salisbury (Darton and others, 1908, p. 14–20). The existence of this lake was discussed by Cook in 1880 (p. 61–64), and a detailed description was given later by Salisbury and Kümmel (in Salisbury and others, 1894; 1895).

Detailed maps and a description of the surface deposits were made for several areas in the State: Passaic quadrangle, N. J.-N. Y. (Darton and others, 1908); Franklin Furnace quadrangle, N. J. (Spencer and others, 1908); Philadelphia district (Bascom and others, 1909a); Raritan quadrangle (Bayley and others, 1914). Only a few reports on the Quaternary geology of New Jersey have appeared since the productive period of 1902–17. They include reports by Leverett (1928) and MacClintock and Richards (1936); other short papers will be mentioned where applicable.

PHYSIOGRAPHY AND GENERAL GEOLOGY

Although a small State, New Jersey has varied and complex physiography and geology. Four physiographic provinces are represented, Valley and Ridge, Highlands, Piedmont, and Coastal Plain (fig. 1). More than 60 formations and all periods of geologic time, except the Carboniferous, Permian, and Jurassic, are represented. The formations generally trend northeast; this trend is evident in the orientation of the physiographic provinces, drainage patterns, and direction of major lines of communication.

Each physiographic province is characterized by rocks of similar origin, type, or age. The Valley and Ridge province, in the northwest part of the State (fig. 1) is characterized, as its name suggests, by a parallel series of valleys and ridges. The ridge farthest northwest is Wallpack Ridge, which is parallel and nearly adjacent to the Delaware River from Port Jervis to Wallpack Bend (fig. 2). The ridge is held up by Upper Silurian and Lower Devonian limestones and sandstones and is flanked on both sides by less resistant shale, siltstone, and calcareous shale. Closely paralleling Wallpack Ridge to the southeast is Kittatinny Mountain (fig. 3). This is the highest and most prominent ridge in the State.

It is underlain by predominantly hard Silurian Shawangunk Conglomerate which forms a partly bare-crested ridge, with a steep southeast face, from the New York border to Delaware Water Gap (fig. 2). A subordinate secondary ridge on the dip-slope (northwest) of Kittatinny Mountain is underlain by red and gray Silurian sandstone and shale (High Falls and Bloomsburg Formations). Wallpack Valley lies between Wallpack Ridge and Kittatinny Mountain and is underlain by Devonian and Silurian limestones and shales. A broad valley, Kittatinny Valley, lies southeast of Kittatinny Mountain and is underlain largely by Kittatinny and Jacksonburg Limestones and Martinsburg Shale of Cambrian and Ordovician age. Lower Cambrian Hardyston Quartzite crops out locally in thin bands along the base of the Kittatinny Limestone.

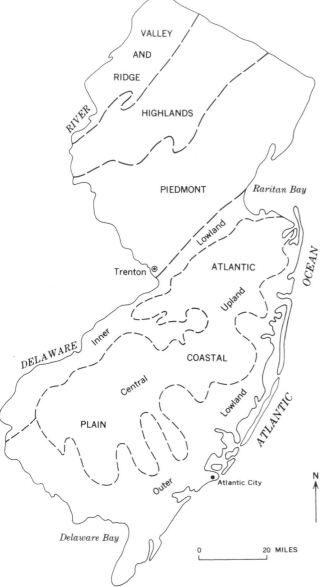

Figure 1. Physiographic provinces of New Jersey.

Nepheline syenite is exposed between the Shawangunk Conglomerate and Martinsburg Shale along the base of Kittatinny Mountain in northern Sussex County (figs. 2, 3).

The Highlands province (Reading Prong), in the central part of northern New Jersey, is underlain chiefly by Precambrian gneisses which form an irregular hilly terrane. Several long narrow valleys trend northeast across the province. These are underlain by fault-block inliers of Paleozoic Kittatinny Limestone and Martinsburg Shale, outcrop belts of Precambrian Franklin Limestone, and infolds of Devonian sandstone and shale. Prominent ridges of Devonian and Silurian quartzite con-

glomerates border one valley in northern Morris and northwestern Passaic Counties (fig. 2).

The Piedmont lies southeast of the Highlands province and comprises the east half of northern New Jersey. This province is underlain by shale, siltstone, sandstone, conglomerate, basalt, and diabase of the Triassic Newark Group. The sedimentary rocks of this group form a broad southeast-sloping piedmont plain. Above this surface are conspicuous ridges of basalt (Watchung Mountains in Bergen, Passaic, Essex, Morris, Union, and Somerset Counties) and diabase (Palisades along the Hudson River and Sourland Mountain to the southwest a few miles north of Trenton – fig. 2). These three provinces, Valley and Ridge, Highlands, and Piedmont, constitute nearly half the land area of New Jersey. The remainder of the State is in the Coastal Plain, a broad sandy plain sloping gently southeast to the Atlantic Ocean. In this report, the Coastal Plain is divided into three subprovinces, an inner lowland, a central upland, and an outer lowland (fig. 1). The boundary of the cen-

Figure 2. Map of New Jersey showing locations of terminal and recessional moraines, and counties and towns mentioned in the text.

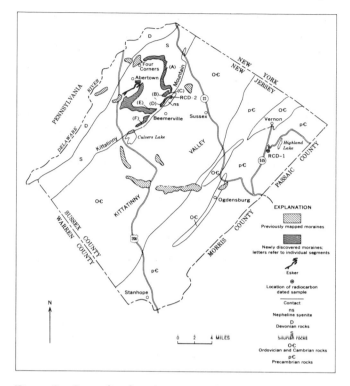

Figure 3. Generalized geologic map of Sussex County, N. J., showing location of end moraines and eskers on Kittatinny Mountain and their relations to moraines in the valleys on either side.

tral upland lies approximately at an altitude of 50 feet, except for a short segment east of Trenton. The Coastal Plain province is underlain almost entirely by unconsolidated marine and fluviatile clay, silt, sand, and gravel of Late Cretaceous and Tertiary age. An abundance of glauconite is a distinctive characteristic of the formations underlying the inner lowland and inner edge of the central upland. Much of the Coastal Plain province is covered by Quaternary deposits (largely fluviatile), particularly in the inner and outer lowlands and the southern one-third of the central upland.

QUATERNARY GEOLOGY OF NEW JERSEY

The Quaternary deposits of New Jersey are in two broad regions; the northern glaciated part of the State and the southern unglaciated part. The northern region includes the area north of the Wisconsin terminal moraine and the area south of the moraine which is north of the latitude of Perth Amboy (fig. 2) and which probably was invaded by pre-Wisconsin glaciers. The southern region is that part of the State south of the latitude of Perth Amboy which lay in the periglacial zone. Two or possibly more stages of glaciation probably are represented in New Jersey. Glaciation of northern New Jersey during Wisconsin time is accepted without ques-

tion. Opinions differ, however, as to the extent and duration of an ice sheet or sheets of pre-Wisconsin age in New Jersey.

GLACIATED NORTHERN REGION

THE WISCONSIN GLACIATION

Glacial deposits of Wisconsin age cover large areas of northern New Jersey north of the terminal moraine. This moraine forms a nearly continuous ridge across the State from Perth Amboy north to Dover and west to Belvidere (fig. 2). The entire area north of the terminal moraine apparently was overridden by ice during Wisconsin time; the crests of the highest ridges, including Kittatinny Mountain, show glacial scour and striae. The general direction of ice advance was southwest (Salisbury and others, 1902, pl. 8). There were local variations: the ice moved nearly due west over Kittatinny Mountain; due south between Belvidere and Netcong; and southeast across the Hudson River onto Long Island (fig. 2). The thick broad band of the terminal moraine across the State indicates that the ice front remained relatively stagnant for a considerable period of time.

After deposition of the terminal moraine, the ice front apparently withdrew for some miles where it again remained stationary and a recessional moraine formed. This moraine, in Union and Essex Counties, is about 7 miles long and is about 10 miles north of the most southerly loop of the terminal moraine. Apparently simultaneously, a discontinuous ridge or a series of linear deposits was laid down across nearly the entire width of Sussex County (figs. 2, 3), about 20 miles north of the terminal moraine across Warren County. After formation of this moraine the ice front withdrew farther to the north and once again remained relatively stationary. Small segments of an end moraine were deposited in Bergen County about 20 miles farther north than the moraine in Union and Essex Counties. In Sussex County the ice front withdrew only about 4 or 5 miles and was then stationary while another or second recessional moraine was deposited north of the earlier or first recessional moraine (figs. 2, 3).

Study of aerial photographs indicates that several small segments of end moraines are in northern Morris and Passaic Counties lying between the southern morainal belt in Sussex County and the small end moraines in northern Bergen County. The presence of such a nearly continuous morainal belt across northern New Jersey, well north of the terminal moraine, suggests that the ice front was stagnant for a considerable time, possibly after a readvance representing a major subdivision of Wisconsin time.

Radiocarbon dates (RCD) from two localities pertain to these recessional moraines. One date (RCD-1) was obtained by Meredith Johnson from a peat bog in the northeastern part of Sussex County (fig. 3) several miles north of the recessional moraine (Minard, 1959, p. 1208). The sample was dated at 10,890 ± 200 years B. P. A second date (RCD-2) was obtained from a sample of peat taken by the senior author from a bog, on Kittatinny Mountain, which is partly dammed by segment (D) (figs. 3, 22) of the newly discovered end moraines (Minard, 1961). This sample, obtained from a depth of 9 feet, dated at 6,260 ± 300 years B. P. (Meyer Rubin, written commun., October 1968, U. S. Geological Survey Radiocarbon Laboratory sample #W-2200). A third date was obtained from a sample taken (with the help of Cornelia Cameron) from a depth of 17 feet in the same bog as sample #W-2200. This sample dated at 7800 ± 650 yrs. B. P. (Meyer Rubin, written commun., December 9, 1968, U. S. Geological Survey Radiocarbon Laboratory sample #W-2236).

Much of the surface north of the terminal moraine is blanketed by a sheet of till of variable thickness.

In addition sizable deposits of stratified drift (kames, kame terraces, kame fields, eskers, and deltas) cover large parts of the valley bottoms. Classic examples of these features can be seen in many places, particularly in the northwest and north-central parts of the State. Flint 1947 (p. 269) suggests that two subdivisions of the Wisconsin are represented in New Jersey, both probably early Wisconsin in age (table 1). Because the glacial deposits in New Jersey are more weathered than those in northern New England they have been assigned an early Wisconsin age. The continuous terminal moraine from Perth Amboy to Belvidere (fig. 2) may represent the southernmost advance of the glaciers during Wisconsin time. The discontinuous recessional moraine(s) across the State well to the north of the terminal moraine (fig. 2) may represent the static front(s) during later Wisconsin time.

EARLY DRIFT

Drift that is probably pre-Wisconsin in age is present, in a belt 15 to 25 miles wide, south of the terminal moraine in the area west of Morristown (fig. 2). This drift was referred to by Salisbury (1893, p. 60, and Salisbury and others, 1894, p. 73; 1902, p. 187) as extra-morainic drift and by Bayley

TABLE 1. HISTORY OF CORRELATION OF THE MAJOR PLEISTOCENE DEPOSITS OF NEW JERSEY (TIME DIVISIONS ARE THOSE GIVEN BY FLINT, 1947, FOR THE MISSISSIPPI RIVER BASIN).

New Jersey					Mississippi Basin
Salisbury and others (1902)	Lewis and Kümmel (1915)	Salisbury and Knapp (1917)	Kümmel (1940)	MacClintock and Richards (1936); MacClintock (1940)	Flint (1947)
Post Kansan Glacial (Wisconsin)	Cape May	Cape May		(Probably not represented)	Mankato (Wisconsin)
					Cary
				Later Wisconsin	Tazewell
				Earlier Wisconsin (Budd Lake drift)	Iowan
Early-Drift(?)	Pensauken		Cape May	Cape May fm.	Sangamon
Early Drift(?)				Illinoian till | Pensauken fm. (in part)	Illinoian
	Bridgeton(?)			Pensauken fm.	Yarmouth
			Jerseyan	Kansan till | Bridgeton fm. (in part)	Kansan
		Pensauken	Pensauken	Bridgeton fm.	Aftonian
	Jerseyan	Bridgeton	‹Bridgeton›		Nebraskan
				Beacon Hill gravel	[Pliocene] Citronelle fm.

and others (1914, p. 16) as Jerseyan Drift although it was shown on the surficial map of that report as extra-morainic drift. Lewis and Kümmel (1912) show it on the State geologic map as Early Drift and refer to it in their accompanying report (1915, p. 73–74) as Jerseyan (Early) Drift and describe it as follows: "Glacial drift, both stratified and unstratified, greatly antedating the moraines of the last, or Wisconsin, stage of glaciation, occurs in discontinuous patches south of the latter to a maximum distance of 23 or 24 miles. In the Highlands it is thicker and more continuous in the wider valleys than on the ridges, while on the Triassic (Piedmont) plain it caps isolated and somewhat flat-topped hills in relations that indicate prolonged erosion since its deposition.

"The great age of at least some portions of this drift is indicated by the fact that since it was formed the main streams have sunk their channels 100 foot and opened wide valleys on extremely gentle gradients. The complete oxidation and leaching that it has undergone and the disintegration of even large bowlders of gneiss and granite deep within its mass are other evidences of its great age." Lewis and Kümmel believed the drift was probably equivalent in age to the Nebraskan of the Mississippi Valley (p. 74). Leverett (1928) correlated part of the Jerseyan with the Illinoian and part probably with the Kansan. Kümmel (1940, p. 153) states that "the drift older than the Wisconsin is not all of the same age." He notes (p. 151) "In New Jersey the glacial deposits are now believed to belong to three widely separated epochs, or stages, Jerseyan (oldest), Illinoian, and Wisconsin (youngest)."

Although Kümmel (1940, p. 153) gives Leverett credit for assigning an Illinoian age to part of the Jerseyan Drift, Salisbury and others (1902, pl. 28) much earlier mapped two different sets of glacial deposits south of the terminal moraine. They classified the first belt south of the terminal moraine as "Extra-morainic glacial drift of undermined [undetermined] age probably in part antedating the moraine." He classified the scattered deposits over a broad area to the south of this belt as "Bodies of extra-morainic drift mainly glacial, much older than the moraine." This same belief in two separate sets of deposits and ages for the Jerseyan Drift was repeated by Bayley and others (1914, p. 16). That the early drift is older than the Wisconsin is attested to by its more advanced weathering and patchy distribution. Large masses of early drift (including all pre-Wisconsin glacial drift) near Basking Ridge (fig. 2) were regarded by Lev-

erett as Illinoian but as Jerseyan (Kansan) by MacClintock (1940, p. 113) on the basis of a statistical analysis of the weathered boulders. Kümmel (1940, p. 154) points out that MacClintock (1940, p. 113) "has determined that if the amount of weathering of gneiss stones in the Wisconsin drift is represented by the figure 42, that of the Illinois drift in New Jersey is 64, and the Jerseyan (Kansan) is 75."

An identifiable terminal moraine is not associated with the Jerseyan Drift. The southern margin of the Wisconsin till sheet is marked by a prominent, nearly continuous, conspicuous morainal ridge (fig. 4), whereas the front of the Jerseyan Drift is an indefinite zone of patchy thin deposits. The lack of a terminal moraine may, in part, be due to subsequent erosion but is more probably due mostly to the fact that the ice front did not remain stationary long enough for a conspicuous end moraine to form. Part of the Jerseyan Drift near the Wisconsin terminal moraine, and at nearly the same altitude, may be, in part, material deposited when the ice front briefly overran the Wisconsin terminal moraine. But this does not seem to be true for all deposits in front of the terminal moraine, because many are completely separated from the moraine and are at considerably higher altitudes (up to several hundred feet higher).

UNGLACIATED SOUTHERN REGION

The southern region is that part of the State south of the latitude of Perth Amboy at the head of Raritan Bay (fig. 2). Although this region was not glaciated it was close enough to the ice mass to have been strongly effected by it (figs. 5A, 5B). The deposits in this part of the State contrast sharply with those in the northern part. The hummocky, kettle and kame topography of the glacial deposits contrasts markedly with the relatively flat topography of the flood plain, natural levee, channel, and fan deposits of the south.

The deposits in the northern part of the State are chiefly ice-contact deposits; those in the south were deposited chiefly by melt water flowing from the ice during periods of glaciation, or by streams during interglacial periods. Some deposits in the southern region, particularly along the coast and Delaware Bay, accumulated in a marine environment.

HISTORY OF INVESTIGATIONS

Interpretation of the environments of depositions and of the age and facies relations of the Quaternary formations of southern New Jersey is

Figure 4.   Vertical aerial photograph (scale 1:20,000) of the Wisconsin terminal moraine, in an arc just north of Budd Lake, N. J., showing conspicuous kettle-kame topography.   Contrast glacially scoured, jointed, gneissic bedrock in upper left with the same unglaciated saprolitic rock in lower left.   Photograph by Aero Service Corporation, winter 1939.

A   B

Figure 5. A, Congeliturbate structures in the upper part of the Magothy Formation of Late Cretaceous age, beneath horizontally stratified Pleistocene sand and gravel at Parsonville in the periglacial area near Trenton. B, Sand-blasted block of ironstone from the Beacon Hill Gravel of Pliocene(?) age. This and similar blocks are near the top of the highest hill in the Roosevelt quadrangle. Note the flutes and grooves, and faceted and polished surfaces separated by sharp ridges.

not easy. The discontinuous, irregular, unconsolidated glacial deposits in the northern part of the State are distinctly different and readily separated from the continuous, regular, consolidated, underlying bedrock formations. Outcrops of bedrock are abundant, large, and fresh. In contrast, in the southern part of the State unconsolidated sediments, of Cretaceous and Tertiary age, which underlie the Quaternary deposits, closely resemble them in composition, internal structure, and weathering, and it is often difficult therefore to distinguish Quaternary deposits from the older sediments.

The difficulty of mapping these deposits was recognized early by Salisbury (Salisbury and Knapp, 1917, p. 1). Field work for this report was begun in 1891 and essentially finished in 1903. Salisbury notes on page 1 that Knapp was responsible for most of the details in the report and that "only those who have tried to unravel the history recorded in such doubtful terms, as in the surface sands and gravels of this region, can appreciate the painstaking effort devoted by Mr. Knapp to the study."

The first major report on the Quaternary of southern New Jersey was by Salisbury (1893, p. 151–166) who, in his description of the "yellow gravel," included nearly all material considered to be of Quaternary age. There was no attempt in this report to divide the "yellow gravel" into units or assign names. The following year, however, the

"yellow gravel" was divided into four units (Salisbury, and others, 1894, p. 39–52).

The oldest of these units was called the first phase (Salisbury and others, 1894, p. 39) and included three types of material (p. 41): (1) coarse gravel, (2) fine gravel, and (3) sand. The coarse gravel (p. 42) is what is now mapped as the Beacon Hill Gravel, named for deposits on Beacon Hill in the northwest part of the Marlboro quadrangle, midway between the Roosevelt and Sandy Hook quadrangles (fig. 6). It also occurs on the hills west and northeast of Clarksburg in the Roosevelt quadrangle (fig. 6). The base of the gravel is at an altitude of about 355 feet on Beacon Hill and ranges from 340 to 310 feet in altitude in the Roosevelt quadrangle (Minard, 1964). This gravel is now considered to be of Pliocene (?) age. The fine gravel occurs on Juliustown Ridge and Arney's Mount in the southern part of the Columbus quadrangle and on Stony Hill in the north-central part of the New Egypt quadrangle (fig. 6). It is now included in the Cohansey Sand of Miocene (?) and Pliocene (?) age and the basal Kirkwood Formation of Miocene age (Lewis and Kümmel, 1912; Owens and Minard, 1962; and Minard and Owens, 1962). The sand, which occurs at Clarksburg in the Roosevelt quadrangle (fig. 6), is now mapped as Cohansey Sand and Kirkwood Formation ("fluffy sand" of Salisbury and others, 1894, p. 43) (Lewis and Kümmel, 1912; and Minard, 1964). Shortly after publication, Salisbury apparently became aware that the interpretation of the first phase was in error, and in later publications these units were mapped correctly (Bascom and others, 1909b; Lewis and Kümmel, 1912).

Salisbury and others (1894, p. 57) called the second unit of the "yellow gravel" the second stage and named it the Pensauken Formation for exposures in Hyltons pits on the south side of Pensauken Creek (Burlington-Camden County line, fig. 2) near where it empties into the Delaware River. The unit was named for what later was designated the arkose phase of the Pensauken. This phase included sand and gravel derived from the Beacon Hill Gravel; clay, greensand, "firesand," and quartz sand from the Cretaceous beds and possibly from the Miocene; arkose from pre-Cretaceous Triassic sandstone, shale, and trap, as well as granite, schist, conglomerate, and gabbro (Salisbury, and others 1894, p. 58).

Salisbury and others (1894, p. 63) called the third unit the third "yellow gravel" formation and named it the Jamesburg loam (1894, p. 63) in reference to the silt and loam in the vicinity of Trenton and to the north, and probably would have named it Trenton except that name was preoccupied. Therefore he assigned the name Jamesburg from a locality northeast of Trenton where the unit was similar. He noted that Jamesburg was but a local name for the Columbia formation, or a part of it, which was the name given to the old drift lying south of the terminal moraine (Salisbury and others, 1894, p. 67). Later the Jamesburg was considered correlative with the younger Cape May Formation and the term was discontinued (Wilmarth, 1938, p. 1039).

Salisbury and others (1894, p. 70–71) called the fourth unit the fourth "yellow gravel" formation and considered it younger than the Columbia Group and possibly correlative with the last glacial period, or younger.

Salisbury considered the Pensauken trough (Perth Amboy-Salem, fig. 2) to have been submerged during the deposition of the material within it, and that the main sources for the material was from the Delaware River, Schuylkill River, and Raritan River (then flowing southwest instead of into Raritan Bay), and Hudson River or Hackensack valley stream(s) in that area (Salisbury, 1895, p. 115). At that time he considered the Pensauken Formation to be equivalent in age to the Lafayette Formation (of previous usage) and to pre-date any known glaciation on the North American continent (1895, p. 117). Later Salisbury and Knapp (1917, p. 78) suggested that the Pen-

sauken may or may not be older than the older glacial drift. MacClintock and Richards (1936, p. 299) state that Salisbury and Knapp (1917) tried to correlate the Pensauken with the Jerseyan Glaciation but that the presence of warm water flora casts grave doubts on a glacial age (p. 300). Fuller (1914, p. 223) correlated the Pensauken with the Jameco Gravel of Long Island, which suggests a pre-Yarmouth age (table 2). Table 1 shows the different ages and stratigraphic assignments for the Pensauken by earlier authors.

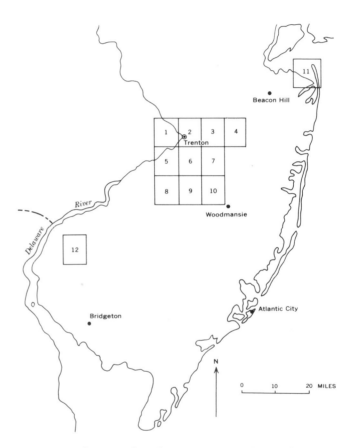

Figure 6. Index map of southern New Jersey showing locations of 12 recently mapped 7½-minute quadrangles and localities mentioned in the text. The field trip on the second day is in the central block of 10 quadrangles. Geologic maps for all but Trenton West have been published by the U. S. Geological Survey:

1.	Trenton West	J. P. Owens (unpub. data)
2.	Trenton East	Owens and Minard (1964b)
3.	Allentown	Owens and Minard (1966)
4.	Roosevelt	Minard (1964)
5.	Bristol	Owens and Minard (1964c)
6.	Columbus	Owens and Minard (1962)
7.	New Egypt	Minard and Owens (1962)
8.	Mount Holly	Minard and others (1964)
9.	Pemberton	Owens and Minard (1964a)
10.	Browns Mills	Minard and Owens (1963)
11.	Sandy Hook	Minard (in press)
12.	Woodstown	Minard (1965)

TABLE 2.   CORRELATION OF PLEISTOCENE TIME DIVISIONS (FROM MAC CLINTOCK AND RICHARDS, 1936, P. 335).

Mississippi Valley		Long Island	New Jersey	
Wisconsin	Mankato	Wisconsin	Wisconsin	
	Cary	Harbor Hill		
	Tazewell	Ronkonkoma		
	Iowan	Manhasset		
Sangamon		Jacob Gardiners	Cape May	
Illinoian			Illinoian	
Yarmouth		Part of Gardiners (?)		Pensauken-Bridgeton Complex
Kansan		Jameco (?)	Jerseyan (?)	
Aftonian				
Nebraskan				

In their final report (1917), Salisbury and Knapp subdivide the Pleistocene deposits of southern New Jersey into three formations. These are, from oldest to youngest, the Bridgeton, Pensauken, and Cape May. These units are still the standard subdivisions of the Pleistocene of southern New Jersey (table 2). In some of the early reports it is difficult to determine what ages were given to these formations; within a single report, an age or correlation is definitely given for one unit, suggested or inferred for a second unit, and not given for a third unit. It seems that the three formations were once intended to be correlated either with the glacial periods, or partly with glacial periods and partly with interglacial periods, or entirely with interglacial periods. Probably any modern interpretation would agree wholly or in part with that of MacClintock and Richards (1936) and MacClintock (1940) (table 1).

MAJOR FORMATIONS

Bridgeton Formation

The Bridgeton Formation was probably deposited by streams flowing from the north and northwest. At present it covers large areas of the surface in the south-central and southwestern parts of the State, and small isolated areas at least as far north as the Trenton vicinity (Bascom and others, 1909b, p. 15).

The formation was divided by Salisbury and Knapp (1917, p. 18–25) into two phases, the Glassboro and Woodmansie. They considered the deposition of the two phases to be more or less contemporaneous but that the Glassboro was characterized by material brought in largely from the crystalline and older rocks to the north and deposited in the trough from Perth Amboy to Salem, whereas the Woodmansie phase, was deposited on the coastal plain southeast of the trough and north of the Mullica River, and was derived from the local unconsolidated coastal plain formations. Salisbury and Knapp (1917, p. 1–10) considered the Bridgeton to be the oldest and highest (at any particular locality) of the three Quaternary formations of southern New Jersey. Earlier, part of the Pensauken may have been considered Bridgeton in the Trenton area and to the northeast. What finally was mapped as Bridgeton (Bascom and others, 1909b, and Lewis and Kümmel, 1912) is the gravel that caps the higher hills below the Beacon Hill Gravel. The altitude of the base of the Beacon Hill Gravel gradually declines south and southwest from about 355 feet at Beacon Hill to 340 and 310 feet near Clarksburg in the Roosevelt quadrangle, and about 180 feet at Woodmansie (fig. 6).

The base of the Bridgeton Formation, however, is at an altitude of about 200 feet in the eastern part of the Columbus quadrangle and at 150 feet in the southeast corner of the Browns Mills quadrangle. The altitude of the base of the formation declines quite uniformly towards the south where near its southern limit at Bridgeton, it is only at an altitude of about 60 feet.

The deposits mapped as Bridgeton have internal structures suggesting fluviatile deposition: horizontal gravel beds, cross stratification in coarse beds, and lenses of gravel. Feldspar and chert are weathered and the deposits are quite silty. In most exposures the formation ranges in thickness from a few feet to about 20 feet. Locally, gravelly parts of the underlying Cohansey Sand were previously included in the Bridgeton. The fluviatile, arkosic Bridgeton is, however, usually more silty and weathered than the clean, quartz sand of the Cohansey.

Pensauken Formation

The Pensauken Formation of Salisbury and Knapp (1917) is more restricted than the Pensauken of Salisbury (1894). The bulk of what is presently mapped as Pensauken (Lewis and Kümmel, 1912) lies within the Perth Amboy-Trenton-Salem trough. It occurs below the older Bridgeton Formation and above the younger Cape May Formation and is believed to have been deposited by a stream system which may have been the southwest extension of the Hudson River, joined by the Raritan, Delaware, and Schuylkill from the northwest. This deposition was believed by Salisbury and Knapp (1917, p. 68) to have followed a long period of erosion after Bridgeton deposition. They favor an age for the Pensauken that ranges from older than to nearly contemporary with the old drift (1917, p. 78).

The altitude of the base of the Pensauken is at about 100 feet in the Trenton area and declines to the southwest so that at Swedesboro, in the northern part of the Woodstown quadrangle (fig. 6), it is only a few feet above sea level. In the type area, at the mouth of Pensauken Creek, the base of the Pensauken is also nearly at sea level (Lewis and Kümmel, 1912). Some of the material mapped as Pensauken at lower altitudes may be younger sediments reworked from the Pensauken.

Campbell and Bascom (1933), in an attempt to explain the source of the formation, considered the increasing abundance of chert toward the northeast and suggested some major shifts in drainage. They referred primarily to the black chert which was believed to have come from the Kittatinny Limestone (p. 304–305). To transport the chert, and other large volume of sediment, to the trough between Perth Amboy and Trenton, they suggested that the Delaware River flowed southeast across Kittatinny Mountain through Culvers Gap (23 miles northeast of Delaware Water Gap, fig. 2). From there it flowed across the northern part of the State into the upper Raritan River drainage basin where it was joined by the Hudson River which flowed approximately parallel to its present course but to the northwest. From their juncture the two rivers flowed as one to the southwest and deposited the coarse Pensauken sediments in the broad trough (Campbell and Bascom, 1933, p. 310–312) of the Inner Lowland (fig. 1).

Howell and Hale (1946) support the Campbell and Bascom theory because of the presence of fossiliferous chert of Devonian age in the Pensauken at Princeton, 10 miles northeast of Trenton (fig. 2).

The Pensauken is arkose, quartzose, and glauconitic (Salisbury and Knapp, 1917, p. 79–83). Where arkosic it is relatively nonglauconitic, where glauconitic it is strongly quartzose and does not contain arkose. The boundary between the arkosic and glauconitic phases nearly corresponds with the inner edge of the glauconite belt (Salisbury and Knapp, 1917, p. 83). The formation (1917, p. 88) is from 10 to 70 feet thick. Minard augered a hole through the formation near New Canton in the central part of the Allentown quadrangle. Seventy-two feet of Pensauken was penetrated; the base was at an altitude of 85 feet.

The Pensauken is at a lower altitude than the Bridgeton and has a more varied composition. The Pensauken and Bridgeton both contain quartz, feldspar, and silt sized material, but only the Pensauken contains glauconite, ironstone, and abundant rock fragments from Triassic bedrock and quartzites from the older formations (High Falls, Shawangunk, and Hardyston). The Pensauken is poorly to well stratified and is both horizontally bedded and crossbedded. These features indicate the generally fluviatile origin of the formation.

Cape May Formation

The Cape May Formation is the youngest Quaternary formation defined by Salisbury and Knapp (1917). Its present distribution, although less widespread than previously, is very similar to its original distribution. The Cape May occurs along most present streams and rivers as bordering terraces, natural levees, and overbank deposits. It is thickest and most continuous along the Delaware River from Trenton south to the Atlantic Ocean and along the lower reaches of all the major streams flowing into the Delaware River, Delaware Bay, south side of Raritan Bay, and the Atlantic Ocean. The formation almost completely blankets Cape May County, its type locality, at the southern tip of New Jersey.

At any one locality it is the lowest in altitude of the three Quaternary formations of southern New Jersey. Its base is mostly at an altitude of less than 50 feet except along the upper reaches of some streams. It is generally from 10 to 30 feet thick but is more than 150 feet thick in Cape May County (Gill, 1962, p. 22–31). Salisbury and Knapp (1917, p. 162) correlated the Cape May with the Wisconsin Glaciation but MacClintock and Richards (1936, p. 317) correlate it with the Sangamon Interglaciation.

Of the three Quaternary formations in southern New Jersey, the Cape May is the only one about which a recent comprehensive report has been written. Gill (1962) made a thorough study of the formation in its type locality in Cape May County. He divided it into four facies. From oldest to youngest, these are: estuarine sand facies, estuarine clay facies, marine sand facies, and deltaic sand facies. Gill (1962, p. 26) correlates the estuarine clay facies with the Gardiners Clay of Long Island, the Philadelphia Clay of Richards (1960), and the Fish House Clay of Woolman (1897, p. 201–244). The radiocarbon date for the overlying deltaic sand facies is in excess of 35,000 years B. P.

The Cape May is readily differentiated from the two older Quaternary formations. It is at a lower altitude at any one locality, and shows little weathering. The sand and gravel is mostly clean and hard. The formation lacks the weathered chert and feldspar and soft Triassic rock fragments of the Bridgeton and Pensauken. Along the rivers flowing into the ocean and Delaware Bay, the Cape May Formation was derived largely from the quartz sands of the Miocene formations and from the older Quaternary units. Only the hard materials survived intact and the soft weathered material was comminuted and deposited as silt and clay, chiefly in distinct layers. Along the Delaware River and streams flowing into it, a greater variety of material is present in the Cape May Formation, but it is still fresh and relatively unweathered as compared with the Bridgeton and Pensauken Formations. Much glauconite is

present locally along the inner edge of the coastal plain (Inner Lowland, fig. 1).

The Cape May is well stratified, crossbedded and horizontally bedded, to nearly massive. Both fluviatile and marine structures are present. Marine fossils are common in Cape May County. The marine topography of the Cape May was discussed by MacClintock (1943).

PRESENT INVESTIGATIONS

In 1957 a mapping project was begun in the lower Delaware Basin by James P. Owens and the senior author, both of the U. S. Geological Survey. Geologic maps were made of twelve 7½-minute quadrangles; eleven have been published (fig. 6). The Quaternary units were mapped in the Woodstown, Roosevelt, and Sandy Hook quadrangles, but only the pre-Quaternary units were shown on the other maps. In addition to the quadrangle maps many short papers were published and several final papers are in preparation. Owens and Minard have prepared a map of the Quaternary deposits of nine quadrangles in the Trenton block (all but the Roosevelt; fig. 6) at a scale of 1:48,000 (J. P. Owens and J. P. Minard, unpub. data). The Quaternary strata are mapped according to lithology, not according to altitude and inferred age relations as was done in most previous reports. Figure 7 shows the nine quadrangles with their major drainage net and several geographic locations in each as reference points. Figure 8 is a generalized geologic map of the pre-Quaternary lithic units in these quadrangles. Figure 9 shows part of the same area with the Quaternary as previously mapped. Figures 10–12 show the distribution of the Quaternary by lithologies. These three maps are modified from the 1:48,000 scale map cited above. Concurrently with the lower Delaware Basin geologic mapping project, the junior author was engaged in a geologic-hydrologic study in the Woodmansie area just southeast of the block of quadrangles shown in figure 6, which included deposits of the Bridgeton Formation.

Distribution of Quaternary units in the Trenton area

The Delaware River and its tributaries constitute the drainage system in the Trenton area (fig. 7). Comparison of the distribution of the pre-Quaternary (fig. 8) and Quaternary units (figs. 10–12) indicates the strong local association of Quaternary deposits with source formations in all but the northwest part of the nine-quadrangle

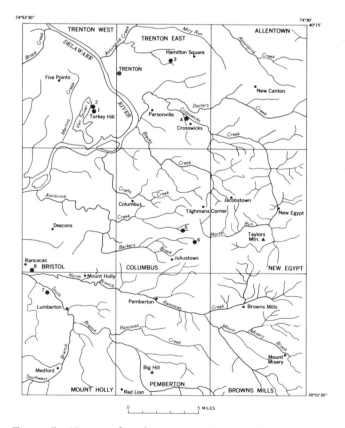

Figure 7. Nine-quadrangle area near Trenton showing major towns, drainage, and localities cited in the text. Numbered points are field-trip stops for the second day.

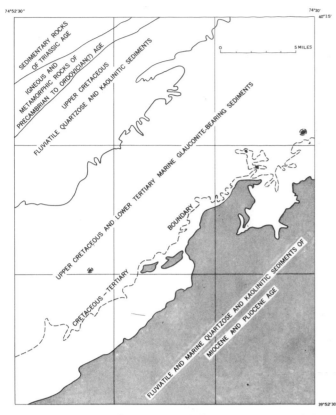

Figure 8. General distribution of the pre-Quaternary lithic units in the nine quadrangles near Trenton.

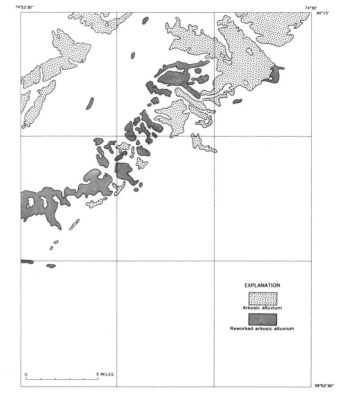

Figure 10. Distribution of arkosic alluvium and reworked arkosic alluvium near Trenton.

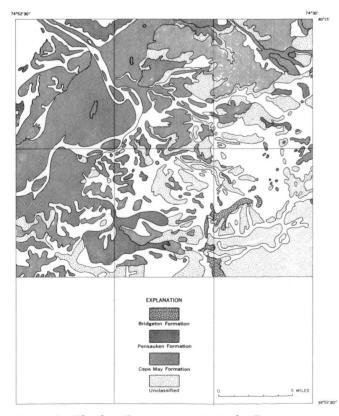

Figure 9. The four Quaternary units in the Trenton area as shown by Bascom and others (1909b).

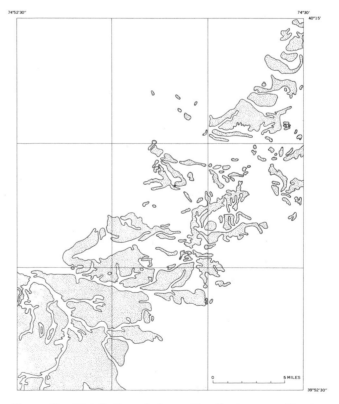

Figure 11. Distribution of glauconitic alluvium near Trenton.

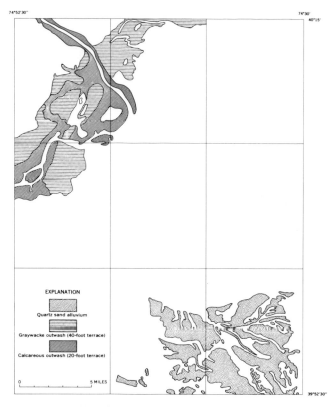

Figure 12. Distribution of quartz sand alluvium south of Trenton, and of graywacke and calcareous outwash along the Delaware River near Trenton.

area. The distribution of the arkosic alluvium, reworked arkosic alluvium, graywacke outwash, and calcareous outwash in the northwest part of the area (figs. 10 and 12) resulted from deposition of material from streams of the Delaware River and former Pensauken trough drainage system which flowed from the north and northwest. The glauconitic and quartz alluvium in the southeast part of the area (figs. 11, 12) were chiefly derived locally, from pre-Quaternary and older Quaternary formations, and deposited by streams flowing north and northwest.

Arkosic alluvium. The distribution of this unit is shown on figure 10. The arkosic alluvium was probably deposited throughout much of the length of the Pensauken trough.

It is the unit most closely associated with the Pensauken of Salisbury and Knapp (1917, p. 67–159). The source of the sediment was from the north-northwest and east. Its distinctive mineralogy is derived from the weathered igneous and metamorphic rocks of the Piedmont and Highlands and the arkosic Stockton Formation of Triassic age of the Piedmont (fig. 1) which lie to the north and

northwest. The base of the arkosic alluvium is generally above an altitude of 80 feet. The distribution of the arkosic alluvium within a long narrow trough, the discontinuous lenses of gravel, the clear-cut cross- and horizontal-stratification, and its channel-shaped cross profile, all indicate deposition in a fluviatile environment. The composition of the alluvium is varied and includes material and rocks from many older formations. Recognizable rocks are Precambrian gneiss, Hardyston Quartzite of Cambrian age, High Falls Formation and Shawangunk Conglomerate of Silurian age, and rocks from the Brunswick, Stockton, and Lockatong Formations, and conglomerate, basalt, and diabase, all of Triassic age. Erosion by the Delaware River and its tributaries in the vicinity of Trenton has reduced the arkosic alluvium to a remnant of its former extent (fig. 10).

Reworked arkosic alluvium. The distribution of this unit (fig. 10) is largely a result of the redeposition of the arkosic alluvium at lower levels. The unit is very similar in composition to the arkosic alluvium and has similar internal structures, indicating its fluviatile origin. It is differentiated from the arkosic alluvium chiefly by its lower altitude, somewhat less weathered appearance, and presence, locally, of rounded iron-oxide-cemented rock fragments. It was derived largely from the arkosic alluvium as streams cut down through it. In addition, there was an influx of other material, as is evident on Turkey Hill (fig. 7). The deposit on Turkey Hill contains boulders and large pieces of Triassic siltstone and sandstone, undoubtedly fresh material derived from upriver. The Turkey Hill deposit is a remnant of an areally larger deposit.

A radiocarbon data of 26,800 ± 1,000 years B. P. was obtained on peat from a bog in the base of the unit in the southeast part of the Trenton East quadrangle (Meyer Rubin, U. S. Geological Survey, written commun., 1962, sample W-1193). The altitude of the bog was about 60 feet.

Glauconitic alluvium. The distribution of the glauconitic alluvium in the Trenton vicinity is shown in figure 11. When compared with the distribution of the glauconite-bearing older sediments shown in figure 8, a close correlation is evident. Sediments constituting this unit are probably some of the most distinct and unique in the area. They are nearly everywhere characterized by conspicuous thin horizontal layers of sand (fig. 13). The stratification is emphasized by the marked

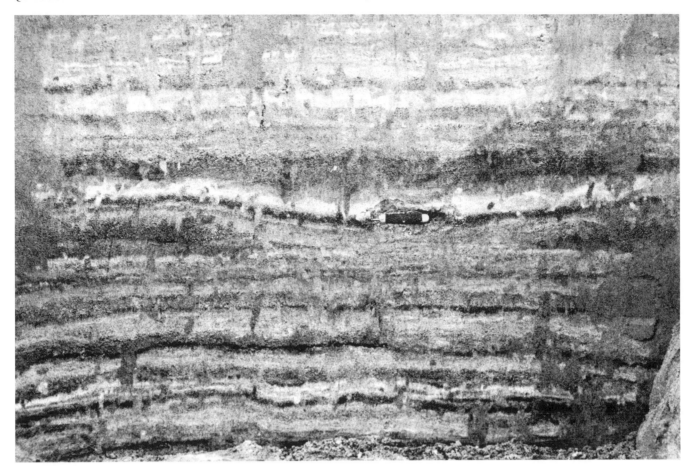

Figure 13. Thin alternating light and dark layers of sand and granules characteristic of the glauconitic alluvium. Note the numerous vertical borings filled by material from the layer above.

alternation of fine and coarse layers and layers rich in green glauconite as compared with layers less rich in that mineral. Except locally in basal layers, there is nearly a complete absence of material coarser than pebbles. The deposits range in altitude from near sea level to 200 feet above. In time and altitude the deposits correlate with both the Cape May and Pensauken Formations.

The uniform stratification and narrow range of sediment size (mostly fine sand to granules) suggest a fluviatile depositional environment of remarkable uniformity for most of this unit. Such uniform bedding can be seen in natural levee and bar deposits. The narrow size range of the sediment can be explained by the narrow sediment size range in the underlying and nearby coastal plain formations, from which the bulk of the material was derived.

Large areas of the glauconitic alluvium were mapped by Salisbury and Knapp (*in* Bascom and others, 1909b) as unclassified deposits (fig. 9) because of their similar lithology and also because of

the difficulty in attempting to correlate them in time and distribution with the other three major units, Bridgeton, Pensauken, and Cape May Formations. Different interpretations of the origin of their unclassified deposits (Salisbury and Knapp, *in* Bascom and others 1909b) include broad alluvial fans and aeolian deposits. The alluvial fan origin has merit for some of the slightly inclined uniformly bedded material, but the aeolian theory does not seem applicable because of the coarseness of some layers and the lack of much distinct cross-stratification. The glauconitic alluvium seems to have been deposited, most probably, by streams flowing from the sandy Coastal Plain areas.

Some of the glauconitic deposits, particularly in the New Egypt, Browns Mills, and Pemberton quadrangles (fig. 7, 8, and 11) lie on non-glauconitic pre-Quaternary quartz sands where the drainage seems to be from areas underlain by these same sands. Part of this is probably explainable by reversals or changes in drainage. A reversal seems evident in the upper reaches of Crosswicks

Creek in the New Egypt quadrangle (fig. 7). If most of the segment of Crosswicks Creek in the quadrangle flowed south, as is indicated by the tributaries south of Jacobstown, glauconite from the older formations could have been carried south and west and deposited on the quartz sand units. Subsequent headward erosion through the cuesta north of New Egypt could have resulted in the capture of the south-flowing drainage. The glauconitic alluvium then was left isolated on the Miocene quartz sands.

Quartz sand alluvium. Figure 12 shows that the quartz sand alluvium is limited to the southeast part of the area. Streams carrying the alluvium from the quartz sand provenance to the east and south, crossed onto glauconite-bearing formations at what is now the northern limit of the quartz sand alluvium. Beyond this point the streams carried a mixed sediment load and deposits of this material are now mapped as glauconitic alluvium. Much of the quartz sand unit would be correlated with the Cape May Formation, largely because of the low altitude of its base, mostly below an altitude of 60 feet, and because it is unweathered. A small area of the quartz sand alluvium near Mount Misery in the southeast part of the Browns Mills quadrangle was previously considered Bridgeton (Lewis and Kümmel, 1912). Its base is at an altitude of about 150 feet and the material is more silty and weathered than the bulk of the quartz sand alluvium.

Graywacke outwash. The distribution of this unit is shown in figure 12. The term graywacke is used to denote the heterogeneous composition of the sediments constituting this unit. It is a mixture of materials derived from the northwest and northeast, which is a source area of diverse bedrock. The graywacke outwash is characterized by clasts of quartz, chert, shale and slate, and gneiss, schist, amphibolite, diabase, and basalt. It is typically gray to reddish gray.

This unit generally underlies the 40-foot terrace level along the Delaware River near Trenton. The graywacke outwash was included in the Cape May of Lewis and Kümmel (1912) because of its altitude and location along the river. A radiocarbon date of $16,700 \pm 420$ years B. P. was obtained from a peat deposit in the unit (Meyer Rubin, U. S. Geological Survey, written commun., 1962, sample W-1151). The peat is in the north-central part of the Bristol quadrangle at an altitude of about 35 feet.

The graywacke outwash seems to have been partly deposited by the Pensauken drainage when the Delaware River was deflected to the southwest several miles above the present deflection point. This interpretation is based on the alinement of deposits to the southwest across the present course of the Delaware River.

Calcareous outwash. The distribution of this unit is shown in figure 12. It is part of the Trenton Gravels of Lewis (1881). The top of this unit is at an altitude of about 20 feet and its base is at least 10 to 15 feet below sea level. The unit forms the broad sandy-surfaced plain along the Delaware River near Trenton. It consists of well sorted gravel overlain by sand and silt.

The calcareous outwash is clean, fresh, and hard. In contrast with all the older deposits, it contains many limestone pebbles and cobbles as well as calcium carbonate as coatings on the underside of pebbles and as a cement between some pebbles. The deposit is typically leached to a depth of 6 to 8 feet; pebbles below this depth are commonly coated by calcium carbonate derived from above.

Except for the upper few feet, most of the material is well-sorted, well-rounded gravel ranging in size from pebbles to small cobbles; large cobbles and boulders constitute only a minor part of the deposit. Pebble lithologies include gneiss, sandstone, conglomerate, chert, argillite, and limestone. The material was probably almost exclusively deposited by the Delaware River and originated from bedrock and glacial deposits in the Delaware River drainage basin.

The Trenton Gravels of H. C. Lewis (1881) were mapped as the Cape May Formation by Lewis and Kümmel (1912). This unit was considered by Peltier (1959, p. 179) to correlate with the Valley Heads Drift of MacClintock and Apfel (1944), which is correlated by them with the Cary Stade of the Wisconsin Glaciation (p. 1162). The overlying sand may be a result of alluvial sand deposition during a rise in sea level following the Valley Heads episode. This sand is correlated by Peltier (1959, p. 184) with Wilson's Two Creek interval, which is considered by Lemke and others (1965, table 1, p. 20) as following the Mankato.

A radiocarbon date of $3,900 \pm 200$ years B. P. was obtained from peat from an auger hole in the 20-foot terrace in the southwest part of the Trenton West quadrangle (Meyer Rubin, U. S. Geological Survey, written commun., 1965, sample W-1539). This date may only indicate the age of the bog which can be considerably younger than the underlying gravel; possibly it accumulated in a younger channel.

ROAD LOG

FIRST DAY — NORTHERN NEW JERSEY

Mileage

0.0	Proceed south on Route 1 from the Holiday Inn Motel at Penns Neck, east of Princeton. Turn right at circle on Washington Street.
0.8	Cross Lake Carnegie, enter Princeton Campus area.
1.7	Left on Nassau Street.
2.1	Right on Route 206.
18.4	Junction 206 and 202 at Somerville circle, bear right.
19.0	Junction 22, bear right then left under 22. Proceed north on 202 and 206.
20.3	Junction Interstate 287. Turn left.
25.3	Road cuts through Watchung Basalt.
29.0	Interstate 287 temporarily ends at Mt. Airy Road in Basking Ridge, just south of junction with Whitenack Road. Proceed north on Mt. Airy Road.
30.6	Right on West Oak Street.
30.9	Cross over Interstate 287.
31.2	Left on Washington Street.
31.5	Right on Ridge Street.
31.7	Left on Finley Street.
32.0	Pass under Interstate 287, park along right side of road just past the underpass. Walk across railroad tracks to exposure. Jerseyan Drift, stop 1, on the southeast end of a hill at the junction of Interstate 287 and the Delaware, Lackawanna, and Western Railroad near the railroad station in Basking Ridge. Exposures are in cuts along the highway and railroad (fig. 14). The north end of the cut is ground moraine consisting mostly of clayey silty sand and gravel. Many pebbles, cobbles, and some boulders of crumbly, rotten gneiss are in this part of the deposit. Recognizable rocks include alaskite, amphibolite, and banded gneiss from the Precambrian Losee, Pochuck, and Byram Gneisses, as well as quartzite conglomerate derived either directly from the Silurian Green Pond Conglomerate to the north, or locally from the Triassic fanglomerate which contains clasts from the Green Pond.

The dark area shown in the upper central part of figure 14 is chiefly stratified sand, probably deposited by melt water. The material shown in the lower center of figure 14 is well-stratified alternating layers of silt to fine sand and gravel. These beds dip westward into the hill. This ex-

posure may represent a segment of end moraine of the early drift. The stratified material was deposited by melt water from the ice margin and the unsorted material accumulated at the ice margin at the base of the hill as melting balanced the rate of advance.

The material in the cutbank shown at the left edge of figure 14 consists mostly of Triassic shale and siltstone fragments.

Continue on Finley Street.

32.7	Junction 202. Turn right.
39.7	Square in Morristown. Continue around square on 202.
42.1	Under Delaware, Lackawanna, and Western Railroad in Morris Plains.
43.4	Junction Route 10. Turn right.
45.1	Junction Interstate 287. Turn left.
52.9	Temporary end of 287 at Montville. Turn right. Proceed on Main Street (202) into Montville.
53.8	Straight ahead on River Road (202 goes right).
54.0	Pass under Interstate 287.
54.3	Left into abandoned pit.

Sand and gravel pit in a kame on the south side of a hill just southeast of Montville, about 13 miles northeast of Morristown, stop 2. Most of the sand and gravel has been removed. The purpose of the stop is to observe the well-preserved glacial scour and striae on unweathered Triassic fanglomerate bedrock (fig. 15). Bedrock was exposed when the overlying well-stratified sand and gravel (fig. 16) was removed for construction use. Striae are distinct, particularly on the upper surfaces. The major direction of the striae is N.55°E. with minor striae 10 to 20 degrees to either side, indicating a predominant northeast to southwest movement of the ice. Beds of gravel are locally cemented by and individual pebbles completely coated with calcium carbonate. Rock types include gneisses, Green Pond Conglomerate, probable Devonian shale, amphibolite, greenstone to greenstone schist, and porphyritic nepheline syenite(?).

The Triassic bedrock is remarkably similar in composition to the overlying glacial deposit. This is one of a few locations where the Triassic contains an abundance of gneiss cobbles and boulders. The fault boundary between the Precambrian rocks of the Highlands and the Triassic rocks of the Piedmont trends northeast through the northern part of Montville about ¾ mile to the northwest. The material constituting the Triassic appears to be a fan containing many boulders derived from the nearby Highlands and

Figure 14. Jerseyan Drift at Basking Ridge, stop 1. Unsorted moraine deposit in far right; stratified silt, sand, and fine gravel in center and left.

deposited on the piedmont by swift-flowing streams from the mountain front. Many of the boulders within the Triassic bedrock have been cut in half and polished by the overriding ice. Turn around; return north along River Road.

Mileage

54.8	Left on 202.
55.1	Right on Taylortown Road at Montville post-office.
55.3	Pass under railroad overpass.
57.0	Right in Taylortown on Boonton Avenue.
62.1	Left on Route 23 at Butler (large pit on left).
79.6	Left on Franklin Avenue (past Franklin Pond).
80.2	Bear right at swimming pool.
80.6	Right on Fowler Street.
80.8	Left over railroad tracks on North Church Road.
81.9	Right at Judson mail box onto oiled road through farmyard.
82.2	Park in large pit.

North Church gravel pit, about 4 miles northwest of Ogdensburg (fig. 3), stop 3. The pit exposure is in the south edge of a flat-topped deposit extending 2 miles in an east-west direction and about ½ mile north-south. Salisbury and others (1902, p. 158) believed this to be an excellent example of a delta. As can be seen in figure 17, long planar foreset beds are covered by flat-lying topset beds. The end moraine across the valley to the south of Ogdensburg and also to the west (fig. 3; Salisbury and others, 1902, p. 158) could have dammed the valley and the delta could have been deposited as an ice-contact feature in the pond between the moraine and ice.

Rock types include shale, quartzite, limestone, and gneiss. Calcium carbonate coats the bottoms of some pebbles and cements some together.

82.5	Return to North Church Road. Turn right.
82.9	Right on 94 at North Church.

On way to stop 4 note kames (fig. 18) in valley south of Hamburg. Excavation in the adjacent valley bottom exposed an ice wedge filling in stratified drift (fig. 19).

85.5	Junction Route 23, continue on 94.

87.0 Right into parking area opposite Wayside Inn. Park buses. Hardyston Sand and Gravel Company, Inc., 1.4 miles north of Hamburg and 6.5 miles north of Ogdensburg, Stop 4. Pit exposure is in a Kame terrace along the valley side. Distinct layers of sand and gravel both horizontal and inclined in different directions are evident (fig. 20). Material is clean, hard, and fresh. Rock types include Franklin Limestone (calcite), hornblende gneiss, pink feldspar granite, siltstone, slate, and shale. Calcium carbonate is present as grain coatings and cement in the sand and gravel. The material probably was deposited in ponded areas between an ice tongue in the valley and the valley wall. Turn around, proceed south on 94.

Mileage

88.5 Right on Route 23 center of Hamburg.

101.7 Entrance to High Point Park.

102.1 Left on Saw Mill Road. During the next 14½ miles the trip route crosses end moraines and a lake bed in the intermontane valley on Kittatinny Mountain. The moraines, shown on figure 3, are those discovered and mapped by the senior author in 1952 (Minard, 1961). The road crosses moraine segment (A) (fig. 3) three times and the lake bed once. Moraine (A) (fig. 21) is the most northerly and longest (8 miles) of the moraines on the mountain (fig. 3).

Just to the south of the southern loop of moraine segment (A) are three more segments (fig. 22). Segment (B) was deposited at the margin of an ice lobe in the intermontane valley before withdrawal of the ice to the north and stabilization and accumulation of segment (A) (figs. 21, 22). Segment (C) was deposited at the end of a narrow tongue of ice which extended westward, through the saddle, from the main glacier filling broad Kittatinny Valley to the southeast. The tongue rested on segment (B) so that moraine (C) rests in an arcuate pattern on (B) (fig. 22). The lithology of (C) is different from (B) in that nepheline syenite and Martinsburg Shale, carried by the ice up from the valley to the east, constitute a large percentage of the deposit. Segment (B) consists mainly of quartzite and sandstone from the

Figure 15. Glacially scoured Triassic fanglomerate which was buried beneath a kame deposit near Montville, stop 2. Striae are well preserved on the upper surface of the bedrock where gneissic and quartzite boulders have been glacially scoured.

Figure 16. Sand and gravel beds were exposed at stop 2 in the early stages of excavation in 1953. These beds covered the fanglomerate bedrock shown in figure 15.

Figure 17. North Church delta, stop 3. Note the conspicuous, long, planar, cross-stratified foreset beds overlain by continuous horizontal topset beds.

Figure 18. Typical kame landform in the valley at the south edge of Hamburg, 5 miles north of Ogdensburg (fig. 3).

Shawangunk Conglomerate and High Falls Formation. Segment (D) was deposited at the margin of the main glacier filling Kittatinny Valley, after withdrawal of the tongue that deposited segment (C) and lies along the crest of the main ridge of Kittatinny Mountain. Figure 23 shows the unsorted nature of (D). It contains much nepheline syenite from the outcrop along the east face of Kittatinny Mountain 200 feet below the moraine (fig. 22). The northern end of this moraine partly dammed a basin scoured in the Shawangunk Conglomerate in the saddle. It was from peat that filled this basin that radiocarbon dates were obtained (RCD-2 and RCD-3, fig. 3 and p. 13).

Mileage
102.2   Right on Ridge Road.
104.4   Road rises on conspicuous ridge of moraine segment (A) and crosses over it.
106.9   Road rises again on ridge of moraine (A) and crosses over it.
107.5   Right on Sawmill Road.

108.7   Right on Deckertown Pike. Road crosses glacial lake bed.
109.1   Road rises on moraine (A) ridge again and crosses it.
110.6   Left on Crigger Road.
114.7   Acute left on Sunrise Mountain Road.
115.5   At Sunrise Mountain overlook. View over Kittatinny Valley to the Highlands (Reading Prong). Lunch stop. Turn around and return to
116.3   Crigger Road. Bear left.
116.6   Park at left side of road.
Walk down Tinsley Trail on right side of road to the valley floor.
East end of moraine (E), along west base of the main ridge of Kittatinny Mountain 4 miles northeast of Culvers Lake (fig. 3), stop 5. This segment of the moraine ridge on Kittatinny Mountain is 6 miles long and 95 feet thick near the western end.

It is composed largely of material from the Shawangunk Conglomerate and High Falls Formation but there are some syenite boulders on the east end. As can be seen in figure

Figure 19. Ice wedge filling in stratified drift deposit near Hamburg. Beds were laterally compressed and crumpled during growth of the wedge and subsequently slumped during melting and filling. Shovel near bottom of filled wedge is nearly 5 feet long.

24, the moraine stands out in sharp contrast to the bedrock. It forms distinct ridges, which loop across bedrock ridges, and the surface at the east end is hummocky and contains kettle holes that are seasonally filled by water. The kettle holes usually become full during late winter and spring and drain internally until they become dry in late summer or fall. The water-filled kettles and porous moraine are an important source of water for the upper drainage of Big Flat Brook during dry periods. This is true for all the moraines and particularly for the lake bed which serves as a reservoir for water when the brook is high and as a source of water when it is low.

Figure 25 shows one of the kettles on the east end of moraine (E) drained dry in late fall. In late winter the kettle is likely to be filled

with water to the level indicated. A boulder field, chiefly of quartzite blocks from the Shawangunk Conglomerate, covers a large area on the east end of the moraine.

As we walk down on the east end of the moraine the trail follows the upper rim of the steep front slope of the moraine. Surface features of the moraines on the mountain are sharp probably, in part, because the moraines are in wooded areas where they have been less modified by erosion than those in the cleared, farmed areas. Return to bus, proceed south.

Mileage

117.3   Crossing center hummocky part of moraine segment (F), the last and southernmost moraine on the mountain (figs. 3, 26). Segment (F) was deposited at the margin of the main glacier which occupied Kittatinny Valley to the southeast. The central part of the moraine, where the series of ridges form an S-shaped loop, was deposited at the end of an ice tongue extending up the inclined valley formed where the bedrock ridge of Shawangunk Conglomerate was shortened longitudinally and crinkled by compressive forces. The material in the moraine consists predominantly of Martinsburg Shale which was derived from valleys below and carried upward as much as several hundred feet. The surface of the moraine, particularly in the central part, is pocked by water-filled kettle holes. Some of these can be seen as we drive past.

120.4   Bear right at junction.

120.6   Right on Route 206.

122.3   Bear left at Tuttles corner (Route 521).

123.3   Bear left at fork.

125.4   Bevans Post Office (Peters Valley).
         Proceed toward Flatbrookville. During the next 10 miles the route follows Flat Brook valley, between Kittatinny Mountain on the left and Wallpack Ridge on the right. The valley bottom is blanketed by glacial deposits which are indicated by kames, kame terraces, and kettle holes. Much of this area will be flooded by waters impounded by the Tocks Island Dam.

130.0   Bear left at fork.

135.2   Left across bridge over Flat Brook in Flatbrookville. Proceed toward Blairstown.

137.5   Millbrook. Bear left, proceed on Millbrook Road.

139.2   Pit in Martinsburg Shale at right side of road just over crest of mountain.

142.8   Hardwick Center. Bear right at fork.

144.7   Junction Millbrook Road and High Street in Blairstown. Enter High Street and immediately turn acute left on Bridge Street.

144.8   Cross Main Street.

145.0   Junction 521 and 94. Turn left.

145.3   Junction 521. Turn right toward Hope.

Figure 20. Pit exposure in a kame terrace northeast of Hamburg, stop 4. Note distinct inclined bedding.

Mileage

151.1 North edge of Hope. Bear left.

151.4 Cross 519. Proceed toward Great Meadows.

156.3 Right on Route 46 in Great Meadows.

156.5 Left on dirt road.

156.9 Drive to far pit.

Terminal moraine at Great Meadows marks the apparent southernmost stillstand of the Wisconsin Glaciation, stop 6. It is interesting to compare the similar internal structures of this deposit (fig. 27) with those of the early drift at stop 1 (fig. 14). Both have stratified deposits near the front and unsorted material overlapping from the rear. Rock types include gneiss, quartzite, sandstone, limestone, and shale. Calcium carbonate coatings and cement are present locally.

157.4 Return to Route 46. Turn left.

164.5 Left at Bridgeville.

168.0 Left on 519.

174.5 Bear left on 519 in Lower Harmony.

176.8 Junction Route 24. Continue on 519.

178.5 Junction Route 22. Bear right then left around circle and proceed south on 22.

179.2 Still Valley circle.

179.4 Bear right on Still Valley Road.

180.6 Turn left into dirt road just before railroad underpass.

180.8 Stop at second road, walk down to pit.

Jerseyan Drift in the southern part of Warren County, 14 miles south of Belvidere, stop 7. The material is weathered, well-stratified sand and fine gravel (fig. 28). The deposit is composed predominantly of flat pieces of rounded shale and siltstone, derived from the Martinsburg Shale, and lesser amounts of red quartzite, black, gray, and white chert, and some rotten gneiss. Quartz pebbles, showing columnar structure are abundant. These are probably from the Martinsburg. Imbricate structure is conspicuous in beds containing platy shale fragments. The deposit is isolated; it occurs at an altitude of 300 feet (150 feet above the nearby Delaware River) and 15 miles south of the Wis-

Figure 21. Stereoscopic pair of vertical aerial photographs (scale 1:20,000) showing the southern part of end moraine (A) where it forms a conspicuous ridge across the valley. Arrows mark the front of the moraine. Before the stream cut through the moraine ridge, a glacial lake one and one-half miles long and about 60 feet deep was impounded behind the moraine. The field areas are on the lakebed deposits which typically consist of 5–6 feet of rust-mottled gray silt-clay over as much as 40 feet of sand and gravel. Photographs from Aero Se·vice Corporation, Winter 1939.

Figure 22. Stereoscopic pair of vertical aerial photographs (scale 1:24,000) showing the southerly loop of end moraine (A), all of (B) and (C), and part of (D) and the peat bog in the saddle. The main ridge is Shawangunk Conglomerate. Photographs from Aero Service Corporation, winter 1939.

consin terminal moraine at Belvidere. There is no evidence of calcium carbonate in the deposit, although it was probably present because the deposit and a considerable area around it is underlain by limestone. Any limestone within the deposit was probably leached during the long period of weathering which has taken place since the material was deposited.

Mileage

181.0 Return to blacktop road. Turn left under the railroad overpass.

181.3 Right.

182.1 Right in Springtown.

182.5 Bear left at fork.

184.7 Right on River Road in Carpentersville.

185.2 Straight ahead (River Road turns left over railroad) to Houdaille Construction Material, Inc. property.

185.9 Drive to rear of pit.
Gravel pit along the Delaware River just north of Carpentersville in southern Warren County,

stop 8. These deposits underlie a flat-topped terrace, the surface of which is about 25–30 feet above the river. The gravel is in long horizontal layers (fig. 29) and is clean, hard, and unweathered. Rock types include sandstone, quartzite, chert, columnar quartz, limestone, and gneiss. Calcium carbonate coats and cements some pebbles. The gravel is an outwash deposit and probably was derived largely from the moraine at Belvidere and transported downstream by the river. These gravels are remarkably similar to those downstream at Trenton which underlie a 20-foot terrace level along the Delaware River. The downstream gravels will be seen at stop 2 tomorrow.

186.7 Return to entrance of pit.

187.0 Left in Carpentersville.

189.2 Right at junction with 519.

191.2 Left.

194.1 Bear right on Route 22. Continue east on 22 and 78 to Route 28.

206.8 Bear right on 28.

Figure 23. Unsorted material in end moraine (D) on the crest of Kittatinny Mountain. In addition to quartzite and sandstone from the Shawangunk Conglomerate and High Falls Formation, there are many cobbles and boulders of nepheline syenite which were carried up from the outcrop 200 feet below (fig. 22).

Mileage

209.0   Somerville circle, junction Routes 28, 202, and 206. Bear right and proceed south on 206.

226.7   Left on Nassau Street in Princeton.

227.1   Right on Routes 526 and 571 (Washington Street).

228.8   Circle at Penns Neck. Bear right around circle and proceed north on Route 1 to motel. End of first day's trip.

SECOND DAY — TRENTON, NEW JERSEY AREA

See figure 7 for location of stops.

Begin trip at Penns Neck traffic circle on Route 1 just south of the motel.

0.0   Proceed south on Route 1.

8.3   Traffic circle at north edge of Trenton. Junction Route 206. Turn 270 degrees around circle and continue south on Route 1 freeway. Cross Delaware River.

12.2   South (right 270°) on Pennsylvania Avenue (second right after toll gate).

14.1   Angle right up ramp to 4-lane highway.

14.8   Left on New Ford Mill Road.

15.6   Right into long pit, U. S. Steel Corporation property. Pit in the east side of Turkey Hill, stop 1. This hill is an isolated inlier of older Pleistocene alluvium rising conspicuously above the broad terrace of younger Pleistocene alluvium on the slip-off slope on the Pennsylvania side of the Delaware River at Trenton. The River here changes its general direction from southeast (across the regional strike) to southwest (parallel to the regional strike). The deflection occurs at the low cuesta along the inner edge of the Coastal Plain, a short distance southeast from the inner edge of the Cretaceous. This cuesta lies along the approximate southeast edge of the Pensauken trough and may partly owe its origin to the erosion of the trough by the former river.

Turkey Hill is underlain by reworked arkosic alluvium (figs. 7, 10). The hill stands isolated amidst younger terrace deposits which lap against its base. Lockwood and Meisler (1960) interpreted the material in Turkey Hill as Illinoian in age, based on altitude, composition, and reddish weathering at the top (p. B-3). It had been mapped as Pensauken Formation by Salisbury (*in* Bascom and others, 1909b) and Peltier (1959, p. 165–171). In this report it is included in the reworked arkosic alluvium unit (fig. 10) and is therefore younger than the Pensauken (arkosic alluvium). Rock types are Triassic shale, siltstone, arkosic sandstone, argillite, and diabase, Precambrian Losee and Pochuck Gneisses, Silurian quartzite conglomerate and sandstone, and amphibolite. Glauconite is present in the upper beds as dark-green botryoidal grains.

Layers of sand alternate with layers and lenses of gravel (fig. 30). Boulders are as much as 4 feet across. Some cobbles and boulders of gneiss are deeply weathered and diabase cobbles and boulders have a soft yellow goethite rind surrounding a hard dark-gray interior.

The material forming Turkey Hill came from at least two directions. The Silurian sandstone and quartzite conglomerate (High Falls Formation and Shawangunk Conglomerate) indicate a source from up the present Delaware River, possibly the Delaware Water Gap area. The glauconite indicates a coastal plain source to the northeast and east, and probably was transported to Turkey Hill by a stream flowing in the Pensauken trough. It is probable that this stream was as large or larger than the Delaware River and that the Delaware River joined it at Turkey Hill or north of it and did not flow around the hill to the east and south as it now does. Later, as the Pensauken trough stream became smaller

Figure 24. Stereoscopic pair of vertical aerial photographs (scale 1:20,000) showing the eastern part of end moraine (E). Conspicuous hummocky kettle-hole topography can be seen at 1 on the east end. Moraine curves north around bedrock ridges at 2 and 3 on the west end, and south in the adjacent valleys. Arrows mark the front of the moraine. The prominent ridge trending northeast at the right is the Shawangunk Conglomerate. Photographs from Aero Service Corporation, winter 1939.

Figure 25. One of the larger kettle holes on the east end of end moraine (E) in December 1952, stop 5. By March 1953, water had filled the kettle to the level indicated by the dashed line, a maximum depth of 8 feet. The pool was nearly 200 feet wide and 300 feet long. The water slowly drains internally and, depending on the amount of precipitation, may not completely drain until late fall. The boulders are covered by blankets of dried pond-surface algae which draped over them as the water level dropped.

as the continental glacier withdrew, the Delaware River extended farther southeast as undercutting and erosion of the material in, under, and bordering the trough progressed. Slip-off slope deposits built up on the inside of the bend south of Trenton, leaving Turkey Hill isolated and surrounded by the younger terrace deposits now flanking its slopes. The base of the gravel and sand of Turkey Hill rests on sand of the Raritan Formation of Late Cretaceous age. This contact was exposed in an excavation in an abandoned pit on the west side of the hill. The top of the Raritan, beneath the hill, is at an altitude of about 20 feet but is generally well below sea level under the surrounding younger terrace deposits which were laid down in channels cut below the protected Cretaceous surface beneath Turkey Hill.

Turn around and return to road.

Mileage

15.8   Left (north) on New Ford Mill Road.

16.2   Left on dirt lane. Cross railroad and walk to edge of lake.

Warners gravel pit, stop 2. Exposure of calcareous outwash (fig. 12), formerly mapped as Cape May (fig. 9). The main exposure in the lake bank is in the 20 foot terrace level. The 2½ mile long Van Sciver Lake was created as the gravel was excavated. The gravel is dug and dredged, and transported by company truck, barge, and railroad. It is fresh, clean, and hard and rock types present are quartzite, sandstone, limestone, chert, gneiss, and diabase. Calcium carbonate, leached from limestone pebbles above, coats and cements pebbles 6–8 feet below the surface.

The gravel underlies broad areas of the lowlands adjacent to the Delaware River near

Figure 26. Stereoscopic pair of vertical aerial photographs (scale 1 : 24,000) showing end moraine segment (F). Note loop in saddle above inclined valley and extension along main ridge in both directions. See text for description. Photographs from Aero Service Corporation, winter 1939.

Figure 27. Wisconsin terminal moraine at Great Meadows, stop 6. Stratified deposits near the front (right) of the moraine are overlapped from the rear (left) by unsorted fine to coarse material. Compare with figure 14.

Figure 29. Pit exposure in river gravels at Carpentersville in southern Warren County, stop 8. Long horizontal strata are typical.

Trenton (fig. 13). It was certainly deposited almost exclusively by the Delaware River. The remarkable similarity, in composition and internal structures, to the gravels along the river to the north was pointed out at stop 8 yesterday (see fig. 29). There is a marked decrease in the amount of Triassic material as compared with Turkey Hill. The gravel shows a gross uniformity in size, ranging mostly between pebbles and small cobbles (fig. 31). The Triassic material probably did not survive the swift uniform flow indicated by the size range. It is also possible that ice rafting of boulders, which would be less likely to be comminuted as readily as pebbles and small cobbles, did not play as dominant a role as it may have during deposition of the Turkey Hill material.

Turn buses around in field.

Mileage

16.5   Return to New Ford Mill Road. Turn left.
17.0   Right on 4-lane highway. Proceed east toward end of highway.
17.8   Merge with Pennsylvania Ave. and proceed north.
19.6   Right on toll bridge (Route 1); cross Delaware River.
20.8   Right on Route 33.
23.8   Right on Klockner Road.
26.1   Right on Hamilton Square Road.
26.5   Right into pit.
        Pit in Pensauken trough east of Trenton, stop 3. Altitude of the upper surface is 100 feet. This is the typical Pensauken Formation (mapped in this report as arkosic alluvium); the material is nearly all sand, which is in contrast to the abundance of gravel at the two previous stops.

Figure 28. Jerseyan Drift in southern Warren County, stop 7. The deposit is 15 miles south of the terminal moraine at Belvidere.

Figure 30. Exposure in a pit on the east side of Turkey Hill, stop 1. Beds of sand alternate with beds of gravel. In this report the deposit is included in the reworked arkosic alluvium unit.

Figure 31. Typical exposure of calcareous outwash gravel in the bank of Van Sciver Lake, stop 2.

Prominent cross-stratification indicates a southwesterly flow (fig. 32). The deposit is chiefly yellow-brown, arkosic, silty quartz sand overlain by gravel and scattered pebbles. A few cobbles are present and weathered sand-size feldspar is abundant. Some layers are stained or cemented by iron oxide. This exposure is similar to most deposits of arkosic alluvium in the Pensauken trough northeast to Perth Amboy.

Mileage

26.9 Return to Hamilton Square Road. Turn left.
27.3 Right on Klockner Road.
29.9 Cross Route 130.
32.7 Right on Main Street in Crosswicks.
33.0 Right into pit.
Stratified glauconitic alluvium is a wide-spread unit (fig. 11) blanketing the glauconite

belt (fig. 8), stop 4. Salisbury (*in* Bascom and others, 1909b) mapped much of it as unclassified deposits (fig. 9). The altitude of the deposit is about 70 feet. Several feet of massive, silty sand overlies 10 feet of well-stratified (horizontally) highly glauconitic alluvium (fig. 33). This in turn overlies less glauconitic cross- and horizontally-stratified sand alternating with lenses of gravel. The lower material is part of the reworked arkosic alluvium. A sample of peat from a bog beneath this material was dated at 26,800 ± 1,000 years B. P. (Meyer Rubin, U. S. Geological Survey, written commun., 1962, sample W-1193).

33.4 Return to Main Street. Turn left. Continue through Crosswicks.
34.0 Right on Crosswicks-Chesterfield Road.
35.0 Left at fork.
35.6 Chesterfield. Proceed south.
38.2 Georgetown.
38.4 Right.
38.8 Cross S39 (68).
39.0 Bear left on Gaunts Bridge Road.
40.9 Right on Route 537 and immediately left across bridge over creek. Park at side of road. Peat bog beside road, stop 5. Stream cuts through peat and logs underlying the flat meadow area on the east side of the road. The bog surface is between 80 and 90 feet above sea level. A radiocarbon date of 10,770 ± 300 years B. P. was obtained from a sample of wood from the base of the bog (Meyer Rubin, U. S. Geological Survey, written commun., 1964, sample W-1411). Numerous similar bogs are present at altitudes of 50 to 100 feet, indicating a ponding possibly caused by rising sea level. Considerable downcutting has taken place subsequently.

Continue south.

Figure 32. Exposure in a pit in arkosic alluvium in the Pensauken trough east of Trenton, stop 3. Conspicuous cross-stratification dipping to the left (southwest) indicates direction of flow.

Figure 33. Exposure in a pit in stratified glauconitic alluvium, stop 4. Thin horizontal layers characterize all exposures of glauconitic alluvium.

Mileage

41.2   Left.

42.7   Right on S39 (68).

43.7   Left at traffic light at the entrance to Fort Dix.

43.9   Park by service station. Walk down abandoned road into old pit, stop 6.

Pit exposes 12 feet of thin horizontal beds of glauconitic alluvium (silty at top), overlying 6–10 feet of pebbly irregular beds of silty sand. This latter deposit is a non-glauconitic unit and is correlated with the quartz sand alluvium. Because of lithology and, particularly, its altitude of 180 feet (the upper ground surface here is about 200 feet altitude), it is correlative with Salisbury's Bridgeton Formation. Beneath this unit is the contrasting clean yellow to yellowish-gray, fine to very coarse sand of the Cohansey Sand of late Tertiary age. The Cohansey is also non-glauconitic. The middle gravel unit (Salisbury's Bridgeton) probably was deposited before erosion had cut deep enough to expose the underlying glauconite-bearing formations which contributed to the overlying glauconitic alluvium.

The glauconitic alluvium is everywhere characterized by its distinctive bedding and uniform texture. The bedding is undoubtedly accentuated by the concentrations of dark-green glauconite in some layers and concentrations of light-gray quartz sand in others. The narrow grain-size distribution probably is a result of the source material being largely sand-sized sediment of the Coastal Plain formations to the east.

Turn around; go back to the traffic light at the entrance to Fort Dix.

44.1   Left into Fort Dix.

45.2   Traffic circle; bear right on Pemberton Road. Pass through Pemberton. Lunch stop.

Figure 34. Large pit in glauconitic alluvium along the south side of the South Branch Rancocas Creek, stop 7. Note the horizontal beds at the left edge of the photograph.

49.7   Right on Route 38 (just across North Branch Rancocas Creek).

52.5   Cross Route 206.

53.9   Road crosses a depression, one of those which are present by the hundreds in the Coastal Plain. Most are present on Quaternary deposits, largely of quartz sands, and seem oriented along broad, former drainage lines.

55.8   Left on Lumberton Road. Pass through Lumberton and cross South Branch Rancocas Creek.

57.5   Right on Creek Road.

58.6   Right at scales into pit.

59.1   Stop in pit.

Large pit operated by Asphtar Inc., stop 7. Uniform horizontal beds in the glauconitic alluvium underlie the broad areas bordering the river (fig. 34). These deposits extend for miles along the three branches of Rancocas Creek and their tributaries. The deposits are all characterized by uniform texture (mostly sand to granule size), long, thin, horizontal beds, and high glauconite content. The deposits are probably, in large part, natural levees. Since deposition, sea level has dropped enough to cut down through 40 feet of the glauconitic alluvium and 10 feet into the underlying Cretaceous and Tertiary formations.

59.6   Return to Creek Road; turn right.

60.6   Bear left on Route 38.

61.4   Bear right on Masonville Road.

62.2   Angle right across Pennsylvania Highway toward Rancocas.

63.0   Angle left; cross over New Jersey Turnpike.

64.7   Right on Main Street in Rancocas.

65.8   Right into lane to Rancocas Valley Farm.

66.5   G. F. Haines farmyard; walk to pit.

Large pit, mostly in reworked arkosic alluvium, stop 8. The bottom of the pit is 30–40 feet above sea level. The predominant material is arkosic sand; lower beds contain layers and lenses of gravel. Long, planar cross strata are common. Cobbles and boulders are abundant in the basal beds and are mostly siltstone, sandstone, and arkose conglomerate of Triassic age.

A unique feature of this pit is the presence of cobbles and boulders of glauconite, the largest of which is 4 feet thick and 6 feet long (fig. 35). These blocks are unconsolidated and crumbly at present, as is true typically in the parent formation. Swift flowing streams must have deposited the coarse gravel which constitutes this deposit. The glauconite blocks could not have survived transport in such swift water unless they were at one time more cohesive. If the blocks were frozen when they dropped into the stream, conceivably from an undercut bank, they could have survived rolling some distance along the stream bottom until

Figure 35. Exposure of reworked arkosic alluvium in a pit in the upper terrace along the north side of the North Branch Rancocas Creek, stop 8. The large dark mass in the center of the photograph is an unconsolidated block of glauconite 4 x 6 feet across. The overlying beds are cross-stratified arkosic sand. Gravel beds abut the lower part of the block. Several smaller blocks of glauconite are near the shovel. The large block probably is from the lower part of the Navesink Formation of Late Cretaceous age.

lodged and covered by sand and gravel. Some of the blocks of glauconite have a thin rind of iron oxide. The large block probably is from the lower part of the Navesink Formation of Late Cretaceous age. Blocks from the Hornerstown Sand of Paleocene age are also present.

On the far (west) side of the pit iron-oxide cemented forms include "pipes" 6 inches in diameter and 6–12 feet long. In the near (east) side of the pit are bedded glauconitic sands. These beds are in sharp contrast with the yellowish-brown arkosic sand. The glauconitic sand consists of cross and horizontally stratified predominantly light-gray to yellow-gray quartz sand alternating with beds of predominantly dark-green glauconite sand.

The pit is underlain at a shallow depth by the Englishtown Formation of Late Cretaceous age. The channel-shaped base of the Quaternary deposits, with its pavement of cobbles and boul-

ders formerly was exposed in the lower part of the pit.

Last stop. Return to motel.

Mileage

67.2 Return to blacktop road. Turn right.
68.1 Cross over New Jersey Turnpike.
68.6 Left.
69.2 Right on Woodlane Road.
72.7 Bear left on Route 537.
74.6 Left (north) on Route 206. Continue north on 206 into Trenton.
89.7 Turn right on South Clinton Street (last street before railroad track across road).
90.3 Circle just south of Pennsylvania Railroad.
90.5 Around circle onto freeway (Route 1).
92.9 Circle at junction with Route 206. Proceed north on Route 1.
101.2 Circle at Penns Neck.

END OF TRIP

REFERENCES CITED

Bascom, Florence, and others, 1909a, Description of the Philadelphia district [Pa.-N. J.-Del.]: U. S. Geol. Survey Geol. Atlas, Folio 162.

Bascom, Florence, and others, 1909b, Description of the Trenton quadrangle [N. J.-Pa.]: U. S. Geol. Survey Geol. Atlas, Folio 167.

Bayley, W. S., Kümmel, H. B., and Salisbury, R. D., 1914, Description of the Raritan quadrangle [N. J.]: U. S. Geol. Survey Geol. Atlas, Folio 191.

Campbell, M. R., and Bascom, Florence, 1933, Origin and structure of the Pensauken gravel: Am. Jour. Sci., v. 26, 5th ser., no. 153, p. 300–318.

Cook, G. H., 1877, Exploration of the portion of New Jersey which is covered by the glacial drift: New Jersey Geol. Survey Ann. Rept. 1877, p. 9–22.

Cook, G. H., 1878, On the glacial and modified drift: New Jersey Geol. Survey Ann. Rept. 1878, p. 8–23.

Cook, G. H., 1880, Glacial drift: New Jersey Geol. Survey Ann. Rept. 1880, p. 16–97.

Darton, N. H., Bayley, W. S., Salisbury, R. D., and Kümmel, H. B., 1908, Description of the Passaic quadrangle [N. J.-N. Y.]: U. S. Geol. Survey Geol. Atlas, Folio 157.

Flint, R. F., 1947, Glacial geology and the Pleistocene epoch: New York, John Wiley and Sons, Inc., 589 p.

Fuller, M. L., 1914, The geology of Long Island, New York: U. S. Geol. Survey Prof. Paper 82, 231 p.

Gill, H. E., 1962, Ground-water resources of Cape May County, New Jersey-Salt-water invasion of principal aquifers: New Jersey Dept. Conserv. and Econ. Devel. Div. Water Policy and Supply Spec. Rept. 18, 171 p.

Howell, B. F., and Hale, H. E., II, 1946, Fossiliferous pebbles in "Pensauken Gravel" at Princeton, New Jersey: Jour. Geology, v. 54, no. 6, p. 386–390.

Kümmel, H. B., 1940, The geology of New Jersey: New Jersey Dept. Conserv. Geol. Ser. Bull. 50, 203 p.

Lemke, R. W., Laird, W. M., Tipton, M. J., and Lindvall, R. M., 1965, Quaternary geology of Northern Great Plains *in* Wright, H. E., and Frey, D. G., eds., The Quaternary of the United States: Princeton, N. J., Princeton Univ. Press, p. 15–27.

Leverett, Frank, 1928, Results of glacial investigations in Pennsylvania and New Jersey in 1926 and 1927 /abs./: Geol. Soc. America Bull., v. 39, p. 151.

Lewis, H. C., 1881, The antiquity and origin of the Trenton gravels: Salem, Mass., 31 p. (extract from Primitive Industry . . . , by Charles C. Abbott).

Lewis, J. V., and Kümmel, H. B., 1912, Geologic map of New Jersey, 1910–12: New Jersey Geol. Survey, scale 1:250,000 (rev. ed. 1950).

Lewis, J. V., and Kümmel, H. B., 1915, The geology of New Jersey; a summary to accompany the geologic map (1910–12): New Jersey Geol. Survey Bull. 14, 146 p.

Lockwood, W. N., and Meisler, Harold, 1960, Illinoian outwash in southeastern Pennsylvania: U. S. Geol. Survey Bull. 1121-B, p. B1–B9.

MacClintock, Paul, 1940, Weathering of the Jerseyan till: Geol. Soc. America Bull., v. 51, p. 103–116.

MacClintock, Paul, 1943, Marine topography of the Cape May Formation: Jour. Geology, v. 51, no. 7, p. 458–472.

MacClintock, Paul, and Apfel, E. T., 1944, Correlation of the drifts of the Salamanca re-entrant, New York: Geol. Soc. America Bull., v. 55, no. 10, p. 1143–1164.

MacClintock, Paul, and Richards, H. G., 1936, Correlation of Late Pleistocene marine and glacial deposits of New Jersey and New York: Geol. Soc. America Bull., v. 47, no. 3, p. 289–338.

Minard, J. P., 1959, Recent Saprolite: Science, v. 129, no. 3357, p. 1206–1209.

Minard, J. P. 1961, End moraines on Kittatinny Mountain, Sussex County, New Jersey: U. S. Geol. Survey Prof. Paper 424-C, p. C61–C64.

Minard, J. P., 1964, Geology of the Roosevelt quadrangle, New Jersey: U. S. Geol. Survey Geol. Quad. Map GQ-340.

Minard, J. P., 1965, Geologic map of the Woodstown quadrangle, Gloucester and Salem Counties, New Jersey: U. S. Geol. Survey Geol. Quad. Map GQ-404.

Minard, J. P., 1969, Geology of the Sandy Hook quadrangle, in Monmouth County, New Jersey: U. S. Geol. Survey Bull. 1276 (in press).

Minard, J. P., and Owens, J. P., 1962, Pre-Quaternary geology of the New Egypt quadrangle, New Jersey: U. S. Geol. Survey Geol. Quad. Map GQ-161.

Minard, J. P., and Owens, J. P., 1963, Pre-Quaternary geology of the Browns Mills quadrangle, New Jersey: U. S. Geol. Survey Geol. Quad. Map GQ-264.

Minard, J. P., Owens, J. P., and Nicholas, T. C., 1964, Pre-Quaternary geology of the Mount Holly quadrangle, New Jersey: U. S. Geol. Survey Geol. Quad. Map GQ-272.

Owens, J. P., and Minard, J. P., 1962, Pre-Quaternary geology of the Columbus quadrangle, New Jersey: U. S. Geol. Survey Geol. Quad. Map GQ-160.

—————— 1964a, Pre-Quaternary geology of the Pemberton quadrangle, N. J.: U. S. Geol. Survey Geol. Quad. Map GQ-262.

—————— 1964b, Pre-Quaternary geology of the Trenton East quadrangle, N. J.-Pa.: U. S. Geol. Survey Geol. Quad. Map GQ-341.

—————— 1964c, Pre-Quaternary geology of the Bristol quadrangle, N. J.-Pa.: U. S. Geol. Survey Geol. Quad. Map GQ-342.

—————— 1966, Pre-Quaternary geology of the Allentown quadrangle, N. J.: U. S. Geol. Survey Geol. Quad. Map GQ-566.

Peltier, L. C., 1959, Late Pleistocene deposits, *in* Willard, Bradford, Geology and mineral resources of Bucks County, Pennsylvania: Pennsylvania Geol. Survey, 4th ser., Bull. C 9, p. 163–184.

Richards, H. G., 1960, The dating of the "Subway Tree" of Philadelphia: Pennsylvania Acad. Sci. Proc., v. 34, p. 107–108.

Salisbury, R. D., 1893, Surface geology; report of progress: New Jersey Geol. Survey Ann. Rept. 1892, p. 37–166.

—— 1895, Surface geology; report of progress: New Jersey Geol. Survey Ann. Rept. 1894, p. 1–149.

—— 1896, Surface geology; report of progress: New Jersey Geol. Survey Ann. Rept. 1895, p. 1–16.

—— 1898, Surface geology; report of progress: New Jersey Geol. Survey Ann. Rept. 1897, p. 1–22.

Salisbury, R. D., 1899, The soils of New Jersey and their relation to the geological formations which underlie them: New Jersey Geol. Survey Ann. Rept. 1898, p. 1–41.

Salisbury, R. D., and Knapp, G. N., 1897, Surface geology; report of progress: New Jersey Geol. Survey Ann. Rept. 1896, p. 1–23.

—— 1917, The Quaternary formations of southern New Jersey: New Jersey Geol. Survey, v. 8, Final Rept. State Geologist, 218, p.

Salisbury, R. D., and Kümmel, H. B., 1895, Lake Passaic, an extinct glacial lake: Jour. Geology, v. 3, p. 533–560.

Salisbury, R. D., and others, 1894, Surface geology; report of progress: New Jersey Geol. Survey Ann. Rept. 1893, p. 33–328.

—— 1902, The glacial geology of New Jersey: New Jersey Geol. Survey, Final Rept., v. 5, 802 p.

Spencer, A. C., and others, 1908, Description of the Franklin Furnace quadrangle [N. J.]: U. S. Geol. Survey Geol. Atlas, Folio 161.

Wilmarth, M. G., 1938, Lexicon of geologic names of the United States: U. S. Geol. Survey Bull. 896, pt. 1, p. 1–1244.

Wilson, L. R., 1932, The Two Creeks Forest Bed, Manitowoc County, Wisconsin: Wisc. Acad. Sci. Trans., v. 27, p. 31–46.

Woolman, Lewis, 1897, Stratigraphy of the Fish House black clay and associated gravel, New Jersey Geol. Survey Ann. Rept. 1896, sec. 4, pt. 3, p. 201–254.

LATE TRIASSIC NEWARK GROUP, NORTH CENTRAL NEW JERSEY AND ADJACENT PENNSYLVANIA AND NEW YORK

F. B. VAN HOUTEN

Department of Geological and Geophysical Sciences, Princeton University, Princeton, New Jersey 08540

GENERAL GEOLOGY

INTRODUCTION

General statement. The Late Triassic Newark Group is preserved in a southwest-trending basin that reaches from Rockland County, New York, to northeast Lancaster County, eastern Pennsylvania. This is the largest of six major (and several minor) Triassic rift valleys in a sinuous belt more than 1,000 miles long, from Nova Scotia to North Carolina (and in subsurface as far south as Florida).

The Newark Basin is the largest of three lobes (Newark, Gettysburg, Culpeper) connected by narrow corridors in a 300-mile arcuate tract from southeastern New York to northeastern Virginia. It is about 140 miles long and a maximum of 32 miles wide along the Delaware River.

The Newark Group consists of 16,000–20,000 feet (4,880–6,100m) of nonmarine sedimentary rocks and associated intrusive and extrusive basic rocks (fig. 1). Their strike generally parallels the trend of the basin and they dip 10–20°NW. Along the northwestern margin these rocks are bounded by Precambrian and Paleozoic rocks of the Reading Prong of the New England Upland (the Highlands). Most of this boundary is a system of high-angle faults, but intermittently the Triassic deposits overlap on rocks of the upland terrane. Within the basin Newark strata lie on Paleozoic and subordinate Precambrian rocks of the Blue Ridge and Piedmont provinces (fig. 2). Along the southeastern margin Newark deposits overlap on Precambrian and Paleozoic rocks of the Piedmont Province, including soda-rich schist and gneiss. Newark strata, in turn, are overlapped by Cretaceous and younger deposits of the Coastal Plain Province to the southeast.

In addition to their general northwest dip Newark strata are locally warped into open anticlines and synclines (fig. 11) and are broken by many small normal faults which complicate attempts to estimate thickness of the formations. In the central part of the basin a few large faults with several thousand feet of vertical displacement may be part of a transcurrent system involving those of the northwest border as well (Sanders, 1962, p. 40). This faulting has nearly doubled the width of the basin.

SEDIMENTARY ROCK UNITS AND TYPE AREAS

Newark Group (System in early reports)—Redfield, 1856, p. 357; Newark, east-central New Jersey.

Stockton Formation—Kümmel, 1897, p. 35–40; vicinity of Stockton, west-central New Jersey.

Lockatong Formation—Kümmel, 1897, p. 40–47; along Lockatong Creek between Stockton and Frenchtown, west-central New Jersey.

Brunswick Formation—Kümmel, 1897, p. 47–55; along Raritan River, vicinity of New Brunswick, central New Jersey.

Hammer Creek Formation—Glaeser, 1963, p. 184; along Hammer Creek in Richland Quadrangle, Pennsylvania.

The common succession of rocks in the Newark Group (fig. 1), as in most of the Triassic basins, consists of a lower, locally conglomeratic arkose, the Stockton Formation, a maximum of 6,000 feet (1,830m) thick, grading upward into a reddish-brown mudstone deposit, the Brunswick Formation, estimated to be as much as 16,000 feet (4,880m) thick. As in other basins, this sequence is thickest

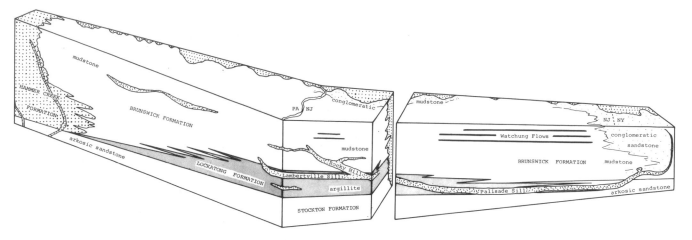

Figure 1. Schematic block diagram of formations and lithofacies of the Newark Group in the Newark Basin in southeastern Pennsylvania, central New Jersey and adjacent New York. Restored as before final faulting, warping and tilting. Thicknesses relative. Looking toward northwest border of the Basin.

near the faulted northwest border where it interfingers with the Hammer Creek Conglomerate.

In the central part of the Newark Basin in New Jersey and adjacent Pennsylvania the two widely-occurring formations are separated by and interfinger with the dark gray and reddish-brown Lockatong Argillite, a maximum of 3,750 feet (1,150m) thick (fig. 1, 3). Compared with many nonmarine basin sequences these three central deposits are remarkably uniform and persistent.

A thick diabase sill intrudes the lower part of the Newark Group in central and northeastern New

Jersey. At its northeast end in New York and its southwest end in western New Jersey (fig. 1) the sheet cuts upward into higher beds. In Pennsylvania less extensive, discordant, basin-shaped sheets of diabase are common. Thick dikes are associated with these intrusions, especially in Pennsylvania. Numerous slender dikes cut across deformed Triassic strata and some extend beyond the basin. Three thick, multiple basaltic lava flows are interbedded with the upper part of the Brunswick Formation in the northeastern half of the basin.

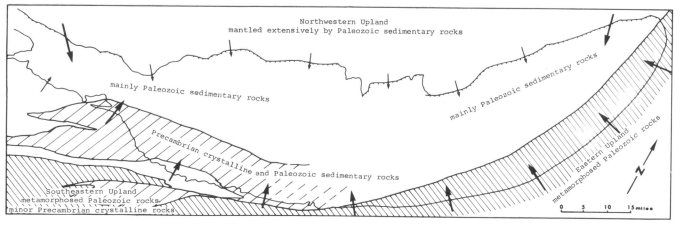

Figure 2. Sketch map of major rock units beneath the Newark Basin and exposed in bordering uplands in Late Triassic time. Basin outlined in thin line, dotted along faulted northwest border.
Terrane predominantly Paleozoic sedimentary rocks of the Highlands and Blue Ridge province — unmarked; metamorphosed Paleozoic rocks of the Piedmont province — closely spaced diagonal lines; Precambrian crystalline rocks of the Blue Ridge province — widely spaced diagonal lines.
Major sources and estimated directions of dispersal — large arrows; local sources and dispersal — small arrows.

Distinctive hornfels developed from the three major rock types in the Newark Group were produced by the intrusive sheets. In contrast, the lava flows had insignificant metamorphic effects.

Stratigraphic nomenclature. The gradational and interfingering relations of the three formations in the central part of the Newark Basin introduce stratigraphic complexities difficult to express with simple nomenclature. Admittedly, some reports used the formation names erroneously in a time-stratigraphic sense, but more commonly the lensing, interfingering formations have been considered to be lithic units, as pointed out clearly by Kümmel (1898, p. 41). Once the stratigraphic complexity is recognized and previous misconceptions corrected, the relations between the units can be described and discussed by using the basic lithostratigraphic terms, formation and member. Substitution of such words as lithofacies (McLaughlin and Willard, 1949) and lithosome (Glaeser, 1963; 1966, p. 186–188) is not necessary.

Fossil content. A variety of fossils has been recovered from the Newark Group, but except for reptilean footprints fossils of major groups are rare, and so far none is diagnostic of specific correlation with Triassic strata of Europe.

A general list, extracted from Baird, 1954; Boch, 1959; Colbert, 1965, 1966; Colbert and Gregory, 1957; and Wherry, 1959, includes:

Stockton Formation: Ferns, equisitales, conifers, ginkgos, cycads; mollusks; labyrinthodont amphibians and phytosaur reptiles. Most of the fossil plants are from uppermost Stockton or basal Lockatong deposits at Carversville, Pennsylvania.

Lockatong Formation: Elasmobranch, coelocanth and palaeoniscid fish; labyrinthodont amphibians; phytosaurs and other reptiles, and rare dinosaur footprints; crustacean estheriids, ostracodes and phyllocarids, and molluskan unios; small burrowers, coprolites and traces of plants. The fossil fish occur only locally, but are very abundant and well-preserved. Fossil plants, in contrast, are remarkably rare.

Brunswick Formation: Coelocanth and higher bony fish; reptiles such as *Chirotherium, Stegomus* and *Hypsognathus,* and abundant dinosaur footprints; traces of plants, including most of the groups found in uppermost Stockton deposits.

Age and correlation. Nonmarine Triassic rocks throughout eastern North America are generally equivalent to the Keuper Formation of northern Europe, and assigned a Karnian, Norian and possibly Rhaetian (Late Triassic) age (see McKee and others, 1959, table 1). Igneous rocks in the upper part of the Newark sequence have been dated at about 190–200 m.y., although Rodgers (1968, p. c-o, 2) points out that "because of alteration of the dated basalt and related rocks, the true age is probably nearer 225 m.y. (Holmes, 1965, p. 361, assigned

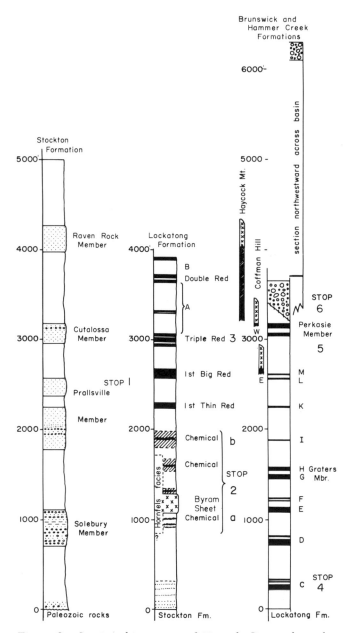

Figure 3. Stratigraphic section of Newark Group along the Delaware River from Stockton northwestward to 3 miles west of Milford, N. J. After McLaughlin, 1943, 1944, 1945, 1946; Van Houten, 1964.

Subordinate units in gray Lockatong Formation are reddish-brown, in reddish-brown Brunswick Formation are dark gray. Shows stratigraphic position of Lockatong chemical cycles, Byram diabase, and Lockatong hornfels, Coffman Hill and Haycock Mt. diabase sheets in easternmost Pennsylvania, and field trip stops.

this date to the *beginning* of the Triassic Period—180 to 225 m.y.). Duration of Late Triassic time is generally estimated at about 15 m.y., or one-third of Triassic time. But the fact that Late Triassic time includes 3 of the 6 Triassic ages suggests that it may have been at least 20 m.y. long. In contrast, Abdel-Monem and Kulp (1968, p. 1232) state that "the sedimentary rocks of the Newark Group of the New Jersey Triassic basin were deposited over a short time interval (<5 m.y.)" without citing any supporting evidence for this abbreviated span.

Apparently a Late Triassic tetrapod fauna of phytosaurs and pseudosuchians (chirotheres) was replaced at about the time of the first lava flow in New Jersey by a younger fauna dominated by bipedal dinosaurs (Colbert and Gregory, 1957). The stratigraphic distribution of these forms suggests a general correlation of the three Watchung lava flows high in the Brunswick Formation with the three flows (Talcott, Holyoke, Hampden) in the Connecticut Valley. By this correlation, Triassic deposition apparently began earlier, perhaps before Karnian time, in the Newark Basin. De Boer (1968, p. 620) has concluded on the basis of magnetic inclination data that all three Watchung flows, as well as the Palisade Sill, correlate with the Holyoke volcanic unit in the Connecticut Valley. The fact that the main Watchung flows are separated by deposits just about as thick as those between the Talcott, Holyoke and Hampden flows implies about equal time of formation and makes De Boer's correlation seem unlikely.

The three sedimentary formations in the Newark Basin are remarkably like three units in the Deep River Basin in North Carolina, and a scanty fauna suggests they may be generally correlative (Colbert and Gregory, 1957, table 1).

Source area. Deposition in the Newark Basin began with influx of arkosic detritus from an uplifted eastern and southern provenance not far from the present outcrop margin (Glaeser, 1966, p. 17, 26–27). Downfaulting along the northwestern margin of the basin also began at this time. Abundant feldspar in the Stockton and Brunswick Formations and their intimate interfingering with the soda-rich Lockatong Formation, imply derivation of all three from the soda-rich crystalline terrane that lay to the east and south at a time when the northwestern upland was still mantled with Paleozoic sedimentary rocks (fig. 2). Radiometric ages of Newark detritus also point to a continuous supply from a metamorphosed Paleozoic source east and south of the basin (Abdel-Monem and Kulp,

1968, p. 1234–1237) while short, vigorous torrents eroding the Paleozoic mantle of the northwestern upland filled the northeast and southwest ends of the basin with wedges of thick conglomerates (figs. 1 and 2).

Former extent. Two contrasting interpretations of the former extent of the Newark deposits have been proposed:

1. The present extent of outcrop is about that of the original basin-fill—the local basin hypothesis.

2. Newark deposits originally covered the "broad terrane" between the Newark Basin and the Connecticut Valley and its southern extension. This reconstruction requires that abundant feldspathic detritus in New Jersey and Pennsylvania was derived from the east side of the postulated broad graben.

Evidence that the detritus in the Newark Basin was derived largely from nearby eastern and southern sources contradicts that "broad-terrane" hypothesis (Van Houten, 1965, p. 826–828). In fact, Glaeser (1966, p. 4, 17, 26–27) has shown that the shape and extent of the original Newark depositional basin were much like the present outcrop area. Although radiometric dating indicates that most of the Triassic detritus in New Jersey was eroded from metamorphosed Paleozoic rocks to the east, this evidence does not necessarily prove a nearby source as claimed by Abdel-Monem and Kulp (1968, p. 1237), for such sediments could have been drived from similar rocks east of the Connecticut valley.

Dispersal. A specific pattern of dispersal within the long Newark Basin is difficult to reconstruct. Clearly, tongues of conglomerate distributed southwestward from the northeast end and northeastward from the southwest end (figs. 2 and 8), and confinement of the Lockatong lacustrine facies to the more central part of the basin, point to an axial distribution of sediments imposed on their initial dispersal from the flanking uplands. But the path of exit of the main drainage from the basin is not known. Regionally, drainage to the southwest seems somewhat more reasonable; or the main drainage may have gathered waters from the northeast and southwest ends and flowed eastward through a valley in the eastern upland in a pattern like that of the Great Valley of California draining into San Francisco Bay. Distribution of the Atlantic continental blocks undoubtedly exerted a significant control on development of the regional drainage.

Rate of accumulation. General estimates of rate of deposition, the role of bypassing, uniform rate of basin sinking, rate of permanent accumulation, and duration of Stockton, Lockatong and Brunswick episodes have been based on data from the stratigraphic section along the Delaware River. The numerous assumptions entailed are the following:

Thickness: Stockton—1,830 m, Lockatong—1,150m, Brunswick—1,900m.

Rate of initial deposition (based on estimated Lockatong rate):

Stockton: 600+ mm/1,000 yr. Most rapid.

Lockatong: 215 mm/1,000 yr, derived from varve-counts in short cycles (Van Houten, 1962, p. 572). Slowest rate.

Brunswick: 305+ mm/1,000 yr. Intermediate rate.

Length of Triassic Period: 45 m.y. 225–180 m.y. ago.

Uniform rate of basin sinking: About 215 mm/1,000 yr. during entire time. This is the rate of accumulation of the Lockatong Formation which presumably recorded a neat balance between sinking and supply with essentially no bypassing.

A. Duration of depositional episodes with uniform basin sinking:

Stockton: 8.50 m.y., of which about 5.5 m.y. were involved in cumulative bypassing.

Lockatong: 5.50 m.y. with no bypassing.

Brunswick: 8.75 m.y. with about 2.50 m.y. of cumulative bypassing.

Newark episode: 22.75 m.y. with about 8.0 m.y. of cumulative bypassing. This duration, about one half of the Triassic Period is consistent with the fact that that Newark Group is assigned to three (Karnian, Norian, and Rhaetian) of the six Triassic stages.

B. Duration of depositional episodes assuming a changing rate of basin sinking to match that of initial deposition of each formation, with no time used up in bypassing:

Stockton: 3.0 m.y.

Lockatong: 5.5 m.y.

Brunswick: 6.25 m.y.

Newark episode: 14.75 m.y. This duration agrees with the commonly estimated 15 m.y. for Late Triassic time. Nevertheless, it requires the unreasonable assumption that no time was involved in bypassing during the entire episode.

C. The conventional method of calculating rates of deposition by plotting thickness against assumed durations produces a similar range of results, according to the duration assumed. If the Newark episode was half of Triassic time (22.5 m.y.)

then the average rate of basin sinking and accumulation was about 215 mm/1,000 yr. If the Newark episode is assumed to have been a third of Triassic time (15 m.y.) then the average rate was 325 mm/1,000 yr.

D. In spite of the numerous assumptions and uncertainties in this exercise, the calculations do provide a useful estimate of the minimum (14.75 m.y.) and maximum (22.75 m.y.) duration of the Newark episode, as well as a minimum (215 mm/1,000 yr) and maximum (325 mm/1,000 yr) average rate of basin sinking and accumulation. Significantly, these rates are more than an order of magnitude slower than those of many flysch and molasse troughs (Fischer, 1969). Even if the estimated thickness of the Newark strata is nearly doubled (9,500 m) the resulting minimum and maximum rates (430–650 mm/1,000 yr) are still slower than those of some flysch and molasse troughs.

Conditions of deposition. Newark deposits like Triassic sequences in most of the rift valleys in eastern North America, comprise the major facies of a piedmont-valley flat environment.

Following deep erosion of the Appalachian region streams from eastern highlands spread arkosic sands across the basin, forming extensive floodplain deposits in sheets commonly bounded by erosional contacts. Streams of low sinuosity that swept entire floodplains carried much of the muddy sediment beyond the basin.

Meanwhile local deposits of debris flows, sheet floods and more distal steep-gradient streams built up a belt of alluvial several miles wide fans along the faulted northwest border.

The episode of sand accumulation ended with central ponding of the longitudinal drainage to form the Lockatong lake with a narrow marginal floodplain facies. The cause of ponding is not known, but it may have been related to especially active influx of alluvial fan deposits from the northwest border at both ends of the basin, combined with continued central sinking.

Once established, conditions in the long lake varied only within narrow limits for several million years. In this stable setting, shielded from tectonic influences, cyclic variations in climate exerted a major control on the lacustrine sedimentation. Similarities between Lockatong detrital cycles and upper Stockton fluvial deposits and between grayish-red chemical cycles and broad Brunswick mudflat deposits indicate that Lockatong deposits accumulated in a shallow lake. In fact, in its later stage it probably was a playa.

With waning of the lake, broad mudflats with

wandering watercourses and weak external drainage prevailed during accumulation of the muddy Brunswick deposits. In this episode the northwest upland may have supplied an increased amount of sediment. Expression of short cycles was suppressed, but long climatic cycles continued to produce episodes of a dry, oxidizing environment and thick sequences of ferric-oxide-rich mud alternating with moister periods and accumulation of thin sequences of dark gray mud.

Although the conspicuous reddish-brown color of these deposits has commonly been cited as evidence of the climate, it yields no particular information about temperature or rainfall. At best, the abundant ferric-oxide pigment in the mud delivered to the basin points to considerable weathering in the source area. Flora and fauna preserved in Newark strata do suggest a warm climate, however.

Thus the physical characteristics of this nonmarine rift valley sequence were determined to a large extent by their geographic environment (table 1).

Paleomagnetic data. A paleomagnetic study by Opdyke (1961) including analyses of samples from the Stockton, Lockatong and Brunswick Formations, as well as from the Watchung flows and Palisade Sill, shows that:

1. The directions of natural remanent magnetization in the sedimentary and in igneous rocks are closely grouped.

2. Magnetization in sedimentary and in igneous rocks has essentially the same inclination.

3. No reversals of magnetization were found.

4. The pole position at about 63°N, 108°E is in good agreement with that obtained from other North American Triassic deposits.

Major references. Most of the observations summarized in this review have been drawn from published reports as follows:

General: Darton 1902; Darton and Kümmel 1908, 1909; Glaeser, 1966; Johnson and McLaughlin, 1957; Kümmel 1897, 1898, 1899, 1914; Lewis 1907a, 1909; MacLachlin 1959; McLaughlin 1943, 1944, 1945, 1946b, 1959; Savage 1968; Van Houten and Savage 1968; Wherry 1931.

Stockton Formation: Rima, Meisler and Longwill 1962.

Lockatong Formation: Van Houten, 1962, 1964, 1965.

Brunswick Formation: Drake, McLaughlin and Davis 1961; McLaughlin 1933, 1946a; Sturm 1956.

Hammer Creek Formation: Carlston 1946; McLaughlin 1939; Glaeser 1963, 1966.

Hornfels: Andreae and Osann 1893; Lewis, 1908; Van Houten and Savage, 1968.

Igneous rocks and zeolites: Fenner 1908; Lewis 1907b, 1908; Schaller 1932; Walker 1940; Bucher and Kerr, 1948; Mason, 1960.

Structure: Drake and Woodward, 1963; McLaughlin 1941, 1959, p. 126–141; Sanders, 1962, 1963.

STOCKTON FORMATION

The Stockton Formation crops out along most of the eastern and southeastern margin of the Newark Basin where it overlaps on basement rocks of the Piedmont Province (fig. 4). Stockton strata also lie along the southern base of the two large fault-blocks in west-central New Jersey and adjacent Pennsylvania.

At the northeast end of the basin, near Piermont, New York, exposures of the formation ends where the Hudson River curves northwestward, cutting across the regional strike (Savage, 1967, p. 34). In the northern fault-block in west-central New Jersey Stockton deposits interfinger with the Hammer Creek Conglomerate near the northwest border fault. At the southwest end of the basin, in an area of poor exposure (see Glaeser, 1966, pl. 3) the Stockton Formation may either have been removed by

TABLE 1. DISTRIBUTION OF SELECTED FEATURES IN NEWARK DEPOSITS

		Lockatong			
	St	L	M	U	Br
Redox potential					
Oxidizing	X	Xx		x	X
Reducing		xX	X	X	xg
Drainage					
Through	X	Xd	xd	xd	X
Closed		xc	Xc	Xc	
Current Action					
Channeling	X	xd	xd	xd	x
Cross-bedding	X	Xd	xd	xd	x
Shrinkage cracks					
Subaerial	x	X	x	x	X
Syneresis		x	X	X	
Salt Casts			xc	Xc	X
Calcareous pellets, nodules					x
Fauna					
Swimmers		xd	xd	xd	xg
Burrowers	x	Xd	xd	xd	X
Tracks	x	x	x	x	X

X—dominant
x—subordinate
d—detrital short cycle
c—chemical short cycle
g—gray units
St—Stockton
L—lower detrital Lockatong
M—middle detrital and chemical Lockatong
U—upper predominantly chemical Lockatong
Br—Brunswick

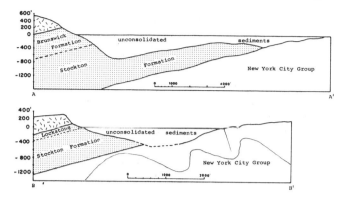

Figure 4. Generalized geologic sections across the Hudson River, showing the contact of the Stockton Formation with metamorphic rocks of the New York City Group.

A-A¹: Between Palisade Sill at South Nyack and Tarrytown, N. Y. After Worzel and Drake, 1959, fig. 5.

B-B¹: At site of George Washington Bridge, between Palisade Sill at Fort Lee, N. J., and upper Manhattan, N. Y. After Berkey, 1948, fig. 8A.

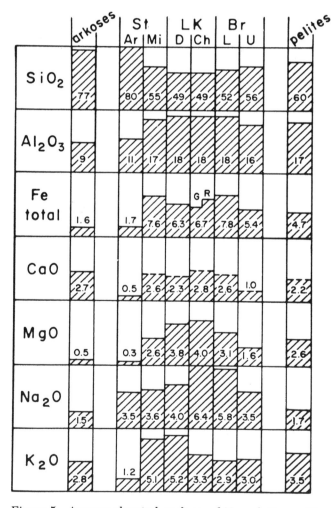

Figure 5. Average chemical analyses of Newark Group (Van Houten, 1965, fig. 3). Arkoses—average for arkoses, Pettijohn, 1963, table 12; pelites—average for pelites, Shaw, 1956, table 9.

erosion before deposition of the overlapping Hammer Creek Formation or it may interfinger with the lowest part of the conglomeratic deposit.

The Stockton Formation is about 5,000 feet (1,525 m) thick along the Delaware River in the northern fault-block. Fifteen miles south, near the Montgomery-Bucks County line, it is as much as 6,000 feet (1,830 m) thick (Rima and others, 1961, p. 9). From this central position the formation thins both to the northeast to about 1,300 feet (396 m) along the Hudson River (fig. 26) and to the southwest where it thins to about 1,000 feet (305 m) in 35 miles.

In the central part of the basin Stockton deposits grade upward into the conformably overlying Lockatong Formation. In the northeastern and southwestern parts Stockton strata are conformably overlain by the Brunswick Formation.

The Stockton Formation consists principally of light yellowish-gray to pale reddish-brown fairly well-sorted arkose and subordinate poorly-sorted conglomerate and reddish-brown mudstone distributed in rather distinct units (fig. 3). In the central part of the basin these units commonly are as much as 50 feet thick and can be traced for several miles. Persistent groups of beds several hundred feet thick have been identified as members (fig. 3; McLaughlin, 1959, p. 74–75). In eastern Pennsylvania the Stockton Formation comprises a lower conglomeratic arkose member, a middle fine- to medium-grained arkose member, and an upper mudstone member (Rima and others, 1962).

Yellowish-gray conglomeratic deposits consist of relatively dispersed, moderately rounded clasts averaging about an inch in diameter and only locally more than 3 inches across, set in a poorly-sorted arkosic matrix. Most of the clasts are quartz, some are quartzite, a few are feldspar, metamorphic rock and shale fragments. Near the base of the formation along the southeastern margin many of the clasts are quite angular. In southeastern Pennsylvania where the basal conglomerate overlies Paleozoic limestone, clasts of limestone are abundant.

Reddish-brown feldspathic mudstone and associated sandstone are well-bedded, very micaceous, and extensively disturbed by burrowers. The mudstone contains abundant illite and muscovite, very little kaolinite, and Na-feldspar predominates over K-feldspar. But K_2O exceeds Na_2O because of the abundant illite and muscovite (fig. 5).

The distinctive well-sorted, medium- to fine-grained arkose occurs in thick well-bedded units only locally interbedded with stringers of conglomerate and mudstone. Bedding, commonly is marked by films of reddish-brown mud or, locally by concentrations of grains of specular hematite (after magnetite). Cross-bedding and channeling are common. Most of the arkose has a texture of interlocking grains (sutured contacts) produced by pressure solution which probably resulted from burial below an estimated 10,000 feet (3,050 m). Authigenic feldspar occurs locally as overgrowths and void-fillings. Silica and calcite are the common cements. Much of the fine-grained arkose is speckled with small yellowish-brown intergranular patches. In the least weathered specimens these consist of a carbonate (ankerite?) extensively stained by goethite.

Stockton arkose contains abundant quartz (50–70 percent) and feldspar (15–40 percent) which generally decreases stratigraphically upward and northward, subordinate muscovite, chert, and metamorphic rock fragments, and a minor amount of biotite and chlorite. Albite-oligoclase commonly is more abundant than K-feldspar in the finer-grained facies, and is less altered. Compared with an average arkose (fig. 5) the Stockton arkose is rich in Na_2O, containing several times as much Na_2O as K_2O.

Relatively common opaque grains are mostly hematite with residual cores of magnetite. Hematite also occurs in intergranular areas as tiny randomly oriented platelets, as well as wispy elongate patches arranged crudely parallel to bedding but generally absent between tight grain contacts.

LOCKATONG FORMATION

The Lockatong Formation is a huge lacustrine lens in the middle of the Newark Group in central New Jersey and adjacent Pennsylvania. It crops out near the eastern border from the vicinity of the George Washington Bridge to Staten Island above and below the Palisade Sill, and along the southeastern border from Kingston, New Jersey, to Phoenixville, Pennsylvania, a distance of about 110 miles. Lockatong strata are also exposed in the lower part of the two large fault-blocks in the central part of the basin.

The Lockatong Formation is about 3,750 feet (1,150 m) thick in the northwestern fault-block (fig. 3) in west-central New Jersey, and thins laterally to the northeast and southwest. At its southwest end 10 miles west of Phoenixville (Schuylkill River), Pennsylvania, Lockatong deposits interfinger with Brunswick Mudstone and wedge out between the Stockton Arkose and Hammer Creek Conglomerate

(Glaeser, 1966, pl. 3). Toward the northeast the metamorphosed Lockatong Formation thins to 500–750 feet in subsurface west of Staten Island. It interfingers with the uppermost part of the Stockton Formation along the Hudson River at least as far north as the George Washington Bridge where it is represented by dark gray hornfels above and below the Palisade Sill (fig. 26). The Formation also thins southward to 2,000 feet (610 m) near Trenton and to 1,500 feet (458 m) near Princeton.

Lockatong deposits conformably overlie the uppermost reddish-brown and dark gray silty, micaceous mudstone of the Stockton Formation. The lower 400–500 feet consists of rather similar micaceous mudstone with subordinate ripple-marked and mudcracked fine-grained sandstone. The Lockatong Formation grades upward into the Brunswick Formation through a thick section of rather regularly alternating reddish-brown and dark gray units (figs. 3 and 7) recurring at 350-foot intervals (McLaughlin, 1933, 1944, 1948, 1949). Gray units above the main body of the Lockatong Formation are successively thinner upwards and laterally, whereas reddish-brown units are successively thicker. The lower 1,600 feet of the interbedded sequence along the Delaware River in the northern fault-block has been assigned to the Lockatong Formation because (1) argillite predominates, (2) the thicker interbedded units are gray, and (3) analcime not only is present but is unusually abundant in the reddish-brown units. These reddish-brown units are thickest and most analcime-rich in the northern fault-block; they extend more than 15 miles to the southwest, and 8 to 10 miles to the northeast where they interfinger with the Hammer Creek Conglomerate.

Throughout most of its extent the Lockatong Formation is arranged in short "detrital" and "chemical" cycles (fig. 6), averaging about 15 feet thick. Based on assumed varve-counts these cycles are interpreted to be the result of expansion and waning of an extensive lake, controlled by the 21,000-year precession cycle (Van Houten, 1962, p. 568, 572–573).

Detrital short cycles averaging 14–20 feet thick comprise several feet of black pyritic shale succeeded by platy dark gray carbonate-rich mudstone in the lower part and tough, massive gray calcareous argillite in the upper. The argillite has a small-scale contorted fabric produced largely by crumpled shrinkage-crack and burrow casts, and is rarely ripplemarked. Thicker detrital cycles commonly contain a 2–5 foot layer or lens of thin-bedded ripplebedded siltstone and very fine-grained sand-

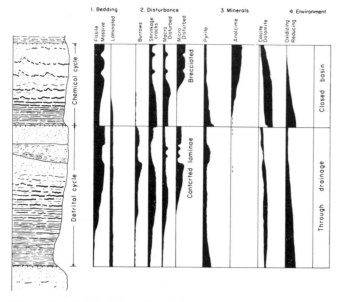

Figure 6. Model of detrital and chemical short cycles of Lockatong Formation, showing distribution and qualitative estimates of prevalence of sedimentary structures, selected minerals and sedimentary environments. Chemical cycles average 10 ft thick, detrital cycles average 17 ft thick.

stone with small-scale disturbed bedding (fig. 6). On average, the detrital cycles contain abundant Na-feldspar, some K-feldspar, illite and muscovite, chlorite and calcite, and a little quartz.

The more common and unique chemical cycles average 8–13 feet thick. Lower beds 1–8 cm thick are alternating dark gray to black platy dolomitic mudstone and marlstone broken by shrinkage cracks. Locally basal beds contain long lenses of crystalline calcite and pyrite as much as 2 cm thick layers of gray marlstone extensively disrupted by syneresis cracking. The upper part of a chemical cycle is tough gray analcime- and carbonate-rich argillite. The massive mudstone is brecciated on a microscopic scale, apparently the product of syneresis. Tiny slender crumpled shrinkage cracks irregularly filled with crystalline dolomite and analcime produce a "birdseye" fabric. Some thinner chemical cycles in the upper part of the formation are grayish-red to reddish-brown (fig. 17). Thinner, darker red beds are disrupted by shrinkage cracks and broken into mosaic intraformational breccia with patches of analcime and dolomite concentrated in the cracks. Thicker massive beds are speckled with tiny lozenge-shaped pseudomorphs of dolomite and analcime after gypsum? or glauberite?, and locally marked by long intricately crumpled crack-filling.

Argillite in the upper part of chemical cycles contains as much as 7 percent Na_2O and as little as 47

percent SiO_2 (fig. 5), and is composed of a maximum of 35–40 percent analcime together with albite, dolomite and calcite, and illite and chlorite. This mineral assemblage apparently resulted from flocculation and precipitation from carbonate-silicate lake water rich in cations, as well as from subsequent authigenesis of the alumino-silicate gel.

Chemical cycles are limited to the central 60 miles of the basin. Beyond this inner area the formation consists of detrital cycles and silty argillite. Clusters of 20–25 short cycles, either predominantly chemical or detrital, occur in long cycles about 325–350 feet thick. These coincide with the pattern of alternating reddish-brown and gray units in the upper part of the Lockatong and throughout most of the overlying Brunswick Formations (figs. 3 and 7).

BRUNSWICK FORMATION

The Brunswick Formation is widely but poorly exposed throughout the Newark Basin. In the central part where Lockatong deposits are thickest the conformable Brunswick Formation is at least 6,000 feet (1,830 m) thick. To the northeast, where the Lockatong Formation is thin or absent Brunswick deposits are several thousand feet thicker; to the southwest they have been estimated to be more than 16,000 feet (4,880 m) thick (fig. 1).

Throughout the central part of the basin the Brunswick Formation consists of a monotonous succession of reddish-brown mudstone and siltstone, with local beds of claystone and fine-grained sandstone. Unlike many other nonmarine basin deposits, this central facies is devoid of well-developed coarse-grained channel-fill deposits. Commonly two kinds of mudstone occur in alternating sequences (fig. 19): one is bright reddish-brown crumbly to fissile mudstone to claystone with persistent bedding; the other is more resistant, massive silty mudstone with broad channel-deposits of fine-grained sandstone and abundant traces of burrowers (Picard and High, 1963).

In addition to evidence of extensive burrowing, Brunswick deposits contain a variety of surface markings, notably shrinkage cracks and reptile footprints. Complete molds of glauberite locally filled with calcite or barite are abundant in scattered 3–12 inch layers. Calcite casts of skeletal glauberite crystals also form rosettes several inches in diameter.

Persistent thin units of dark gray mudstone are distributed through the formation at 350–400-foot intervals matching those in the upper part of the Lockatong Formation (figs. 3 and 7). Eight such units have been identified in deposits in the northern fault-block (McLaughlin, 1959, p. 102–104). Two of these are thicker and more widespread. The lower, Graters Member is about 1,420 feet (433 m) above the base of the formation; the upper, Perkasie Member is about 3,025 feet (923 m) above the base (fig. 3).

Near the northwestern margin Brunswick deposits contain abundant sandstone and conglomerate, and several miles from the border fault they interfinger with the Hammer Creek Conglomerate. At the southwest end of the basin the Brunswick Formation interfingers with the Hammer Creek Formation in its type area (Glaeser, 1966, p. 10), pointing to an important influx of detritus from the northwest upland (fig. 8). Similarly, in northeastern New Jersey the Brunswick Formation contains lenticular deposits of reddish-brown, poorly-sorted arkose and tongues of conglomerate with clasts of quartz, quartzite, limestone and feldspar as much as 5 inches in diameter. In adjacent Rockland County, New York, Brunswick deposits are increasingly conglomeratic (Savage, 1968). Here the formation consists of: (1) a lower arkosic facies (750–1,000 feet) including interbedded burrowed micaceous mudstone locally containing abundant intraclasts and pellets, and lenses of conglomerate with clasts (some of which are Paleozoic sedimentary rocks) as much as 2½ inches in diameter; (2) burrowed silty mudstone and well-bedded arkose with scattered clasts as much as an inch in diameter (1,000–1,500 feet; 305–458 m); (3) conglomeratic sandstone and conglomerate (at least 3,000 feet; 915 m) with clasts as much as 12 inches in diameter composed in part of Paleozoic sedimentary rocks and rare clasts of gneiss and gabbro. This succession grades upward and northwestward into the Hammer Creek Conglomerate.

The widespread feldspathic mudstone and micaceous siltstone in the Brunswick Formation consist primarily of quartz (50–75 percent in siltstone, 10–30 percent in mudstone), less than 15 percent feldspar (except in the arkosic lower part), and relatively rare lithic fragments in deposits in the central and southeastern part of the basin. The fact that feldspar is common within several miles of the northwest border points to a significant influx from the southern and eastern upland composed of metamorphic rocks.

Both Na-feldspar and K-feldspar are present; the former commonly predominates. Illite is the major clay mineral, with chlorite subordinate (Sturm, 1956). As the base is presently delimited in the central part of the basin the lowest 150 feet as-

signed to the Brunswick Formation contain a minor amount of analcime. Hematite is the pigment mineral coating grains and staining the clay fraction; it is also the common opaque mineral. There is no reliable way to determine whether the hematite pigment was inherited directly from red soil or produced after deposition (Van Houten, 1968, p. 404–405).

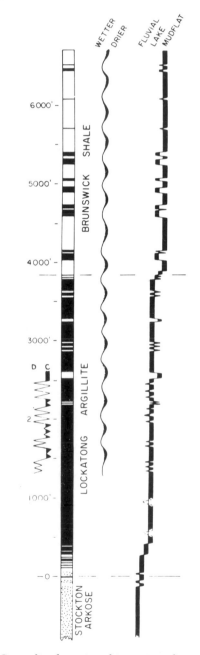

Figure 7. Generalized stratigraphic section of uppermost Stockton, Lockatong and Brunswick formations. Black—gray units; white—reddish-brown to grayish-red units. Left of column—distribution of short chemical (C) and detrital (D) cycles in pattern of intermediate and long cycles. Postulated long climatic cycles of wetter and drier phases and sequence of alternating geographic environments sketched in column at right.

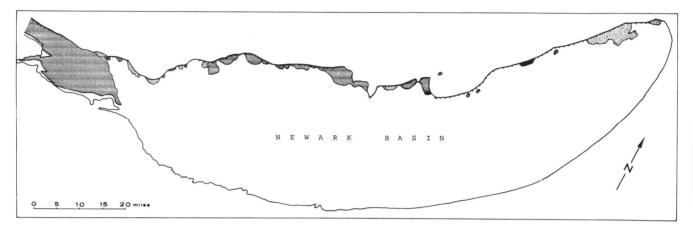

Figure 8. Sketch map of outcrops of Hammer Creek Formation in Newark Basin. Dotted heavy line is faulted northwest border. Predominantly Paleozoic quartzite clasts—curvilinear pattern; predominantly Paleozoic limestone clasts—dotted pattern; predominantly Precambrian crystalline clasts—black. Data mostly from Carlston, 1946; Glaeser, 1966.

Compared with the average fine-grained mudstone (fig. 5) Brunswick mudstone generally is rich in Na_2O and contains a little less K_2O. Most of the Na_2O apparently is in detrital Na-feldspar; only a minor amount is in authigenic feldspar. The K_2O is in illite and sericite; relatively little is in K-feldspar.

HAMMER CREEK FORMATION

Along the entire length of the northwest border patches of very coarse conglomerate project several miles into the basin (figs. 3 and 8). Some of these deposits interfinger with Brunswick strata; others interfinger with lower deposits of the Newark Group (fig. 1). The clasts in this very coarse border facies are almost entirely lower and middle Paleozoic sedimentary rocks (fig. 2); Precambrian gneissic debris predominated only very locally. Throughout most of its extent the conglomeratic facies is separated from its northwestern source area by the border fault, but locally the conglomerate overlaps on rocks of the northwest upland.

At the narrow southwest end of the basin a thick deposit of coarse conglomerate that interfingers with the Lockatong and Brunswick formations was named the Hammer Creek Formation by Glaeser (1963; 1966, p. 11). The name is here extended to all the predominantly conglomeratic deposits in the Newark Group composed largely of Paleozoic and very locally of Precambrian clasts derived from the northwest upland, and concentrated mainly along the northwest margin except where the basin narrows at its northeast and southwest ends.

In the central part of the basin Newark deposits

uplifted on the northern fault-block reveal that the Stockton, Lockatong and Brunswick Formations interfinger with the marginal conglomeratic facies (Kümmel, 1898, p. 57; McLaughlin, 1946, p. 90). Reasonably, this relationship prevails in subsurface along much of the northwest border.

At the northeast end of the basin *exposed* Hammer Creek deposits are about 1,000 feet (305 m) thick with clasts as much as 4 feet in diameter, extend about 3.5 miles across the basin (Savage, 1968, p. 56), and interfinger with the upper part of the Brunswick Formation to the southwest.

Throughout its lateral extent the Hammer Creek Formation varies from a poorly-sorted, rather well-rounded conglomerate to markedly angular breccia. The clasts range in size from common 4–8 inch cobbles to less common fragments 1–4 feet long. The composition also differs along the extent of the formation (fig. 8). In the type area in the southwest and in west-central New Jersey along the Delaware River, the clasts are predominantly Paleozoic quartzite. In eastern Pennsylvania, locally in New Jersey and in Rockland County, New York, angular Paleozoic limestone clasts predominate. Clasts of Precambrian gneiss predominate and fragments of basalt are present in only a very few patches. In outcrops along the Delaware River the Hammer Creek Formation consists of lenticular units of fining-upwards detritus (figs. 20 and 21). The sandstone fraction is lithic-rich, but it does contain Na-feldspar sand grains.

Although the roundness of the tough quartzite clasts suggests prolonged abrasion, the character and distribution of the deposits point to local accumulation near the foot of an upland terrane by debris flow and short steep-gradient streams from a source within 10–25 miles (Carlston, 1946). Presumably the northwest border of the basin lay essentially along the present outcrop margin. Moreover, faulting along it and erosion of the upland block prevailed throughout much of Late Triassic (Newark) time. The most abundant supply of coarse detritus was from the northwest uplands adjacent to the narrow northeast and southwest ends, and these sources became especially active after accumulation of the Stockton Arkose.

NEWARK HORNFELS

Sedimentary rocks of the Newark Group have been metamorphosed by both extrusive and intrusive basic igneous rocks. Metamorphic effect of the thick Watchung lava flows on reddish-brown Brunswick mudstone amounted to little more than alteration of its color to greenish gray several inches from the contact and of its illite-chlorite matrix to a non-expanding 14Å mineral.

Metamorphic rocks produced by the intrusions reflect the effects of temperature, controlled largely by distance from intrusion; and of composition, controlled by the sedimentary facies involved. Most of the published attempts to assign these hornfels to metamorphic facies have failed to recognize the role of the basically different rock types.

Stockton Arkose intruded by the Palisade Sill along the Hudson River shows a marked effect only within a foot or two of the intrusion. This produced megascopic diopside and sphene, together with minor apatite and magnetite, and locally green hornblende. Vermiculite and talc are also present in minor amounts. Additional Na-feldspar was also produced whereas both K-feldspar and quartz were diminished. In finer-grained silty arkosic K-feldspar and quartz commonly were eliminated and abundant diopside colors the rock a pale greenish gray. Walker (1961, p. 13) states that arkose along the lower Palisade contact was rheomorphically mobilized and injected into adjacent hornfels in a lit-par-lit fashion, but no locality is cited.

Hematitic mudstone and micaceous siltstone of the Brunswick Formation were only mildly responsive to thermal metamorphism by the sills. Hornfels developed in these deposits occur along the border of the Palisade Sill in northeastern New Jersey and adjacent New York, in the contact zone of the Cushetunk Mountain, Rocky Hill and Lambertville sheets in west-central New Jersey, and of the Coffman Hill, Haycock Mountain and similar sheets in adjacent Pennsylvania. Metamorphic

effects occur through several hundred feet of section, but marked alteration is limited to the inner few hundred feet. The least altered Brunswick Mudstone is a dark reddish-brown to grayish-lavender rock with ½- to 1½-inch nodules of epidote and chlorite. Nearer the sill the matrix is chlorite-sericite with large shot-like patches of coarsely crystalline chlorite-muscovite or epidote and magnetite pseudomorphic after cordierite. Veinlets and patches of feldspar, hornblende, pyroxene and tourmaline, as well as scattered sphene and specularite produce local variations. The characteristic pale gray to pale lavender color is due largely to conversion of aphanitic hematite pigment to fine-grained specularite. Magnetite is developed only in the dark gray hornfels 50 feet or less from the intrusion. This inner facies is also marked by the presence of tourmaline and cordierite.

The Lockatong Formation was the most susceptible to thermal metamorphism and has been extensively reconstituted to varieties of very fine-grained biotite, Na-feldspar hornfels marked by the absence of quartz, a paucity of SiO_2, and an abundance of Na_2O. These hornfels occur above and below the Palisade Sill from the vicinity of the George Washington Bridge to west of Staten Island, and along the border of intrusive sheets in west-central New Jersey and adjacent Pennsylvania. In spite of the marked mineralogical alteration of these rocks they still exhibit their characteristic primary sedimentary structures and fossils diagnostic of this cyclic lacustrine facies. Distinctive metamorphic minerals developed in hornfelsed Lockatong cycles include (figs. 29 and 31):

Upper chemical argillite: nepheline, cancrinite, thomsonite and calcite within 80 feet of the intrusion; cancrinite, thomsonite, and increasing unaltered analcime within 440 feet of the intrusion. Locally sodalite, muscovite, green-brown hornblende, and pale green augite are developed.

Upper detrital argillite: K-feldspar, scapolite, aegirine, diopside, clinozoisite, chlorite and magnetite.

Middle carbonate-rich mudstone: grossularite, andradite, diopside, prehnite, and calcite and idocrase in carbonate layers.

Lower black shale: epidote, magnetite, cordierite, andalusite.

Hornfelsed xenoliths of Lockatong Argillite a few 10's of feet long occur in the basal part of the Palisade Sill (figs. 27, 28), and of its offshoot at North Bergen (fig. 30). In the uppermost part of the diabase at Mt. Gilboa (Brookville) a block of nepheline and analcime syenite (fig. 15) apparently was produced by reaction of a soda-rich Lockatong Argillite xenolith with the granophyric differentiate of the diabase (Barker and Long, 1969).

IGNEOUS ROCKS

The intrusive and extrusive basic igneous rocks associated with Newark sedimentary rocks are tholeiitic olivine-poor basalt and diabase characteristic of rift-valley sequences. These igneous rocks and their counterparts in the other Triassic basins record the emplacement of basic magma on an enormous scale, 1,000 miles in length and about 200 miles wide. Significantly, they correlate broadly with the vast Karoo dolerites and equivalent basic rocks in the Southern Hemisphere, and like them, apparently were emplaced during an episode of tension attending the widening of the Atlantic Basin.

Within the Newark Basin igneous rocks of rather varied form are distributed in two distinct provinces generally separated by the two major intrabasin faults. The thick Palisade Sill and its offshoots and the three extensive Watchung lava flows center in the northeastern half of the basin, whereas less extensive basin-shaped sheets and thick associated dikes are limited to the southwestern part of the basin. Thus the lateral extent of intruded sheets is far greater than that of the flows.

The Palisade quartz diabase sill was intruded into the uppermost Stockton Formation and basal Lockatong Formation as seen along the Hudson River in eastern New Jersey. It is a distinctly layered sheet as much as 1,000 feet (305 m) thick with a marked concentration of olivine 15–20 feet thick about 40–50 feet above its base. Both locally and regionally the sill transects enclosing strata. Locally it cuts as much as 100 feet across beds with a ragged contact. Because lower and upper contacts are very irregular, tongues of trap intrude enclosing beds and blocks of sedimentary rock have commonly been caught up as xenoliths. To the northeast in Rockland County, New York, and to the southwest in west-central New Jersey and adjacent Pennsylvania the sill intrudes the Brunswick Formation (fig. 1). Contrary to Carey's view (1961, p. 167) the higher the stratigraphic position of this sill the more markedly it cross-cuts enclosing strata. Minor offshoots from the top of the Palisade Sill form the thin sheet at North Bergen (Granton Quarry), New Jersey, and the Snake Hill plugs at Secaucus, northeast of Newark.

Local evidence of a steep cross-cutting contact along the west edge of the outcrops suggests that the sill may not extend down dip to the west (Lowe, 1959; see also Geologic Map of New Jersey, 1940). On the contrary, drilling in the Hackensack Meadows west of the Palisades reveals that in subsurface some parts of the sill dip about 10°W with the enclosing strata (Widmer, 1959). A gentle, west-dipping sheet is also suggested by the presence of high-grade hornfels less than 500 feet below the surface as much as 2 miles west of the west edge of the Palisades on Staten Island. Moreover, Lewis' (1907, p. 205) review of the available evidence suggests that the sill apparently is conformable with the enclosing strata down dip to the west.

The Palisade Sill has long been cited as an outstanding example of simple gravitational differentiation by settling of early-formed olivine (fig. 9) and later pyroxene crystals (Walker, 1940). Actually, such a mechanism apparently fails to account for associated features and more recently other processes have been suggested:

1. Fractional crystallization by ionic diffusion and convective circulation of the magma, supplemented by minor gravitational settling of large olivine crystals (Hess, 1956, p. 1960).

2. Fractional crystallization with late-stage injection of a new pulse of magma into the partially crystallized mush. Large olivine crystals were concentrated by gravitational settling (K. R. Walker, 1962; [1969] in press).

3. Mechanical or hydrodynamic flow differentiation, with central concentration of early-formed olivine crystals during intrusion. The crystal mush then settled as a result of fluctuation of velocity and pressure (Simkin, 1965, p. 102; Bhattacharji, 1967).

West of its spectacular exposure along the Hudson River the Palisade Sill and associated slender dikes reappear in central and western New Jersey in the Lockatong and Brunswick formations as the Rocky Hill sheet (1,500 feet (458 m) thick), and the Sourland Mountain and Lambertville Sheet (1,800 feet (549 m) thick) in the Hopewell fault-block. Additional faulted masses include Pennington Mountain, Baldpate Mountain, Solebury Mountain and its feeder dike of Bowman Hill, Belle Mountain, Gilboa Mountain and the Byram sheet along the Delaware River, as well as the isolated Cushetunk Mountain and Round Mountain high in the Brunswick Formation.

In Pennsylvania, west of the major fault-blocks, diabase sheets that thicken toward their axis and have distinctive ringlike outcrop patterns, intruded different levels of the Brunswick mudstone (Hotz,

1952). Some are very near the top of the formation, yet they are coarse-grained and markedly differentiated. Apparently the amount of overburden was not a simple control of the level of intrusion or the extent of differentiation. Associated with the basin-shaped sheets in the southwest are thick, rather irregular dikes. One of the thickest of these is 2,500 feet (760 m) wide and extends from the southeastern to the northwestern border west of the Schuylkill River through 10,000 feet of Newark strata.

Long, slender steep-dipping, north- to northeast-trending dikes are common throughout the basin. These are discordant to enclosing rocks and are especially common across the west-trending Cornwall-Kelvin Displacement (fig. 12), are much straighter than the sinuous basin trends, and commonly extend beyond the basin border but are not offset along its faulted margin. Thus these dikes are younger than the sills and flows, their subsequent tilting, and the major faulting of the basin.

Tholeiitic lava flows in the upper part of the Brunswick Formation in the eastern half of the Newark Basin consist of three multiple Watchung flows about 50 miles long, and their 400-foot-thick Sand Brook and New Germantown remnants to the west. Each is composed of at least two thick flow units. The stratigraphic sequence of flows comprises:

Top:
    Brunswick Mudstone
    3rd Watchung Mountain—about 300 feet thick
    Brunswick Mudstone—about 1,200 feet thick
    2nd Watchung Mountain—700–900 feet thick
    Brunswick Mudstone—500–600 feet thick
    1st Watchung Mountain—600–650 feet thick
    Brunswick Mudstone

Outpouring of the Watchung basalt began without preliminary explosive activity and spread fluid lava at least 50 miles in single flows as much as 150–175 feet thick. In a general way the distribution of the lava records the axial position of the Newark Basin, thus indicating its displacement toward the northwest border in late Brunswick time. If DeBoer's (1968, p. 619–620) correlation of these three flows and the intrusions as well with the middle, Holyoke volcanic event in the Connecticut Valley is correct, the igneous episode in New Jersey was of very short duration indeed.

The lower two flow units of the 1st Watchung Mountain, the lowest one markedly jointed, the overlying one ropy and pillow lava (fig. 10), exhibit many of the characteristic features of nonmarine basaltic flows and are much like successive flows of the Columbia River basalt (see Mackin, 1961, p. 14–17) in south-central Washington.

In the vicinity of Paterson, New Jersey, the uppermost vesicular part of the lowest flow unit and the overlying pillow and ropy lava of the second flow unit of the 1st Watchung Mountain (fig. 24) are mineralized with spectacular assemblages of zeolites and associated secondary minerals. Fenner (1908) and Schaller (1932) concluded that the zeolites were formed when the second unit (mistakenly believed by Fenner, 1907, p. 312–325, to be

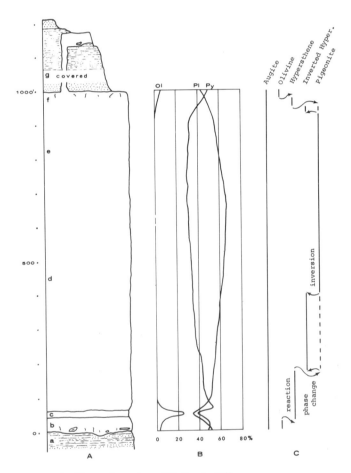

Figure 9. Vertical section of Palisade Sill.
  A. Composite section in vicinity of Edgewater and North Bergen, N. J.
    a. Lockatong hornfels and Stockton Arkose.
    b. Chilled border of sill with white veins and xenoliths.
    c. Olivine diabase 3 to 20 ft thick.
    d. Normal diabase, coarser and more acidic toward top.
    e. Coarse diabase with pegmatitic schlieren.
    f. Chilled border with white veins, xenoliths rare.
    g. Lockatong hornfels and Stockton Arkose intruded by subsidiary sill at North Bergen.
  B. Percentage by weight of olivine, plagioclase and pyroxene (Walker, 1940).
  C. Model of crystallization sequence (Jacobeen, 1949).

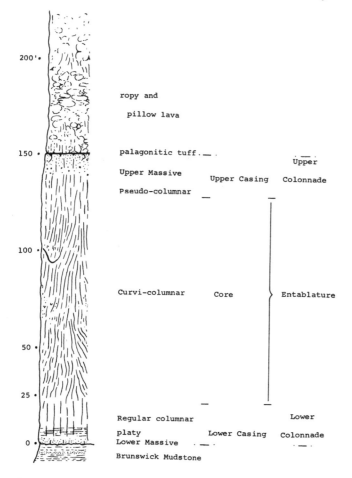

Figure 10. Vertical section of first and second flow units of 1st Watchung Mt., northeastern N. J.; and terminology commonly applied to subdivisions.

STRUCTURE

The Newark rift valley is a post-orogenic half-graben formed during the Palisade Disturbance that terminated the Appalachian orogenic cycle during Late Triassic time (Rodgers, 1967, p. 419–421). The basin follows the regional grain of Appalachian structures (30–50° NE) and its rocks dip 10–20 NW. A conspicuous line of faults with as much as 18,000 feet of stratigraphic displacement marks its northwestern border. *Total* displacement along the faults, including removal of detritus from the source area, may have been several thousand feet more. Many thousands of feet of differential displacement also occurred along the flanks of the eastern and southeastern source area. This, too, presumably was accompanied by faulting not now clearly expressed in exposed terrane or rock relations.

Within the basin both faulting and warping are common (fig. 11). Repetition of stratigraphic sequences by the major faults produced the wide central part of the basin. The north to northeast trending Hopewell Fault (figs. 11 and 14) with 6,000–10,000 feet (1,830–3,050 m) of stratigraphic displacement along at least 20 miles of trace in west-central New Jersey may continue northward to the northwest border fault. It continues west-southwest across the Delaware River as the Pidcock Creek Fault and a branch southwest as the Buckmansville Fault which continues toward the west-trending Chalfont Fault. At both ends of the Hopewell Fault major diabase sheets, which are older than the last movement on the Fault, pinch out at the surface before reaching the rupture.

The northeast to north-trending Flemington Fault (figs. 11 and 14) with as much as 10,000 feet (3,050 m) of stratigraphic displacement at the Delaware River, extends to the border fault, and as the Furlong Fault continues southwest to the Chalfont Fault.

Sanders (1962, 1963, p. 512–514) has speculated that displacement on the Hopewell and Flemington faults includes right-lateral strike-slip movement of as much as 12 miles, now expressed in the marked offset of the northwest border fault in west-central New Jersey.

The west-trending Chalfont Fault in Pennsylvania, lying in the middle of the west-trending part of the Newark Basin, may be a surface expression of the deepseated Cornwall-Kelvin Displacement (fig. 12). At the west end of the basin a southeast trending fault extending at least 12 miles from Reading to west of Phoenixville is the locus of a thick dike and has been downthrown about 2,500

an upfaulted part of the first unit) flowed into a saline lake on the Brunswick mudflats. More reasonably, the first columnar flow deranged the regional drainage. Then the overlying second flow was extruded and became pillowed where it encountered local lakes. Development of pillows locally and at many levels testifies to the complexity of flow emplacement. But extrusion into the lakes probably did not produce the zeolite mineralization. Comparison with other nonmarine lava flows reveals that *unburied* Cenozoic and Recent pillow lavas contain no abundant zeolites, and that spectacular zeolite mineralization has occurred in jointed, nonpillowed flows (as in Triassic basalt in Nova Scotia). This evidence suggests that the pillow lava, together with the underlying vesicular lava, at Paterson, New Jersey, was mineralized after burial by circulating ground water, perhaps derived largely from compaction and dewatering of the underlying Brunswick mudstone.

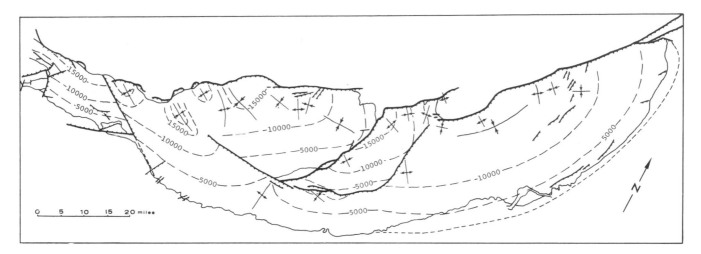

Figure 11. Generalized structure map of Newark Basin. Major faults with dots on downthrown side; anticlines and synclines; isopach lines with thickness in feet.

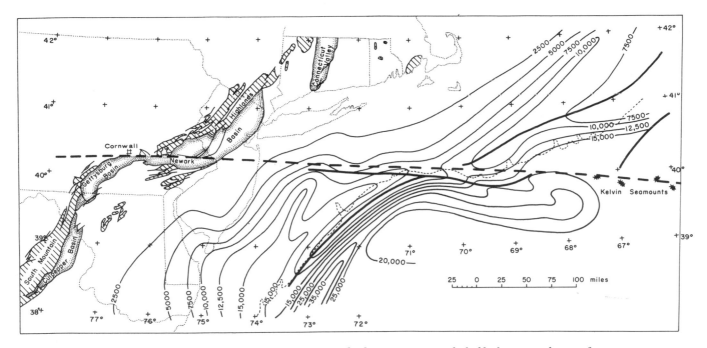

Figure 12. Map of central Atlantic states and adjacent continental shelf, showing relation of postulated Cornwall-Kelvin Displacement at about 40°N to trend of Triassic rift valleys and major Precambrian masses, and to basement topography and magnetic anomaly belt of the shelf. After Drake and Woodward, 1963.

　　Cornwall-Kelvin Displacement with estimated 75–100 miles dextral movement—dashed heavy line.

　　Magnetic anomaly belt—solid heavy line.

　　Contour lines on buried basement—solid thin lines, depth in feet.

　　Triassic rift valleys—stippled.

　　Exposed Precambrian basement—oblique lines.

feet (760 m) on the east side. Other conspicuous dikes, especially in Pennsylvania, apparently record a pattern of faults. Many minor faults, mostly northeast-trending, with less than 500 feet of displacement occur throughout the Basin. They are most obvious in resistant sills and lava flows.

Newark rocks have also been warped into low anticlines and synclines (fig. 11), most of which are near major faults. Commonly their axes are transverse to trends of the faults and of the basin-axis. This relation suggests that development of the folds was related to movement along the major faults (Wheeler, 1939).

Deformation of the Newark Basin involved the following events:

1. Uplift of an east-southeast borderland during the entire episode, supplying the distinctive Na-rich detritus to the basin.

2. Essentially continuous downfaulting along the northwest border producing a regional northwest-sloping gradient in the basin, and a northwest upland that shed coarse Paleozoic detritus along its steep scarp throughout most of Late Triassic time. Locally the alluvial fan deposits overlapped

earlier faults as the active faulting shifted northwestward. Toward the end of this episode extensive sills and lava flows were emplaced.

3. Widespread, marked deformation of all major faults and folds and tilting of the basin occurred after accumulation of the thick basin-fill and major igneous activity.

4. Emplacement of swarms of dikes in tensional fractures, many of which cut across regional structures.

The Newark Basin and its counterparts in eastern North America may have been part of a fracture system initiated by and ultimately destroyed by widening of the Atlantic Basin in Mesozoic time. Moreover, the striking eastward trend of the west end of the basin and of adjacent Appalachian structures (fig. 12) may reflect late Paleozoic right-lateral faulting along the Cornwall-Kelvin Displacement (Drake and Woodward, 1963; Drake and others, 1963). Movement on this wrench fault probably produced the offset of magnetic anomalies under the continental shelf in the vicinity of the Kelvin Seamounts.

FIELD TRIP

ROUTE AND OBJECTIVE

This two-day trip across the Newark Basin in New Jersey and New York (fig. 13) consists of 11 stops at the main rock types. The first day's traverse northward along the Delaware River (fig. 14) crosses three major fault-blocks. Stops in the northernmost block are at the successively younger soda-rich Stockton Arkose (1), the cyclic analcime-rich Lockatong Argillite and the Byram diabase and associated nepheline-cancrinite hornfels (2 and 3), lowermost transitional deposits of the Brunswick Formation (4), characteristic Brunswick Mudstone (5), and the Hammer Creek Conglomerate (6) at the northwest border.

The second day's traverse is northeastward from a stop at the columnar first flow of the 1st Watchung Mt. northwest of Newark (7), to its pillowed second flow with zeolites and associated minerals at West Paterson (8), then to the northeastern end of the basin in Rockland County, New York, to the coarser, lenticular, feldspathic facies of the lower Brunswick Formation at Nyack (9). Return southward on the Palisade Sill affords a stop at its famous olivine layer and lower scapolite and cordierite hornfels (10) and at the Lockatong prehnite, grossularite, diopside hornfels above the sill (11).

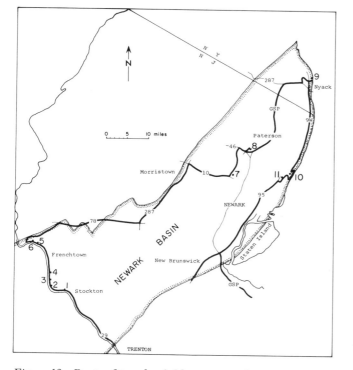

Figure 13. Route of two-day field trip across the Triassic rocks of Newark Basin, central and northeastern New Jersey and adjacent New York.

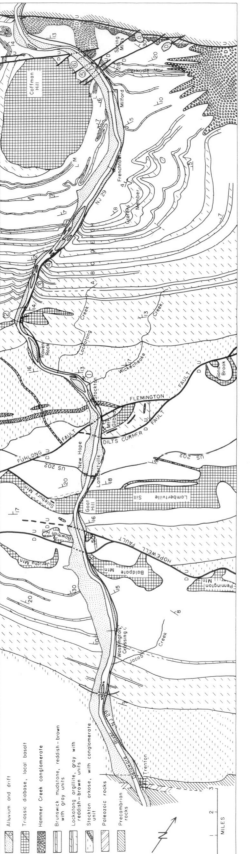

Figure 14. Geologic map of Triassic rocks along Delaware River, west-central New Jersey and adjacent Pennsylvania. Mainly after Johnson and McLaughlin 1957, pl. 1; and Drake, McLaughlin and Davis, 1961. Shows location of field trip stops.

ROAD LOG

Mileage

0.0   Calhoun St. cloverleaf, Trenton. Rocks below the Coastal Plain mantle and Triassic basin-fill are exposed in a 1.5 mile-wide belt. These basement rocks consist of the Baltimore Gneiss (about 1,100 m.y., Hopson, 1964, p. 28), the soda-rich Wissahickon Schist (latest Precambrian or earliest Cambrian; Hopson, 1964, p. 207), and Early Cambrian Chickies Quartzite. A similar sequence underlies much of the southeastern side of the Newark Basin. Traverse is along *Rt. 29 and the river road the entire first day.*

0.7   Crossing approximate contact of Stockton Arkose and basement rocks. Southern fault block is about 9 miles wide.

4.7   Scudder Falls Bridge. West portal in Pennsylvania is in uppermost Stockton Arkose, with basal platy black mudstone of Lockatong Formation exposed at N end of northern access. East portal is about on the contact. Lockatong Formation in southern fault-block is about 1,500–1,700 ft thick, suggesting marginal thinning.

5.1   Analcime-bearing reddish-brown argillite of

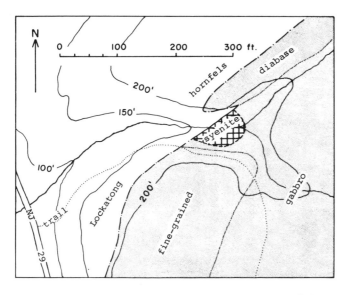

Figure 15. Geologic sketch map of Lockatong hornfels, diabase and gabbro, and nepheline and analcime syenite, Mt. Gilboa area, 0.3 mile south of Brookville, N. J. After Barker and Long, 1969, fig. 2.

upper Lockatong Formation was exposed in road-side excavations.

5.6   Analcime-rich reddish-brown argillite crosses road. Short cycles well-displayed in old quarry on west side of river.

6.1   Jacob Creek. Analcime-rich reddish-brown uppermost Lockatong Formation crops out in deep ravine. Belt of Brunswick Formation to N is about 4 miles wide. Local exposures on E side of road.

7.5   Washington Crossing. Intersection with Rt. 546. Gray tongues in Brunswick exposed locally.

10.7   Old quarry in Baldpate Mt. diabase. This intrusion in the Brunswick Formation, as the others along the traverse, apparently is a faulted part of the Palisade Sill.

11.0   Moore Creek. Southern Branch of Hopewell Fault crosses river and continues west as the Buchmanville Fault.

11.5   Mercer County Workhouse quarry in small Belle Mt. diabase mass. Across the river this sheet forms Bowman Hill (marked by stone tower), and like several others, has cross-cutting contacts.

12.0   Crossing northern branch of Hopewell Fault. Middle fault-block to north is about 4–5 miles wide along the river.

12.8   Lambertville Sill rises to Goat Hill (altitude, 450 ft) at south end; crosses river and forms Solebury Mt. Brunswick hornfels just above the contact south of town includes a tourmaline-rich facies; south of New Hope, Pennsylvania, on the west bank, metamorphosed cyclic drab deposits include grossularite-prehnite-muscovite hornfels.

13.8   Crossing approximate top of Lambertville Sill (1,800 ft thick).

14.1   NW-dipping Brunswick hornfels in roadcut to E just south of Rt. 518.

14.4   Junction Rt. 20 and U. S. 202. Turn west into Lambertville; turn north at first traffic light; continue on Rt. 29 across Brunswick Formation.

16.1   Crossing southern branch of Flemington Fault which repeats Lockatong Formation.

16.3   Mt. Gilboa quarry in very coarse-grained diabase (gabbro). Lockatong hornfels on south edge of quarry at entrance to property.

16.9   Lockatong nepheline-cancrinite hornfels above diabase intrusion. Nepheline and analcime syenites crop out along small creek to E (fig. 15). Barker and Long (1968; 1969 in press) have proposed that these syenites formed by re-action of Lockatong Argillite with a grano-phyric differentiate of diabase.

17.2 Crossing Flemington Fault at Brookville; it continues west as the Furlong Fault along which uplifted Cambro-Ordovician limestone is over-lain by the Stockton Formation. The fault block to the north is about 17 miles wide.

17.3 Arkosic conglomerate of Solebury Member (385 ft thick) of Stockton Formation. Pebbles of quartz and quartzite are as much as 3 inches in diameter; clasts of feldspar are rare.

17.75 Stockton. Schoolyard with exposure of Sole-bury Member.

18.6 Bear left after crossing Wickecheoke Creek at junction with Rt. 519.

STOP 1.

Mileage

18.9 Quarry in Stockton Arkose, upper Prallsville Member, 200 ft thick; 2,500 ft above base of the formation.
   Items:
   1. Well-sorted, medium-grained gray to yel-lowish gray arkose, with relatively little associated mudstone.
   2. Na-feldspar to K-feldspar is commonly 2:1.
   3. Channeling marked by thin reddish-brown mudstone layers.
   4. Cross-bedding, both large and small scale.
   5. Micro-placers of specular hematite (after magnetite) grains locally outline thin bed-ding. Implications for detrital remanent magnetization?
   6. Small yellowish-brown spots widely dis-tributed through arkose. Mainly goethite here; an iron-stained carbonate mineral in less altered outcrops.
   7. Thin layers of micaceous reddish-brown, well-bedded mudstone.
   8. Isolated burrows in arkose. Abundant hori-zontal burrows preserved on sole of arkose units. Mudstone extensively burrowed. Implications for red beds, detrital remanent magnetization, and rates of accumulation?
   9. Intraformational mud-chip conglomerate, commonly distributed on cross-bedding foreset slopes.
   10. Very coarse-grained kaolinized arkose near top of quarry.

20.1 Lockatong Creek. Quarry in Stockton Arkose high on west bank of the river.

21.3 Raven Rock. Old quarry to E in massive fine-grained arkose of the Raven Rock Member 300 ft thick, just S of road to Rosemont.

21.6 Well-bedded reddish-brown sandy mudstone and sandstone in uppermost Stockton Formation.

22.0 Crossing estimated Stockton-Lockatong contact.

22.8 Southernmost road outcrop of Lockatong horn-fels, about 250 ft below Byram intrusion.

STOP 2.

22.9 Roadcut in Lockatong Formation and Byram diabase sheet; 1,000–1,500 ft above base of the formation. Traverse 2a, 0.2 mile long, through Lockatong hornfels below and above the Byram intrusion (fig. 16).
   Items:
   1. Vague cyclic pattern of major ledges of albite-biotite hornfels; Nepheline occurs about 100 ft below sill.
   2. Lower contact of chilled border of diabase with sheared, weathered Lockatong pelitic hornfels.
   3. Coarse-grained diabase, extensively frac-tured. Joint minerals include calcite, epi-dote, prehnite, tourmaline and amphibole. Conspicuous lineations expressed in con-centration of joint minerals have no apparent relation to compositional layering in the diabase.
   4. Complex upper contact (enlargement), essentially undisturbed along curved con-tact; faulted and sheared diabase and horn-fels on north side. Nepheline, cancrinite, pyroxene and amphibole present in both blocks of Lockatong albite-biotite hornfels.
   5. Relation of 70-foot sequence from fault to outcrops north of creek is obscure. Iso-lated exposures and sheared zones at road-side topographically below position of northern lobe of diabase (see sketch map).
   6. O marks beginning of undisturbed, continu-ous sequence of Lockatong Formation, esti-mated to be about 70 ft above the faulted contact with diabase. Nepheline absent above the 11-foot mark (81 ft above diabase); cancrinite and albite diminish and analcime and thomsonite increase out to the 370-foot mark. Analcime and albite present above this level. Analcime and thomsonite also occur as joint minerals.

23.1 7. Metal disc at 10-foot mark 6 ft above road records level of flood water in September, 1955.
   Conspicuous white sprays of cancrinite, albite and calcite at 21-foot mark apparently developed from analcime-dolomite sprays as seen at Stop 3.

23.3 Old quarry in Lockatong cancrinite-rich horn-fels. Thomsonite is a common joint mineral along with analcime, calcite, and rarer ilmenite. Scapolite in the hornfels and laumontite on

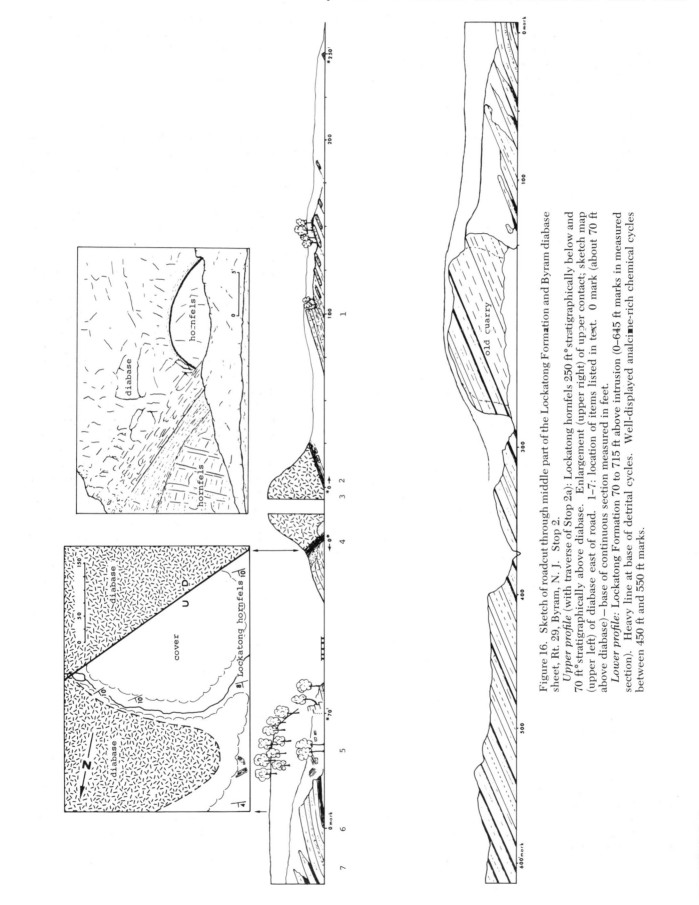

Figure 16. Sketch of roadcut through middle part of the Lockatong Formation and Byram diabase sheet, Rt. 29, Byram, N. J. Stop 2.

Upper profile (with traverse of Stop 2a): Lockatong hornfels 250 ft° stratigraphically below and 70 ft° stratigraphically above diabase. Enlargement (upper right) of upper contact; sketch map (upper left) of diabase east of road. 1–7: location of items listed in text. 0 mark (about 70 ft above diabase)—base of continuous section measured in feet.

Lower profile: Lockatong Formation 70 to 715 ft above intrusion (0–645 ft marks in measured section). Heavy line at base of detrital cycles. Well-displayed analcime-rich chemical cycles between 450 ft and 550 ft marks.

joints reported by Lewis (1909, p. 96) have not been confirmed. This dark gray hornfels produced crushed stone megascopically almost indistinguishable from fine grained trap.

23.6 Traverse 2b through typical analcime-rich chemical cycles of Lockatong argillite (450–550-foot mark; fig. 16).

Detrital cycles 15–20 feet thick (fig. 6) have easily eroded black shale at the base and a feldspathic siltstone to very fine-grained sandstone unit in the upper part.

Chemical cycles 10–15 feet thick (fig. 6) have 1) a platy lower part marked by alternating black mudstone and tan-weathering dolomitic layers commonly disrupted by shrinkage cracks, 2) a more massive middle part marked by extensively disrupted dolomitic beds, and 3) an upper more uniform analcime-rich part, commonly speckled with analcime and dolomite patches in a birds-eye fabric.

23.9 First Thin Red unit. Lowest of several reddish-brown units mapped by McLaughlin. These are the most analcime-rich of the Lockatong deposits. In a 325–375 foot pattern of long cycles, reddish-brown units coincide with repeated thick reddish-brown units of the Brunswick Formation.

24.15 First Big Red unit, 110 ft thick.

STOP 3.

Mileage

24.4 Quarry in Triple Red unit (fig. 17); 3,000 ft above base of the formation. Lockatong Argillite is now used only locally as building stone.

Items:

1. Top of lowest gray detrital cycle containing quartz.

2. Lowest complete gray cycle with quartz and marked shrinkage disruption of dolomitic layers in lower part. Upper third is analcime-rich with no quartz.

3. Third and fourth cycles have lower platy dark lavender argillite with conspicuous shrinkage cracks, mosaic intraformational conglomerate and large white analcime-dolomite patches. Upper massive part is reddish-brown, speckled with tiny patches locally arranged in rosettes several inches in diameter.

4. Fifth cycle at top of east wall has extensively brecciated upper part composed of disrupted layers of dolomitic gray argillite, with crinkled crack-casts and injection filling of dark reddish-brown argillite.

25.0 Double Red unit exposed in creek is uppermost reddish-brown analcime-rich unit assigned to Lockatong Formation.

25.3 Top of Lockatong Formation, uppermost gray argillite unit B.

Base of the Brunswick Formation placed at the top of gray unit B and below the lowest thick reddish-brown unit is convenient for field work. But the lower half of this reddish-brown unit contains analcime and dolomite and resembles argillite in the upper part of the Lockatong Formation more than it does typical mudstone of the upper Brunswick Formation.

STOP 4.

25.5 Ravine and side road to E in middle of lowest reddish-brown unit assigned to Brunswick Formation (about 200 ft above base of formation).

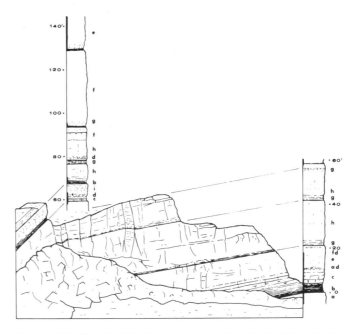

Figure 17. Sketch of Lockatong chemical cycles in Triple Red unit, Busik Quarry, Rt. 29, N. J. looking south, Stop 3. Heavy lines on columns—gray deposits.

a. dark gray quartz-bearing detrital argillite

b. black pyritic shale and platy mudstone

c. platy dark gray to black mudstone with thick crumpled dolomitic shrinkage-crack casts

d. upward-concaved planes and thin zones of shearing in massive dolomitic argillite

e. dark gray analcime-bearing argillite

f. reddish-brown analcime-rich argillite

g. platy blackish-red argillite, extensively cracked, with mosaic type intraformational conglomerate and thin tan dolomitic pullapart fragments. Conspicuous birdseye fabric with large analcime and dolomite patches (rarely with minor K-feldspar) in shrinkage cracks

h. massive reddish-brown argillite speckled with tiny patches of analcime and dolomite commonly filling skeletal crystal casts in rosettes and radiating patterns

i. extensively disrupted greenish gray argillite with breccia immersed in grayish red argillite. Complex pattern of crumpled and injected shrinkage-crack casts.

Traverse 0.2 mile long to lowest Gray unit C about 300 ft above the base (fig. 18).

Massive reddish-brown deposits commonly burrowed and characterized by repeated shrinkage-cracking of two kinds of layers:

1. very thin greenish-gray beds; locally wafers are scattered through reddish-brown argillite.
2. layers of brown-weathering dolomite-rich mudstone ½–1½ inches thick, commonly as pullaparts; locally fragments are widely dispersed.

Mileage

25.7   Reddish-brown argillite above Gray unit C contains scattered tiny flecks of pyrite and scattered brown-weathering irregular dolomitic nodules as much as 1½ inches long. Many of them contain irregular patches of sparry calcite or are small septarian nodules with cracks filled with calcite. Some are crudely concretionary and may be of algal origin.

Gray unit C is the lowest of eight recurring

gray units centered at about 400-foot intervals. McLaughlin traced Unit C northeast to its end in the Hammer Creek Conglomerate, and 16 miles southwestward where it thins and ends in reddish-brown mudstone. The unit consists of black, platy carbonate-rich beds with abundant pyrite, below massive gray argillite commonly with long, slender crumpled crack casts. The platy part contains two very distinctive deposits.

1. black mudstone layers about 1–1½ inches thick, with wide complexly crumpled and injected shrinkage crack casts of brown-weathering muddy dolomite and minor calcite.
2. Long lenses of pyritic pellet dolomite 6–12 inches thick, with silty round and flattened burrow casts and thin arcuate calcitic skeletal debris (ostracodes?) in the upper part. The intraclasts are 1/16- to ¼-inch mudstone and massive and crudely concretionary dolomitic pellets, some of which have internal shrinkage cracks filled with sparry calcite.

26.35   Gray unit D exposed in ravine of Warford Creek.

26.9   Gray unit E and F exposed in ravine and along side road to E. Gray units G and H, 220 feet higher, are included in the Graters Member, 135 feet thick. It has been traced for 40 miles by McLaughlin, but is not exposed along Rt. 29.

30.3   Frenchtown. Junction with Rt. 12. Turn W then first N and continue N on river road. North of town, Brunswick strata in broad Hunterdon syncline dip W under diabase sheet of Coffman Hill intruded high in the Brunswick Formation. High cliff on W side of river displays patterned sequence of more and less resistant mudstone and claystone units of Brunswick Formation.

32.3   Estimated position of Gray unit I.

33.9   Milford. Junction with Rt. 519. Turn W at traffic light, then N just W of drive-in bank. Continue to NW on narrow river road. WATCH FOR TRUCKS.

34.5   Channeling in sandy units of Brunswick Formation.

34.6   Gray units L and M. Strata nearly horizontal.

35.0   Position of first of 4 small faults, down on E, extending NE into Paleozoic and Precambrian rocks of the Highlands and SSW into the diabase sheet. Marked jointing in roadcut. Most easterly fault is downthrown about 60 ft on east side; middle two have combined throw of about 200 ft (Drake and others, 1961).

35.2   Estimated position of second fault.

STOP 5.

35.3   Middle Brunswick Mudstone in patterned sequence, about 2,900 ft above base of formation. Dip 6–10°N.

Items along 0.2 mile traverse:

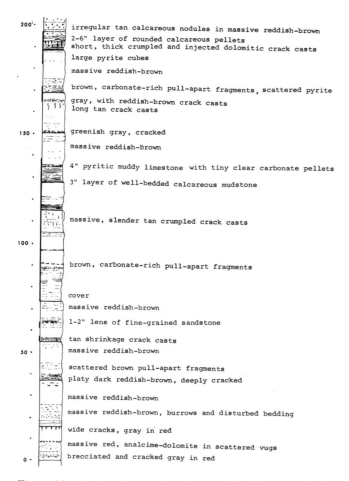

irregular tan calcareous nodules in massive reddish-brown
2-6" layer of rounded calcareous pellets
short, thick crumpled and injected dolomitic crack casts
large pyrite cubes
massive reddish-brown
brown, carbonate-rich pull-apart fragments, scattered pyrite
gray, with reddish-brown crack casts
long tan crack casts
greenish gray, cracked
massive reddish-brown
4" pyritic muddy limestone with tiny clear carbonate pellets
3" layer of well-bedded calcareous mudstone
massive, slender tan crumpled crack casts
brown, carbonate-rich pull-apart fragments
cover
massive reddish-brown
1-2" lens of fine-grained sandstone
tan shrinkage crack casts
massive reddish-brown
scattered brown pull-apart fragments
platy dark reddish-brown, deeply cracked
massive reddish-brown
massive reddish-brown, burrows and disturbed bedding
wide cracks, gray in red
massive red, analcime-dolomite in scattered vugs
brecciated and cracked gray in red

Figure 18. Stratigraphic section of upper 75 ft of lowest reddish-brown unit and overlying lowest gray unit C in lower part of Brunswick Formation. High roadcut along Rt. 29, N. J. Stop 4.

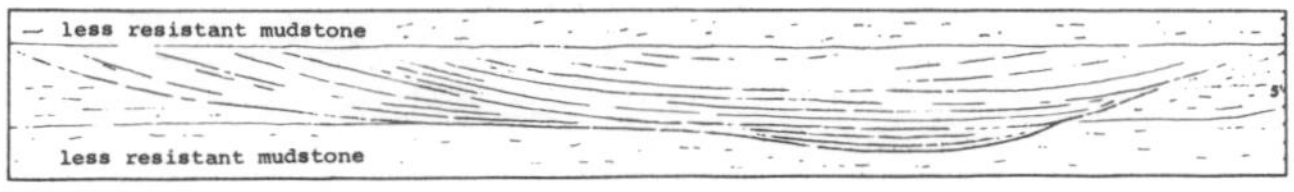

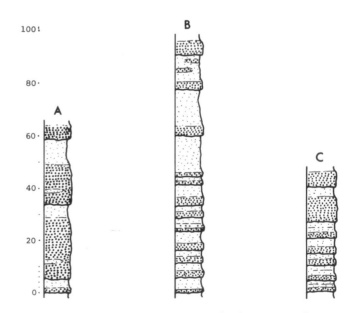

Figure 19. Sketch of channel-fill deposit in more resistant member of rhythmic alternation in the middle part of the Brunswick Formation; river road, about 1 mile west of Milford, N. J. Stop 5.

Deposit consists of successive and overlapping persistent beds 1–3 inches thick, commonly grading upward from fine-grained feldspathic sandstone to mudstone. Burrowing occurs both within the beds and on bedding surfaces with shrinkage cracks.

1. Third fault well-exposed at E end of exposure.
2. Sequence of 29 or more alternations averaging 8.5 ft thick (Picard and High, 1963).
3. Role of burrowing in destruction of lamination and in reflecting rate of burial.
4. Channel-fill deposits in silty Brunswick Mudstone (fig. 19). Well-bedded 1–3 inch layers — fine-grained sandstone in lower part; some are graded; many are burrowed; top of layer is marked by burrows and shrinkage cracks.
5. Feldspar in siltstone and fine-grained sandstone points to derivation from crystalline source area.

35.6   Estimated position of fourth fault at ravine and side road to E.

STOP 6 (MILEPOST 37).

Mileage

35.9   Southward projecting lobe of quartzite-rich Hammer Creek Conglomerate about 100 ft above its base. Dip generally 10–15°NW. Fault in ravine between the two major outcrops.

Most clasts are less than 3 inches in diameter, a few are 5 inches across; the largest are 9 inches long. The detritus is arranged in a succession of fining-upwards sequences (fig. 20) scoured at the base and gradational upwards (fig. 21B, C). The upper poorly-sorted mudstone is locally cross-bedded in SE outcrop, but essentially unbedded in NW outcrops. Calcareous nodules and patches in upper part of each unit are suggestive of cornstone in fossil soils of the upper Old Red Sandstone (Burgess, 1961).

This border facies exhibits many of the characteristic features of alluvial fan deposits which differ markedly from regionally extensive, rather well-sorted and rounded conglomerates of many molasse sequences. Some of its significant features are:

1. Located less than five miles from the border fault.
2. Source area probably less than 25 miles away.
3. Abundant matrix and associated poorly-sorted sandstone indicating rapid deposition and little by-passing.
4. Conspicuous lensing and channeling.
5. Rapid gradation into a more distal, finer-grained facies.

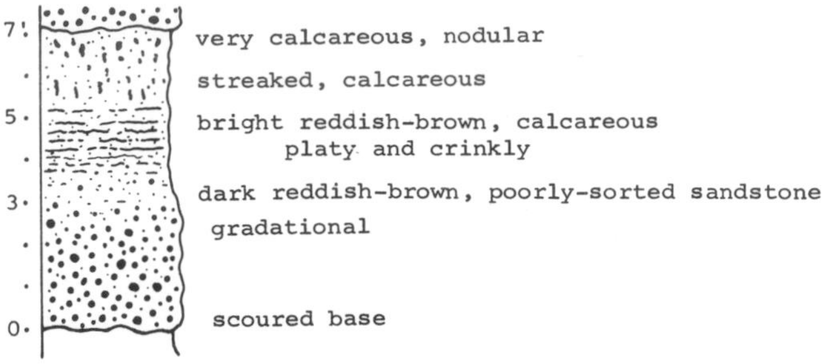

36.4   NW end of Pebble Bluffs roadcut. Fining-upwards Hammer Creek units are about 25–30 ft thick (fig. 21A). The largest clasts are about 13 inches long.

37.7   Turn N on Phillips Road. Gilbert Generating Station in valley to south. Steep cliff on far S side of river shows patterned sequence of upper Brunswick deposits. Uppermost beds are gray hornfels beneath Coffman Hill diabase.

Road to north crosses Brunswick mudstone between lobes of Hammer Creek Conglomerate composed entirely of Paleozoic clasts.

Figure 20. Model of fining-upwards conglomerate-sandstone unit, showing sequence of elements commonly present. Pebble Bluffs, river road, about 2 miles west of Milford, N. J. Stop 6. See fig. 21.

very calcareous, nodular
streaked, calcareous
bright reddish-brown, calcareous
    platy and crinkly
dark reddish-brown, poorly-sorted sandstone
gradational
scoured base

Figure 21. Vertical sections of lenticular fining-upwards conglomerate-sandstone units. Pebble Bluffs, river road, about 2 miles west of Milford, N. J. Stop 6. A is 0.5 mile northwest of C. Maximum clast size: A—about 13 inches long; B—about 9 inches long.

Mileage

38.6 Turn E on Church Road running along border fault. Precambrian terrane of Musconetcong Mt. to N. Sliver of lower Paleozoic rocks faulted against Precambrian gneiss and overlapped by the Hammer Creek Conglomerate.

40.0 Quarry in Hammer Creek Conglomerate to S. Paleozoic (mainly Silurian) quartzite clasts are extensively fractured by crushing in fault zone.

41.6 Spring Mills. Turn S then E on Rt. 519, in Brunswick Formation south of narrow belt of limestone-pebble border conglomerate.

42.9 Take left fork to Little York.

43.6 Patch of older drift crossed in next 0.4 mi.

44.0 Road turns N, joined by road from S.

44.3 Little York. Patch of quartzite-rich Hammer Creek Conglomerate faulted against a fault-sliver of Cambrian limestone. Turn E to Pattenburg.

45.7 Traverse enters major area of quartzite-rich conglomerate which to E and SE interfingers with the Brunswick, Lockatong and Stockton formations. Relations of this sort indicate movement along the border fault during much of Late Triassic time.

46.7 Cross Rt. 579 on border fault. Road follows sliver of Paleozoic rocks in border fault zone.

48.2 Pattenburg, on northern edge of Hammer Creek Conglomerate lying on Cambro-Ordovician basement rocks which are faulted against gneisses of the Highlands. Continue NE to join U. S. 22.

49.3 U. S. 22. Head ESE across Cambro-Ordovician basement for 0.5 mi, then across Hammer Creek Conglomerate.

50.8 Norton intersection. Cross Ordovician shale overlapped by border conglomerate.

51.8 Exposure of dark gray Ordovician shale to S.

52.0 Outcrop of red and gray Ordovician shale with thin beds of chert and limestone to S. These rocks yield Deepkill and Normanskill (Early to Middle Ordovician) graptolites.

52.4 Join Interstate 78E.

53.6 Clinton exit. Interstate 78E traverses contact of Paleozoic rocks and overlapping Hammer Creek Conglomerate.

55.6 Complexly faulted Cambro-Ordovician strata and Precambrian gneisses that project S into Newark Basin are bounded on the E by the Flemington Fault.

57.3 Crossing NE end of the Flemington Fault at junction with border fault. Precambrian gneiss to S is faulted against a syncline in the Cushetunk Mt. diabase which encloses the Round Valley reservoir.

61.3 Syncline in New Germantown flow (400 ft ± thick) forms low hills a mile N.

67.7 Intersection of Interstate 78 and 287. Arcuate NW-trending Watchung Mts. rise to E. 1st Watchung ridge ends half a mile to SE. Turn N on IS 287 which runs parallel to the 2nd Watchung ridge for 1.5 mile.

69.4 IS 287 turns E and crosses 2nd Watchung Mt.

71.4 At Moggy Hollow (crossing Mine Brook Road) road heads NE between the recurved 2nd Watchung Mt. to the NW and the 3rd (Long Hill) Watchung to the E and SE.

Glacial Lake Passaic dammed between the Highlands and the Watchung Mts. drained southwestward through Moggy Hollow to the North Branch of the Raritan River when outlets to the northeast were blocked by the ice sheet.

73.4 Narrow gap between 2nd and 3rd Watchung ridges.

74.4 Bernardsville to NW on border fault; Basking Ridge to SE. To E the Great Swamp is in broad syncline of 3rd Watchung Flow.

76.4 IS 287 crosses Passaic River. In its last stage Glacial Lake Passaic was drained northeastward through gaps at Little Falls and Paterson by the Passaic River system. A small lake was left in the area of the Great Swamp.

The border fault 1,500 ft to NW continues as the remarkably straight Ramapo Fault for about 50 miles to the NE end of the basin.

77.4 Narrow gap between border fault and west end of anticlinal outlier of 3rd Watchung Flow (New Vernon Ridge).

78.4 Patch of Hammer Creek Conglomerate in low hill to E.

80.6 Northern horn of New Vernon anticline in 3rd Watchung flow.

82.3 Crossing Rt. 510 east of Morristown.

85.4 Intersection Rt. 10. Leave IS 287, turn SE on Rt. 10. Traverse crosses upper Brunswick terrane above 3rd Watchung Mt.

86.8 Whippany intersection. Continue SE across Black Meadows. Cross Passaic River.

91.4 Morehousetown intersection. Riker Hill ahead is part of 3rd Watchung Mt.

92.85 Livingston intersection. Top of 2nd Watchung Mt. 0.3 mi ahead.

95.10 Base of 2nd Watchung Mt.

95.50 Cross Valley between 1st and 2nd Watchung ridges; top of 1st Watchung Mt.

96.25 Intersection with Prospect Ave. (Rt. 577) on crest of 1st Watchung Mt. Turn N.

96.75 Access road to Interstate 280 deep cut through first flow of 1st Watchung Mt. (fig. 22).

STOP 7.

97.20 Lower part of first flow and underlying Brunswick Formation, deep roadcut on Interstate 280 (fig. 23; see also fig. 10).

Items:

1. Sandy facies of feldspathic Brunswick mudstone and interbedded arkosic sandstone. Generally well-bedded, micaceous: mud-

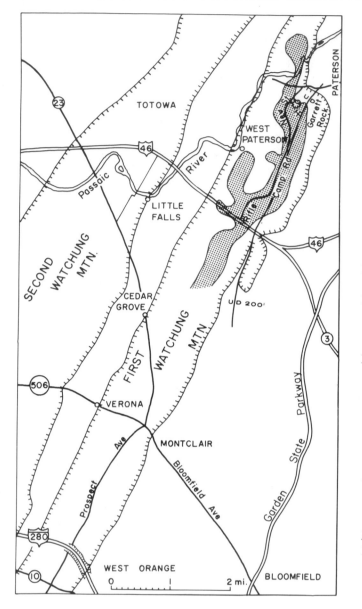

Figure 22. Sketch map of part of 1st and 2nd Watchung Mountains, central New Jersey, showing location of deep roadcut through lowest flow on Interstate Highway 280, West Orange (Stop 7) and New Street Quarry in lower part of second flow unit of 1st Watchung Mountain, West Paterson (Stop 8).

Shaded area—distribution of pillow lava in second flow unit as mapped by Fenner, 1906. Pillow lava undoubtedly extends farther west under cover and was present to the east where the 1st Watchung flows have been removed by erosion.

cracking in finer-grained beds; thin layers of mud-chip conglomerate in sandstone; burrows very rare; trails on some bedding surfaces.

2. Very thin hornfels zone. Small patches of reddish-brown mudstone locally caught up in basal lava.

3. Lower vesicular zone with inclined pipe amygdules. Most amygdules are quartz and calcite. Lining of anhydrite and gypsum found in a vesicle about a half inch in diameter.

4. Well-developed lower colonnade of thick vertical columnar joints.

5. Main body of flow is curvi-columnar zone with markedly radiating slender joints. (Manspeizer, 1969)

6. No clear evidence of flow units within this 150–175 foot flow.

97.7   Exit from IS 280. Turn E on Rt. 577 along crest of 1st Watchung Mt.

99.9   Cross Bloomfield Ave. to Rt. 23 heading NW.

101.9   Cedar Grove, top of 1st Watchung Mt. Cross Brunswick Formation.

102.9   Base of 2nd Watchung Mt. Entering broad gap at Little Falls occupied by Passaic River which drained Glacial Lake Passaic.

103.6   Turn N onto Montclair Road 0.1 mi past traffic light.

104.2   Turn E on Main St., then N at first traffic light onto Union Ave. Cross Passaic River; view of Little Falls to E and W. Continue on Union Ave. to Rt. 46.

105.0   Turn E on U. S. 46.

106.1   Top of 1st Watchung Mt. on east side of valley.

106.6   Lower part of second flow of 1st Watchung Mt. in roadcut. Ropy and pillow lava to S (right), local columnar joints on N (left).

106.8   Turn S at Great Notch exit, then N on overpass to Rifle Camp Road. Long reservoirs lie along Garrett Mt. Fault (fig. 22).

109.1   Bear left on narrow New St., West Paterson.

STOP 8.

Mileage

109.5   New St. Quarry (abandoned in upper vesicular unit of 1st Watchung Mt. (fig. 24; see also fig. 10). These units and the palagonitic tuff between them are well-displayed in Vandermade's Prospect Park (Sowerbutt) Quarry, Paterson; it is not open to large groups.

Items:

1. West-dipping vesicular unit 10 feet thick at top of lowest flow of 1st Watchung Mt. exposed along base of east wall.

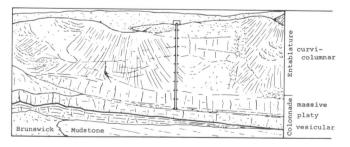

Figure 23. Sketch of lower part of first flow of 1st Watchung Mountain, west side of roadcut on Interstate Highway 280, 0.5 mile south of Rt. 577, N. J. Stop 7. Pipe is 100 ft high.

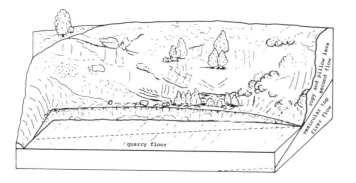

Figure 24. Sketch of east wall of New Street Quarry, West Paterson, N. J. Stop 8. Vesicular top of first flow unit of 1st Watchung Mountain crops out in basal ledge a maximum of 10 ft high. Above it is ropy and pillow lava of the second flow unit which contains abundant zeolites and associated secondary minerals.

2. Pale reddish-brown altered tuff a few inches thick locally at top of and churned into vesicular unit. The tuff is now largely non-expanding 14Å and 7Å clay minerals.

3. Pillowed and columnar second flow unit. Irregular development of pillows high in the flow suggests a complex process of extrusion into ponds dammed by the flow itself.

4. Large patches of zeolites and other minerals in voids between pillows as well as in vesicles. Thin layer of tuff and basaltic glass of pillows have not been zeolitized.

5. On west wall pillows in lower half are elongate N–S, in upper half they are elongate E–W.

6. Large concentrations of zeolites in jointed vesicular basalt in Triassic flows in Nova Scotia and in Cenozoic flows in Iceland suggest that the pillow lava is not a basic control of zeolite mineralization at West Paterson as contended by Fenner (1908) and Schaller (1932).

112.2 Return to Great Notch, take U. S. 46E.

112.3 Columnar jointing in lower flow of 1st Watchung Mt. on S.

112.4 Crossing Garrett Mt. Fault (fig. 22). First flow of 1st Watchung Mt. is downfaulted 200 ft or more on E. Continue on U. S. 46.

112.6 Brunswick Formation below 1st Watchung Mt. on S.

114.2 Join Garden State Parkway at Interchange 154 N. Travel 19 miles N to N. Y. Turnpike at Spring Valley. Sandy Brunswick roadcut just past Exit 155 (0.8 mi); conglomeratic Brunswick roadcut 0.5 mi past N. J.-N. Y. border (17 mi).

133.2 N. Y. Turnpike E.

137.1 Roadcut in upper part of Palisade Sill.

138.1 Gigantic columnar joints, characteristic of Palisade Sill.

138.5 Exit 11. Take Rt. 59E.

139.0 Traffic light at intersection with 9W. Turn N.

139.9 Turn E at Rockland Lumber Mart, downhill on Birchwood Ave.

140.5 Turn N on Broadway.

141.1 South end of Verdreitege Hook ahead. Spectacular talus at very steep angle of repose.

West of Nyack the Palisades are as much as a mile west of the Hudson River because the sill intruded a higher part of the west-dipping Brunswick Formation. It swings back to the river as Verdreitege Hook at Nyack Beach State Park, cutting somewhat lower Brunswick deposits. At its recurved north end the sheet is a dike intruding successively higher Brunswick beds; its trace turns sharply to the west almost perpendicular to strike.

A distinct olivine-rich zone is absent at the northern end of the Palisade intrusion in higher Newark strata, and also at its southwestern end in west-central New Jersey.

STOP 9.

Mileage

141.5 Nyack Beach State Park. Road E down steep hill to Hudson River. Lower part of Brunswick Formation, about 500–1,000 ft above base.

The Park authorities have made this stop available to us. Please do not disturb the foliage or the outcrops. Collect specimens only in those areas of recent blasting or rock fall. Depending on the season watch for copperhead snakes, yellow-jacket wasps, poison ivy and poison sumac.

Items:

1. Very rapid lateral changes in thickness and characteristics of units throughout traverse.

2. Within vertical sequences units of same grain-size differ markedly in thickness.

3. Bright reddish-brown micaceous mudstone commonly contains silty burrow fillings, and abundant rounded intraclasts 1/16- to 1/4-inch in diameter. Most are calcitic mudstone, a few are aphanitic limestone with shrinkage cracks and scattered silt. These pellets are locally concentrated in cross-bedded strata 6–12 inches thick, indicating at least local transport.

4. Burrowing is common in beds of mudstone, and especially evident in the upper part and on soles of overlying sandstone with scour "pods."

5. Many of the mudstone beds directly overlie a marked scour surface on sandstone.

6. Soft-sediment deformational features, such as contorted bedding and mudstone dikes.

7. Arkose ranges from very well-sorted medium to fine-grained varieties with channels marked by basal intraformational mud chips,

and cross-bedding in thicker units generally directed to SW-W, to poorly-sorted coarse-grained sandstone with scattered clasts of Paleozoic sedimentary rocks indicating derivation of some detritus from the N or NW.

Many of the features suggest a braided stream environment with relatively steep slope, abundant sediment supply, and highly variable discharge (Smith, 1968, p. 285). These include: cut and fill, irregular bedding surfaces, variable bed thickness, marked change in grain-size laterally and vertically, large-scale planar cross-bedding, local concentrations of intraclasts, and a variety of structures representing bed-roughness forms.

Traverse along Hudson River. Distances on footpath estimated from north end of parking area.

0–895 ft: Mostly cover.

895–950 ft: Picnic tables and grills.

1,450–1,550 ft: Outcrop exhibits marked channeling and lensing of mudstone with abundant intraclasts of calcite and mud, fine-grained intraformational conglomerate, and an 18-inch layer of intricately contorted mudstone (fig. 25).

1,570–2,215 ft: Much talus.

2,215–3,745 ft: Very large talus blocks; well-developed columnar jointing in sill. Pass fireplaces.

3,745–3,925 ft: Tree-clad slope.

4,040–4,400 ft: Outcrop 150 ft south of open stone shelter, consists of following units from top to bottom:

10 ft: Massive light gray coarse arkose, locally cross-bedded.

5 ft: Thin-bedded, reddish-brown arkosic siltstone and very fine-grained sandstone, mudstone in upper foot. Channeling and lensing in 6–12 inch units. Locally disturbed bedding and isolated burrows.

7 ft: Massive, locally cross-bedded light gray conglomeratic arkose. Clasts consist of angular quartz as much as 2½ inches in diameter, feldspar as much as an inch long and intraformational mudstone.

Path level.

Continue on path. Conspicuous jointing in sill to W. Sing-Sing prison across river to E.

Enormous talus blocks along path between 4,400 to 5,500 ft.

5,570–6,080 ft: Base of sill about at path level north to retaining wall on W side of path. N of retaining wall 10 feet of cross-bedded, light gray arkose which apparently overlies reddish-brown mudstone seen in next outcrop to N.

6,230–6,420 ft: Cover.

6,575–7,185 ft: Continuous exposure of beds of mudstone and arkosic units exhibiting the intricate variations and facies changes characteristic of these fluvial deposits.

Return to parking area.

142.0 Return to Broadway; head S.

143.0 Turn W on Birchwood Ave., return to U. S. 9W.

143.6 U. S. 9W. Turn S.

144.5 Traffic light, continue S on Palisade Sill (fig. 26). Scattered outcrops of diabase.

148.4 Glacial deposits on diabase to E.

148.9 Sparkill Gap. Sparkill Creek is the only stream that flows eastward across the Palisade Sill to the Hudson River. According to Johnson (1931, p. 76–131) the Hudson River originally flowed

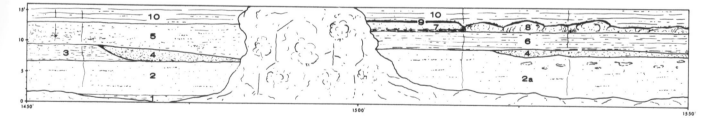

Figure 25. Sketch of outcrop of lower Brunswick Formation, Nyack Beach State Park, N. Y. Stop 9. 1,450–1,550 ft north of parking area.
  1. Fine-grained arkose.
  2. Hackly, micaceous mudstone with abundant scattered calcareous clasts and pellets, commonly ⅛–¼ inch in diameter; also in thin stringers.
  2a. As in 2, with irregular, mostly horizontally elongate, gray calcareous nodules 3–6 inches long.
  3. Parallel-bedded fine-grained muddy arkose.
  4. Intraformational calcareous pellet conglomerate, and mudstone loaded with calcareous pellets and intraclasts. Low-angle, large-scale cross-bedding.
  5. Thin-bedded fine-grained arkose; abundant small-scale cross-bedding and channeling.
  6. Pale pinkish gray fine-grained arkose; parallel bedding.
  7. Homogeneous micaceous silty mudstone, thin-bedded in lower part.
  8. Thin-bedded fine-grained arkose and mudstone, extensively contorted.
  9. Discontinuous layer of bright reddish-brown micaceous claystone 2–3 inches thick.
  10. Bedded arkose.

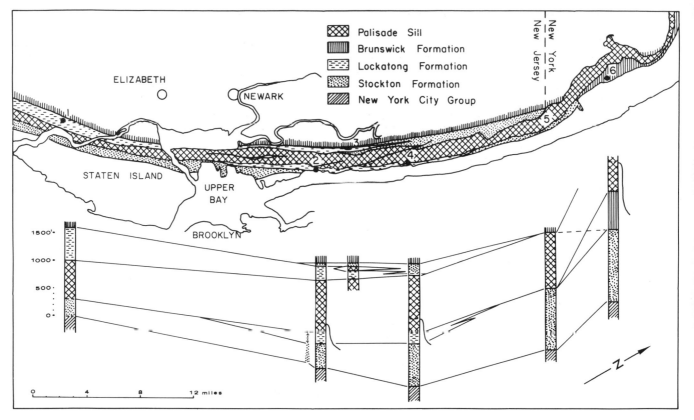

Figure 26. Generalized geologic map and stratigraphic sections of Newark rocks along the northeastern margin of the Newark Basin, showing interfingering of Stockton and Lockatong strata and cross-cutting of the Palisade Sill. See fig. 4.

    1. 500-foot cores, Sewaren, N. J. Van Houten, private report.
    2. Cores and excavations, Lincoln Tunnel. Berkey, 1948.
    3. Granton Quarry, North Bergen, N. J. Van Houten and Savage, 1968, p. 98, fig. 22.
    4. Cores and excavations, George Washington Bridge, Berkey, 1948.
    5. Estimated section at N. J.–N. Y. boundary.
    6. Outcrops and geophysical data, South Nyack, N. Y. Worzel and Drake, 1959.

westward on the Schooley erosion surface above the position of the sill, was superposed on it and cut a water gap (see Thornbury, 1954, p. 234–240 for discussion of this explanation).

149.4   Tallman Mountain Park, entrance to E. Olivine zone is present in sill.

150.6   Intersection with Oak Tree Road.

    Stockton arkose exposed along Hudson River at Sneden's Landing to E contains abundant zircon, rutile, and anatase, whereas Brunswick constituents such as tourmaline and garnet are rare, and metamorphic rock fragments are absent (Savage, 1968, p. 56–62).

151.5   N. Y.–N. J. border. Entrance to Lamont-Doherty Geological Observatory of Columbia University to E.

158.0   Glaciated diabase on crest of Palisades to E.

161.8   Overpass of Palisade Interstate Parkway. Continue S on Lemoine Ave. (Rt. 67).

162.3   Cross west portal of George Washington Bridge (fig. 26). Lockatong hornfels at upper contact in roadcut of Interstate 80 a mile W includes grossularite-andradite, prehnite, diopside varieties.

Clepsysaurus (*Rutiodon*), a phytosaur, was found in Stockton strata below the sill to the E. Continue S on Rt. 67, then 67 and 5.

164.0   Turn E on Rt. 5E; Palisade Amusement Park to S. Descend to base of sill. Olivine layer in lower part is at road level at second left-hand bend. Fault in sill just S of main switchback.

165.0   River Road, Edgewater. Continue S.

165.15   Traffic light. Turn W up Demsey Ave. to Undercliff Ave.; turn N.

STOP 10.

Mileage

165.3   Palisade quartz diabase sill and Lockatong hornfels along abandoned trolley route. Park opposite small park with benches (fig. 27; also see fig. 9).

    Items:

    1. Well-bedded, dark gray Lockatong biotite-albite hornfels and arkose along creek bottom.

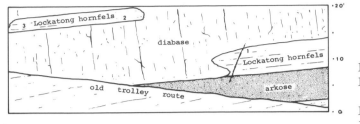

Figure 27. Sketch of contact zone at base of Palisade Sill, old trolley route, Edgewater, N. J. Stop 10. Shows irregular contact with Lockatong hornfels, minor fracture zone, tongue of arkose, and 2-foot thick xenolith of Lockatong hornfels.
1. Scapolite, aegirine, Na-feldspar hornfels
2. Cordierite, biotite, Na-feldspar hornfels
3. Very coarse-grained biotite, Na-feldspar hornfels with minor pyroxenes.

    2. Pale greenish-gray chert-like albite-diopside hornfels.
    3. Very irregular diabase-hornfels contact. Scapolite-aegirine variety in innermost few inches showing rheomorphic effects.
    4. Long hornfels xenolith 2 ft thick with original stratification preserved, 15 feet above contact. Cordierite-rich and coarse-grained plagioclase varieties.
    5. Olivine zone about 15 ft thick, 50 ft above base of sill at switchback on trolley route. Hypotheses proposed to explain the differentiation and account for the olivine-rich layer include gravitational settling, fractional crystallization, late-stage injection of a new pulse of magma, and mechanical flow differentiation.

165.5   Proceed N on Undercliff Ave. to next corner, turn E and descend to River Road. Turn S.
166.2   Stockton and Lockatong hornfels behind Edgewater sewage treatment plant and Edgewater Welding Co.
166.5   Fault in sill north of R. R. tunnel entrance (fig. 28).
166.7   Large xenoliths exposed to W in excavation behind Virginia Lee Lace Co. A 15 x 30 ft and a 4 x 6 ft xenolith of biotite-plagioclase hornfels with pyroxene-rich and muscovite-chlorite-rich varieties.
166.8   Long NNE-trending fault in sill at re-entrant in cliff (fig. 28), down 275 ft on E.
166.9   Cyclic Lockatong hornfels in excavation W of Mobil Service Center (fig. 29).
        Items:
        1. Base about 90 ft below sill.
        2. Three complete cycles below tongue of arkose 18 ft thick.
        3. Grossularite-prehnite-idocrase varieties in lower and middle carbonate-rich part of cycle.
        4. Analcime-scapolite-diopside varieties in upper part of cycle.
167.0   Intersection with Gorge Road. Turn N. Road-

cut in upper-most 60 ft of Lockatong hornfels with vague pattern of 4–7 foot cycles which include distinctive layers of grossularite and prehnite. Road follows fault-line ravine (fig. 28).
167.2   Base of sill to W is at road level.
167.9   Intersection with Edgewater Road (Rt. 50). Turn W and continue on Edgewater Road.
169.1   Traffic light at Shaler Blvd. Turn SW.
169.6   Traffic light at U. S. 9 and 1 (Broad Ave.) Turn S.
170.6   Knob of Stockton Arkose to W; contains interbedded 2–3 inch limestone unit (Kindle, 1944, p. 4).

STOP 11.

Mileage
171.1   Granton Quarry in Lockatong hornfels and overlying subsidiary sill, W. side of U. S. 1 and 9, North Bergen. Park in rear of Diana Stores Corp. (fig. 30, 31). Base of outcrop about 125–

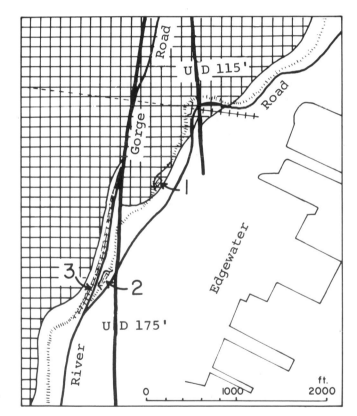

Figure 28. Sketch map of faulted Palisade Sill above interfingering Stockton and Lockatong hornfels, along River Road, Edgewater, N. J. Hachures outline base of the Palisades. After Kummel, 1898, fig. 14.
1. Xenoliths of Lockatong hornfels about 15 x 30 ft and 4 x 6 ft, 10 to 20 ft above floor of excavation behind Virginia Lee Lace Company plant.
2. Excavation in cyclic Lockatong hornfels on west side of River Road, opposite Mobil Service Center.
3. Road cut on west side of Gorge Road, from tongue of Stockton Arkose through cyclic Lockatong hornfels to the base of the Palisade Sill.

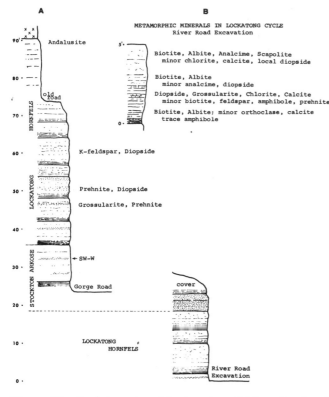

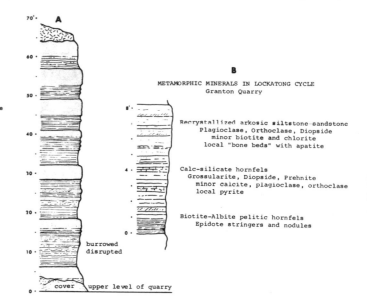

Figure 31. Stratigraphic section (A) and model of cyclic Lockatong hornfels (B) showing distribution of principal metamorphic minerals, Granton Quarry, North Bergen, N. J. See fig. 30.

Figure 29. Lockatong hornfels, junction of River Road and Gorge Road, Edgewater, N. J. See fig. 28.

A. Stratigraphic section of uppermost cyclic Lockatong Formation with tongue of Stockton Arkose, showing distribution of some distinctive metamorphic minerals and dispersal direction.

B. Model of cyclic Lockatong hornfels in River Road excavation showing distribution of principal metamorphic minerals.

150 ft above Palisade Sill; subsidiary sill is about 40 ft thick.

Items:

1. North wall displays six complete cycles of Lockatong hornfels, 8–14 ft thick.
2. West end of cycle 5 and 6 broken and injected by diabase. Small xenolith 15 ft to W in lower part of wall.
3. Well preserved fossil fish and reptiles with recrystallized apatite bones, as well as

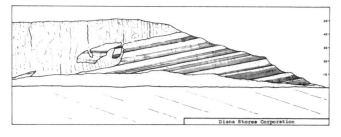

Figure 30. Sketch of cyclic Lockatong hornfels and xenoliths, north wall of Granton Quarry, Diana Stores Corp. property, west side U. S. 1 and 9, North Bergen, N. J. Stop 11. Base is about 125–150 ft above top of Palisade Sill.

estheriids, occur on bedding surfaces mainly in lower and middle platy part of cycles.

4. Lockatong hornfels include calc-silicate virieties in the middle carbonate-rich part and extensively feldspathized and recrystallized diopside-rich fine-grained arkose in the upper part.

171.8 Return to U. S. 9 and 1. Continue S.

172.4 Discordant upper contact of Palisade Sill and Stockton Arkose behind Ward's Way Shipping Co. to E. Stockton strata dip 15°NW; the surface of the diabase dips 45°NW. Cross-cutting relations here and at other places at the top of the sheet suggest that the sill may have been fed by a dike along its present western border.

Borings in the Hackensack Meadow to the W indicate that at least locally the diabase is concordant with the west-dipping Newark strata (Widmer, 1959). At Sewaren, SE of Newark and 2 miles W of the west edge of the sill on Staten Island, high-grade nepheline, sodalite, grossularite and diopside-rich varieties of hornfels 300–500 ft below the surface suggest that the diabase dips gently westward from Staten Island to lie less than the 750 ft below the surface at Sewaren.

174.1 Exit to N. J. Turnpike; head W.

175.4 Enter N. J. Turnpike S.

177.2 Little Snake Hill diabase plug to S.

177.4 Snake Hill diabase plug to NW.

182.8 Newark Airport, exit 14.

END OF TRIP

REFERENCES CITED

Abdel-Monem, A. A., and Kulp, J. L., 1968, Paleogeography and the source of sediments of the Triassic Basin, New Jersey, by K-Ar dating: Geol. Soc. America Bull., v. 79, p. 1231–1242.

Andreae, A., and Osann, A., 1893, Tiefencontacte an den intrusiven diabasen von New Jersey: Naturh-med. Ver. Heidelberg, Verh (NF) 5, p. 16–27.

Baird, D., 1954, *Chirotherium lulli*, a pseudosuchian reptile from New Jersey: Bull. Mus. Comp. Zoo., v. 111, N. 4, p. 165–192.

Barker, D. S., and Long, L. E., 1968, Feldspathoidal syenite formed by assimilation in a quartz diabase sill: Geol. Soc. America Sp. Paper 115, p. 12–13.

———— (1969), Feldspathoidal syenite in a quartz diabase sill, Brookville, New Jersey: Jour. Petrology, (in press).

Berkey, C. P., 1948, Engineering geology in New York City; Geol. Soc. America, Guidebook 61st Ann. Mtg., p. 51–66.

Bhattacharji, S., 1967, Mechanics of flow differentiation in ultramafic and mafic sills: Jour. Geology, v. 75, p. 101–112.

Boch, W., 1959, New eastern American Triassic fishes and Triassic correlations: Geol. Center Res. Ser., v. 1, 184 p.

Bucher, W. H., and Kerr, P. F., 1948, Excursion to the 1st Watchung Basalt at Paterson, New Jersey: Geol. Soc. America, Guidebook 61st Ann. Mtg., p. 109–119.

Burgess, I. C., 1961, Fossil soils of the Upper Old Red Sandstone of South Ayrshire: Trans. Geol. Soc. Glasgow, v. 24, p. 138–153.

Carey, S. W., 1961, Relation of basic intrusions to thickness of sediments: *in* Dolerite, Hobart., Tasm., p. 165–169.

Carlston, C. W., 1946, Appalachian drainage and the Highland border sediments of the Newark Series: Geol. Soc. America Bull., v. 57, p. 997–1,032.

Colbert, E. H., 1965, A phytosaur from North Bergen, New Jersey: Amer. Mus. Nov. 2230, p. 1–25.

———— 1966, A gliding reptile from the Triassic of New Jersey: Amer. Mus. Nov. 2246, p. 1–23.

Colbert, E. H., and Gregory, J. T. 1957, Correlation of continental Triassic sediments by vertebrate fossils: Geol. Soc. America Bull., v. 68, p. 1,456–1,467.

Darton, N. H., 1902, Juratriass Rocks: *in* U. S. Geol. Survey, Atlas Folio 83, p. 6–10.

Darton, N. H., and Kümmel, H. B., 1908, Triassic System: *in* U. S. Geol. Survey, Atlas Folio 157, p. 7–13, 21, 2.

———— 1909, Triassic System: *in* U. S. Geol. Survey Atlas Folio 167, p. 6–11, 17–19.

De Boer, J., 1968, Paleomagnetic differentiation and correlation of the Late Triassic volcanic rocks in the central Appalachians (with special reference to the Connecticut Valley): Geol. Soc. America Bull., v. 79, p. 609–626.

Drake, A. A., McLaughlin, D. B., and Davis, R. E., 1961, Geology of the Frenchtown Quadrangle, New Jersey-Pennsylvania: U. S. Geol. Survey Quad. Map GQ Map 133.

Drake, C. L., Heirtzler, J., and Hirshman, J., 1963, Magnetic anomalies off eastern North America: Jour. Geophys. Res., v. 68, p. 5259–5275.

Drake, C. L., and Woodward, H. P., 1963, Appalachian curvature, wrench faulting and offshore structures: Trans. N. Y. Acad. Sci., Ser. II, v. 26, p. 48–63.

Fenner, C. N., 1908, Features indicative of physiographic conditions prevailing at the time of the trap extrusions in New Jersey: Jour. Geology, v. 16, p. 299–327.

Fischer, A. G., 1969, Geological time-distance rates: the Bubnoff unit: Geol. Soc. America Bull., v. 80, p. 549–552.

Glaeser, J. D., 1963, Lithostratigraphic nomenclature of the Newark-Gettysburg Basin: Penn. Acad. Sci., Proc., v. 37, p. 179–188.

———— 1966, Provenance, dispersal and depositional environments of Triassic sediments in the Newark-Gettysburg Basin: Pennsylvania Geol. Survey, 4th Ser., Bull. G. 43, 168 p.

Hess, H. H., 1956, Discussion of Walker, F., The magnetic properties and differentiation of dolerite sills—a critical discussion: Am. Jour. Sci., v. 254, p. 446–451.

Holmes, Arthur, 1965, Principles of Physical Geology: Ronald Press Co., New York, 1,288 p.

Hopson, C. L., 1964, The crystalline rocks of Howard and Montgomery Counties, *in* The Geology of Howard and Montgomery Counties: Maryland Geol. Survey, p. 27–215.

Hotz, P. E., 1952, Form of diabase sheets in southeastern Pennsylvania: Amer. Jour. Sci., v. 250, p. 375–388.

Jacobeen, F. H., 1949, Differentiation of Lambertville Diabase: sen. thesis, Department of Geology, Princeton University, 22 p.

Johnson, D. W., 1931, Stream Sculpture on the Atlantic Slope: Columbia Univ. Press, New York, 142 p.

Johnson, M. E., and McLaughlin, D. B., 1957, Triassic formations in the Delaware Valley; Geol. Soc. America Guidebook for field trips, No. 2, p. 31–59.

Kindle, C. H., 1944, A discovery of limestone in the Newark Series: Geol. Rev., City College Geol. Soc., v. 4, p. 3–4.

King, P. B., 1961, Systematic pattern of Triassic dikes in the Appalachian region: U. S. Geol. Survey Prof. Paper 424, p. B93–B95.

Kümmel, H. B., 1897, The Newark system; A report of progress: New Jersey Geol. Survey Ann. Rpt. 1896, p. 25–88.

———— 1898, The Newark system or red sandstone belt: New Jersey Geol. Survey, Ann. Rpt. 1897, p. 23–159.

———— 1899, The Newark or new red sandstone rocks of Rockland Co., N. Y.: N. Y. State Geol. Survey, Ann. Rpt. 18, (1899), p. 9–50.

———— 1914, Triassic Rocks: *in* U. S. Geol. Survey, Folio 191, p. 13–16, 20.

Lewis, J. V., 1907a, The origin and relations of the Newark rocks: New Jersey St. Geol. Survey, Ann. Rept. 1906, p. 97–129.

Lewis, J. V., 1907b, Structure and correlation of Newark trap rocks of New Jersey: Geol. Soc. America Bull., v. 18, p. 195–210.

—— 1908, Petrography of the Newark igneous rocks of New Jersey: New Jersey Geol. Survey Ann. Rept. State Geol. (1907), p. 97–167.

—— 1909, Building stones of New Jersey: Ann. Rpt. New Jersey State Geol. Survey Ann. Rept. St. Geol. 1908, p. 53–124.

Lowe, K. E., 1959, Structure of the Palisades intrusion at Haverstraw and West Nyack, New York: N. Y. Acad. Sci., Annals, v. 80, p. 1,127–1,147.

Mackin, J. H. 1961, A stratigraphic section in the Yakima Basalt. and the Ellensburg Formation in south-central Washington: Washington Div. Mines and Geol., Rpt. Inv. 19, p. 1–45.

MacLachlin, J. C. 1959, Paleotectonic maps of the Triassic System: in McKee, E. D. and others, U. S. Geol. Survey Misc. Geol. Inv. Map I-300, p. 12–13.

Manspeizer, Warren, 1969, Radical and concentric joints, First Watchung Mountains, New Jersey: Geol. Soc. America, 4th Ann. Mtg. N.E. Sect. (Abstract).

Mason, B. H., 1960, Trap rock minerals of New Jersey: N. J. Bur. Geol. Topo., Bull. 64, 51 p.

McKee, E. D., Oriel, S. S., Kepner, K. B., MacLachlin, M. E., Goldsmith, J. W., MacLachlin, J. C., and Mudge, M. R., 1959, Paleotectonic maps of the Triassic system: U. S. Geol. Survey Misc. Geol. Inv. Map I-300, 33 p.

McLaughlin, D. B., 1933, A note on the stratigraphy of the Brunswick formation (Newark) in Pennsylvania: Mich. Acad. Sci. Papers, v. 18, p. 421–435.

—— 1939, A great alluvial fan in the Triassic of Pennsylvania: Mich. Acad. Sci., Papers, v. 24, p. 59–74.

—— 1941, The distribution of minor faults in the Triassic of Pennsylvania: Mich. Acad. Sci., Arts, Letters, v. 27, p. 465–479.

—— 1943, The Revere well and Triassic stratigraphy: Pa. Acad. Sci., Proc. v. 17, p. 104–110.

—— 1944, Triassic stratigraphy in the Point Pleasant district, Pennsylvania: Pa. Acad. Sci., Proc., v. 18, p. 62–69.

—— 1945, Type sections of the Stockton and Lockatong formations: Pa. Acad. Sci., Proc., v. 19, p. 102–113.

—— 1946a, The Triassic rocks of the Hunterdon Plateau, New Jersey: Pa. Acad. Sci., Proc., v. 20, p. 89–98.

—— 1946b, Continuity of strata in the Newark series: Mich. Acad. Sci. Papers, v. 32, p. 295–303.

—— 1959, Mesozoic Rocks: in Geology and mineral resources of Bucks Co., Pa., Pennsylvania State Geol. Survey Bull. C9, p. 55–162.

Opdyke, N. D. 1961, The paleomagnetism of the New Jersey Triassic: a field study of the inclination error in red sediments: Jour. Geophys. Research, v. 66, p. 1941–1949.

Pettijohn, F. J., 1963, Chemical composition of sandstones—excluding carbonate and volcanic sands: in Fleischer, Michael, ed., Data of Geochemistry; U. S. Geol. Survey Prof. Paper 440-S, 21 p.

Picard, M. D. and High, L. R., 1963, Rhythmic alternations in the Triassic Chugwater and Brunswick Formations, Wyoming and New Jersey: Contr. Geol., v. 2, p. 87–99.

Redfield, W. C. 1856, On the relations of the fossil fishes of the sandstone of Connecticut and other Atlantic States to the Liassic and Oölitic periods: Amer. Jour. Sci., v. 22, p. 357–363.

Rima, D. R., Meisler, H., and Longwill, S., 1962, Geology and hydrology of the Stockton Formation in southeastern Pennsylvania: Pennsylvania Geol. Survey, Bull., W-14, 111 p.

Rodgers, John, 1967, Chronology of tectonic movements in the Appalachian region of eastern North America: Amer. Jour. Sci., v. 265, p. 408–427.

—— 1968, Introduction to the Triassic of Connecticut: Conn. Geol. and Nat. Hist. Survey, Guidebook 2, p. C-O, 1–2.

Sanders, J. E., 1962, Strike-slip displacement on faults in Triassic rocks in New Jersey: Science, v. 136; p. 40–42.

—— 1963, Late Triassic tectonic history of northeastern United States: Amer. Jour. Sci., v. 261, p. 501–524.

Savage, E. L., 1968, The Triassic rocks of the Northern Newark Basin: N. Y. St. Geol. Assoc., Guidebook for field excursions, Trip c., p. 49–68.

Schaeffer, B., 1952, The Triassic Coelocanth fish, Diplurus, with observations on the evolution of the Coelocanthini: Amer. Mus. Nat. Hist. Bull., v. 99, p. 31–78.

Schaller, W. T., 1932, The crystal cavities of the New Jersey zeolite region: U. S. Geol. Survey Bull. 832, 90 p.

Shaw, D. M., 1956, Geochemistry of pelitic rocks. Pt. 3. Major elements and general geochemistry: Geol. Soc. America Bull., v. 67, p. 919–934.

Simkin, Tom, 1965, The piritic sills of northwest Trotternish, Isle of Skye, Scotland: Ph.D. thesis, Dept. of Geol., Princeton Univ., 139 p.

Smith, N. D., 1968, Criteria for recognition of the braided stream environment with examples from the lower Silurian of the North-Central Appalachians: Geol. Soc. America, abstracts, p. 285.

Sturm, Edward, 1956, Mineralogy and petrology of the Newark Group sediments of New Jersey: Ph.D. thesis, Rutgers Univ.

Thornbury, W. D., 1954, Principles of Geomorphology: John Wiley and Co., New York, 618 p.

Van Houten, F. B., 1962, Cyclic sedimentation and the origin of analcime-rich upper Triassic Lockatong Formation, west-central New Jersey and adjacent Pennsylvania: Amer. Jour. Sci., v. 260, p. 561–576.

—— 1964, Cyclic lacustrine sedimentation, Upper Triassic Lockatong Formation, central New Jersey and adjacent Pennsylvania: Pennsylvania Geol. Survey Bull. 169, p. 497–531.

—— 1965, Composition of Triassic Lockatong and associated formations of Newark Group, central New Jersey and adjacent Pennsylvania: Amer. Jour. Sci., v. 263, p. 825–863.

———— 1968, Iron Oxides in red beds; Geol. Soc. America Bull., v. 79, p. 399–416.

Van Houten, F. B., and Savage, E. L., 1968, The Triassic Rocks of the northern Newark Basin: N. Y. St. Geol. Assoc. Guidebook for field excursions, Trip c road log, p. 69–100.

Walker, F., 1940, Differentiation of the Palisade diabase, New Jersey: Geol. Soc. America Bull., v. 51, p. 1,059–1,105.

Walker, K. R., and Poldervaart, A., 1962, The Palisade Sill: Internat. Miner. Assoc. 3rd Gen. Cong., North. Field Excur. Guidebook, Washington, D. C., p. 5–7.

Walker, K. R., 1969, The Palisade Sill, New Jersey. A re-investigation: Geol. Soc. America Spec. Paper III.

Wheeler, G., 1939, Triassic fault-line deflections and associated warping: Jour. Geology, v. 47, p. 337–370.

Wherry, E. T., 1959, Triassic life, p. 114–124, *in* Geology and mineral resources of Bucks Co., Penn.: Pennsylvania State Geol. Survey Bull. C 9, 243 p.

Widmer, Kemble, 1959, Bedrock map of the Hackensack Meadows: New Jersey Geol. Survey, Geol. Rpt. No. 1, 26 p.

Worzel, J. L., and Drake, C. L., 1959, Structure section across the Hudson River at Nyack, N. Y., from seismic observation: N. Y. Acad. Sci., Ann., v. 80, p. 1,092–1,105.

ENGINEERING GEOLOGY OF THE YARDS CREEK HYDRO-ELECTRIC PUMPED STORAGE PROJECT

BENNETT L. SMITH

Department of Geology, Rutgers University, New Brunswick, New Jersey 08903

GENERAL NATURE OF THE PROJECT

The Yards Creek Project (fig. 1) has involved the construction of a reservoir having a capacity of 4,900 acre-ft on top of Kittatinny Mountain about five miles northeast of Delaware Water Gap and a second reservoir of somewhat larger capacity on the south side of the mountain in the valley of a small creek at an elevation some 700 feet below the upper reservoir. The two ponds are connected by a combination of penstock and tunnel through a powerhouse capable of pumping water from the lower to upper reservoir and, by reversing the flow, capable of generating 330,000 kilowatts of electricity. In effect, the system is a huge hydraulic storage battery. Greatly oversimplifying the economic aspects, it might be said that the system ordinarily is operated so that relatively inexpensive power is used during nights and weekends to store energy in the upper reservoir while the power that is generated is used to meet peak demands and emergencies and thus is relatively valuable power. About 70 percent of the energy used in pumping is recovered in the generating mode but the average difference in value between pumping and generating energy is more than enough to offset this loss of energy.

GEOLOGIC SETTING

As shown in fig. 2, the area of the upper reservoir is underlain by the Shawangunk Formation (Silurian in age and approximately equivalent to the Tuscarora Formation in Pennsylvania). The Martinsburg Formation (Ordovician) is the bedrock in the lower reservoir area; the contact between the two formations is well exposed in an excavation just above the lower portal of the tunnel (fig. 3) that is part of the system which connects the two reservoirs.

For local engineering geology purposes the Shawangunk Formation can be subdivided into three members: a lower member consisting of relatively thickbedded and massive quartzite and conglomerate with only a couple of thin shale beds, a middle member made up of quartzite and slatey shale thinly interbedded in about equal amounts and an upper member consisting mostly of quartzite but with thin beds of shale making up six percent of the member. The lower and middle members are each about 200 feet thick, the upper member is at least 1,000 feet thick and above it is the High Falls Formation (with which we are not concerned in the Yards Creek project). The upper reservoir location is athwart the axis of a syncline in the Shawangunk Formation which plunges southwesterly at 10 to 12 degrees (fig. 2).

EARLY GEOLOGIC INVESTIGATIONS

Studies of pumped storage possibilities in the general area had been made by Public Service Electric and Gas Company engineers as early as 1947 but it was not for another 10 years that the idea of pumped storage became sufficiently attractive to investigate in detail. There have been more than 20 pumped storage projects operating in Europe since the 1930's so the idea is not a new one but all of these older plants have separate pumps and turbines and some were for seasonal rather than daily storage. The development of an efficient reversible pump-turbine coupled with a motor-generator is an important reason for a rather sudden growth of interest in pumped storage in recent years.

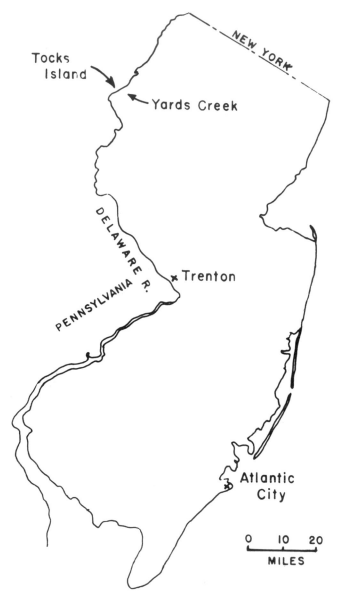

Figure 1. Location of the Yards Creek and Tocks Island Project.

In the first engineering geology study of the Yards Creek site made by this writer for New Jersey Power and Light Company in 1957, attention was focused on the need for quantitative studies of the permeability of the upper reservoir quartzite bedrock because an unlined reservoir was contemplated, impounded by conventional dikes. All other topographically comparable upper reservoir sites known to this writer have some kind of artificial bituminous or other lining.

In 1961, after the geology had been mapped in considerable detail, a test boring program was begun in the upper reservoir area. A total of 42 diamond drill borings were put down around the rim of the proposed reservoir, including 14 which were drilled for test grouting purposes. Most of the borings ranged in depth from 40 to 70 ft with some much deeper; nearly all were drilled at angles of 60 to 75 degrees from the horizontal and at bearings calculated to intersect fracture or joint systems at specified angles. It was considered essential to learn as much as possible about the behavior of joints with depth because the quartzite rock itself was essentially impervious, and any leakage from the reservoir under the dam essentially would be along joints.

All borings were hydraulically pressure tested in 10-foot sections (using a 10-foot double or zone packer) and the results were calculated in terms of a permeability factor. The average of all of the permeability tests for each increment of depth when plotted against depth showed an apparently meaningful statistical relationship with the indicated permeability of the rock decreasing exponentially by an order of magnitude with each 30 ft of depth. While individual tests at any given depth varied considerably from boring to boring, an apparent statistical fit seemed to make the results quantitatively meaningful. A test grouting program was carried out in two local areas selected as represent-

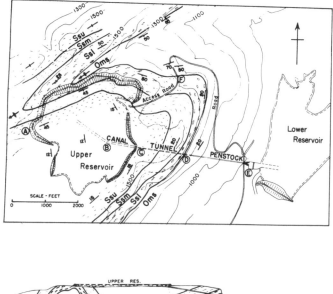

Figure 2. Geologic map and cross section, Yards Creek Project. The symbol Oms represents the Ordovician Martinsburg Formation, Ssl represents the lower member of the Silurian Shawangunk Formation, Ssm and Ssu are the middle and upper members respectively of the Shawangunk Formation. Strike and dip of bedding and cleavage are shown by conventional symbols. In the cross section, the bedding of the Martinsburg Formation is indicated by dotted lines while the cleavage directions are shown by the unbroken lines.

Figure 3. Lower portal of the 20 ft diameter pressure tunnel. The contact between Ordovician Martinsburg Formation and Silurian Shawangunk quartzite is at about the middle of the picture and about midway between the two horizontal rows of planks. Note presplit walls. Location D of figure 2.

ing two widely prevalent geologic conditions in the foundation.

The results of the grout test and of the permeability data indicated that, for the most part, a single line grout curtain could be effective in controlling leakage. On the basis of these results, and simple flow net analyses, the design engineers prepared foundation grouting specifications aiming for a foundation permeability in the order of 10^{-5} cm/sec.

Meanwhile, other geologic investigations were having a bearing upon the design of the structures. The original conceptual design for the upper reservoir dikes had involved a rock fill structure with a concrete upper face and a grout curtain at the upstream toe. On the mountain there is a thin mantle of glacial till, rarely more than 5 or 10 feet thick, and certain exploratory work in the upper reservoir area indicated that enough relatively impervious glacial till could be obtained to permit consideration of a thin earth core structure. The ultimate design for the upper reservoir dikes called for a core 8 ft wide at the crest sloping 1:7 on either side. The design also called for four ft of fine filter and an additional four ft of coarse filter on both upstream and downstream sides. Outside of the filter zone of either side was dumped rock shell sloping 1.75 to 1. The material for the rock shell was obtained from a quarry inside the reservoir and from a canal which leads across part of the basin to the upper portal of the tunnel. (The canal was needed to minimize dead storage since the lowest original ground in the area of the reservoir was on that side of the reservoir nearly opposite from the tunnel portal).

Sand and gravel deposits in the valley of Paulins Kill a couple of miles south of the project area were examined as a potential source of filter materials. Most of these deposits, however, had an undesirable excess of fine filter as compared to coarse. At Columbia on the Delaware River about 8 miles from the project, a source was located where, by a simple crushing and screening operation, and without washing, suitable gravel and sand could be produced in equal quantities.

Exploration in the lower reservoir area involved mainly that of learning enough about conditions in the Yards Creek channel to permit an engineering judgement as to the optimum alignment for the main dam. Abundant silty till was available nearby for the construction of a relatively simple earth dam having a central zone of silty glacial till and an outer zone upstream and downstream of slightly more pervious sandy till.

Bedrock in the center of the channel at the lower reservoir main dam was as much as 50 feet below the original ground surface. This overburden consisted of organic silt at shallow depth underlain in turn by varved silts and sand, gravel and boulders and till. The so-called slurry trench method (in one of its early applications for this purpose) was used to effect a cut off or impervious barrier beneath the dam. This method consists of excavating a narrow trench with nearly vertical sides down to an impervious formation. The trench is held open by the pressure exerted by a slurry mixture of water and bentonite. After excavation the trench is backfilled with impervious compacted earth material. In view of the presence of these pervious and unstable sands and gravels, the alternative method would have involved a very large multistage excavation dewatered by well points.

One particular feature of the rock excavations that were made on the project involved the widespread use of "pre-splitting" for control of overbreak, particularly wherever concrete was to be placed against excavated rock. In the powerhouse excavation there was, for example, a total contact area in the order of 80,000 sq ft between concrete and rock where the rock served as the outer form for placement of the concrete. Because the cost of concrete was in the order of $2.00 per cubic foot the advantages of controlling overbreak are obvious. The Yards Creek Project was one of the first to employ this technique of pre-splitting or controlled blasting, although it has come into widespread use since that time. It was difficult to control overbreak with a burden of less than 10 or 15 feet of rock but below such depth, pre-splitting worked very nicely in both the Martinsburg and the Shawangunk Formations (fig. 4). In addition to the powerhouse, the technique was used for all of the penstock footings, in both the lower and upper portals of the tunnel, along the canal walls in the upper reservoir and along the walls of the spillway.

In addition to geologic notes on the contract drawings prepared by the design engineers, other geologic information which was made available as bidding documents included (1) a comprehensive geologic report, (2) detailed reports of the test grouting program with particular reference to procedures that were used and rates and amounts of grout injection, (3) a report of a test-drilling program involving pneumatic drills (having in mind that drilling in the quartzite for rock fill material and for excavation of the canal and tunnel represented a very large part of upper reservoir construction costs), (4) a special report on the geology of the tunnel route forecasting in detail the amount of excavation that would have to be made in the Martinsburg Slate on the one hand and each of the three members of the Shawangunk Formation on the other and asserting that these units on the tunnel route were the same as those exposed on the east bank of the Delaware River at the Water Gap.

THE UPPER RESERVOIR GROUT CURTAIN

The Yards Creek upper reservoir is of some special interest because (1) of the way it is perched on Kittatinny Mountain and, without any artificial lining, is enclosed by nearly two miles of conventional dikes and grout curtain, (2) percussion drilling was used for most of the grouting and (3) hydraulic pressure testing of drill holes served as a major test for effectiveness of grouting and for the prediction of seepage losses through the bedrock. The

Figure 4. Excavation for powerhouse in an early stage of construction. Note presplit walls in the Martinsburg Formation. From each of the three pump turbine units, a 10 ft diameter penstock leads through a tunnel in the back wall of the powerhouse to the surface about 250 feet away, where the three pipes are joined to form a 19 ft diameter penstock. Two of the three tunnels can be seen in the photograph. Location E of figure 2.

grouting procedures will be described in greater detail in a separate publication but a brief description is given below.

Following excavation of overburden and weathered or loose and detached rock from a trench under the core and filter portions of the dam, exploratory diamond core holes were drilled at intervals of about 200 ft to about twice the depth anticipated, on the basis of prior knowledge, for the ultimate depth of grout curtain in any particular area. On the basis of the information, including pressure testing, obtained from these borings, decisions would be made as to spacing, depth and orientation (azimuth and inclination) of primary percussion drilled grout holes. Following grouting of these, secondaries would be drilled as needed on the basis of two principal criteria: the grout take and the permeability as indicated by hydraulic pressure testing. For the most part three or four passes through an area were sufficient to accomplish such closure but locally additional passes were required and in three short reaches, totaling about 500 ft of dike alignment, three or more rows of grout holes were needed. Directions to drill each boring were given in the field with a particular purpose or a geologic objective in mind; no drilling was done in a merely arbitrary manner. Grouting was mostly simple packer grouting but some staging was done. In any given area, after it appeared that a satisfactory curtain had been effected, a final (ordinarily limited) set of diamond borings was put down to check on the effectiveness of the percussion work and in a few cases additional sets of diamond hole were needed to achieve the desired degree of impermeability.

Used with discretion and having the local geologic conditions in mind, the hydraulic pressure tests indicated on a statistical basis that the design permeability in the order of magnitude 10^{-5} cm/sec was being achieved. The need for a tight foundation was dictated in large part by the fact that the contributing watershed for the pumped storage system is a little less than five square miles. Allowing for the legally required minimum release of water at the main lower reservoir dam and allowing for evaporation from the reservoir surfaces, a loss of as little as one cfs through the dikes in the western part and southern part of the reservoir, where such loss would escape from the system, might have to be replaced in some way.

In the 9,000 ft long curtain, a total of 1,311 percussion and 397 diamond borings were grouted. Meaningful comparisons of the performance of the percussion and diamond borings depend upon local geologic conditions but in general it was found that if as much as 1.0 cu ft of solids could be injected in a boring, there was little difference in the behavior of the two types. However, essentially no grout (0.1 cu ft of solids or less) could be injected in 36 percent of the percussion holes compared to only 10 percent of the diamond borings. In those parts of the formation where even slightly weathered shale layers occurred interbedded with the quartzite, percussion drilling locally was so ineffective that the cutoff had to be made mostly by diamond borings. Because the percussion drilling cost was about one-third that of diamond work and because of the greater speed and therefore greater flexibility of the operation, percussion work was done to the fullest extent practicable. Percussion borings were 2⅞-inch diameter, diamond borings were of AX size. Approximately 17,000 cu ft of solids were injected in some 55,000 linear feet of borings. Fly ash was used wherever possible, partly to replace cement by as much as 50 percent and partly to reduce rate of grout injection where large flows were encountered. Pressures were very carefully monitored to avoid using more than 1 psi per ft of depth except for very shallow packer settings where pressures of 5 to 10 psi were used with great caution and usually only after low pressure grouting had been completed at a setting of three to five feet below the surface.

The amount of seepage under the dams is estimated at about 100 gpm for the 9,000 ft total length of dike and this is in close agreement with the permeability that was sought.

Although the initial investigations were carried out by New Jersey Power and Light Company, the project was built by and is operated by Jersey Central Power & Light, which is a sister company in the General Public Utilities organization. Public Service Gas and Electric Co. have a 50 percent interest in the project. Ebasco Services, Inc. were design engineers; Burns and Roe, Inc. were construction supervisors. The author was retained independently by the power company from initial exploration through construction.

POSSIBLE RELATIONSHIP TO THE TOCKS ISLAND PROJECT

Some time ago Jersey Central Power & Light Co. and Public Service Electric and Gas filed an application with the Delaware River Basin Commission to build additional pumped storage generating capacity in connection with the Tocks Island project which thus would combine pumped storage, flood control, maintenance of minimum downstream flow in the Delaware, recreation values and future water supply. The ultimate generating capacity from pumped storage would be in the order of 1,300 kilowatts and would involve building additional reservoir capacity on the mountain contiguous to the existing upper reservoir, connected by shaft and tunnel to an underground powerhouse near the Tocks Island dam. The Tocks Island reservoir would be used as the lower reservoir.

Much of the land required for the Yards Creek project had been owned by the State of New Jersey. The contract negotiated with the State for this property involved an exchange for other land which the company owned or could acquire, as well as a sum of money (used by the State in the purchase of land for Wawayanda State Park in the northeastern part of the state) and certain commitments regarding water supply. With regard to the latter, the contract commits the company, at such time as (1) the State needs the water and acquires the legal right to take it, (2) Tocks Island reservoir is built, and (3) the Delaware-side pumped storage plant is built, to transfer water from the Tocks Island reservoir to the Yards Creek lower reservoir at the bare cost for the net pumping lift. In other words the State would reimburse the company for the cost of lifting the water some 300 feet from the level of the Delaware to the level of Yards Creek but would make no payment for the investment in the facilities required for such service.

Possible advantages of this plan would include the fact that the water could flow by gravity in a tunnel or pipeline from Yards Creek to metropolitan New Jersey or to storage facilities such as Round Valley reservoir or those in the Pequannock River system. The state would also avoid the cost of diversion facilities. Ultimately a definitive comparison will have to be made with the cost and other factors involved in diversion schemes at other possible downstream locations on the Delaware River.

ROAD LOG

The field trip route is through the Water Gap on the Pennsylvania side, crossing the Delaware at Portland and following route 94 to Walnut Valley (about 3.5 miles west of Blairstown), where a turn is made left to the Yards Creek project. After visiting the project the return trip is made via route 94 and route 80 through the Gap on the New Jersey side.

Mileage

0.0   The road log begins on route 611 opposite the toll gate at the north end of the route 80 bridge at Delaware Water Gap.

2.0   Brief stop to examine the Shawangunk Formation.

6.0   Turn on toll bridge over the Delaware and follow route 94 toward Blairstown.

12.1   Turn left off route 94, following signs to the Yards Creek Project. At 13.5, cross Yards Creek. 14.6 gate to project.

Proceeding up the access road to the top of Kittantinny Mountain, one passes in turn (1) the main lower dam and spillway, (2) the powerhouse on the right and the 19 ft diameter penstock on the left, (3) road cuts in Martinsburg slate and, near the top, cuts in Shawangunk quartzite. At the outlook point at the summit, one can see — with decent visibility — in addition to the reservoir, Kittatinny Mountain extending northeasterly toward High Point in the extreme northern part of New Jersey, the Catskill Plateau to the northwest and the New Jersey Highlands to the southeast.

The outlook spot is approximately on the axis of a syncline, plunging 10 to 12 degrees southwesterly. The upper portal of the pressure tunnel is about 2,000 feet southerly. The highest dam is about due northwest from the observation point. Rock fill for embankment construction was obtained from quarry operations within the reservoir. Earth materials for the core of the dikes were obtained from borrow pits inside the reservoir and from other locations on the mountain.

The trip will conclude with (1) a visit to the lower portal of the pressure tunnel to examine the Martinsburg-Shawangunk contact, (2) a visit to a series of ponds that have been built to study the effects of fluctuating water levels on fish propagation and (3) a brief examination of samples, cores and models. The group will proceed by bus to the Delaware Water Gap Bridge and continue on to the Tocks Island damsite.

GEOLOGY OF TOCKS ISLAND AREA AND ITS ENGINEERING SIGNIFICANCE

A. J. DEPMAN AND D. G. PARRILLO
U. S. Army Engineers District, Philadelphia, Pa. 19106

The Tocks Island dam site will be located on the Delaware River approximately two miles upstream from Shawnee-on-the-Delaware, Pa. The center-line will be about 100 feet downstream from the lower end of Tocks Island (see figure 1). The reservoir formed will extend 39 miles upstream at

Figure 1. General view of proposed Tocks Island project.

Figure 2. Artist's conception of dam and Tocks Island reservoir.

flood pool and have a capacity of nearly 1,100,000 acre-feet.

As proposed, the dam will be a combination earth and rockfill type with an impervious core and upstream impervious blanket (see figure 2). Some pertinent statistics are:

Elevation Top of Dam	455'
Elevation River Bottom	295'
Spillway Crest Elevation	392'
Water Supply Pool Elevation	410'
Flood Control Pool	432'
Spillway Design Flood Max. Pool	445'
Width of Dam @ Crest	30'
Width of Dam @ Base	3,260'
Length of Dam (excluding Spillway)	2,600'
Spillway Width	383'
Outlet Works	26' dia tunnel 1,835' long
Power Intake Structure	2–24' dia tunnels 1,260' long

HISTORICAL INTEREST

The site lies in what was known in Colonial times as Minisink Valley, a Lenni Lenape Indian name with the obscure meaning of "the land from which waters disappeared"!

Early in the 18th Century Dutch traders, explorers and miners visited this part of New Jersey from the Hudson Valley, to be followed later by the British.

A monument to these hardy adventurers is the existence of the Old Copper Mine (Pahaquarra Mine), 1646–1650, located in Kittatinny Mountain about two miles upstream of the site. There is some evidence to indicate these are the oldest copper mines in New Jersey, if not in the United States and certainly among the oldest mines of any kind to be worked by Europeans on this continent.

GEOLOGY AT THE DAM SITE

At the dam site the Delaware River flows in a broad U-shaped valley with a gentle gradient. The left abutment (New Jersey side) is formed by the western flank of Kittatinny Mountain (crest elevation 1,500 feet MSL datum). The Silurian High Falls formation (New Jersey equivalent of the Pennsylvania Geological Survey Bloomsburg Formation) is at or within 10 feet of the surface in most areas, with the exception of one area where a thick morainal ridge strikes east-west up the slope. Outcrops along Old Mine Road at the base of the left abutment show the red and green beds of the High Falls with a N65°E strike dipping into the river about 45° northwest. Riverward the bedrock surface plunges sharply to depths of 135 feet (figure 3) near where the High Falls formation grades into the overlying Poxono Island formation. At this juncture the attitude of the beds is near vertical. The bedrock surface rises gently beneath the river sediments and on the Pennsylvania side of the valley the slope steepens and rocks of Devonian age crop out above elevation 450 MSL.

The major structure in the valley is a broad SW plunging asymmetric syncline with a SE dipping axial plane. Minor undulations and cross folds are found within the larger structure. The east limb forms Kittatinny Mountain (in New Jersey); the west limb forms Wallpack Ridge (in Pennsylvania). The trough of the syncline is located under the deep Pleistocene deposits on the Pennsylvania bank of the river.

A series of glacial deposits, as much as 250 feet thick, capped by a veneer of alluvium, overlies the bedrock surface beneath the river and most of the right abutment (Pennsylvania side).

STRATIGRAPHY

Materials encountered at the dam site range from Recent unconsolidated sediments to hard, quartzitic Silurian rocks. Detailed stratigraphic descriptions and engineering significance of the consolidated formations are shown in table 1.

PLEISTOCENE GEOLOGY

Deposits of Wisconsin age have been recognized at the dam site. Four recessional moraines have been identified in the Delaware valley and there is evidence that a fifth occurred in the vicinity of Tocks Island. Borings show a hummocky belt of till above bedrock and below ice-contact glacio-fluvial and glacio-lacustrine deposits. Low morainal ridges diagonally ascend both sides of the valley. These features can be recognized topographically, particularly on the Pennsylvania side approximately one-half mile upstream of the power line crossing. As the ice edge waned from the Tocks Island position, melting of the crevassed marginal zone left behind isolated ice blocks. Outwash was deposited around, between and over these masses. Fine-grained sediments accumulated in local ponded areas. Some of these sediments show slump and collapse features indicating melting of buried ice masses after sediment deposition.

Leaching of carbonate rock debris near the surface and precipitation from ground water at depth has resulted in many cemented lenses within the valley train (outwash) material. Fluvial erosion has dissected the valley train to its present level. The stream course seems to have followed, for the most part, a series of longitudinal ice-block depressions in the valley train.

The postulated sequence of events from the last stages of glaciation through the present land form at Tocks Island is shown in figure 4 (A-E). Detailed descriptions and engineering classifications of the glacially derived sediments are shown in table 2.

GEOLOGICAL CONDITIONS AFFECTING DESIGN

One major problem in selecting the dam site was locating the best foundation within the glacial deposits. The gravel "highs" associated with the end moraine provided this locale. Geophysical investigations and a few borings narrowed the site of the proposed dam to a zone of the river approximately 3,000 feet long located downstream of the lower end of Tocks Island. Detailed investigations within this zone to locate the most suitable foundation conditions was accomplished by means of borings on a basic 200' grid pattern and approximately 1,200 undisturbed Shelby-tube samples taken in the glacio-lacustrine sediments. Results of these detailed investigations confirmed the unique geological setting, which afforded the best foundation conditions for an earth and rockfilled dam.

The permeability of the glacial deposits in the Pennsylvania bluff are believed to present no major problem since weathering has progressed to depths of ten feet, yielding decomposed shale products and a naturally impervious blanket. In sections where this natural blanket is breached, it will be re-established during construction of the dam.

Location of the spillway was governed by the

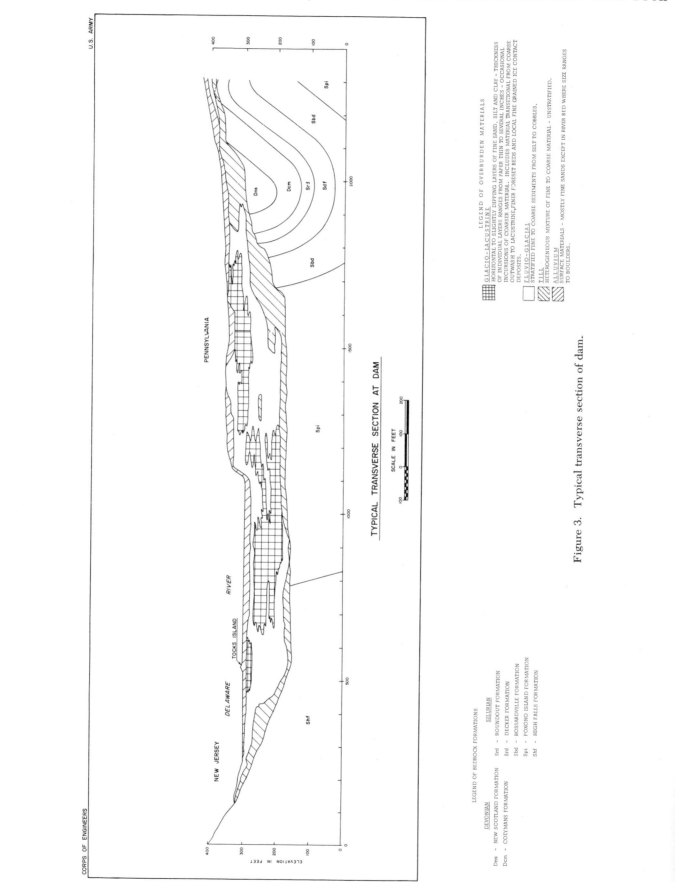

Figure 3.   Typical transverse section of dam.

TABLE 1. STRATIGRAPHIC COLUMN AT DAM SITE AND ENGINEERING SIGNIFICANCE OF ROCK TYPES

Age	Formation	Member	Thickness of Unit at Dam Site	Description	Engineering Significance & General Remarks
Devonian	Marcellus		Not Present at Site	Black, fissile carbonaceous SHALE.	*Marcellus* — forms steep walls and extensive talus slopes along right rim of reservoir from Wallpack Bend to Matamoras. Intense wave action and fluctuating water levels may cause local slides & slumping.
	Onondaga			Dark, thin-bedded SHALE and med.-bedded cherty LIMESTONE.	
	Oriskany			White, f. to c. conglomeratic SANDSTONE.	
	Port Ewen			Gray, calcareous & siliceous silty SHALE.	*Onondaga through Bossardville* — used mainly for stratigraphic & structural correlation. These formations form the left rim of the reservoir from Wallpack Bend to Port Jervis, N. Y., and the right rim from Wallpack Bend to Tocks Island. All limestones are susceptible to solution activity & potential reservoir leakage; however, no cavernous conditions or areas of large scale solution activity are known to exist. At the dam site these formations will not be involved in any major excavation.
	Minisink			Gray, argillaceous LIMESTONE & SHALE.	
	New Scotland	Flatbrookville	35′±	Med. dk. gy. calcareous fossiliferous SHALE & argillaceous LIMESTONE with nodules and layers of chert; cleavage.	
	Coeymans	Total	82′–89′	Med. gy. quartzose conglomeratic SANDSTONE w. calcareous cement. Local limestone beds & cherty zones.	
		Stormville Disconformity		Massive fossiliferous LIMESTONE with reef structure 25′ thick overlain by sandy LIMESTONE & black SHALE; mud cracks.	
		Shawnee I.	42′		
		Peters Valley	5′–7′	Lt. gy. f.–c. calcareous SANDSTONE.	
		Depue I.	20′	Med.-thin bedded clayey LIMESTONE.	
Silurian	Rondout	Total	31′–34′	Thin bedded dk. gy. calc. SHALE & sandy LIMESTONE; bedding undulatory; mud cracks common. Buff, platy weathering	
		Mashipicong	11′–12′		
		Whiteport	5′–7′	Med. bedded to massive DOLOMITE.	
		Duttonville	15′	Interbedded calc. SHALE & LIMESTONE.	
	Decker Ferry	Wallpack Center	55′	Med.-thin bedded sandy LIMESTONE & DOLOMITE, calc. SILTSTONE, SANDSTONE & SHALE. Fossiliferous, reef structures.	
	Bossardville		112′	Laminated to thin bedded clayey, v.f. crystalline LIMESTONE; calc. SHALE & SILTSTONE, some beds massive & dolomitic.	*Poxono* — underlies valley fill in the reservoir area & at the dam site. Several zones exhibit extensive solution activity & much rehealing.
	Poxono Island		675′	Green, red & mottled calc. SHALE, MUDSTONE, dolomitic SHALE, DOLOMITE & LIMESTONE.	*High Falls* — forms left rim of reservoir from Wallpack Bend to the dam site. All structure in left abutment will be founded in this formation.
	High Falls (Bloomsburg)		1,500′	Red & green mottled SHALE, MUDSTONE, SILTSTONE, SANDSTONE & QUARTZITE; cross bedding common in coarser grained rock.	*Shawangunk* — will not be encountered in dam structures; possibly will be encountered if underground powerhouse is constructed in pumped storage scheme. Is not exposed in reservoir. Potential source of excellent quality concrete aggregate and dimension stone.
	Shawangunk		1,500′	Cross-bedded QUARTZITE & CONGLOMERATE, with interbedded black ARGILLITE.	
	—Unconformity or fault—				
Ordovician	Martinsburg		Not present at site	SLATE & GRAYWACKE.	Not of significance.

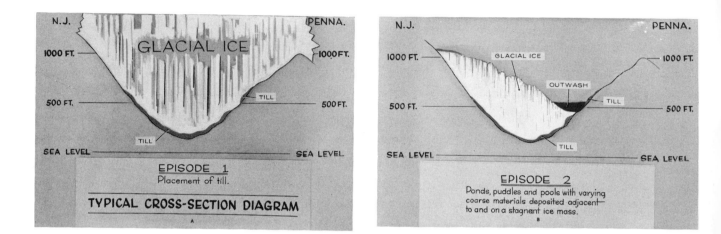

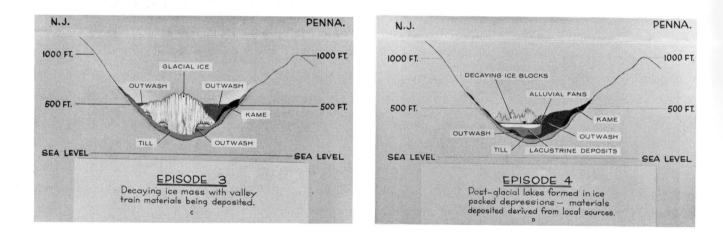

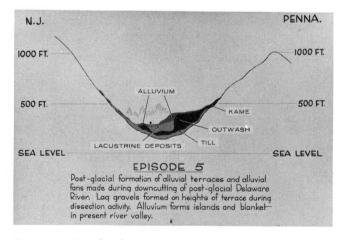

Figure 4.   Postulated sequence of glaciation at Tocks Island.

TABLE 2. ENGINEERING CLASSIFICATION OF PLEISTOCENE SEDIMENTS

Type	Section Symbol	Overburden Materials	Permeability	Foundation	Borrow Pervious	Borrow Impervious
Alluvium	VI	River Bed—fine–coarse reworked Pleistocene	High	Good–Excel	Good	
Glacio-Laucustrine Deposits	I	Flood Plain—fine–med. silty sands	Low	Weak		
		Horizontal gray silts, very fine sand; red clay up to 50% of section	Low	Weak		
	IIa	Horizontal to gently dipping layers of fine sand, silt, clay. Thickness of individual layers range from paper thin to several inches. Clay content 10–20% in occurrences below El 300 under river & up to 50% above El 300 under bluff	Low	Weak		
	IIb	Same as IIa, but bedding disturbed locally by slumping or sliding, contains local incursions of coarser material	Low	Weak		
Glacio-Fluvial Deposits	III	Silt, fine to med. sand, rare fine gravel. Includes material transitioned from coarse outwash to type II, finer foreset beds and local fine grained ice contact deposits	Low–Medium	Weak–Good		
	IV	Stratified fine to coarse sediments from silt to cobbles	High–Low	Good–Excel	Good–Fair Below wthrd zone	Top 10' Fair–Excel
Till	V	Heterogeneous mixture of fine to coarse material unstratified	Low	Good–Excel		Good

type of rock in both abutments—limestone in the right; sandstone, siltstone, quartzite in the left. The stability of slopes are a major concern, since the maximum cut will be over 300 feet. Several bedding plane slips are intersected in the proposed intake structure and spillway cut, all dipping into the excavations at angles varying from 17° to 35° (zones 10, 11, 12, figure 5). Two of these (zones 10 & 12) occur at or near the base of laminated shales and siltstones with deep desiccation cracks. In both structural features there are varying degrees of weathering and decomposition.

Since these zones will be controlling factors in design of slopes, a detailed exploration program was initiated to ascertain the physical conditions of zones 10 and 12, as they will affect design. Three 36-inch diameter calyx holes were drilled to intersect this lithologic unit in critical areas to examine the in-situ conditions. In one hole, zone 12 was rock bolted and overcored to obtain samples to evaluate its strength. To further investigate the geometry of the bedding plane slip, 4-inch cores were obtained behind proposed slopes for testing

and piezometers set to measure hydrostatic pressures.

The 5 x 7 ft adit begins at the riverward side of the right spillway wall following the centerline of the spillway crest at approximately the base of excavation for this structure for a distance of 600 feet. Cross cuts at zone 12 and the end of the tunnel are each 100' in length. In-situ and laboratory tests have been performed to establish the following primary parameters: the approximate shear strength of the more critical rock zones and along the more critical weak planes; the stress intensity and direction present within the various rock types, and the general deformation characteristics of different rock types.

An attempt was also made to correlate seismic field velocities with sonic core velocities so that information obtained from testing in the tunnel adit may be extended to other parts of the site.

The design of the rock slopes will be determined on the basis of data obtained from all of the above mentioned tests and several slopes will be constructed in a proposed test quarry.

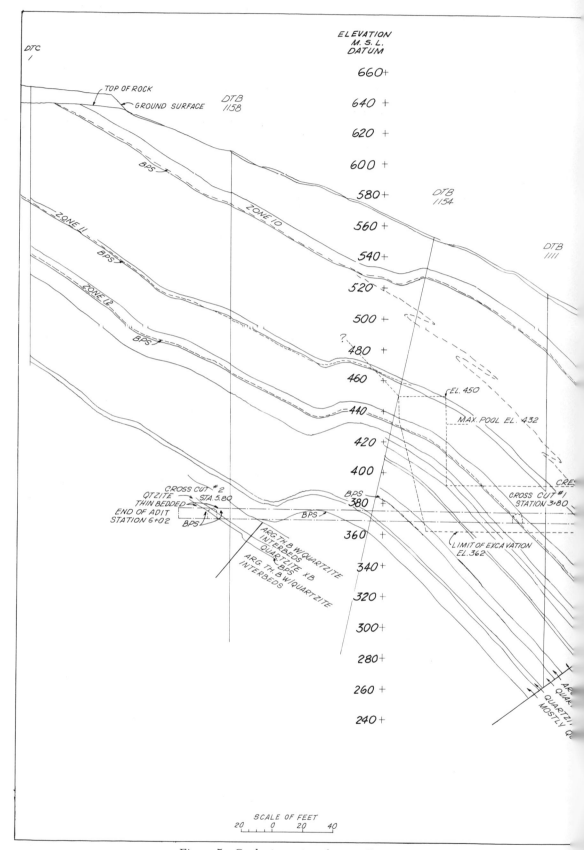

Figure 5. Geologic section along spillway crest.

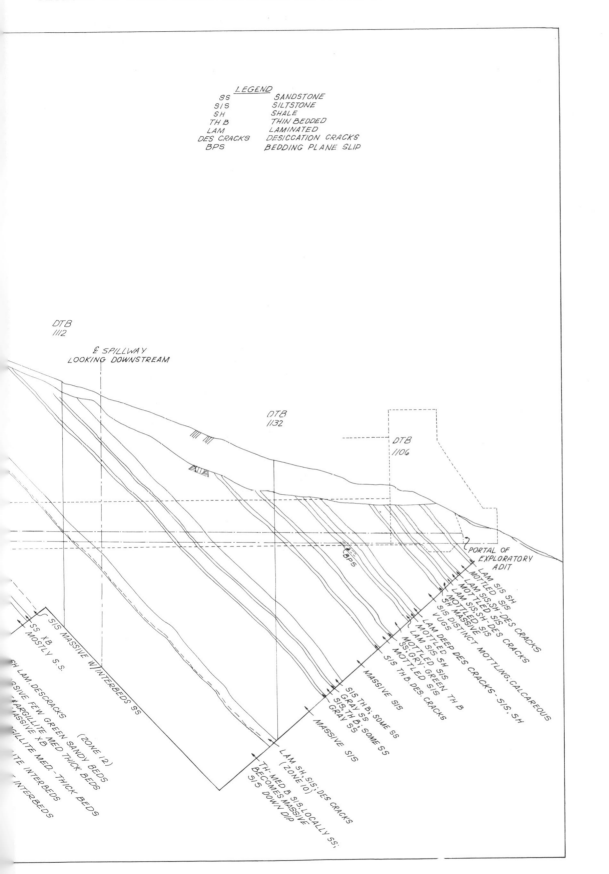

LEGEND
SS — SANDSTONE
SIS — SILTSTONE
SH — SHALE
TH B — THIN BEDDED
LAM — LAMINATED
DES CRACKS — DESICCATION CRACKS
BPS — BEDDING PLANE SLIP

£ SPILLWAY
LOOKING DOWNSTREAM

The seven million yards of rock obtained from excavation for structures in the left abutment will be used in the dam. Impervious core and blanket material is available from the uppermost weathered part of the outwash and from the till. Most of the unweathered outwash and alluvium will provide pervious fill, but processing may be required because of the significant percentage of siltstone and shale particles. Concrete aggregate can be obtained from suitable parts of the High Falls within the excavation, the lower and upper Shawangunk, or from commercial sources in the area.

ROAD LOG

Mileage

0.0	Delaware Water Gap Bridge, New Jersey.
0.5	Pass toll gate at Pennsylvania end of bridge—note road cut on left side—Silurian Shawangunk—High Falls (Bloomsburg) formations contact.
1.0	Leave Interstate 80 at cloverleaf—take county road north from village of Delaware Water Gap, Pa. toward Shawnee-on-the-Delaware.
1.5	Road cut on right side—Wisconsin esker.
2.5	Continuous road cut and outcrop on left exposes the Devonian New Scotland, Coeymans, Rondout and Decker formations.
2.8	Fred Waring's Shawnee Inn and golf course on alluvial Shawnee Island on the right.
3.0	Waterfalls on left provide excellent exposure of Bossardville limestone and Decker formation. Mud (desiccation) cracks in the Bossardville in creekbed are as much as 20 feet long.
5.5	Arrive approximate centerline of proposed Tocks Island Dam at Coutz' Farm. Observe biohermal Shawnee Island member of the Coeymans formation in ridge rock exposed along the power line clearing opposite Tocks Island. Observe rock and soil samples representing right (Penna) abutment and foundation. 45 minute stop here. Return via same route to Water Gap bridge.
11.0	Delaware Water Gap National Recreation Area Information Center.
	Lunch stop. One hour.
12.0	Observe anticline in road cut—Silurian Shawangunk—High Falls formation contact—follow anticlinal structure from east to west limb.
12.5	Continuous outcrop of High Falls formation—note cleavage, joint patterns and dip of red and green beds. Dip @ 45°NW Strike @ N65°E.
15.9	Arrive at centerline of proposed Tocks Island Dam, N. J. Form two groups—Group No. 1 climb to site of calyx hole excavations and cores; Group No. 2 enter test adit to observe rock instrumentation zones, cross cuts and stratigraphy representative of the left abutment structures of the project. Stop estimated to last approximately 1½ to 2 hours. Depart Tocks Island Area and return to Atlantic City.

ILMENITE DEPOSITS OF THE NEW JERSEY COASTAL PLAIN

FRANK J. MARKEWICZ

Bureau of Geology and Topography, New Jersey Department
of Conservation and Economic Development, Trenton, New Jersey 08625

INTRODUCTION

At the 1957 Geological Society of America meeting in Atlantic City, a paper entitled "Preliminary Report on Ilmenite-Bearing Sands From the Coastal Plain of New Jersey," was presented by the author and his colleague, Daniel Parrillo, presently with the U. S. Army Corps of Engineers. In this paper it was indicated that four potentially economic ilmenite sand type ore bodies had been delineated in the Coastal Plain. After the official news release on November 23, 1956 a number of companies interested in ilmenite optioned thousands of acres and drilled hundreds of holes in the potential areas delineated by the New Jersey Geological Survey (fig. 1). Today (1968), three companies, Glidden-Durkee Division of SCM Corp., American Smelting and Refining Corp., and American Cyanamid; collectively hold approximately 8,500 acres in Ocean and Burlington Counties. The Glidden Company has produced ilmenite concentrate from their modern plant at Lakehurst since 1962. Of the four privately held ore bodies, three were discovered by the New Jersey Survey. In addition to the above, there is approximately 2,500–3,500 acres of potential ilmenite ore bearing land owned by the Federal and State Governments. Localities in which high ilmenite concentrations were found, but not considered economic because of residential development, included areas west of Laurelton and east of Lakewood. A large area stripped of its overburden for a school foundation near Laurelton exposed a two foot thick black ilmenite pavement approximately 2.5–4.5 feet below the surface. Some of the material was so laden with ilmenite that it was slightly indurated by iron leached from the heavy mineral concentration. Other less continuous heavy mineral concentrations occur in the Medford Lakes area. Work by the New Jersey Survey in the Medford Lakes region revealed local economic heavy mineral concentrations, however, the pattern proved to be erratic, with local high percentages bordered immediately by very low values. Subsequent detailed deep drilling on a reasonably large area by a commercial company substantiated the erratic deposition.

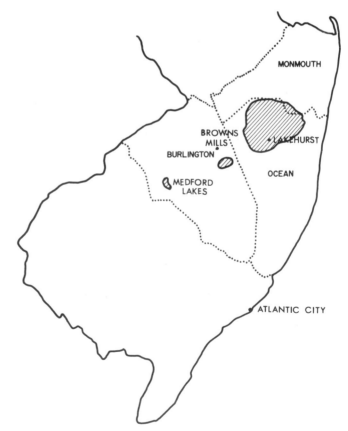

Figure 1. Outline map showing areas of ilmenite deposits.

363

Some recent work by an independent concern indicates interesting heavy mineral percentages in Silver Bay, which lies north of Toms River. No work has been done by the Survey at this location.

ACKNOWLEDGMENTS

The author is grateful to the Glidden Co., American Smelting and Refining Corp., and American Cyanamid Company for their sincere cooperation. Special thanks go to Mr. S. Schworm and Mr. J. Gijlstra of the Glidden Company. Gratitude is extended to members of the New Jersey Geological Survey for their assistance and to Mr. Meredith E. Johnson, former State Geologist who critically reviewed the manuscript.

HISTORY

Since the original announcement made by the New Jersey Geological Survey on the occurrence of commercial ilmenite bodies, there have been numerous requests asking how these deposits were found and what initially prompted the investigations for commercial concentrations of heavy minerals. Ilmenite was known to occur as part of the heavy minerals in the silica sands of the southern Coastal Plain and also in sand and gravel pits, but there was no mention of potentially economic concentrations.

This is an appropriate time to relate the historical and geological background associated with the discovery of these deposits. During the early to mid 1950's there developed in New Jersey, as everywhere else, the search for radioactive mineral deposits. The Survey was almost swamped by requests from professional and amateur uranium seekers, in addition to farmers, homeowners, and even persons poaching on their neighbor's building lot in the middle of the night with a geiger counter. Investigation of these requests coupled with our own basic field work and research on radioactive minerals, rocks, and deposits led to the location of several monazite-bearing rock and sand type deposits near Chester, New Jersey. While investigating the alluvial sands for their monazite content in river valleys of Morris County, especially near Chester, a significant amount of black opaque heavy minerals were found in Black River Valley. Although the highest percentage of the metallic heavies were magnetite, a considerable percentage of ilmenite or ilmeno-magnetite was noted. Knowing that perhaps thousands of feet of Precambrian rock had been eroded from the Highlands of northern New Jersey, this thought provoking question presented itself; where was the more stable suite

of heavy minerals finally deposited that had been eroded from the Precambrian rocks? Surely there is not enough residual sandy material left in northern New Jersey to account for the enormous amount of rock that was eroded. Logically, much or most of the residual sandy material containing the intermediate and more stable heavy minerals were transported to the Cretaceous and Tertiary seas that had invaded southern New Jersey. This posed another question: what ancient drainage transported these materials and in what locality would heavy mineral concentrations occur; and where could these concentrations be found in the present day Coastal Plain? In addition, if any commercial deposits existed they must be found at the surface or at shallow depth. The widespread (Pliocene?) Cohansey Formation composed principally of quartz sand appeared to be the most logical formation to investigate because it probably contained sands that were reworked from shore deposits of the underlying Tertiary and Cretaceous formations. All of the seventeen Coastal Plain formations had near shore alluvial deposits, beach sands, and dune deposits, and the material from these deposits was in part reworked by the advancing and retreating seas from one sequence to the next. In as much as the Cohansey is the latest wide spread Coastal Plain deposit, it appeared that there might be sufficient concentrations of heavy minerals somewhere in the Cohansey to justify an exploration program. The problem then was in what area of the Cohansey were the minerals concentrated, if indeed any concentrations existed at all? In addition to investigations of the large number of exposures at the numerous sand, gravel, and glass sand pits, several companies had conducted ilmenite investigations along the coast without turning up any concentrations to encourage further exploration. All of the available reports dealing with the mineralogy of the Coastal Plain reported ilmenite in the heavy mineral fraction, but made no mention of any significant concentrations. The subject was studied for approximately three months by the author, and it was decided that the ancestral drainage patterns of the Delaware and Millstone Rivers provided the best clues for interpreting the history of Cohansey and Post-Cohansey deposition. In addition to transporting sediments from the Paleozoic and Precambrian rocks in the northern part of the state, these two rivers were probably eroding and transporting near shore and beach sands from the exposed formations (Tertiary and Cretaceous) that extended inland beyond their present outcrop limits. It was determined from field studies and well samples that the ancestral

Millstone which now flows northward had at one time flowed southeastward toward the coast. A glance at the New Jersey geologic map shows that the Delaware River, as large as it is, and very probably it was larger, is diverted ninety degrees immediately south of Trenton where it comes in contact with unconsolidated sediments. Why and by what mechanism is a mighty river diverted by materials that can be removed by hand, especially so when upstream a mere 10 to 12 miles from the turn, it has cut a fairly straight course for some 35 miles through diabase, argillite, sandstone, and shale. The plausible reason is that at one time the ancestral Delaware flowed southeastward throughout its entire length and was not diverted below Trenton because of soft sediments, but became a product of stream capture by smaller subsequent stream(s) flowing southwestward parallel with the strike of the basal upper Cretaceous sediments as shown in figure 2. These smaller subsequent streams drained into the Schuylkill River, a water course that is also interpreted to have previously flowed southeastward across New Jersey. The

Schuylkill became a product of capture by smaller subsequent streams that were flowing to the southwest below what is now Philadelphia. By projecting a line southeastward into the Coastal Plain from the Rocky Hill gap of the Millstone River and another line along the Delaware River from Lambertville to Trenton and across the Coastal Plain it focused attention to the northwest projecting lobe of Cohansey in the Collier's Mills-Lakehurst region. A much smaller northwest projecting lobe in the Cohansey exists southeastward of Camden in line with a southeast projected course of the Schuylkill River.

During November, 1956, several shallow depth holes (4–6 ft) dug in Cohansey sand near Collier's Mills revealed heavy mineral concentrations of three to four percent; however, it was a nearby newly dug cesspool hole that revealed a substantial percentage of heavy minerals for the total depth of 12 feet. Laboratory results on the sand from this site showed approximately 10 percent heavy minerals. Petrographic work indicated the heavies consisted of 80 to 85 percent black to brown il-

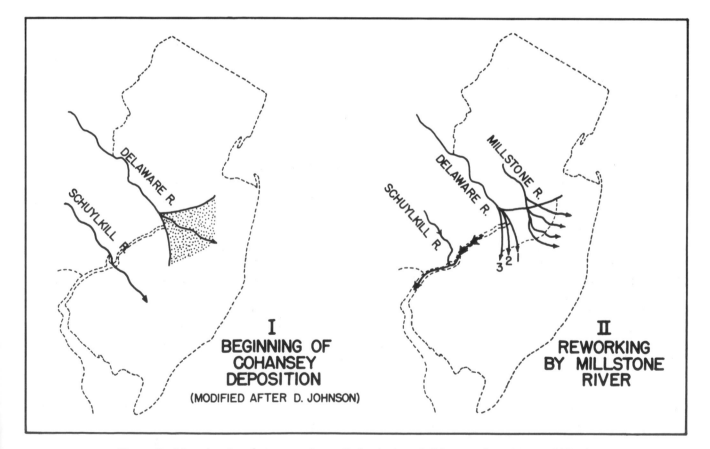

Figure 2. Map showing drainage patterns, I—beginning of Cohansey deposition and II—reworking by Millstone River.

menite and tan to gray highly polished and rounded leucoxene. Early chemical tests indicated a TiO_2 content of 52–58 percent in the ilmenite. An exploration program was immediately initiated near Collier's Mills in the area of the original discovery. Subsurface exploration was accomplished with a 4-inch soil-type auger and all road cuts, stream cutbanks, utility pole and house excavations were checked for heavy mineral content. Even the sand brought to the surface by ants was carefully washed and examined with a hand lens. After approximately 50 holes were augered, it became evident that an area in excess of 1,000 acres contained over 3 percent of heavy minerals to a depth of 20 feet or more. In some holes the heavy minerals averaged 10–12 percent for a depth of 10 to 15 feet. In most road cuts and other exposures the heavy minerals are effectively masked by a thin oxidized clay film or are mixed with soil or organic material in the upper profile. One of the most effective exploration tools proved to be a watch glass and a plastic wash bottle filled with water. Because clay, soil and oxidation all tend to mask the heavy minerals, a small amount of sand from an outcrop was put into the watch glass and water added. By stirring and rubbing between the fingers, most of the clay was loosened and put into suspension. The cloudy water is decanted and the process repeated until the sands are fairly clean. By rotating (panning) and settling the heavies to the bottom, a close approximation (after a sufficient period of experience) of the heavy mineral content could be made by looking at the bottom of the glass.

ILMENITE DISTRIBUTION

During the early phase of the program, exploration was limited to the Cohansey Formation, however, as experience and field work continued, it became evident that the contact of the Cohansey with the underlying Kirkwood Formation it also contained a significant amount of heavy mineral values. The mineral grains in this zone are mixed in size, there being the typical large mineral grains as found in the Cohansey in addition to many finer grains. In tracing the smaller grains stratigraphically downward, it became evident that the Kirkwood contained substantial quantities of very fine heavy minerals (fig. 3).

The Survey exploration work was limited to only those areas that appeared suitable for economic exploitation. Non-exploitable areas were investigated principally for purposes of geologic continuity. In many places the land is worth more as real estate than as mineral producing property.

As work progressed, "patchy" heavy mineral concentrations were observed in sand and gravel of the Cape May, Pensauken, and other scattered surficial deposits of Pleistocene age. At some localities, narrow sinuous channel sediments composed of coarse sand and gravel contain appreciable amounts of coarse Cohansey type heavy minerals. As exploration work progressed to the southeast towards Toms River and several ilmenite areas were delineated, it became apparent that the upper Tertiary sediments in the northern part of the Coastal Plain constituted an ilmenite mineral province.

GEOLOGIC FORMATIONS

The three principal formations in the Coastal Plain that contain commercial quantities of ilmenite are the Cape May, Cohansey, and Kirkwood. In addition to these, the more recent Pensauken contains local heavy mineral concentrations, but these are too limited in extent to be of commercial potential alone, nevertheless, they become a limited source of ore where they overlie any one of the above formations that are commercial.

KIRKWOOD FORMATION

The marine Miocene Kirkwood Formation which immediately underlies the Cohansey was deposited unconformably on the Shark River – Manasquan deposits of Eocene age (fig. 4). The Kirkwood Formation extends from Asbury Park in Monmouth County southwestward to Lower Alloway Creek Township in Salem County and throughout its extent consists of fine micaceous sand, silt, and clay. Toward the southwest, the formation becomes more clayey, especially in Salem County, where a thick bed of olive to chocolate colored clay is known as the Alloway Clay. In Monmouth and Ocean Counties, the Kirkwood is a very fine to locally medium grained sometimes clayey, micaceous sand with layers and lenses of clay that range in thickness from a few inches to more than 20 feet. The upper yellow to locally red silty to clayey fine sand grades downward through progressively darker material to the black, slightly to moderately lignitic, fine, micaceous sand. An effective leaching process has bleached the exposed upper part of the formation and in the process has completely removed the lignite.

The Kirkwood Formation ranges in thickness from about 100 feet along its outcrop to more than 650–750 feet in the subsurface at Atlantic City. Heavy mineral-bearing sand in the northeastern part of the outcrop area is 20 to more than 50 feet thick. The fine grained Kirkwood heavy mineral suite contains less ilmenite and leucoxene than the Cohansey and a larger percentage of minerals such as zircon, kyanite, sillimanite, rutile, and others. In addition, there is a slightly greater amount of completely non-economic heavy minerals. The Kirkwood heavies are concentrated in long, fairly flat and continuous laminae with little structure except for shallow cross-beds and minor undulations (fig. 5).

COHANSEY FORMATION

After withdrawal of the Kirkwood sea, fluvial deposition of the Pliocene(?) Cohansey commenced in the inland region with deltaic and probable restricted shallow estuarine deposits occurring

toward the coast. Basal Cohansey, which consists of very fine to medium sand and some fine gravel, was reworked from the exposed Kirkwood. Sediment laden streams flowing from the northwest also deposited sediments that were eroded from the inland lower Tertiary and Cretaceous formations which extended further inland.

The Cohansey typically consists of poorly sorted fine to coarse grained quartz sand. In many areas the fine and coarse sands predominate; however, locally pebble size particles occur. A thin clay veneer coats the sand and heavy minerals in some areas. Clay beds from a few inches to 10 feet thick

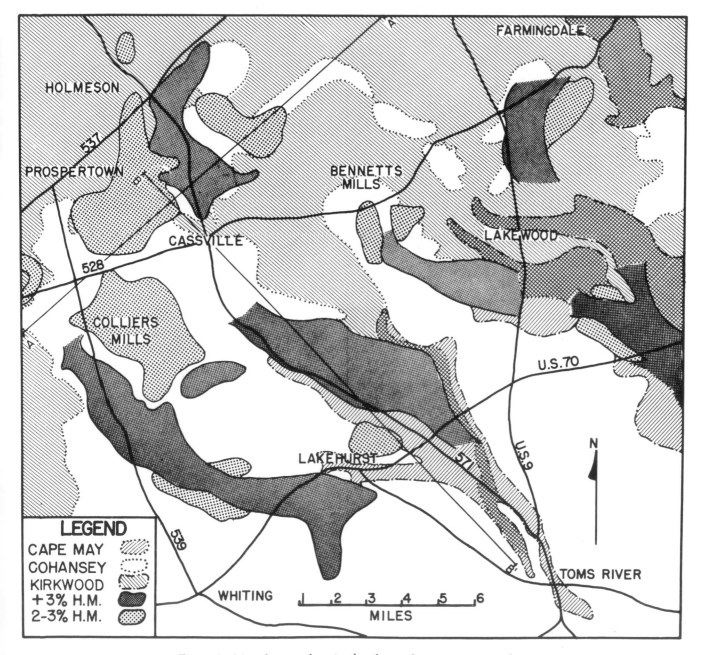

Figure 3. Map showing ilmenite distribution by percentage weight.

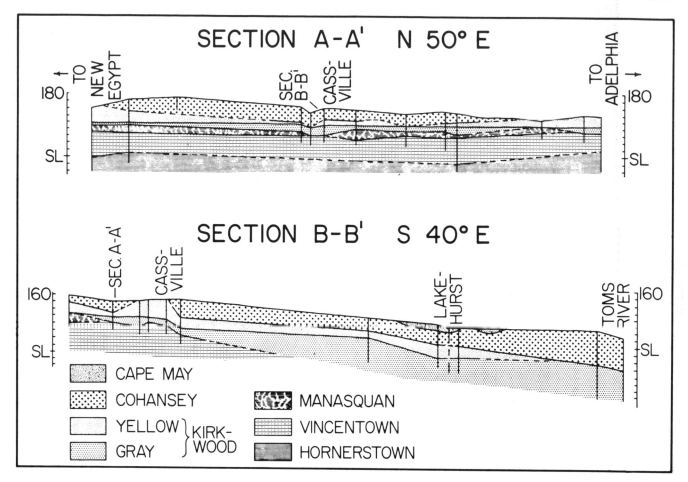

Figure 4. Geologic sections.

are sometimes present, although the ratio of clay to sand is small. These clays sometimes contain lignitic and comminuted vegetable matter. There have been reports of a recognizable fauna found in the clay, but this has never been verified, therefore, exact dating of the formation becomes somewhat questionable. A flora comparable with some upper Miocene localities in Europe has been found near Bridgeton.

From field experience the author believes the Cohansey is roughly divisible into three generalized provinces based upon mineralogic and lithology differences (fig. 6):

Ilmenite Province

This unit is composed of fairly clean quartz sand with some gravel and limited clay. The unit typically contains appreciable amounts of ilmenite.

Quartz Sand and Clay Province

A less clean quartz sand and gravel region with greater amounts of clay in the form of lenses and beds and as clay coatings around quartz grains. Scattered thin lenses and beds of white quartz sand are more common than in the ilmenite province. Heavy mineral content is typically low.

High Silica Sand Province

This province contains moderately dirty quartz sand and gravel to clean quartz sand and gravel mixed with clay beds grading from discontinuous lenses to locally thick beds. Thin to moderately thick (40 foot) beds of very clean high silica sand with a low heavy mineral content. Substantial tonnages of glass and various industrial sands are produced annually from this area.

The Cohansey ranges from 100 feet thick in its outcrop area to more than 250 feet thick in the lower southeast counties of the State. There are no continuous sand members or clay lenses that can be recognized over long distances. Cross-bedding, local cut and fill structures and heterogenity of grain

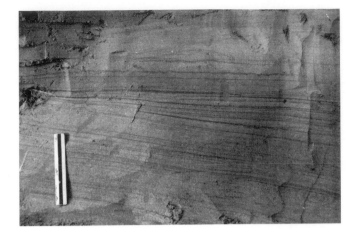

Figure 5. Kirkwood heavy mineral bandlets and laminae. Note the long, straight, laminae in contrast to the irregular wavy cross-bedded Cohansey laminae in figs. 12A and 12B.

size suggests an active stream environment on an extensive alluvial plain in the inland deposits. The type of environment suggested for Cohansey deposition toward the coast is deltaic and estuarine with local swampy areas.

CAPE MAY FORMATION

The Cape May Formation of Pleistocene age is limited in areal extent in the ilmenite province, nevertheless, it is important because it locally contains appreciable amounts of heavy mineral. The work of Salisbury (1898) and Kümmel (1940) suggests that Cape May deposition was complex, and this conclusion is supported by later work. Sand and gravel of the Cape May Formation are found along the present day shore and also bordering the major streams in the Coastal Plain. The Cape May sediments were deposited along low elevation inland drainage basins, as estuarine deposits, and along the margin of the coast.

The Cape May Formation varies from clean, fine, quartz sand to coarse quartz sand with lenses and pockets of fine to coarse, clean to dirty gravel commonly occurring as channel fill deposits. Quartz is the predominant constituent in the sand and gravel, however, it is not unusual to find chert pebbles in the gravel. Cross-bedding is common locally, especially in units containing a mixture of sand and gravel. Channel deposits are a few feet to moderately wide and consist of anastomosing stream channels spread over a large areal extent. Within areas of this type, channels are cut by channels, making cross-bedding most abundant.

The Cape May consists of a mixture of materials derived from the underlying Cohansey and Kirk-

wood and also sediments that were transported from further inland. Heavy mineral concentrations depend primarily upon the source of the sand. Deposits derived predominantly from the Cohansey will contain more heavy mineral over a larger area than a sand from a mixed origin which will generally have an erratic heavy mineral distribution and concentration. At some sites the Cape May contains a greater heavy mineral content than the underlying Cohansey due to reworking of the heavy mineral-bearing Cohansey sand by streams.

In addition to the Cape May Formation, there are other thin, scattered Pleistocene deposits in the ilmenite province. These include the Beacon Hill, Pensauken, and Bridgeton Formations which are composed of sand, gravel and limited clay beds. Quartz is the predominant constituent, but locally a small amount of chert may also be present.

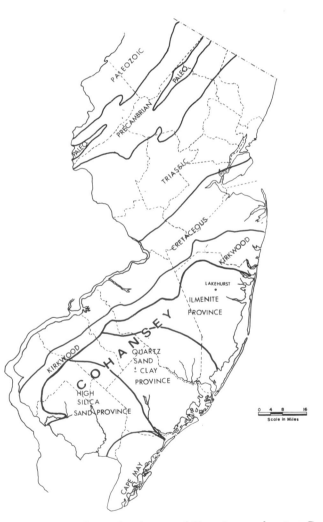

Figure 6. Geologic sketch map of New Jersey showing Cohansey provinces.

ORIGIN OF ILMENITE DEPOSITS

The origin of the heavy minerals and their method of concentration has stimulated discussion since the original announcement of the deposits was made. Several geologists have observed these deposits and each have had different interpretations regarding the origin and mechanics of deposition in the Cohansey Formation. Some geologists interpret the deposits as marine, whereas others state they are fluvial concentrations formed by braided streams. The author and his colleague, Mr. Parrillo, originally interpreted the concentrations to be fluvial in the inner landward deposits and probably deltaic and estuarine nearer to the coast. There appears to be general agreement among geologists that heavy minerals of the Kirkwood Formation were concentrated by marine waters in a semi-restricted near-shore environment. The following briefly outlines the formations and deposition method:

Formation	Age	Environment
Cape May	Quaternary	Fluvial, low lying estuarine and local marine.
Pensauken Bridgeton	Quaternary	Fluvial flood plains, channel type deposits common.
Cohansey	Miocene(?)- Pliocene(?)	Inland fluvial deposits with seaward deposits deltaic with local estuarine influences.
Kirkwood	Miocene	Semi-restricted and open marine.

During Miocene time, the Kirkwood sea may not have extended many miles further northwest than the present outcrop limits of the Kirkwood Formation. The fine grained sediments in the present day outcrop area of the Kirkwood represent a shallow semi-restricted lagoonal or back-bay low energy environment. Much of the fine micaceous sand and silt in the Kirkwood probably represents material eroded from the older sediments of the Coastal Plain lying to the northwest and also sediments brought in from northern New Jersey and eastern Pennsylvania. These were carried by low to moderate energy streams down to the Kirkwood bays and inlets in a setting analogous to the present day streams that carry sediments into Barnegat Bay. The dark, well rounded and polished, fine to very coarse locally pebbly sand found in the subsurface downdip from the Kirkwood outcrop area may be indicative of an offshore bar. Open marine type fossils are found in the downdip facies, whereas the updip clays and very fine grained lagoonal sediments contain a restricted fossil suite. The fine, floury sand typical of the Kirkwood in its outcrop area appears to thin toward the southeast. It is more than 50 feet thick in the northeast and entirely absent in wells to the southeast at Toms River.

During the slow regression of the Kirkwood sea, the gentle shore currents and waves washed across the sandy material removing the lighter grains and concentrating a heavy mineral fraction as shown in figure 5. This process created a 10 mile wide band of fine grained Kirkwood sand which contains two or more percent of heavy minerals.

The continuing uplift of the land mass caused withdrawal of the Kirkwood sea and began a period of erosion. The rise of the land mass increased stream gradients and they began to erode the exposed portions of the inner Coastal Plain. During this period, the ancestral Delaware flowed across New Jersey toward the coast rather than turning at right angles to the southwest as it does today at Trenton. The Delaware transported material eroded from the inner Coastal Plain formations and from rocks further upstream. As the stream flow velocity was reduced southeast of Trenton, the sediment load was deposited over a broad area as alluvial sheet sands. The braiding stream spread and reconcentrated the minerals over an ever widening area.

At the same time that the Delaware was depositing sediments in the northern part of the Coastal Plain, the ancestral Schuylkill River was also flowing across New Jersey and depositing sediments containing heavy minerals. This helps explain the heavy mineral deposits found in the Medford Lakes area. Contemporaneous with the deposition of the above regions, subsequent streams still further to the southwest in Pennsylvania and Delaware were flowing southwest and were gradually eroding headward parallel to the strike of the coastal plain formations. Eventually these subsequent streams captured and diverted the Schuylkill to the southwest. During the Schuylkill capture, the Delaware probably maintained its course to the coast. The process was repeated by subsequent tributaries of the Schuylkill which flowed southwest parallel to strike. Finally, the Schuylkill tributaries pirated the Delaware and diverted it to the southwest. The development of the present drainage is schematically shown in figure 2. The Delaware gradually oriented itself parallel with the strike of the formations and settled in a channel similar to its present day course.

Although the Delaware deposited sand containing heavy minerals, a later stage of drainage events reconcentrated some previously deposited mineral concentrations into higher grade deposits. This later drainage was the ancestral Millstone River flowing to the southeast which not only introduced some new materials from the higher inner portions of the Coastal Plain, but also reworked portions of the previously deposited sheet sands. Terraces and deep channel deposits that border the Millstone River region indicate that during late Pliocene(?) and Pleistocene time the Millstone River flowed southward. These terraces decrease in elevation from north to south, and well samples in the Kingston-Plainsboro-Cranbury area reveal an unusual thickness of gravelly sand that are correlated with the Pensauken Formation. Pensauken deposition by the ancestral Millstone continued during its southeast flowing lifespan, and contributed to and reconcentrated portions of the underlying sheet sands by braided stream action across the broad alluvial deposit.

It is recognized that there are alternate explanations for the origin of these deposits and that some who have studied them believe, for example, that the ilmenite in the Cohansey Formation was concentrated solely by marine action. This would imply, however, that the Cohansey is solely of marine origin, and the evidence against this is rather strong. There are certain horizons exposed in the Glidden pits that are atypical of the Cohansey. Subsequent work may disprove some of the ideas as expressed in this report, however, pursuit of these principles made possible the discovery and exploration of commercial ilmenite deposits.

Heavy Mineral Sources

A major percentage of the ilmenite originated from the Precambrian gneisses and granites with some contribution from the Triassic rocks (fig. 6). This would include southern New York and northern New Jersey granites, gneisses, pegmatites, schists, syenites, diorites, and central Jersey Triassic diabases and sandstones. The low percentage of rutile and monazite in the ilmenite province suggests a Precambrian origin because these rocks are generally low in these specific minerals but contain moderate to locally substantial amounts of ilmenite and ilmeno-magnetite. The mixtures of titanium and iron in the Precambrian rocks ranges from pure ilmenite to ilmeno-magnetite down through mixtures of ilmenite-hematite. Ilmenite also occurs in the Wissahickon Schist and Baltimore Gneiss which crops out from southwest of Philadelphia to north of Trenton. Zircon could have been supplied by both the northern New Jersey rocks and those

metasediments and gneisses lying between Trenton and Philadelphia. Kyanite, sillimanite, staurolite, rutile, and monazite were probably derived principally from the Wissahickon Schist and Baltimore Gneiss. The general path of heavy mineral transport and concentration no doubt includes a series of steps that begins with the removal of a heavy mineral grain by some minor tributary streams of Upper Cretaceous to Tertiary age and subsequent deposition along the coast that existed during this period of geologic time. In addition, some of the very stable minerals may have originated in areas far removed from New Jersey and were added to the Upper Cretaceous and Tertiary sediments by long shore current transportation. Considering that a substantial percentage of the ilmenite grains are sub-angular suggests that there may not have been extreme distances involved in the travel path of this mineral. In addition, a large percentage of the minerals were probably removed from their original setting during the middle to latter part of the Tertiary. During this period of travel and temporary rest, attrition and leaching was eliminating the less stable grains and also removing iron from the ilmenite-magnetite mixtures.

MINERALOGY

During the course of the ilmenite exploration project, more than 350 auger holes were drilled in Ocean, Monmouth and Burlington Counties. The heavy and light mineral fractions of all the auger holes in addition to many drill cutting samples from water wells were studied with a binocular microscope. To maintain a close cross-check on the mineralogy, a number of heavy mineral fractions from different localities were examined with a petrographic microscope. Visual heavy mineral estimates were made of each sample batch in addition to periodic mineral counts which kept a control check on heavy mineral distribution.

Quartz, the most common mineral in the light fraction of the Kirkwood Formation can amount to more than 95 percent. Most grains are generally sub-angular to angular, and are slightly polished while the coarse grains are usually more rounded and exhibit a somewhat greater degree of polish. Feldspar may be from three to seven percent in the finer grained Kirkwood sediment. Thin, transparent muscovite flakes, possibly originating from the Chickies Formation of southeastern Pennsylvania, are common in the fine grained, floury sand. Chert is often present in the coarser Kirkwood, while feldspar is absent. This relationship is especially true in the downdip facies. Lignite is typically present in the darker non-eluviated fine micaceous sand and silt of the Kirkwood.

Quartz comprises a greater percentage of the light fraction in the Cohansey than it does in the Kirkwood. The grains are usually more rounded and tend to be more yellow. Chert, though rare, is sometimes found in the coarser sand, and even where present does not constitute a significant percentage. Occasional mica flakes may be present, but in the basal facies there frequently can be several percent of mica that was reworked from the Kirkwood. Small fragments of "ironstone" are occasionally found in the coarser sands. Glauconite is practically non-existent in the Cohansey and in the updip facies of the Kirkwood in Ocean and Monmouth Counties.

In the light fraction of the Cape May Formation, sub-rounded quartz is the predominant mineral — generally being in excess of 95 percent. Other light minerals include variable amounts of chert which occasionally contain Paleozoic fossils and a few grains of weathered glauconite at some localities.

The percentage weight of heavy minerals in the various screen sizes for the Cohansey and Kirkwood are given in table 1.

The average heavy mineral composition, by grain count, of the Cohansey Formation, Kirkwood Formation, and Cape May Formation is given in table 2.

The word "opaques" as used in the table, includes black and brown ilmenite, and the altered ilmenite, leucoxene. With minor exceptions all of the heavy mineral suites are closely related.

The fine grained Kirkwood minerals contain slightly less leucoxene than the other two formations, but there is an accompanying increase in the

TABLE 2.   HEAVY MINERAL DISTRIBUTION

	Cohansey	Kirkwood	Cape May
opaques	85	76	80
rutile	2	3	1
zircon	5	10	5
staurolite	1	4	1
ky-sil	2	4	2

percentage of zircon, rutile, staurolite, kyanite, and sillimanite. Kirkwood ilmenite is predominantly sub-angular. Leucoxene generally exhibits a greater degree of rounding than the dark ilmenite. No doubt many of the small, angular ilmenite grains found in the Kirkwood have been derived from breakage during transportation. Some ilmenite grains are coated with limonitic and/or other iron compounds, but the majority are clean.

The Cohansey Formation contains the highest percentage of opaques, consistently comprising over 80 percent of the heavy minerals, and in some instances up to 90 percent. The opaques are more altered than in the Kirkwood and the mineral suite is more limited. The heavy minerals, by contrast with those of the Kirkwood Formation, are much coarser. Cohansey ilmenite is predominantly sub-angular, though generally it shows more wear than the Kirkwood ilmenite. As in the Kirkwood, some ilmenite grains are coated with limonite. Many of the larger grains show alteration progressing along parallel planes and some grains with numerous planes of alteration have a foliated appearance. Many large brown ilmenite grains exhibit an etched pattern due to alteration. When broken, the dark ilmenite has a sooty, non-crystalline appearance, probably similar to "Stage 2 ilmenite" examined by Bailey and others (1956).

The Cape May suite is very similar to that of the Cohansey. It is composed of about the same percentage of opaques, but there is slightly more alteration of these minerals. There is an increase in the amount of limonitic aggregate at the expense of the other constituents, except the opaques.

Ilmenite from New Jersey sands exhibits a progressive stage of alteration ranging from black ilmenite, through brown ilmenite or leucoxene to reddish tan, gray, and cream-colored leucoxene, most of the latter varieties having a high luster. Almost all black grains show at least some alteration. The black opaques may be considered as ilmenite, but an X-ray analysis did not give an ilmenite pattern, showing instead a few weak lines of the rutile pattern. Many ilmenite grains are partially or wholly coated with rims of cream-colored leucoxene. In some lighter colored leu-

TABLE 1.   SCREEN ANALYSIS — KIRKWOOD AND
COHANSEY FORMATIONS

KIRKWOOD FORMATION

Screen Size	Percent Weight Retained	Percent H-Min. Fraction	Bulk Percent H-Min.
+16 mesh	1.8	—	—
−16 +32 mesh	2.4	—	—
−32 +60 mesh	2.2	tr.	—
−60 +115 mesh	61.0	2.5	2.10
−115 +250 mesh	28.5	11.9	3.40
−250 mesh	4.0	13.8	0.55

COHANSEY FORMATION

Screen Size	Percent Weight Retained	Percent H-Min. Fraction	Bulk Percent H-Min.
+16 mesh	9.5	—	—
−16 +32 mesh	31.0	tr.	—
−32 +60 mesh	37.4	7.3	2.75
−60 +115 mesh	17.6	24.5	4.30
−115 +250 mesh	3.0	18.3	0.55
−250 mesh	1.5	3.3	.05

coxene grains, alteration has progressed sufficiently to produce a very fine, felted mesh work of what appears to be rutile (fig. 7). This material has been termed "rutilated leucoxene" implying that the meshwork of crystals may be rutile or aggregates of rutile with other titanium minerals. Very fine, tan-colored rutile needles with sharp edges have been noted in heavy mineral concentrations, which may in part be authigenic, or derived from the rutilated leucoxene. Some mica-like grains also possess a very fine meshwork of rutile needles.

Magnetite and many of the less stable heavy minerals constitute only a trace or are non-existent in the heavy mineral suites of the Cohansey Formation and Kirkwood Formations in the ilmenite province. The absence of such minerals as magnetite, hornblende and garnet indicates a moderately long period of leaching probably assisted by significant amounts of organic acids and a fluctuating water table. It is possible that a substantial amount of leaching occurred in the provenance area, especially if a deep regolith was present.

There are significant amounts of staurolite in the lower Upper Cretaceous Raritan and Magothy Formations, however, it generally constitutes two percent or less of the heavy mineral fraction in the ilmenite province, except for the Kirkwood which can contain as much as four percent.

Most of the highly altered ilmenite has a TiO_2 content from 60 to 80 percent or more, and may amount to as much as 13 percent of the total heavy minerals. In a magnetic separation, much of this material would behave like rutile and may be accepted as such because of the high TiO_2 content.

The percentage of TiO_2 given in table 3 is from a composite of 8 samples of both Cohansey and Kirkwood material. It is slightly lower than the average of some individual samples, several of which ran as high as 60 percent. Note the high ratio of TiO_2 to total iron (Fe).

Rutile ranges in color from nearly black, dark red, red or tan as mentioned above. Many elbow twins have been noted in the larger grains found in the Cohansey and Cape May. Some anatase is occasionally included in the rutile fraction, most of it is pale yellow, but a few indigo blue grains were noted.

TABLE 3. CHEMICAL ANALYSIS OF
ILMENITE FROM COMPOSITE SAMPLE

TiO_2	54.7	(Percent)
Fe	23.1	
Mn	0.68	
Ca	0.3	
TiO_2/total Fe ratio	2.3	

Figure 7. Photomicrograph of rutilated leucoxene. Note fine meshwork of twinned crystals. ×275

Zircons occur as nearly euhedral to well-rounded grains and many contain inclusions of magnetite and/or ilmenite. A large percentage of the zircons are clear, though light purple, gray, and tan translucent to almost opaque varieties have been observed. Many of the gray and tan zircons are seamed with microscopic fractures. Rare earth bearing and lightly radioactive zircons, similar to the tan almost opaque type, have been noted in the Precambrian of Morris, Warren, and Sussex Counties of northern New Jersey.

Associated heavy minerals other than those listed in table 3 include tourmaline, andalusite, amphibole, hypersthene, garnet, epidote, monazite and an occasional grain of glauconite. Tourmaline usually is present only in trace amounts, however, at the Glidden Company deposit there are local pockets containing more than 5 percent tourmaline in the mineral fraction. Except for the tourmaline at Glidden, none of the above individual minerals ever amounted to more than 1 or 2 percent and usually the combined total is less than 3 percent.

Because commercial processing of heavy minerals employs magnetic separation in the beneficiation process, several ilmenite samples from the Kirkwood and Cohansey were run to determine how it behaved magnetically. Table 4 indicates that the largest percentage of heavy minerals was extracted with a Memco induction-type magnetic separator when using a current of 0.5 to 1.5 amperes. The heavy mineral percentages are visual estimates.

RESERVES

The total potentially economic reserves in the ilmenite province, exclusive of the Medford Lakes area but including an ore body south of Browns Mills, is estimated to be in excess of 15,000 acres (fig. 3). This estimate includes all private, federal and state lands, a large part of which probably will never be mined. Many acres of rich ilmenite sand lie beneath residential developments such as those about Lakehurst and Lakewood. A total of approximately 8,500 acres is currently held by the three companies mentioned in the earlier part of the report. The acreage figure includes only those areas where heavy mineral values are consistently better than 3 percent for depths considered to be

TABLE 4. MAGNETIC SEPARATION OF HEAVY MINERALS FROM THE KIRKWOOD AND COHANSEY FORMATIONS

KIRKWOOD FORMATION (Composite of 4 samples)

+100 mesh (2.5 percent of Bulk Sand)

Percent H. Min. in Fraction		Principal Constituents
magnetic	trace	—
1.0 amp.	66.0	—
1.5 amps.	10.0	—
2.7 amps.	7.0	—
non-mag.	17.0	—

−100 mesh (97.5 percent of Bulk Sand)

Percent H. Min. in Fraction		Principal Constituents
magnetic	0.4	all black opaques
0.5 amp.	69.9	80 percent black opaques, 15 percent brown opaques
1.0 amp.	9.7 } 82.7 percent	50 percent black opaques, 35 percent brown opaques
1.5 amps.	3.1	20 percent black opaques, 25 percent brown opaques, 25 percent cream opaques, 5 percent staurolite, 5 percent kyanite-sillimanite, 5 percent zircon and rutile
2.7 amps.	3.6	60 percent cream opaques, 10 percent rutile, 15 percent kyanite-sillimanite, 5 percent zircon
non-mag.	13.2	65 percent zircon, 15 percent kyanite-sillimanite, 15 percent cream opaques and rutile

COHANSEY FORMATION (Composite of 6 samples)

+100 mesh (71 Percent of Bulk Sand)

Percent H. Min. in Fraction		Principal Constituents
magnetic	trace	all black opaques
0.5 amp.	24.8	95 percent + black opaques
1.0 amp.	28.5 } 91.3 percent	75 percent black opaques, 20 percent brown opaques
1.5 amps.	38.0	50 percent black opaques, 40 percent brown opaques, 10 percent staurolite, tourmaline, etc.
2.7 amps.	1.6	50 percent brown opaques, 20 percent cream-colored opaques and staurolite, tourmaline, kyanite, sillimanite, rutile, etc.
non-mag.	7.1	60 percent zircon, 15 percent kyanite-sillimanite, 15 percent rutile

−100 mesh (19 percent of Bulk Sand)

Percent H. Min. in Fraction		Principal Constituents
magnetic	trace	all black opaques
0.5 amp.	38.2	85 percent black opaques
1.0 amps.	32.0 } 87.1 percent	70 percent black opaques, 20 percent brown opaques
1.5 amps.	16.9	50 percent black opaques, 30 percent brown opaques 5 percent cream opaques, also staurolite, tourmaline, kyanite-sillimanite, etc.
2.7 amps.	2.7	50 percent cream opaques, 15 percent rutile, 10 percent kyanite-sillimanite, 15 percent zircon
non-mag.	10.3	75 percent zircon, 10 percent rutile, 10 kyanite-sillimanite

a minimum of 20–25 feet. Throughout much of the reserve area ore grade sand lies at grass roots and can go down to depths of 80 feet or more. In the thicker ore zones, concentrations may be present in the Cape May, in the underlying Cohansey, and continue into the Kirkwood Formation. As would be expected, not every foot is ore, there are horizons from less than one foot to more than five feet that may be below 1.5 percent heavy minerals. However, within the next ten feet or so the ore might assay 10 to 15 percent heavy mineral for more than five feet. Throughout much of the reserve area the ore does not form a continuous lateral horizon, but tends to form a discontinuous irregular pattern with very rich, moderate, and low mineral zones which, when averaged out, will assay better than 3 percent heavy mineral.

The following reserve figures, which include all land except that beneath Lakewood and Lakehurst proper, are believed to be modestly calculated based upon an economic depth of 25 feet with an average of 3 percent heavy mineral.

Total Acres	Heavy Mineral Reserves	TiO₂ Reserves
15,000	25,000,000 tons	12,500,000 tons

The tons of TiO_2 are calculated on the basis of an average of 50 percent TiO_2 content in the bulk heavy mineral suite. Considering that there are approximately 8,500 acres currently held as ilmenite reserves and if the total combined production rate averaged 250,000 tons of heavy mineral per year, this would give a life expectancy of 60–65 years for the combined properties.

LAKEHURST ILMENITE MINE – GLIDDEN COMPANY

During the latter part of 1958 and early 1959 the Glidden Company explored several areas in Ocean County in search of an economic ilmenite deposit. The company, properly known as the Glidden-Durkee Division of SCM Corp., produces titanium dioxide at their Baltimore, Maryland plant. Since the bulk of their ilmenite requirement was imported, they were interested in acquiring a domestic source. Their early effort was concentrated on State land near Collier's Mills. A moderate amount of ore land was delineated on State property, but to put together an economic amount of acreage necessitated extension into Federal and private land. Because of land acquisition barriers in this area, Glidden's efforts were diverted elsewhere.

Glidden subsequently acquired several land positions at Legler, located approximately two miles due north of Lakehurst on Route 571. This area had been explored by the New Jersey Survey and constituted one of the better targets considering ore grade and potential economic acreage. The Survey's exploratory work was substantiated by a subsequent exploration program. Glidden conducted additional exploration and constructed a pilot plant which indicated that a suitable ilmenite concentrate could be produced for processing at their Baltimore plant. Construction of the main plant became a reality, and in late 1962 the first ilmenite concentrate was produced. Figure 8 is an aerial view of the Glidden plant.

GENERAL DESCRIPTION OF THE DEPOSIT

The Cohansey ore deposit on the Glidden property is overlain by a mantle of Pensauken sand and gravel from less than one to more than ten feet thick. This material is stripped and stockpiled elsewhere on the property. Cape May sand is present at lower elevations toward the river, however, it is not included in the mining process at this time.

The ore bearing sand is removed by a large capacity drag-line bucket in lieu of a dredge because of the deep and varying depths to the water table. After passing through a wobbler feeder which scalps out plus inch and a quarter objects such as clay balls, wood and ironstone, the sand is pumped through a pipeline to the mill. Following screening, the concentration process includes a combination of circular and spiral concentrators in the wet milling process and high intensity magnetic and electrostatic separators in the dry mill. The ilmenite-leucoxene concentrate which averages 61.5 percent TiO_2 is shipped by rail to Baltimore. Zircon, rutile, kyanite, sillimanite, and the other heavy minerals are stockpiled for further concentration at a later date. An average heavy mineral grain count for the deposit is:

Percent H-Min.	Minerals
85–90	ilmenite-leucoxene
2–4	zircon
1–2	kyanite-sillimanite
tr.–1	staurolite
tr.–1	rutile, anatase
tr.–2+	tourmaline
tr.	garnet, monazite, epidote, andalusite, hypersthene

The ore bearing horizon varies from less than 20 to more than 40 feet thick with 25 feet being the average ore zone. Exploratory drilling into the subsurface Kirkwood indicates another ore horizon with a heavy mineral content of approximately 4.5 percent. Ore zones in the Kirkwood range from a little less than 3 to more than 10 percent heavy minerals. Some holes, 100 feet deep, have pene-

Figure 8. Aerial view of Glidden Plant.

trated more than 40 feet of Kirkwood ore without reaching the bottom contact of the ore zone. Between the bottom of the Cohansey ore zone and the top of the Kirkwood ore there is a barren red sand zone 20 to 40 feet thick. There are no immediate plans for mining the Kirkwood ore.

PENSAUKEN FORMATION

Pleistocene sand and gravel, predominantly the Pensauken Formation, mantles the Cohansey with a blanket of material from a few inches to more than 10 feet thick. The projected thickness and direction of the Pensauken is unpredictable because of erratic deposition and post-Pensauken erosion. The sand and gravel can be a few inches thick and then suddenly form a deeply incised channel in the Cohansey (fig. 9A). Cross beds, as shown in figure 9B, are very abundant and frequently change direction in the course of 10 to 20 feet.

The Pensauken and other Pleistocene deposits consist of generally dirty, silty to clayey sand and sandy gravel which may grade to local beds of gravel devoid of any sand. The sand and gravel is predominantly yellowish brown and can contain an occasional rust-colored lense. Quartz pebbles are sub-angular to well rounded (fig. 9C) and generally fractured internally due to collision during transportation. Most of the quartz has a smooth surface, but many grains possess an irregular corroded surface. Many pebbles are vein type quartz. Flat ironstone fragments with beveled edges are common in the coarser sized gravels. These were eroded from ironstone deposits that capped pre-Pensauken gravelly hills to the northwest. Chert is rare, though it probably was present in the deposits from whence the gravel was derived. Because it was leached and chalky, it did not survive transportation during Pensauken deposition. The occasional chert pebble that is found is medium to light dark gray and was probably introduced as new material mixed in with the older gravels. Stained quartzite pebbles (Hardyston-Chickies?), occasional light pink conglomerate (Silurian?) pebbles, and a few hornfels (?) fragments have been found.

COHANSEY FORMATION

The Cohansey has a wide range of colors including light gray, tan, yellow, brown, red, and any mixture of the above within the same lithic unit. The different facies include a clean, ilmenitic quartz sand, a dirty, slightly clayey, frequently highly ilmenitic brown sand, or a red to reddish brown, stained, barren quartz sand with small fragments of ironstone. Occasional pebbles or limited gravel layers may exist in any one of the facies. The coarse red to reddish brown iron stained barren sand was utilized as a lower Cohansey marker in this region of ilmenite exploration. This unit is rather widespread in the Glidden deposit and also is a marker horizon. The Kirkwood is generally found not too far below this unit.

Cross-bedding, cut and fill, and limited slump features are not uncommon in the Cohansey. Grayish white and less abundant chocolate colored clay

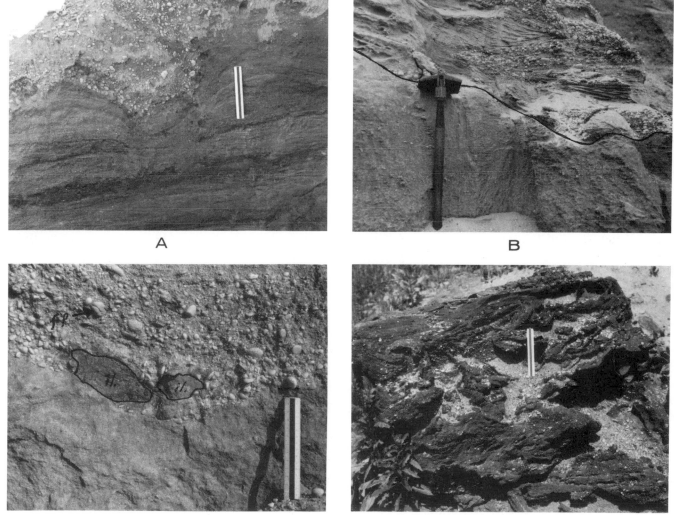

A

B

C

D

Figure 9. A, Narrow Pensauken channel incised in Cohansey mineral bed. Note haphazard position of pebbles. B, Cross-bedded Pleistocene sand and gravel channel incised in finely disseminated ilmenite bed. Line delineates contact. C, Close-up view of gravel channel cut in mineral bed. Note lack of gravel orientation, fractured pebble (fp), and dislodged pieces of ilmenite ore at bottom of gravel (il). D, Large irregular ironstone mass with encased pods and pockets of white very coarse sand and fine gravel. Dredged from depth of 35 ft.

lenses occur randomly in the deposit. Vegetable matter consisting of seed carpels, rootlets, and other bits of organic matter are sometimes found in the chocolate clay. Frequently a sand unit will contain numerous very small to 6″ clay chips scattered throughout the unit. Iron oxide lenses and irregular masses (fig. 9D) both indurated and in the process of cementation occur randomly above and below the water table. In many respects, especially in some local units, the Glidden material is unlike typical Cohansey, but considering the overall variability of the formation, it is not unusual that these variations are present.

The Cohansey consists primarily of fine to coarse angular to rounded quartz with a substantial amount of heavy minerals. The yellow limonitic stained quartz is clear, milky or less frequently, smokey on the interior. Some coarse to pebble size, clear, well polished quartz specimens found at the Glidden pit are of Cape May diamond quality. Quartz occurs as single crystal grains to the more common strained grains seamed with microfractures. One slightly rounded quartz crystal, broken parallel with the long axis, measured nearly two inches long. Exterior surfaces may be clean and smooth or irregularly pock-marked with micro-cavities due to mineral leaching or micro-percussion marks from collision with other grains in rapidly moving water.

The possible sources of the quartz in the Cohansey include all the quartz bearing Precambrian, Paleozoic, and Triassic rocks in eastern Pennsylvania, New Jersey, and southeastern New York.

In addition, long shore currents probably transported a great deal of quartz from other areas and introduced it into the Cretaceous and Tertiary formations whose inland extensions were in part eroded and redeposited as the Cohansey Formation.

In some facies of the light gray, clean, coarse sand there are abundant fragments of very white leached chert to fresher, dense, medium gray, non-leached chert. Leaching is so advanced in some fragments that it is easily powdered between the fingers. The chert ranges from smooth surfaced varieties to irregularly pitted chert containing a few to a myriad of percussion marks. Pitting is due to percussion fracturing, leaching of fossils or dissolving of pyrite, dolomite or calcite crystals. Many chert specimens contain a small fossil brachiopod(s) crinoid columnals, bryozoa, coral or ostracodes. The fossil suite appears to be of lower Paleozoic age and may be in large part derived from Silurian-Devonian limestone. The oolitic and dense, darker chert fragments are probably from the "Kittatinny" dolomite.

CHARACTER OF THE ORE DEPOSIT

The geology of the ore deposit is complex with respect to the continuity of specific ore horizons, certain lithologies, and the depositional processes of some ore zones (fig. 10A). Heavy mineral content tends to be irregular with some zones averaging less than 3 percent, whereas others a few feet distant will exceed 15 percent heavy mineral. The

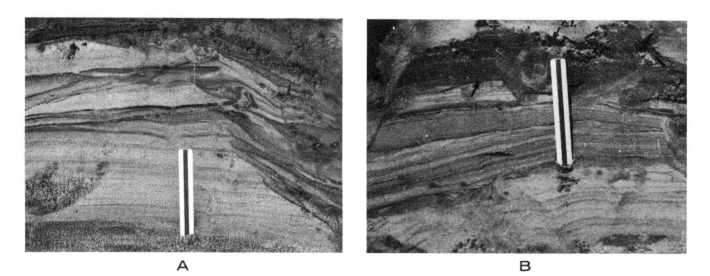

A   B

Figure 10. A, Heavy mineral bands and laminae passing into slump(?) structure. One of many exotic sedimentary structures in the deposit. Disseminated heavies with "salt and pepper" texture at bottom of photograph. B, Very rich mineral band (arrow) overlain by thinner semi-indurated oxide layer (0.1.). Note progressive decrease in heavy mineral from top to bottom.

deposit averages 5 percent heavy mineral for an average thickness of 25 feet. Tracing of specific mineral beds, especially the high concentration bands, is difficult because of sudden pinch, swells or a wispy heavy mineral fade-out. Some very rich concentrations up to 1 foot thick extend laterally or vertically into an indurated iron oxide layer that can be practically devoid of heavy minerals (fig. 10B). Heavy minerals typically form a heterogenous assemblage of different type concentrations diagrammatically shown in figure 11 and described below.

(1) Distinct, straight, wavy, or finely cross-bedded laminae spaced from less than ¼ inch to more than 1 inch apart (fig. 12A). Fade-out of individual laminae very common. Sand between laminae is sometimes low or almost barren of heavy mineral content. Small fold-like structures, faults, and slumps have been noted in this unit (fig. 12B).

(2) Heavy minerals uniformly disseminated throughout the sand exhibiting a coarse salt and pepper texture. Mineral percentages are more uniform for greater thicknesses in this type of occurrence (fig. 12C). Frequent heavy mineral "knots" may be found in this unit (fig. 12D).

(3) Heavy minerals in erratic lenses, bands, or pods that pinch, swell, or terminate by wispy fade-out (fig. 13A).

(4) Rich mineral layers in beds ranging from several inches to 1 foot in thickness (fig. 13B). These tend to pinch, swell, or may grade into, or terminate by passing into a semi-indurated iron layer. Some swells or bulges noted in this unit have a conspicuous light colored core or "eye" of low to almost barren sand.

(5) Coarsely cross-bedded mineral layers spaced from 1 inch to more than 3 inches apart. Sand between layers generally low to almost barren.

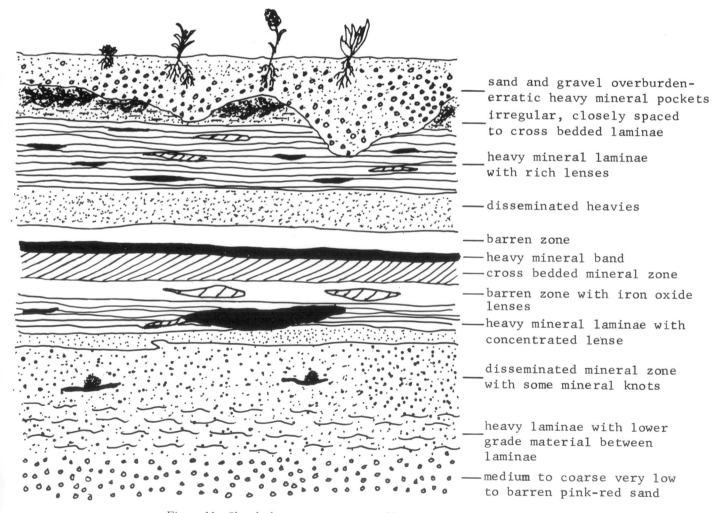

sand and gravel overburden-
erratic heavy mineral pockets
irregular, closely spaced
to cross bedded laminae

heavy mineral laminae
with rich lenses

disseminated heavies

barren zone
heavy mineral band
cross bedded mineral zone
barren zone with iron oxide
lenses
heavy mineral laminae with
concentrated lense

disseminated mineral zone
with some mineral knots

heavy laminae with lower
grade material between
laminae
medium to coarse very low
to barren pink-red sand

Figure 11. Sketch showing various types of heavy mineral concentrations.

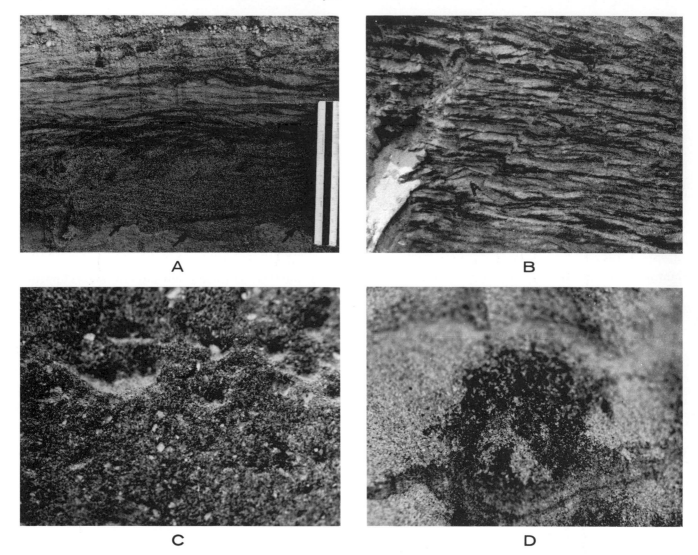

Figure 12. A, Finely cross-bedded and disseminated mineral layers overlain by poorly sorted mineral bearing gravelly unit. Arrows point to irregular lower contact of disseminated band. B, Irregularly spaced, wavy to finely cross-bedded mineral seamlets with micro fault (A-B). C, Close-up of finely disseminated ore with scattered clay bits (arrow). D, Close-up of ilmenite "knot" in disseminated ore zone. Note less concentrated core within "knot" (arrow).

(6) Heavy minerals associated with the gravelly overburden. Concentrations in gravel beds are erratic in mineral percentage and occurrence.

During the latter part of 1967, an unusual clay, silt and lignitic clay and silt unit containing a black, silty, lignite-charcoal band 4 feet thick (fig. 13C) was encountered toward the northwest end of the property. The unit varies somewhat in thickness, but is up to 15 feet thick at some points and extended more than 800 feet in the long dimension during October, 1968. One partially carbonized piece of wood found in this unit measured 3 feet in length, 1 foot in width, and 4 inches in thickness. Ore grade material is found above and below this unit. This unusual deposit is interpreted to be either the remains of a depressed and protected wooded swamp that existed on the fluvial plain during deposition, or was a depression in the alluvial plain in which clay, silt, lignite and woody fragments were deposited. Some of the charcoal and associated red clay lumps in the black layer remind the author of a similar association noted in Alaska in which a lignitic-charcoal deposit was determined to be the site of an ancient fire. Although finely lignitic, chocolate colored clays are found in the Cohansey further south, the age of this unit is unknown, and it may be a large channel deposit of post-Cohansey age.

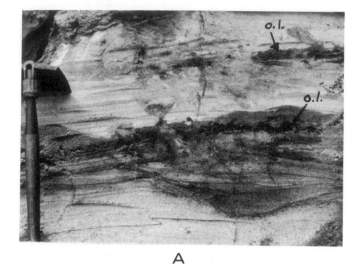

Figure 13. A, Heavy mineral pod with semi-indurated oxide layer (0.1.). Note "fold" in thin ilmenite seam toward bottom of photograph. B, Highly concentrated, dark, heavy mineral band underlain by less concentrated thinner band. Note distorted less mineralized zone (arrow). C, Pit face with clay and black lignitic layer. Overburden stripped off, ore zone (marked ORE) overlying clay layer is approximately 10 feet thick. Ore also occurs below dark layer.

Microscopic examination of the silt and clayey carbonaceous material shows that the organic material is in the form of seed carpels, rootlet remains, and other bits of vegetable matter. It is possible that the woody material burned initially, and after the fire was exhausted, either by burn-out or inundation by sediment, a new growth of grasses and reedy plants developed and this was subsequently covered by silt and sand.

The trip will start at 8:00 A.M. from Haddon Hall, Atlantic City and return to the hotel at approximately 5:00 P.M. Transportation by charter bus. Lunch (dutch treat) will be at a local restaurant in Lakehurst.

Note: This trip is limited to 35 persons on a first come, first serve basis. No students please. Everyone *must bring* their own *safety hat* and have *prescription or safety glasses*. Considering that it is mid-November, warm clothing and suitable footwear should be brought in the event of inclement weather.

Sincere gratitude is extended to Glidden-Durkee Division of SCM Corp., especially Mr. S. Schworm, Gen. Manager, and Mr. Jan Gijlstra for their cooperation in making this field trip possible.

ROAD LOG

Mileage

0.0   Haddon Hall Hotel — departure at 8:00 A.M. Bus will proceed to Atlantic City Expressway, traveling west.

8.5   Enter Garden State Parkway northbound via Exit 38 from Atlantic City Expressway.

20.2   Bridge over Mullica River.

23.0   Bridge over Bass River.

40.1   Note tall stack of the Oyster Creek Nuclear Power Plant to the right.

52.8   Exit at interchange 82 for State Route 37. Proceed west (right) on Route 37.

59.2   Junction of State Routes 37 and 70. Enter Route 70 semi-circle, bearing left or west on Route 70.

59.4   Turn right onto Route 547.

59.6   Bear left at fork in road.

59.8   Lakehurst Naval Air Station on left.

61.4   Stop sign — junction of Routes 547 and 571. Turn west (left) onto Route 571.

63.2   Entrance to Glidden Co. plant on right. Note sign. Proceed down entrance road to Glidden Plant and parking area.

During the A.M. the pits and field operation will be visited. Participants will be assembled in groups for the field excursion. *Safety hats.*

After lunch the ore concentration process will be observed in the plant.

Return to Atlantic City via the same road system.

REFERENCES CITED

Bailey, S. W., Cameron, E. N., Spedden, H. R., and Weege, R. J., 1956, The Alteration of Ilmenite in Beach Sand: Econ. Geology, v. 51, No. 3, p. 268.

Cannon, H. B., 1957, Sand Ilmenites of the Eastern United States: Paper presented at AIME Mining Conference in Tampa, Florida.

Carpenter, J. N., Detweiler, J. C., Gillson, J. L., Weichel, E. C., and Wood, J. P., 1953, Mining and Concentration of Ilmenite and Associated Minerals at Trail Ridge, Florida: Mining Eng., August, 1953.

Gillson, J. L., 1959, Sand Deposits of Titanium Minerals: T. P. 4799H, AIME Trans. Min. Eng., p. 421–429.

Johnson, D., 1931, Stream Sculpture on the Atlantic Slope: Columbia University Press, 1931.

Kümmel, Henry B., 1940, The Geology of New Jersey: Geological Survey of New Jersey, Bull. 50, p. 20–21, 158.

Lynd, L. E., Sigurdson, H., North, C. H., and Anderson, W. W., 1954, Characteristics of Titaniferous Concentrates: Min. Eng. T. P. 3815-H, Trans. AIME, p. 817–824.

Markewicz, F. J., and Parrillo, D. G., 1957, Preliminary Report on Ilmenite-bearing sands from the Coastal Plain of New Jersey: (abs.) Geological Soc. Am. Ann. Mtg., p. 92.

Markewicz, F. J., Parrillo, D. G., and Johnson, M. E., 1958, Titanium Sands of Southern New Jersey, N. J. Geol. Sur. booklet, Paper presented at AIME Annual Meeting, February, 1958.

Martens, J. H. C., 1956, Industrial Sands of New Jersey: Bur. of Min. Research Bull. 6, Rutgers University Press.

Quirk, R., and Eilersten, N. A., 1963, Methods and costs of Exploration and Pilot Testing of Ilmenite-bearing sands, Lakehurst Mine, The Glidden Co., Ocean County, N. J.: U. S. Bur. Mines cir. 8197.

Salisbury, R. D., and Knapp, G. N., 1916, The Quaternary Formations of Southern New Jersey: vol. VII, Geol. Survey of New Jersey.